SPACE SHUTTLE MISSIONS OF THE 80'S PART II

AMERICAN ASTRONAUTICAL SOCIETY

Publications Office, Post Office Box 28130, San Diego, California 92128

Science and Technology

Vol. 1 Manned Space Reliability Symposium, 1965 (112 pages)—$15.00
Vol. 2 Towards Deeper Space Penetration, 1965 (182 pages)—$15.00
Vol. 3 Orbital Hodograph Analysis, S. P. Altman, 1965 (150 pages)—$15.00
Vol. 4 Scientific Experiments for Manned Orbital Flight, 1965 (372 pages)—$25.00
Vol. 5 Physiological and Performance Determinants in Manned Space Systems, 1965 (220 pages)—$15.00
Vol. 6 Space Electronics Symposium, 1965 (404 pages)—$25.00
Vol. 7 Theodore Von Karman Memorial Seminar, 1966 (140 pages)—$15.00
Vol. 8 Impact of Space Exploration on Society, 1966 (382 pages)—$25.00
Vol. 9 Recent Developments in Space Flight Mechanics, 1966 (280 pages)—$20.00
Vol. 10 Space Age in Fiscal Year 2001, 1967 (458 pages)—$30.00
Vol. 11 Space Flight Mechanics Symposium, 1967 (618 pages)—$35.00; Microfiche Suppl.—$6.00 extra
Vol. 12 The Management of Aerospace Programs, 1967 (392 pages)—$20.00
Vol. 13 The Physics of the Moon, 1967 (260 pages)—$20.00
Vol. 14 Interpretation of Lunar Probe Data, 1967 (270 pages)—$20.00
Vol. 15 Future Space Program and Impact on Range and Network Development, 1967 (583 pages)—$30.00
Vol. 16 The Voyage to the Planets, 1968 (184 pages)—$15.00
Vol. 17 Use of Space Systems for Planetary Geology and Geophysics, 1968 (623 pages)—$35.00; Microfiche Suppl.—$4.00 extra
Vol. 18 Technology and Social Progress, 1969 (170 pages)—$15.00
Vol. 19 Exobiology—The Search for Extraterrestrial Life, 1969 (184 pages)—$15.00
Vol. 20 Bioengineering and Cabin Ecology, 1969 (162 pages)—$10.00
Vol. 21 Reducing the Cost of Space Transportation, 1969 (264 pages)—$20.00
Vol. 22 Planning Challenges of the 70's in the Public Domain, 1970 (504 pages)—$30.00; Microfiche Suppl.—$12.00 extra
Vol. 23 Space Technology and Earth Problems, 1970 (418 pages)—$25.00; Microfiche Suppl.—$12.00 extra
Vol. 24 Aerospace Research and Development, 1970 (500 pages)—$30.00
Vol. 25 Geological Problems in Lunar and Planetary Research, 1971 (750 pages)—$40.00
Vol. 26 Technology Utilization Ideas for the 70's and Beyond, 1971 (312 pages)—$25.00
Vol. 27 International Cooperation in Space Operations and Exploration, 1971 (194 pages)—$16.00
Vol. 28 Astronomy from a Space Platform, 1972 (416 pages)—$30.00
Vol. 29 Space Technology Transfer to Community and Industry, 1972 (196 pages [Microfiche only])—$10.00
Vol. 30 Space Shuttle Payloads, 1973 (532 pages)—$35.00
Vol. 31 The Second Fifteen Years in Space, 1973 (212 pages)—$20.00
Vol. 32 Health Care Systems, 1974 (264 pages)—$20.00
Vol. 33 Orbital International Laboratory, Third & Fourth Oil Symposia, 1974 (322 pages)—$25.00
Vol. 34 Management and Design of Long-Life Systems, 1974 (198 pages)—$16.00
Vol. 35 Energy Delta, Supply vs. Demand, 1975 (604 pages)—Paperback—$40.00; Microfiche—$25.00
Vol. 36 Skylab and Pioneer Report, 1975 (160 pages)—$16.00
Vol. 37 Space Rescue & Safety, 1974 (294 pages)—$25.00
Vol. 38 Skylab Science Experiments, 1975 (274 pages)—$25.00
Vol. 39 Environmental Control and Agri-Technology, 1976 (346 pages)—$30.00
Vol. 40 Future Space Activities, 1976 (182 pages)—$20.00 (incl. numerical & author index for all Goddard Memorial Symposia)
Vol. 41 Space Rescue & Safety 1975, (230 pages)—$20.00

Advances in the Astronautical Sciences

Vol. 1-5 AAS Proceedings 1957-60. Available in Microfiche. $20.00 per volume.
Vol. 6   Sixth Annual Meeting Proceedings, 1961 (968 pages)—$35.00
Vol. 7-8 Third West Coast & Seventh Annual Meeting Proceedings, 1961-63. Available in Microfiche. $20.00 per volume.
Vol. 9   Fourth Western Meeting Proceedings, 1963 (910 pages)—$35.00
Vol. 10   Manned Lunar Flight, 1963 (310 pages)—$15.00
Vol. 11   Eighth Annual Meeting Proceedings, 1963 (808 pages)—$35.00
Vol. 12   Scientific Satellites, 1963 (262 pages)—$15.00
Vol. 13   Ninth Annual Meeting: Interplanetary Missions, 1963 (690 pages)—$30.00
Vol. 14   Physical and Biological Phenomena in a Weightless State, 1963 (382 pages)—$20.00
Vol. 15   Exploration of Mars, 1963 (634 pages)—$35.00
Vol. 16   Space Rendezvous, Rescue, and Recovery, 1963; Part I (1028 pages)—$35.00; Part II (380 pages)—$20.00
Vol. 17   Bioastronautics—Fundamental and Practical Problems, 1964 (128 pages)—$15.00
Vol. 18   Lunar Flight Programs, 1964 (630 pages)—$35.00
Vol. 19   Unmanned Exploration of the Solar System, 1965 (1000 pages)—$45.00
Vol. 20   Post Apollo Space Exploration, 1966; Part I (572 pages)—$35.00; Part II (648 pages)—$35.00
Vol. 21   Practical Space Applications, 1967 (508 pages)—$30.00
Vol. 22   The Search for Extraterrestrial Life, 1967 (388 pages)—$25.00; Microfiche Suppl.—$3.00 extra
Vol. 23   Commercial Utilization of Space, 1968 (512 pages, plus 24 microfiches)—$35.00
Vol. 24   Exploration of Space, 1968 (363 pages)—$20.00
Vol. 25   Advanced Space Experiments, 1969 (530 pages)—$35.00
Vol. 26   Planning Challenges of the 70's in Space, 1970 (470 pages)—$30.00; Microfiche Suppl.—$10.00 extra
Vol. 27   Space Stations, 1970 (606 pages)—$35.00
Vol. 28   Space Shuttles and Interplanetary Missions, 1970 (488 pages)—$30.00
Vol. 29   The Outer Solar System, 1971; Part I (618 pages)—$35.00; Part II (740 pages)—$35.00
Vol. 30   International Congress of Space Benefits, 1974 (528 pages)—$35.00
Vol. 31   The Skylab Results, 1975 (1174 pages, Microfiche only)—$45.00
Vol. 32   Space Shuttle Missions of the 80's, 1976 (approx. 1000 pages)—$70.00; Microfiche Suppl.—$60.00
Vol. 33   AAS/AIAA Astrodynamics Conference 1975, (approx. 500 pages)—$35.00; 1976 Microfiche Suppl.—$35.00

Special Volumes

1. Weightlessness—Physical Phenomena and Biological Effects, 1961 (182 pages)—$15.00
2. Lunar Exploration and Spacecraft Systems, 1962 (214 pages)—$15.00

DISTRIBUTED BY **UNIVELT, INC.**, P.O. Box 28130, San Diego, California 92128

AAS PRESIDENT
    Philip H. Bolger   U.S. Department of Transportation

AAS VICE PRESIDENT - PUBLICATIONS
    Dr. Charles Sheffield   Earth Satellite Corporation

SERIES EDITOR
    Dr. Horace Jacobs   Lockheed-California Company

EDITORS
    William J. Bursnall
    Dr. George W. Morgenthaler   Martin Marietta Corporation
    Dr. G. E. Simonson

ART STAFF
    J. C. Speas   Lockheed-California Company

    Thanks are due Marie Heidbreder of the Martin-Marietta Corporation and Diana Law of the Lockheed-California Company for final preparation of the manuscript for the printer.

    Front cover and title sheet, courtesy of Lockheed-California Company.

LARGE SPACE TELESCOPE (LST)

AN AMERICAN *Astronautical* SOCIETY PUBLICATION

Space Shuttle Missions of the 80's

**Volume 32
Part 2**

ADVANCES IN THE ASTRONAUTICAL SCIENCES

Edited by
W. J. Bursnall
G. W. Morgenthaler
G. E. Simonson

Proceedings of American Astronautical Society
Twenty First Annual Meeting

Co-Sponsored by Rocky Mountain Section of the American Institute of Aeronautics and Astronautics, the Denver Section of the Institute of Electrical and Electronic Engineers, and the Rocky Mountain Chapter of Operations Research Society of America

August 26-28, 1976   Denver, Colorado

Distributed by UNIVELT, INC., P.O. Box 28130, San Diego, CA 92128

Copyright 1977

by

AMERICAN ASTRONAUTICAL SOCIETY

AAS Publications Office
P. O. Box 28130
San Diego, California 92128

Affiliated with the American Association for the Advancement of Science
Member of the International Astronautical Federation

Library of Congress Card No. 57-43769

PART I   ISBN 87703-078-2
PART II   ISBN 87703-087-1

Printed and Bound in the U.S.A.

FOREWORD

THEME

The Space Shuttle Transportation System is here. How can we maximize its use for the benefit of mankind?

INTRODUCTION

After several years of intense technical, planning, and fabrication efforts, the era of the Space Shuttle Transportation System is dawning. The aerospace industry and the potential users of this Space Transportation System must now concentrate on Space Shuttle Payloads and the host of experiments and special instruments that will enable society to gain a return on the imaginative investment in this great project.

Over 750 scientists, engineers, political leaders and users, both U.S. and foreign, met at the Brown Palace Hotel in Denver for the 21st Annual Meeting of the AAS to discuss ideas for Shuttle Payload Applications and to plan for future payoff missions. What are the tasks that should be accomplished on these missions? Which instruments will be needed? Who will take the lead in science missions? In Earth resources missions? In Communications experiments? In international collaborative efforts? Even though the U.S., and indeed the major industrial nations, were in a serious economic recession, and the preceding few years had seen budgetary cutbacks in the Nation's space program as a result of loss of public favor for space missions that were purely "spectaculars," a new spirit seemed to prevail at the Denver meeting. The Shuttle would be available and new missions at greater efficiency would be possible. All that was needed was imagination, decisiveness, and dedication. The attitude in Denver was positive - all systems were go!

JOINT NATIONAL MEETING

In keeping with the premise that the Space Shuttle should afford easier access to the use of space for a wide range of users, the meeting was co-sponsored with the local sections of three other major professional societies: The Rocky Mountain Section of the American Institute of Aeronautics and Astronautics; the Denver Section of the Institute of Electrical and Electronic Engineers; and the Rocky Mountain Chapter of TIMS/ORSA (The Institute of Management Sciences/Operations Research Society of America).

MEETING STRUCTURE

After the Call to Order by the General Program Chairman, Dr. George W. Morgenthaler, the attendees were welcomed by the Honorable Richard D. Lamm, Governor of Colorado. Governor Lamm challenged the scientists and engineers who had planned the Space Shuttle and its sophisticated subsystems to apply their skills equally to developing systems to help solve mankind's problems here on Earth.

He was followed by Dr. John Naugle of NASA, who characterized the Shuttle and its Payloads. The opening session also included presentations by leaders of the European consortia who are perfecting the Spacelab for Shuttle Missions, and papers by NASA officials responsible for Shuttle Payload Development and Shuttle Payload Applications.

During the next two days plenary sessions in the Main Ballroom of the Brown Palace Hotel detailed the major payloads already identified for the early Space Shuttle Missions, namely, the Large Space Telescope (LST) and the Atmospheric, Magnetospheric, and Plasmas in Space payload (AMPS). Another topic reviewed was the plan for a Space Tug. The Space Tug would be carried into orbit by the Space Shuttle, along with key payloads; the Tug would then boost these payloads from a lower parking orbit to geosynchronous or other higher energy orbits as needed.

Concurrently with the plenary sessions, technical specialist groups responsible for key Shuttle and Payload subsystems (such as avionics, communications, propulsion, life sciences, payload integration, optics, precision pointing, mechanical systems) met in smaller working groups to hear papers and to discuss problems and progress in these areas. Many other specialist sessions were held on science instrument development, detectors, and experiment data management.

Because of the importance of energy management to man's future and the applications of space technology to this critical area, special sessions were held to focus attention on how the Shuttle System and payloads may help. Great interest was shown in the possibility of an orbital solar power generator with a microwave down-link to Earth.

Two formal luncheons were part of the Symposium. The luncheon on August 26 featured Congressman Don Fuqua of Florida. He reported on the July hearings of the House Sub-Committee on Space Science and Applications. The August 28 luncheon featured Mr. William Dean, Vice President of Rockwell International, the prime contractor for the Shuttle Orbiter, who spoke on "Shuttle Missions - ROI." He was followed by a distinguished interdisciplinary panel which further discussed the nature of the "return on the Shuttle investment" the nation could receive as the actual flights occurred. The Panel included Dr. Richard Goody (Chairman), a scientist; Congresswoman Patricia Schroeder, Dr. Gerard O'Neill, scientist; Capt. Wally Schirra, astronaut; Dr. Robert Anderson, economist; and Mr. Press Layton, engineer.

That evening the traditional AAS Honors Night Banquet included the presentation of the Fellow Award to the new AAS Fellows and the awarding of the annual AAS honors. Senator Frank E. Moss of Utah, Chairman of the Senate Space Committee, spoke on "The Exploitation of Space for the Benefit of Mankind."

INTERNATIONAL ATTENDANCE

The meeting was well attended by international scientists, engineers, and industrialists active in Spacelab and other European Shuttle payload projects. They made many contributions to the planning of future international space missions.

THE SECOND NATIONAL SPACE ART SHOW

The Second National Space Art Show was a strong attraction enjoyed by the public as well as attendees at the Symposium. Mr. Doug Filter was Director. This juried show featured interpretations of space flight, and man's new space environment as perceived by leading artists.

ASTRONAUTICAL EXHIBITS

An extensive Exhibit of space systems featured entries from leading industrial firms and from government agencies. They depicted Shuttle payloads as well as next-generation space instruments for use in exploiting space for mankind.

MEETING COMMITTEE

General Program Chairman	Dr. G. W. Morgenthaler
Deputy	W. L. Kershaw
LST Program Chairman	Dr. A. Hoag
Assistant	J. C. Spencer
Tug Program Chairman	J. H. Disher
Assistant	T. J. Goyette
AMPS Program Chairman	Dr. C. A. Lundquist

Assistant	S. R. Schrock
Shuttle Payloads Chairman	M. E. Turner
AIAA - Rocky Mountain Section	Col. J. P. Wittry
ORSA/TIMS-Rocky Mountain Chapter	A. Anglund
IEEE - Rocky Mountain Section	W. B. Collins
Administration Committee Chairman	A. C. Sellke
Arrangements	S. H. Buzzard
Program & Printing	G. L. Wenner
Mailing	G. E. Simonson
Registration	D. E. Schilling
Publicity	S. Butler, D. L. Yakobson
Preprints	W. J. Bursnall
Finance	W. D. Wiley
Exhibits	H. Bass
Editors	H. E. Nylander, J. H. Nelson, W. P. Pratt
Family Program	K. Bursnall, J. E. Yakobson
Space Art Show	D. C. Filter
Audio Visual	J. Jaumotte
Protocol	S. P. DeJaeger

The enthusiasm and interest shown by the participants was infectious. The perspective gained by the formal presentations and the informal technical and applications exchanges will contribute greatly to sound planning for the future utilization of the Space Shuttle and its payloads for the benefit of mankind.

ACKNOWLEDGMENTS

The cooperation and support of many people is required in the preparation of a book of this scope. The editors would particularly like to thank the authors, who submitted their manuscripts for publication, and Judy Gerlach, Marie Heidbreder, and Kathy Reynolds, who contributed many hours in the transcription of tapes and preparation of manuscripts.

The editors

PREFACE

At this historic meeting which featured Space Shuttle Missions of the 80's - the major space engineering program of the United States for the coming decade- some one hundred and fifty technical papers were presented. Publication of all of these papers in full would have resulted in a bulky set of books consisting of about three thousand pages. The editors therefore decided to produce in this hard-copy proceedings some thirty papers in full and all remaining papers in the form of four or five-page summaries with significant illustrations. Nearly all papers that appear in these volumes as summaries are published in full in a microfiche supplement designated as Volume 25, AAS Microfiche Series. Both hardcopy and microfiche proceedings may be ordered from Univelt Inc., P.O. Box 28130, San Diego, California 92128.

H. Jacobs
Series Editor

CONTENTS

PART I

Page No.

I. INTRODUCTION — 1

    Welcome
        Honorable Richard D. Lamm — 3

    Keynote Address -- The Space Transportation System and Its Payloads (AAS75-121)
        John Naugle — 7

    Department of Defense and Shuttle (AAS75-122)
        Brigadier General Henry B. Stelling, Jr. — 21

    Spacelab Programme - Status Review (AAS75-124)
        Bernard Deloffre — 33

    Spacelab Status/Systems Capability (AAS75-123)
        Hans E. W. Hoffman — 55

    Spacelab Payloads (AAS75-125)
        G. W. Sharp — 80

    Shuttle Payload Development (AAS75-126)
        James Murphy — 117

    Applications Payloads (AAS75-127)
        Charles W. Mathews — 137

    Congressional Views on Long Range Space Plans
        Honorable Don Fuqua — 159

    The Exploitation of Space for the Benefit of Mankind
        Honorable Frank E. Moss (AAS75-136) — 165

    Shuttle Missions - Return on Investment (ROI) Panel Discussion
        William E. Dean, Chairman — 173

II. LARGE SPACE TELESCOPE (LST) PROGRAM — 197

    Potential for Advancement of Space Astronomy (AAS75-128)
        Arthur D. Code — 199

Contents (cont'd)   Page No.

    Concepts of LST Operation (AAS75-129)
        C. R. O'Dell   219

    Large Space Telescope Program Status (AAS75-130)
        James A. Downey III   223

    The LST Scientific Instruments (AAS75-131)
        George M. Levin   235

LST Mirror Development

    Mirror Substrate Material and Manufacturing for the Large
    Space Telescope (AAS75-175)
        William C. Lewis   245

    Fabrication and Test of 1.8-Meter-Diameter, High Quality
    Ule Mirror (AAS75-176)
        Richard J. Wollensak and Clarence A. Rose   249

    Design and Testing with a Reflective Null System (AAS75-177)
        L. Montagnino and A. Offner   261

    Test Results on Homogeneity of Expansion for a 1.8-M Ule
    Lightweight Mirror (AAS75-178)
        G. Friedman and G. Gasser   266

LST Telescope Performance

    Science Performance Considerations for the Design of LST
        Damon D. Ostrander and James C. Tuttle (AAS75-202)   271

    Optical Performance Control (AAS75-203)
        Terence A. Facey   275

    Impact of Focal Plane Dynamics on Image Quality (AAS75-204)
        William J. Pragluski, Peter W. Abbott, Jack F. Eastman   279

    Stray Light from Out of Field Sources (AAS75-205)
        Robert J. Noll   284

    Design of Highly Stable Optical Support Structure (AAS75-206)
        Michael H. Krim   288

LST Instrument and Detector Development

    Large Format SEC Orthicon Integrating Television Sensor
    for LST (AAS75-207)
        John L. Lowrance   293

Contents (cont'd) Page No.

    Thermo/Structural Design Considerations to Achieve the Large
    Space Telescope Line-of-Sight Requirements (AAS75-190)
        Domenick J. Tenerelli   301

    Design of Low-Thermal-Distortion LST Metering Structure
        John R. Lager (AAS75-191)   305

    3 Axis Simulation of the LST Pointing Control Subsystem -
    A Multi-Discipline Activity (AAS75-192)
        W. W. Emsley, T. D. Fehr, D. C. Fosth, and D. L. Knobbs   310

LST Data Management

    LST Data Management and Mission Operations Concept (AAS75-193)
        R. Walker, F. Hudson, and L. Murphy   314

    Data Management for Large Space Telescope (AAS75-194)
        G. R. Hope, Jr., and T. J. Rasser   318

    A Cost Effective Data Management Subsystem for the LST (AAS75-195)
        John A. Dougherty, Thomas D. Patterson, and Albert E. Cole   328

    System Consideration, Design Approach and Test of a Low Gain
    Spherical Coverage Antenna for Large Space Vehicles (AAS75-197)
        Manuel R. Moreno, Thomas D. Patterson, and Richard E.
        Ferguson   333

LST Mission Analysis and Operations

    System Application of the Fault Tolerant Memory (AAS75-296)
        L. J. Murphy   338

    A Scientific Operations Plan for the Large Space Telescope
        Donald K. West (AAS75-198)   342

    LST Operations, A Typical Day (AAS75-199)
        William J. Pragluski and Robert H. Brown   347

    Large Space Telescope Mission Analysis (AAS75-200)
        Frank M. Friedlaender   353

    Automation of the LST (AAS75-201)
        William W. Warnock and C. William Case   357

LST Maintainability and Operations

    Large Space Telescope External Interfaces (AAS75-179)
        Richard E. Collart   361

    LST Refurbishment and Support (AAS75-180)
        John Henschke   365

Contents (cont'd) Page No.

    The Intensified Charge Coupled Device as a Photon Counting Imager (AAS75-208)
        Jack T. Williams 371

    The Infrared Capabilities of the Large Space Telescope
        D. E. Kleinmann (AAS75-209) 374

    Development of an Infrared Spectroradiometer (AAS75-210)
        W. H. Alff and J. G. Thunen 379

    Faint Object Spectrograph (AAS75-211)
        William P. Devereux 382

    High Speed Area Photometer Conceptual Design and Integration
        William Bloomquist and Fred Steputis (AAS75-212) 385

    High Resolution Spectrograph (AAS75-213)
        Keith Peacock 389

    The European Space Agency Study of Photon Counting Imaging for LST (AAS75-294)
        R. J. Laurance 393

LST Precision Pointing and Control Systems

    An Analytical and Experimental Evaluation of Actuator Vibration on LST Image Distortion (AAS75-184)
        A. D. Houston, L. W. Hodge, Jr., and T. J. Kertesz 405

    Development of a Large-Inertia Fine-Pointing and Dimensional Stability Simulator (AAS75-185)
        R. L. Gates, D. H. Wine, R. W. Seiferth, and N. A. Osborne 409

    Evaluation of Communication Antenna Drive System Design Requirements to Allow TDRS Tracking During LST Fine Pointing
        A. J. Besonis and C. J. Chang (AAS75-186) 413

    A Small Instrument Pointing System for Shuttle Sortie Missions (AAS75-187)
        Carl W. Hendrikson and Ewald E. Schmidt 418

    A Fine-Pointing Facility for Spacelab Experiments in the 1980's - The Instrument Pointing Subsystem (AAS75-276)
        H. Heusmann and J. Collin Jones 422

    LST Interferometric Fine Guidance Sensor (AAS75-188)
        A. B. Wissinger and R. H. Carricato 426

    Prismatic Grating Star Tracker (AAS75-189)
        Allen H. Greenleaf 430

Contents (cont'd)   Page No.

    Simulation of the On-Orbit Maintenance Cycle for LST
        J. A. Donnelly (AAS75-181)   435

    LST Power System Long Life Design Techniques (AAS75-182)
        Owen B. Smith, Richard L. Donovan, and James L. Oberg   439

    Large Space Telescope - Orbital Crew EV Maintenance
    Operations (AAS75-183)
        H. T. Fisher   443

III. ATMOSPHERES, MAGNETOSPHERES, AND PLASMAS IN SPACE (AMPS)
     PROGRAM   447

    Challenges in Space Physics (AAS75-132)
        Billy M. McCormac   449

    Challenges in the Atmospheric Sciences (AAS75-133)
        Robert E. Dickinson   463

    AMPS Science Objectives and Philosophy (AAS75-134)
        E. R. Schmerling   471

    Amps Program Status (AAS75-135)
        Rein Ise   476

AMPS Science

    Some Experiments with Energetic Particle Injectors (AAS75-215)
        W. Bernstein and John R. Winckler   487

    AMPS Experiments Involving Gas Releases (AAS75-217)
        Lewis M. Linson   501

    LIDAR Systems for AMPS (AAS75-218)
        Richard D. Hake, Jr.   506

AMPS Engineering Challenges

    Observation of Spacecraft Generated Electrostatic Fields
    In the Vicinity of the ATS 6 Satellite (AAS75-220)
        Elden C. Whipple, Jr.   510

    Measurement of Static Electric Fields at the Surface of
    Satellites (AAS75-222)
        J. E. Nanevicz   519

    Preliminary Shuttle Payload Contamination Assessment (AAS75-228)
        E. B. Ress, R. O. Rantanen, and L. E. Bareiss   539

    AMPS Data Management Concepts (AAS75-229)
        P. N. Metzelaar   543

Contents (cont'd)

Experiment Integration of AMPS (AAS75-230)
   Robert Witholder 549

Ultraviolet Remote Sensing of Atmospheric Ozone from Payloads Using Shuttle Capabilities (AAS75-231)
   A. J. Krueger and D. F. Heath 561

Design of High Voltage Insulation Systems for Aerospace Equipment (AAS75-232)
   W. G. Dunbar 567

PART II

IV. SPACE TUG AND INTERIM UPPER STAGE (IUS) PROGRAMS 571

Space Tug Requirements and Planning (AAS75-137)
   J. Wild and M. D. Kitchens 573

Space Tug Baseline Description and Status (AAS75-138)
   William Teir and Alfred G. Orillion 591

Alternate Tug/IUS Approaches (AAS75-140)
   A. O. Tischler 611

Orbital Transportation in the 1980's and Beyond (AAS75-141)
   Hubert P. Davis 623

Space Tug Operations

An Integrated Approach to Tug/IUS Mission Operations (AAS75-152)
   Billy S. King and Roger A. Chamberlain 645

Tug Ground Operations - An Approach to Early Identification of Operational Impacts (AAS75-153)
   John L. Best 651

Space Tug Mission Operations (AAS75-154)
   Kenneth C. Nuss 656

Spaceflight Tracking and Data Network Support in the Space Tug Era (AAS75-214)
   L. R. Stelter and Robert D. Godfrey 661

Space Tug and Spacecraft Economics

The Space Tug Economic Analysis Study - What We Learned (AAS75-147)
   Charles V. Hopkins 685

A Survey of the Economics of Materials Processing in Space
   B. P. Miller (AAS75-149) 689

Contents (cont'd)   Page No.

    The Economics of Satellite Maintenance (AAS75-150)
        Wilfred L. DeRocher, Jr., and Richard G. Sosnay   711

Interim Upper Stage (IUS)

    Burner II Interim Upper Stage (AAS75-170)
        Henry Kudish   715

    Transtage Interim Upper Stage (AAS75-171)
        Peter B. Teets   719

    Centaur as an Initial Upper Stage for the Space Shuttle
        D. J. Jones and D. A. Heald (AAS75-172)   724

    Delta as an Interim Upper Stage (IUS) (AAS75-173)
        R. P. Dawson and J. F. Meyers   729

    Spinning Solid Perigee Stage (AAS75-174)
        H. A. Rosen, C. R. Jones, and L. M. Bronstein   733

    Agena Interim Upper Stage (AAS75-292)
        J. H. Guill   737

Space Tug Engines

    RL10 Derivatives for IUS/Tug (AAS75-155)
        J. P. B. Cuffe   759

    Advanced Space Engine Component Technology Status (AAS75-156)
        A. T. Zachary   763

    Aerospike Development Status (AAS75-157)
        J. Campbell and H. G. Diem   768

    Solid Rocket Technology Advancements for Space Tug and IUS
    Applications (AAS75-159)
        W. Ascher, R. L. Bailey, J. W. Behm, and W. Gin   772

Space Tug Propulsion Systems

    Requirements and Considerations in Selecting Space Tug
    Propulsion Systems (AAS75-160)
        Christopher J. Cohan   775

    Space Tug Propulsion Systems - Storable Versus Cryogenic
        W. E. Pipes (AAS75-161)   779

    Mixed-Mode Propulsion Systems for Full Capability Space Tugs
        R. Salkeld and R. Beichel (AAS75-162)   785

Contents (cont'd)   Page No.

    A Candidate Mission Using the Shuttle and Solar Electric
    Propulsion (AAS75-163)
        John H. Duxbury and Robert C. Finke   789

    Status Report on Nuclear Electric Propulsion Systems (NEP)
        J. W. Stearns, Jr. (AAS75-164)   799

Space Tug Mechanical and Interfaces

    Tug and Payload-to-Orbiter Interface Requirements (AAS75-165)
        Edward H. Bock   804

    Tug Payload Interfaces (AAS75-166)
        Fritz Runge   808

    Tug Rendezvous and Docking with a Spacecraft - A Remote,
    Manned Approach (AAS75-167)
        Michael J. Hurley   812

    Space Tug Thermal Control (AAS75-168)
        T. L. Ward   817

    NDT for Space Tug Thin Gage Materials (AAS75-169)
        Ward D. Rummel   822

Space Tug Avionics and Communications

    Tug Avionics System Overview (AAS75-142)
        Maurice T. Raaberg and James I. Newcomb   828

    Lightweight Fuel Cell Powerplant for Tug (AAS75-143)
        Lawrence M. Handley   833

    Space Tug Laser Gyro IMU   (AAS75-144)
        Robert F. Morrison   839

    Interferometric Landmark Tracker Applied to Precise Space
    Tug and Payload Navigation System (AAS75-145)
        D. H. Aldrich and W. F. Hubbarth   857

    LSI Computer for 1980's Space Mission (AAS75-146)
        W. A. Clapp, J. E. Saultz, and Dr. J. B. White   861

V.   SPACELAB AND AUTOMATED PAYLOADS PROGRAMS   865

Spacelab Operations

    Spacelab Task Allocation - A Preliminary Determination of
    Onboard Versus Ground Operational Priorities (AAS75-236)
        W. J. Harris and R. E. Holmen   867

Contents (cont'd) — Page No.

NASA/ESA CV-990 Airborne Simulation of Spacelab (AAS75-237)
D. Mulholland, C. Neel, J. DeWaard, R. Lovelett,
L. Weaver, and R. Parker 870

Spacelab Payload Accommodation (AAS75-238)
Donald M. Waltz 872

Concept Verification Test: Evaluation of Spacelab/Payload
Operation Concepts (AAS75-239)
R. O. McBrayer and H. H. Watters 879

Analysis of Extended-Duration Sortie Missions (AAS75-295)
Robert C. Ring and Wilton C. Lide 905

Spacelab Payload Planning

Space Transportation System Payloads Data and Analysis (AAS75-244)
J. D. Peterson and H. G. Craft, Jr. 910

Programmatic Aspects of German Shuttle/Spacelab Utilization
Gottfried Greger (AAS75-245) 915

Future Payload Technology Requirements (AAS75-246)
Howard M. Ikerd and Larry R. Alton 929

Computer-Aided Scheduling of Spacelab Ground Operations
J. K. Willoughby (AAS75-247) 934

Spacelab Integration

The Integration of Commercial Payloads into Spacelab (AAS75-240)
H. L. Bloom and K. R. Taylor 939

Low Cost Integration Techniques for Spacelab Payloads (AAS75-241)
C. A. Braunwarth and T. C. Aepli 943

Spacelab Payload and Program Planning in Germany (AAS75-243)
Horst Schreiber 947

Spacelab Resources (AAS75-297)
Hans M. Kappler 957

Spacelab Science Payloads

Astronomy Spacelab Payloads (AAS75-260)
Richard Ott and Gary Wengrow 962

Contents (cont'd)   Page No.

    Atmospheric X-Ray Emission Experiment for Shuttle (AAS75-261)
      R. A. Goldberg, K. L. Hallam, and J. G. Emming   968

    Spacelab Ultraviolet-Optical Telescope Facility (AAS75-262)
      Murk Bottema   973

    Adaptation of an Existing Cosmic Ray Ionization Spectrometer Experiment to Spacelab (AAS75-263)
      U. R. Alvarado, J. F. Ormes, and C. V. Stahle   977

    A Large Cooled Infrared Telescope Facility for Spacelab (-234)
      Stephen G. McCarthy, Lou S. Young, and Fred C. Witteborn   981

Spacelab Life Sciences and Applications

    Application of Space Shuttle to Fundamental and Applied Microbiological Research (AAS75-255)
      Jerry V. Mayeux   985

    Life Sciences Manned Payloads for Shuttle/Spacelab (AAS75-256)
      Dennis B. Heppner, Goerge L. Drake, and Chester B. May   989

    Shuttle Bioresearch Laboratory Breadboard Simulations
      S. T. Taketa (AAS75-257)   999

    A Life Sciences Spacelab Mission Simulation (AAS75-258)
      John A. Mason, F. Story Musgrave, and Dennis R. Morrison   1000

    Space Shuttle Trace Gas Analyzer (AAS75-259)
      Wallace Dencker   1003

Spacelab Technology Payloads

    An Orbiting Molecular Shield Vaccum Facility: A Materials Laboratory in Space (AAS75-248)
      James W. Youngblood, R. A. Outlaw, Leonard T. Melfi, Jr., and John R. McIlhaney   1007

    Advanced Extravehicular Mobility Unit Technology Experiments
      Gary Wengrow (AAS75-249)   1012

    The Advanced Technology Laboratory (AAS75-250)
      C. Llewellyn and R. Milliken   1018

    Shuttle Entry Technology Payloads (AAS75-251)
      Paul M. Siemers III   1023

    Landmark Tracking Technology (AAS75-278)
      J. D. Welch, W. E. Sivertson, and R. G. Wilson   1045

Contents (cont'd)   Page No.

Communications Payloads

Shuttle Communication Experiments (AAS75-252)
  John J. Woodruff and Donald R. Peters   1049

Large Deployable Antenna Shuttle Experiment (AAS75-253)
  R. E. Freeland, J. G. Smith, J. C. Springett, and K. E. Woo   1053

Availability of a Communications Satellite, Requirement and Feasibility (AAS75-254)
  Garry D. Gordon   1057

Multidiscipline Payloads

Mission Considerations for Multidiscipline Applications Payloads (AAS75-268)
  John M. Macdonald   1061

The Long Duration Exposure Facility - A Shuttle Transported Low-Cost Technology Experiment Carrier (AAS75-269)
  John D. Dibattista   1073

Payload Planning for the First Spacelab Mission - A European View (AAS75-270)
  A. V. Breitenstein and D. Davidts   1077

Spacelab Utilization in Different Fields of Science and Applications (AAS75-271)
  Jacques Collet   1089

Early Space Station User Accommodations (AAS75-291)
  Donald R. Saxton and Harry L. Wobers   1103

Automated Payloads

Automation of Space Processing Applications Shuttle Payloads
  Walter E. Crosmer, Oakley T. Neau, and James Poe (AAS75-264)   1107

Shuttle Benefits for Automated Retrievable Cryogenically Cooled Payloads (AAS75-265)
  John O. Simpson, Thomas M. Spencer, and William H. Follett, Jr.   1111

Retrieval of the HEAO-C Spacecraft with Space Shuttle
  David H. Mitchell (AAS75-267)   1133

The Multimission Modular Spacecraft for the 80's (AAS75-235)
  Robert O. Bartlett and Frank J. Cepollina   1137

Contents (cont'd)   Page No.

Automated Planetary Spacecraft

Shuttle/IUS Performance for Planetary Missions (AAS75-273)
M. J. Cork, J. M. Driver, and J. L. Wright   1171

Possibilities for Reducing High-Energy Performance
Requirements (AAS75-279)
G. R. Hollenbeck   1175

Impact of Space Transportation System on Planetary Spacecraft
and Missions Design (AAS75-274)
Phillip M. Barnett   1180

Using the Shuttle for Future Advanced Planetary Missions
L. D. Friedman and W. Scofield (AAS75-275)   1185

Panel Discussion

Can We Use the STS to Improve the Planetary Program?
Moderator - Daniel Herman, Panel Members - John Niehoff,
Jack Wild, Robert Parks, and Phil Culbertson   1197

VI. ENERGY IN THE SHUTTLE ERA   1221

Energy Research Overview/Alternatives for Energy Development
Thomas J. Vogenthaler (AAS75-280)   1223

The Satellite Solar Power Station - A Focus for Future Space
Shuttle Missions (AAS75-281)
Peter E. Glaser   1239

Nuclear Power in the Shuttle Era (AAS75-283)
S. R. Ross   1263

Availability and Variability of Solar Energy in the Rocky
Mountain Region (AAS75-285)
Richard C. Burriss   1279

Modeling the Western Coal Industry (AAS75-286)
William Ganter, Claude McMillan, and Fred Glover   1283

Minimum Cost Solar Thermal Electric Power Systems: A Dynamic
Programming Based Approach (AAS75-287)
William S. Duff   1288

100 MWe Solar Power Plant Design Configuration and Performance
F. A. Blake (AAS75-288)   1293

Contents (cont'd)   Page No.

    Minimization of Overtime Shift Hours with an L-P Model
        Gerald L. Kaes   (AAS75-289)   1301

    Deposited Today - Consumable Tomorrow Sewage and Solid Waste
    Hydrogenation (AAS75-290)
        Clyde W. LaGrone   1305

INTRODUCTION

SPACE TUG & INTERIM UPPER STAGE (IUS) PROGRAMS

SPACE TUG & INTERIM UPPER STAGE (IUS) PROGRAMS

Program Chairman   John H. Disher, NASA/Headquarters
Assistant   T. J. Goyette, Martin Marietta Corporation

Space Tug Operations
     Chairman   Sidney P. Saucier, NASA/MSFC
     Co-Chairman   Robert Mohling, Beech Aircraft Corporation

Space Tug and Spacecraft Economics
     Chairman   Joel S. Greenberg, Princeton University
     Co-Chairman   Aaron N. Silver, Martin Marietta Corporation

Interim Upper Stage
     Chairman   Lt. Col. Joel Rosenzweig, USAF/SAMSO
     Co-Chairman   Lt. Col. E. J. Bauman, USAF Academy/DFACS

Space Tug Engines
     Chairman   Robert D. Culp, University of Colorado
     Co-Chairman   Preston Layton, Princeton University

Space Tug Propulsion Systems
     Chairman   Preston Layton, Princeton University
     Co-Chairman   Robert D. Culp, University of Colorado

Space Tug Mechanical and Interfaces
     Chairman   Hugh P. Davis, NASA/JSC
     Co-Chairman   John Sterrett, Consultant

Space Tug Avionics and Communications
     Chairman   M. T. Raaberg, General Dynamics, Convair Div.
     Co-Chairman   Capt. A. C. Baer, USAF Academy/DFACS

AAS 75-137

SPACE TUG REQUIREMENTS AND PLANNING

J. Wild, M. D. Kitchens[*]

A brief background of the Shuttle upper stage program is provided describing its evolution to the current two phase program. This program consists of the Interim Upper Stage (IUS) to be followed later by the full capability Space Tug. The requirements imposed upon the Space Tug and their origin are discussed. The mission and payload requirements indicate that about 50% of the automated payloads require an upper stage. Other factors that influence the Tug design, such as Shuttle interfaces, launch operations and flight control are presented. Key driver requirements as known today are summarized.

The relationship between the IUS and Tug programs and their schedule milestones are discussed. The NASA activities that are necessary for the IUS program are addressed as are those study and simulation/demonstration activities planned for Space Tug.

---

[*] NASA Headquarters, Washington, D. C.

BACKGROUND

The Space Tug is an unmanned propulsive stage used to carry automated payloads to and from orbits beyond those which can be obtained by the Shuttle alone. It is an essential part of the Space Transportation System (STS).

The Space Tug has been actively studied since 1969. NASA and DoD conducted exploratory analyses during 1969 and 1970. NASA, DoD and the European Launcher Development Organization conducted system feasibility studies during 1970 and 1971. Detailed stage concept and analyses and applications studies were conducted in 1971 and 1972. All of these activities were supported by technology studies and advanced development laboratory tests. During 1973, NASA and DoD combined funds and jointly conducted extensive Space Tug System analyses. The data base thus formed covered a broad range of stage configurations, program concepts, operational characteristics, and technical approaches.

Both NASA and DoD conducted program assessments and came to the same conclusions. An advanced technology stage designed specifically for the projected 1980-1991 mission requirements had the potential for lowest operational cost and provided the highest performance capability. Additionally, this stage offered the best long term economic benefits because it provided lowest transportation costs and, ultimately, lower payload costs by allowing on-orbit payload refurbishment or payload retrieval with subsequent ground refurbishment.

The data also indicated that development of a Tug based on current technology and available for use on initial Shuttle flights required resources considerably beyond those expected to be available to NASA prior to the Shuttle initial operational date. The DoD was in a similar posture. Furthermore, in examining the transition of NASA and DoD payloads from current expendable launch system to Shuttle operations, two factors became predominant. The first was that the extent of modification to high priority, operational DoD payloads in this period must be minimized in order to provide uninterrupted capability during the early years of Shuttle operations when a back-up capability with current expendable vehicles may be required. The second was the lack of a hard requirement for high altitude payload servicing or payload retrieval during these early years. The considerations led to a two-phase program concept in which an interim upper stage of limited capability would be developed from a modification of an existing propulsive stage for early use in the Shuttle--to be followed later when requirements dictate and resources permit, by the new, more capable Tug.

NASA and DoD have agreed that DoD will develop the Interim Upper Stage (IUS) and NASA will plan for the more capable Space Tug to be available later. The IUS is to be available for civil and military missions when the Shuttle becomes operational. The time of introduction of the Space Tug will be determined by analyses of the requirements and economics of Tug traffic, including consideration of on-orbit servicing of payloads and payload retrieval. These capabilities will be provided by the Space Tug with its introduction to the STS inventory to meet the needs of all STS users during the Shuttle era.

REQUIREMENTS

The Space Tug is an integral part of the Space Transportation System and as such has requirements that originate from the interfaces with the Space Shuttle, the STS ground systems, the payloads and from the missions and operations. The Tug is a high performance, fully reusable stage that is carried into and returned from, low earth orbit by the Space Shuttle. After low earth orbit insertion by the Orbiter, the Tug and its payload are checked out prior to deployment. The Tug is then deployed and readied for main engine ignition to deliver one or more payloads to their final higher energy orbits. If payload retrieval or on-orbit servicing is required, the Tug will perform the rendezvous maneuvers and dock with the payload. For on-orbit servicing, the Tug will carry a service module which will perform the on-orbit servicing function. After completing this phase of the mission, the Tug will return to low earth orbit (with its payload on retrieval missions) where the Orbiter will rendezvous with and recover the Tug (and payload) after safing. The Tug will then be returned to Earth for maintenance and refurbishment in preparation for its next mission.

The Space Tug is required to operate in several mission arenas. The most energetic missions are the outer planet exploration missions which occasionally require expenditure of the Tug, or the addition of an auxiliary stage, to achieve the necessary velocity. The geosynchronous missions constitute the highest traffic rate. Missions requiring polar, sun-synchronous, and highly elliptical intermediate inclination orbits complete the basic mission arenas in which the Tug must operate. The different arenas impose varying requirements on the Space Tug in terms of thermal environment, navigation, communications and propulsion.

The Tug requirements include payload delivery and services capabilities beyond present day launch vehicles and the IUS. For example, multiple payloads will be deployed and retrieved in a single flight, thus reducing user transportation cost and allowing ground refurbishment of payloads. Recent analyses indicate a high potential cost saving for on-orbit servicing of payloads. This capability imposes rendezvous and docking requirements on the Tug similar to those required by payload retrieval missions, except that additional equipment for exchanging payload modules must be accommodated as well as the modules themselves. The increased payload delivery capability also offers the potential for lower spacecraft cost by easing the requirement for very lightweight designs.

The improved payload services planned for the Tug require that the Tug be designed to accommodate provisions for supplying payload power, telemetry, command and control, checkout, and possibly stimuli for calibration, during delivery to the payload's final orbit or trajectory. Final precision orbital placement and payload pointing are also required of the Tug.

Safety also imposes design and operational requirements on the Space Tug. The stage must be designed so that it can be safely processed for launch and be handled and refurbished after return from a normal mission. Return from an aborted mission imposes additional requirements on ground equipment at the launch site and at contingency landing sites. For example, removal of residual propellants and contents of high pressure containers may be required. Transport of the Tug in the Shuttle bay and orbital flight in the vicinity of the Orbiter impose requirements such as the ability to dump propellants during aborts, withstand crash landings, reduce operating tank pressures while in the bay, and safe a deployed Tug by command override from the orbiter crew. Caution and warning signals will be required to alert the crew in case of difficulties. Physical and functional interfaces with the Shuttle, the STS launch facilities and the flight control facilities also place requirements on the Space Tug.

In order to lower operating costs, the Tug must be reusable. This requirement results in further design and operations requirements such as easy rapid ground turnaround after each mission, design for maintainability, and long life highly reliable subsystems. Tug subsystems must withstand many cycles of operation with minimum maintenance costs. For example, the engine is expected to be capable of ten hours of operation, including 300 starts and stops between major overhauls and still deliver very high, predictable performance on each mission.

Mission and Payloads

Analyses have consistently shown that approximately 50% of the STS payloads must be delivered to orbits that are above the Shuttle operating altitude. One recent analysis indicates that 72 automated payloads will use the IUS and 245 will use the Tug, from a total number of 608 to be delivered by the STS. Of the 245 Tug payloads, 109 are DoD and the remaining 136 are NASA and commercial payloads. Twenty five per cent of the DoD payloads are delivered to geosynchronous orbits with the remainder to various altitude and inclination orbits. Sixty three per cent of the NASA and commercial payloads require geosynchronous orbits, 24% are planetary payloads, 9% are polar and sun-synchronous payloads, and the remaining 4% are to a variety of highly elliptical intermediate inclination orbits. Figure 1 shows this distribution.

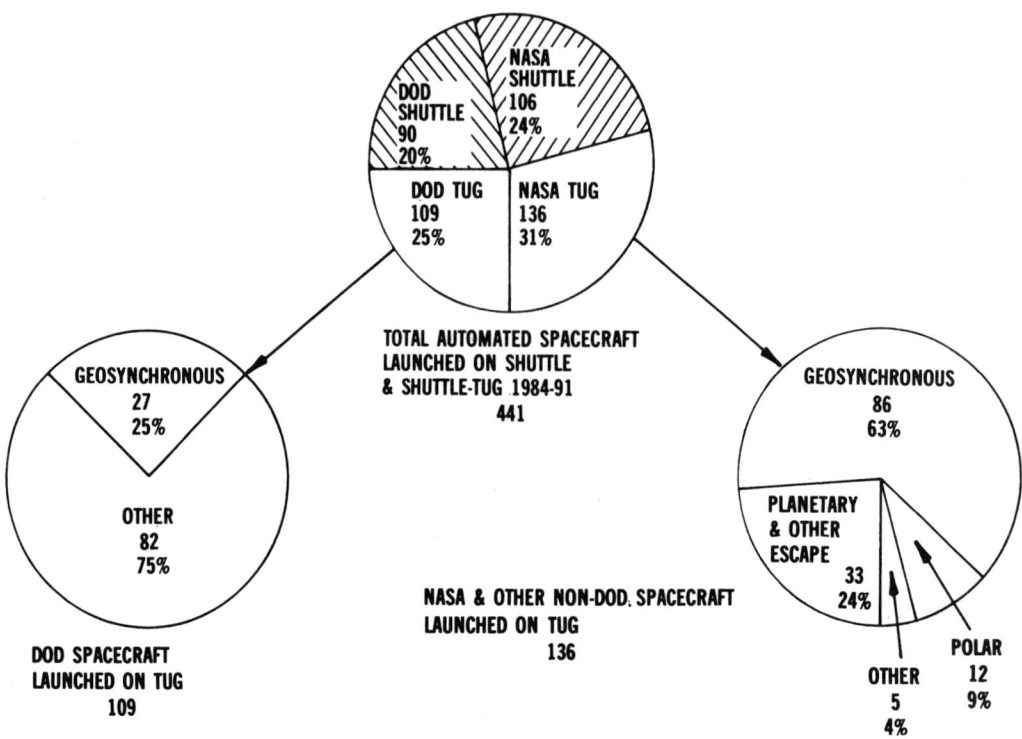

Fig. 1 Representative Tug Payload Distribution

A wide range of payload weights are to be delivered by the Tug to an equally wide range of velocities. Figure 2 shows payload weights and ΔV requirements for the different mission arenas during the Tug's operational phase. Note that the planetary missions require both the heaviest payloads (10,000 lbs at $\Delta V=12000$ ft/sec) and the highest velocities (1,000 lbs at $\Delta V=28,000$ ft/sec). Also, note the large number of payloads at geosynchronous velocities, 14,000 ft/sec.

The performance capability of the current Titan III E/Centaur with a TE-364-4 is also shown on Figure 2 to illustrate a representative energy boundry. Missions below and to the left of the curve can be captured, those to the right and above cannot. It is evident that the planetary missions are a strong Tug performance design driver. For geosynchronous missions, the Tug must deliver the indicated payload ($\Delta V=14,000$ ft/sec) and still have the necessary propellants to return itself, and possibly a retrieved payload, to the Shuttle orbiter for return to the ground. The very high velocity planetary missions require expenditure of the Tug possibly as a final flight after several less energetic missions. Alternatively, an expendable auxiliary stage may, in some cases, be more economically employed with the Tug for these missions and the Tug returned for reuse.

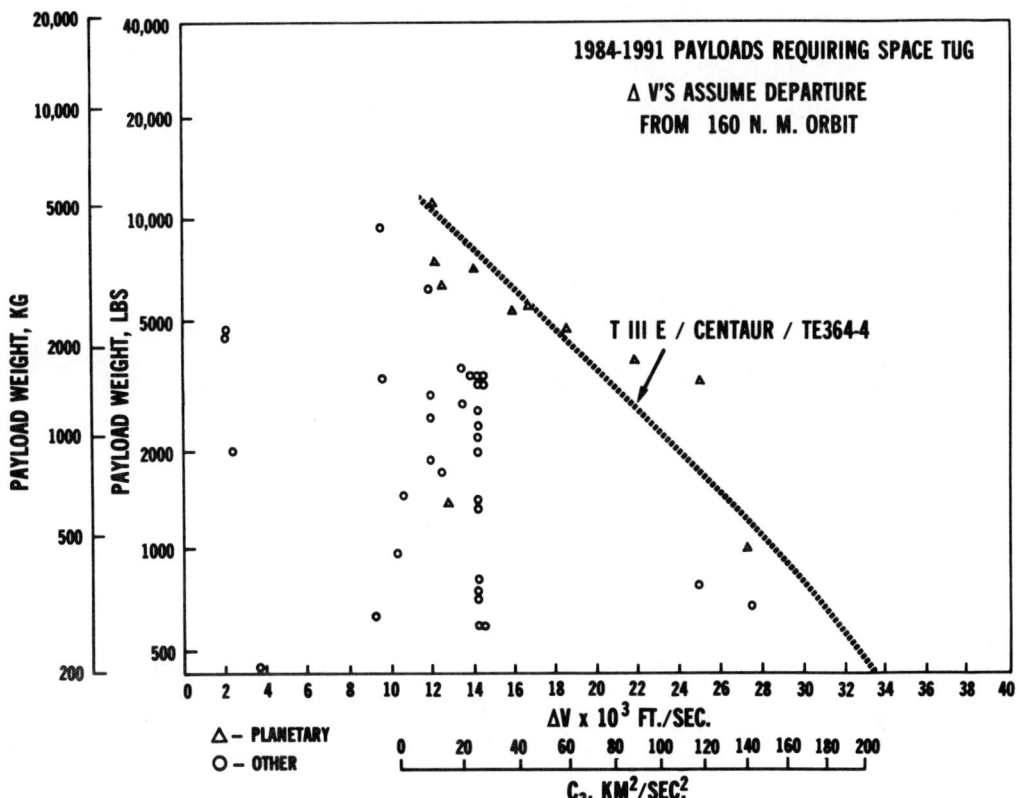

Fig. 2 Representative NASA Mission Model

Mission and payload requirements are summarized in Figure 3.

- **MISSION ARENAS — GEOSYNCHRONOUS, PLANETARY, SUN-SYNCHRONOUS, POLAR, INTERMEDIATE INCLINATION**
- **PAYLOAD WEIGHTS — FROM ZERO TO >10,000 LBS**
- **MULTIPLE PAYLOAD DELIVERY CAPABILITY**
- **PAYLOAD RETRIEVAL AND ON-ORBIT SERVICING**
- **PAYLOAD INTERFACES**
  - **STANDARDIZED MECHANICAL, FLUID, ELECTRICAL**
  - **TELEMETRY, COMMAND, TDRSS**
  - **INSERTION ACCURACIES**
  - **ENVIRONMENTAL**
- **AUXILIARY STAGE FOR SOME PLANETARY MISSIONS**

Fig. 3 Payload Requirements

Space Transportation System Interfaces

Since the Tug is processed at the launch site, carried to low Earth orbit in the Orbiter bay, and controlled from a flight control center, there are strong physical and functional interfaces between these STS elements and the Tug.

Shuttle

Requirements imposed by the Shuttle are highlighted in Figure 4. The Shuttle induced flight loads of about 3g strongly influence the design of the Tug to Shuttle attachments and the resulting load distribution paths through the Tug. The current baseline design provides for Tug support in the Orbiter bay via a cylindrical adapter that fits between the aft ring of the Tug and the rear longeron fittings of the cargo bay. In addition, forward fittings between the Tug and orbiter are used to aid in carrying vertical and side loads. This structural arrangement is similar to that of many other Shuttle payloads except that the aft adapter is also used to rotate the Tug to a position where the Orbiter manipulator arm, operated by the crew, can safely remove the Tug and position it above and clear of the Orbiter. A manipulator attach point is required on the Tug. Other means of deployment from the Orbiter have been and will continue to be examined. The Tug must be designed to withstand a 9g crash landing without breaking apart and causing further damage to the Orbiter and crew.

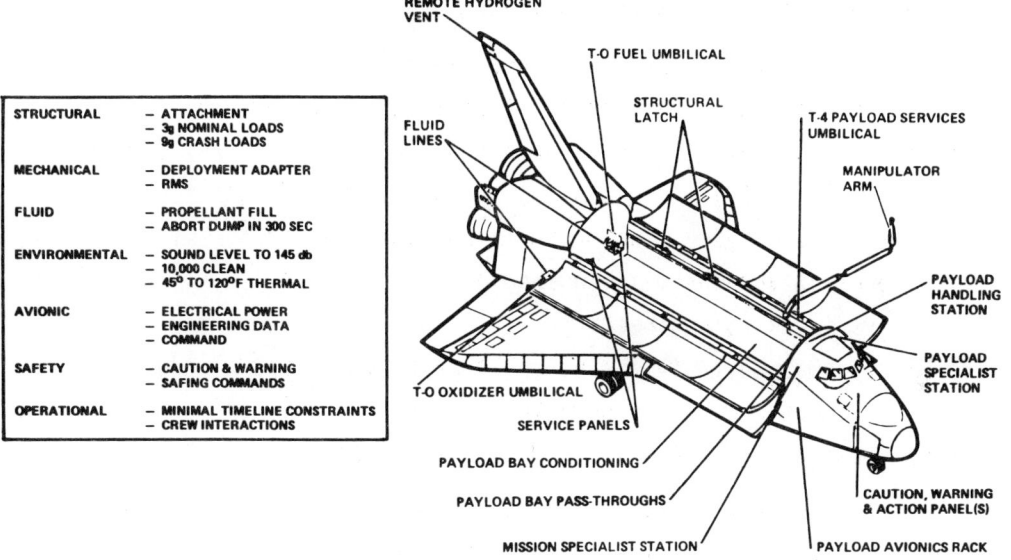

Fig. 4 Shuttle Interface Requirements

The fluid interfaces include means for filling and draining the Tug main propulsion tanks while in the bay on the launch pad. Propellant vent and stage purge lines are also required. These latter interfaces must be disconnected in space and later reconnected when the Tug is returned to the cargo bay for return to earth. One of the more demanding interface requirements is the ability to dump in flight all of the Tug propellants. Return to launch site aborts require that all propellants be dumped in about 300 seconds. Aborts to orbit permit more time for a propellant dump but have the additional requirement that the propellants being dumped provide the necessary acceleration for propellant settling.

The primary Shuttle environmental interfaces that result in Space Tug requirements are the sound pressure level of about 145 db and the cleanliness requirements. The cleanliness requirement is fundamentally a payload requirement. The Tug must be designed and operated in a manner that will not degrade payload cleanliness requirements. Some payloads may require 10,000 class cleanliness. It appears that this requirement can be met with a Tug that is processed in a factory clean environment if proper precautions are taken, such as careful control of the Shuttle's cargo bay ground purge during prelaunch operations and installation of barriers between the payload and Tug during Tug prelaunch processing. The Shuttle cargo bay thermal environment is expected to range from 45 to 120°F during entry and after landing.

Avionic and electrical interface requirements exist in the area of Tug engineering and status data transmitted to the Orbiter crew, and crew commands to the Tug. Safety information and command capability are required while inside the cargo bay and while outside the bay in the near vicinity of the Orbiter. Both visual and audible signals will be provided to the pilot and the mission specialist. The mission specialist will also be provided Tug engineering data. Combined payload and Tug electrical power while in the cargo bay can approach 3kw under some conditions, thus requiring operation of the Tug fuel cell during ascent to supplement the Orbiter supplied electrical power.

The Tug must be operationally compatible with the Shuttle. It must operate with acceptable crew capabilities and minimize timeline constraints. The Tug must conform to the 65,000 lb (Tug and payload) Shuttle cargo weight limits and to the 15 x 60 foot cargo bay volume allowing maximum payload volume consistent with good Tug performance, design practice, and low costs.

Launch Site

One fundamental launch site requirement is that Tug processes cause minimum impact on Shuttle ground operations. This is achieved by Tug pre-launch processing off-line to Shuttle processing. A Tug processing facility will be required to receive, checkout prior to launch, and mate and integrate spacecraft with the Tug. The Tug, with its payload, will be mated with the Orbiter at the launch pad thus requiring compatibility with the STS payload changeout room. Facilities will be required for Tug maintenance after return from a mission and possibly for storage between missions.

The facilities, crews, and operational processes must be adequate for about 20 missions per year on the average. The Tug launch requirements are not fundamentally different from those of the IUS, except for the safing and handling after a return mission and the refurbishment and maintenance requirements. Special requirements may exist for safing a Tug and its payload after return from an aborted mission. Figure 5 illustrates the basic launch site process and functions.

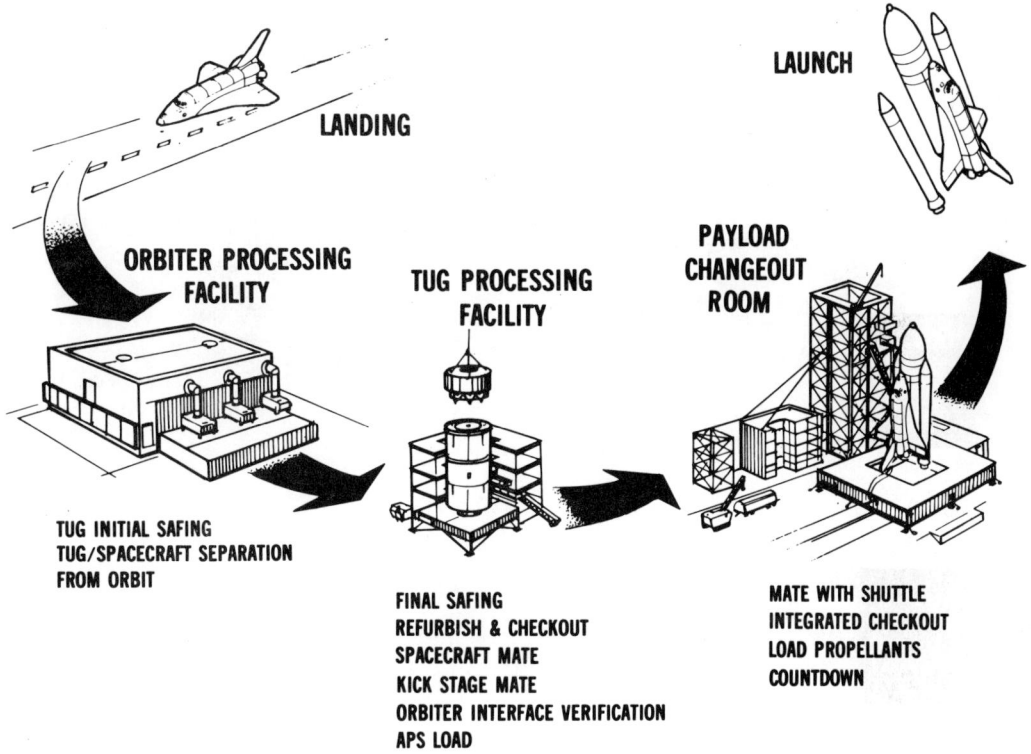

Fig. 5   Launch Site Operations

Mission Control

Mission control and flight monitoring generate requirements primarily in the areas of communications and computation. Figure 6 shows the communications links envisioned. The different locations in the figure are shown to illustrate the functions and interfaces; some of these may, at times, be co-located. For example, Tug mission control may be physically located at the Shuttle mission control site for NASA missions, but may be located elsewhere for DoD Tug missions.

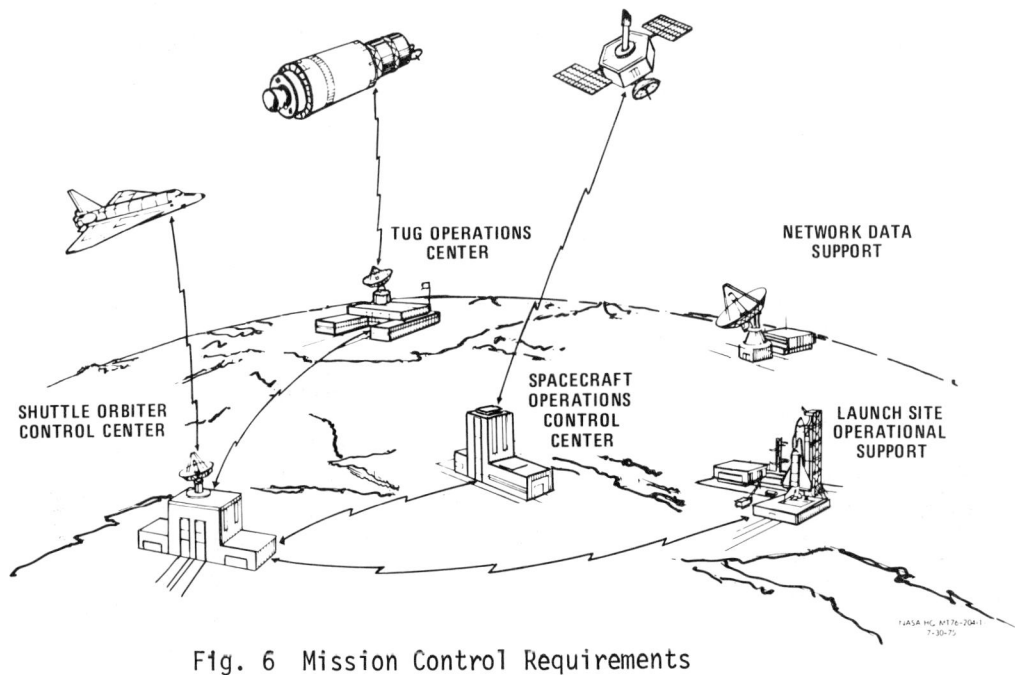

Fig. 6 Mission Control Requirements

The Tug operations center functions are primarily systems checkout and monitoring, trajectory coordination and control, systems command and contingency support as necessary, and payload requirements accommodations during transit. The functions of the Orbiter control center are Shuttle and Tug launch monitoring, Shuttle trajectory management, Shuttle systems monitoring and, as necessary, contingency support and rescue operations. The required spacecraft operations functions are prelaunch payload readiness verification, trajectory and ephemeris requirements, payload systems checkout and real-time monitoring, and long term on-orbit operations. Requirements for launch site "mission control" operations are related to preparation of the systems for launch, system readiness interfaces with mission control sites, analysis of prelaunch data, and countdown and launch operations. Communication network requirements include scheduling of the network, acquisition and distribution of real-time and dumped data, and processing support.

These mission control functions result in requirements for facilities, manpower, software, and mission planning and integration activities for the Tug that are not yet well defined. These requirements should, however, constitute relatively minor additions to those already existing for Shuttle and IUS if appropriate planning is accomplished early.

Key Requirements

Analyses of the known requirements placed on the Space Tug have identified the key, or driver requirements as shown in Figure 7. The most energetic missions are the planetary ones. The Pioneer Saturn/Uranus/Titan missions require ΔV's of about 28,000 ft/sec above the Shuttle's orbit with a payload weighing about 1,000 lbs. Another demanding planetary mission is the Mariner Jupiter Orbiter requiring a ΔV of around 22,000 ft/sec above the Shuttle's orbit with a payload weighing about 3,800 lbs. As noted earlier, these high energy requirements suggest the use of auxiliary stages to supplement the basic performance of the reusable Space Tug.

A second key driver is the high traffic rate. About 20 flights per year are anticipated with around 10 flights per year to geosynchronous orbit. This means that the Tug must be refurbished and missions planned and

- PLANETARY MISSION ENERGY LEVELS
  - $C_3 = 142$ KM²/SEC² WITH 500Kg PAYLOAD
  - $C_3 = 90$ KM²/SEC² WITH 1800Kg PAYLOAD
  - AUXILIARY STAGES
- MISSION FREQUENCY
  - ~20 LAUNCHES PER YEAR
  - ~10 GEOSYNCHRONOUS MISSIONS PER YEAR
- SHUTTLE AND GROUND SYSTEMS INTERFACES
  - 65 K LBS; 15' x 60' BAY
- SAFETY
  - MANNED COMPATIBILITY
  - PROPELLANT DUMP FOR ABORTS
- RETRIEVAL OF PAYLOADS AND/OR ON-ORBIT PAYLOAD SERVICING
- STAGE REUSE
- LOW COST
  - MINIMUM TUG OPERATIONS COST

Fig. 7 Driver Requirements

accomplished, including payload retrieval and multiple payload delivery, at a rate of about once a month for geosynchronous flights alone.

The Tug must, of course, be compatible with the Shuttle and ground systems. The Tug size and performance must be carefully matched to allow high performance and maximum payload volume with the 65,000 lb Shuttle capability and the 15 X 60 foot cargo bay.

The Tug must be designed and routinely operated in a manner that minimizes hazards and risks to personnel, the Shuttle and ground equipment and the environment. Manned compatibility is required when inside or in the vicinity of the Shuttle Orbiter. Propellant dump is required for aborts.

Payload retrieval from orbit and on-orbit payload servicing are key design requirements. Automated rendezvous and docking, and exchange of payload systems or modules appear to be technically feasible, but require further development and demonstration. The intent is to provide improved or updated payloads by ground refurbishment or in-space maintenance and systems updates, which should contribute to lower user costs.

If the objective of providing low user costs is to be realized, a key Tug requirement will be low operating costs. This requirement dictates that a reusable Tug be provided at a low procurement cost, that it be capable of deploying multiple payloads on a single flight so that users can share the cost per flight, and that it be capable of payload retrieval and/or on-orbit servicing to reduce the number of payloads to be procured.

A number of these key requirements are similar to those required of the IUS. The IUS to Tug transition must not impose undue interface changes to payloads, the Shuttle Orbiter, or launch and flight control ground systems.

The basic Tug requirement is to extend the STS benefits of low cost transportation to all NASA, DoD, and commercial payloads whose orbits are beyond the Shuttle operating regime. These requirements are based on the current NASA and DoD mission models which represent the most realistic projection of payload requirements available. As the IUS capabilities evolve and the payload requirements change and mature, these effects will be reflected in the Space Tug planning.

TUG PLANNING

Tug planning is closely related to IUS characteristics, capabilities, and utility. NASA's near term IUS planning consists basically of continued analyses of non-DoD mission and payload requirements; definition of the Shuttle, launch site, and flight control requirements and interfaces; auxiliary stage requirements and definition; and IUS mission capture and utility analyses, including the impact of the planning for transition from current expendable launch vehicles to the IUS. The IUS configuration selection and definition, along with the NASA and DoD agreements on operations responsibilities will allow definition of the NASA unique IUS hardware and software requirements so that these developments can be initiated in parallel with DoD IUS stage validation and full scale development phases. IUS processing facilities must be developed and activated by 1979 for a mid-1980 IUS launch at KSC. These facility developments should allow later introduction of the Tug with minimum impact.

The schedule relationship between the IUS and Tug programs is shown in Figure 8. DoD IUS development, NASA unique IUS developments, auxiliary stage design and development (phase C/D) and Tug design and development, all occur, or are initiated in the 1978 and 1979 time period, with significant funding required for all four activities in 1979. It is necessary that the NASA unique IUS developments be conducted in parallel with the DoD IUS development to support the mid-1980 first flight. The auxiliary stage development must also parallel the IUS development because its first flight is anticipated only about a year later. The integration of the auxiliary with the IUS must be carefully coordinated during IUS development. The need for early introduction of the Space Tug requires that Phase C/D be initiated in 1979. This Phase C/D initiation is also appropriate for good interaction with the auxiliary stage development since it is a goal to develop an auxiliary stage that can operate with both the IUS and Tug.

Other key NASA milestones are the definition of NASA unique IUS requirements and program definition (Phase B) for the auxiliary stage and the Tug. These three activities will be accomplished during 1976 and 1977. The approximate two year lead time between Congressional budget submissions and initiation of these activities requires near-term Agency

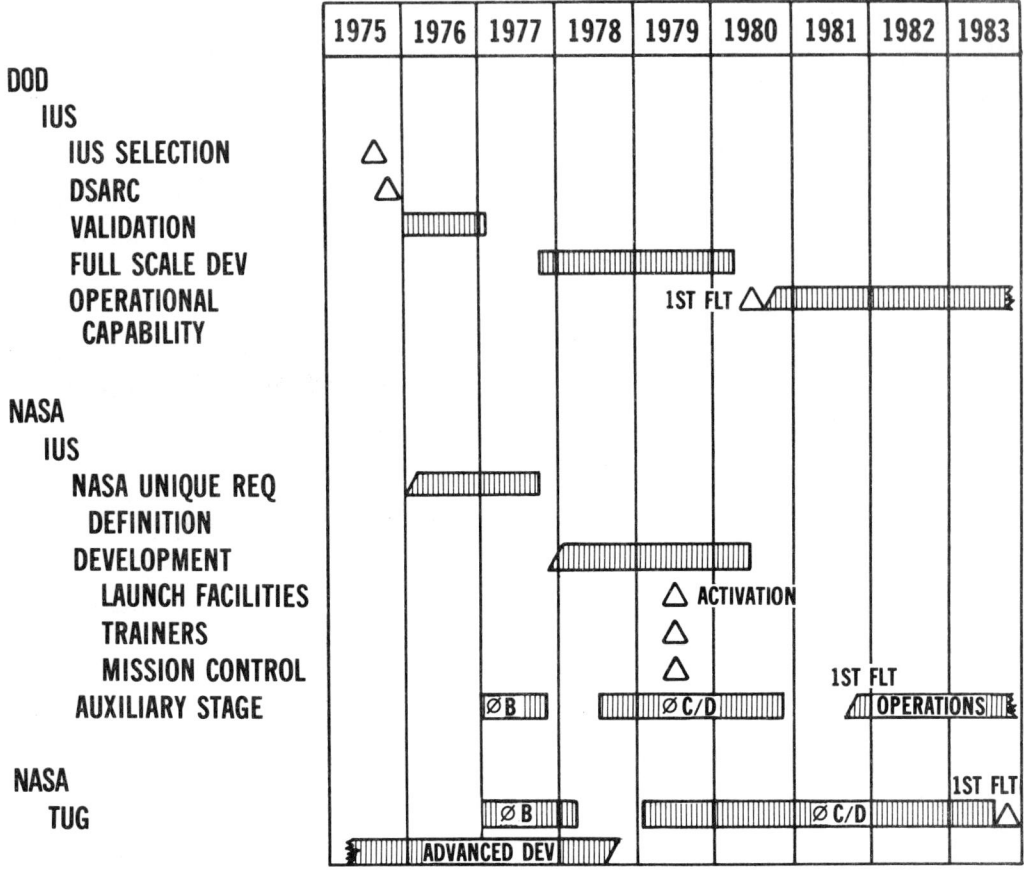

Fig. 8   IUS and Tug Milestones

decisions and firm commitments.

Specific NASA IUS and Tug plans for the near-term cover four general areas: definition of NASA unique IUS requirements and preparation for development, auxiliary stage feasibility analyses, Space Tug requirements and stage definitions in preparation for Phase B, and simulations and demonstrations of long lead-time, critical Tug systems and subsystems.

The NASA unique IUS efforts include further definition of payload and Shuttle requirements and interface definitions; of on-board flight software necessary for NASA missions, e.g., planetary; and of overall operational interfaces with ground systems, payloads and the Shuttle.

The KSC IUS software, launch processes, and facilities will receive increased emphasis as will the real-time Shuttle and IUS flight control interactions. Crew training requirements, including task procedures, and simulation requirements, need early definition.

Auxiliary stage planning includes completion of the current requirement study this fall. The requirement for a concept feasibility effort early next year will be examined in light of the IUS stage characteristics and the need for a program definition study to be initiated later next year.

Tug studies will examine the mission and payload requirements, including retrieval and on-orbit servicing, and Tug systems and their integration. Tug refurbishment requirements and techniques will be examined as will the IUS and Tug program interactions. These activities will be oriented toward preparation for Tug Phase B initiation in 1976 (FY 1977).

Tug simulations and demonstrations will provide laboratory data on hardware and software to help support preparation for Phase B and to identify potential problems and problem solutions for critical long lead-time Tug Systems. The performance of a high area ratio nozzle will be demonstrated on the RL-10 engine and firing of the Lewis Research Center Advanced Space Engine thrust chamber with its high area ration nozzle will be demonstrated. An aerospike thrust chamber will be tested. Simulations of Tug to payload docking and Tug to Shuttle deployment and retrieval are planned. These simulations will provide data on requirements, capabilities and limitations of sensors, of docking mechanisms, and on the dynamics to be expected in zero-g. Analyses of operations and techniques for checkout of the Tug while in orbit are planned. Selected avionics subsystems may also be demonstrated.

The baseline Tug program, the stage definition, capabilities, schedules and relation to the IUS are the results of work that was initiated as far back as 1969. It is based on a foundation of stage feasibility and design analyses, requirements and utilization analyses, complete system analyses, and program and economic analyses. The viability of the baseline program is dependent upon the firmness and accuracy of the payload and mission requirements, the nature of the IUS, and the funds available for development.

After the IUS configuration selection, the baseline Tug program will be reassessed and Phase B commitment decisions made for budget submissions. NASA and DoD agreements, with respect to both IUS and Tug must be finalized.

SUMMARY

The Space Tug is an integral part of the STS, being required for about 50% of the missions during its operational era. It is the second phase of the two phase upper stage program agreed upon by NASA and DoD as the most viable program approach. NASA and DoD have agreed that DoD will develop the IUS and NASA will plan for Tug development.

The baseline Tug program has been formulated on the basis of a long-term analysis of the mission and payload requirements, stage designs and capabilities, and resource requirements and availabilities. Key requirements are the high energy of the planetary missions, the high traffic rate of geosynchronous missions, and the capability for payload retrieval and/or on-orbit servicing.

The IUS and Tug programs are highly interrelated in terms of capability, funding and schedule. The urgency of Tug initial operational capability is dependent on the IUS capabilities and the funds available for development. The baseline Tug program is to be assessed after IUS configuration selection.

Based on the best data available now, the present baseline Tug program in conjunction with the IUS will provide the best national upper stage program for the STS operational era.

AAS 75-138

SPACE TUG BASELINE
DESCRIPTION AND STATUS
William Teir and Alfred G. Orillion*

INTRODUCTION

The highly versatile, reusable and space operational stage that is the economical extension of the Space Shuttle Orbiter operating regime, known as the Space Tug, is still in the conceptual design phase. When one speaks of a "Baseline Definition," it represents that concept that is configured to best meet defined or projected requirements at the time of planned operational capability. Consequently, the current Baseline Tug description discussed herein is that necessary to meet current payload mission model requirements in the mid-1980s for the least risk and development investment. Requirements are always dependent on national (commercial, scientific and military) and international space transportation needs.

CURRENT BASELINE SPACE TUG OVERVIEW

The Baseline Space Tug that is used as a point of reference today is a nominal 15' dia. by 30 ft. long LOX, LH_2 stage with a main propellant capacity of about 50,000 lbs. As shown in Figure 1, it is made up of subsystems and components that are for the most part being developed and are within the current state-of-the-art. Little new technological development is required for this stage, although some of the desired operational techniques will require development efforts. The Space Tug, as currently defined, is strictly an earth based space operational stage. Once deployed from the Shuttle in orbit, it is designed to automatically deliver spacecraft to a predetermined position and velocity and/or

---

* IUS/Tug Task Team, NASA George C. Marshall Space Flight Center

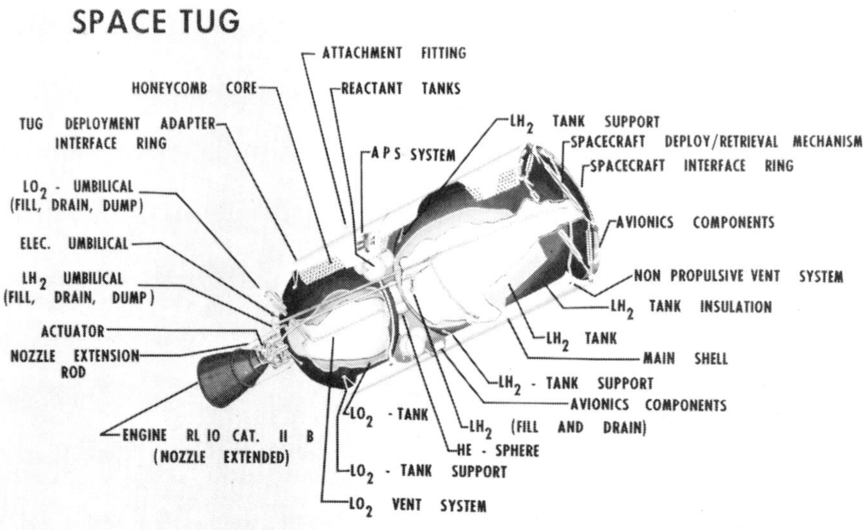

SPACE TUG

FIG. 1

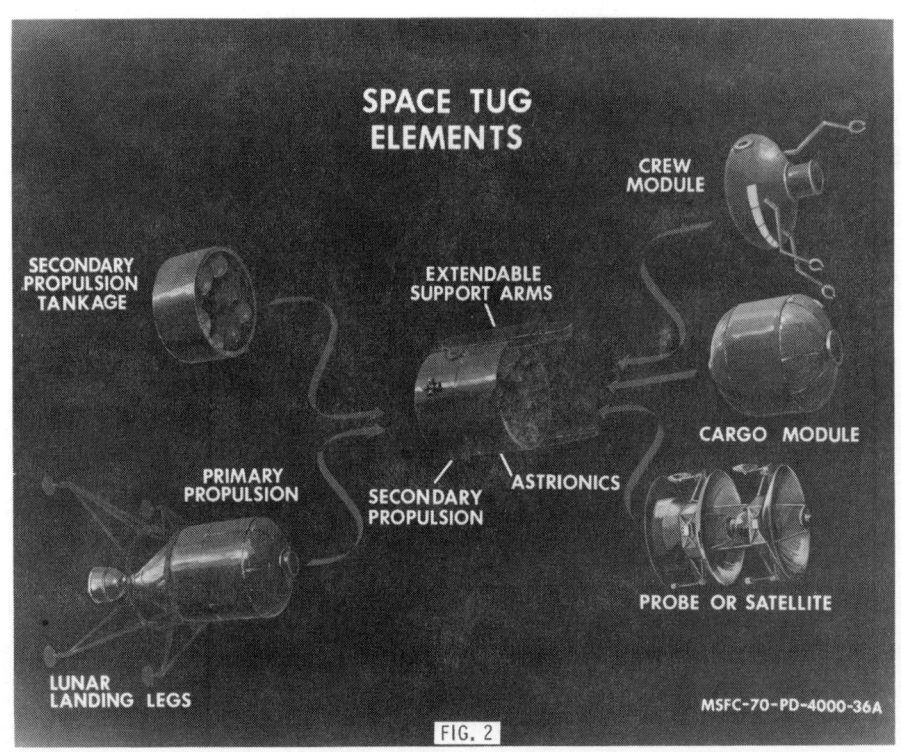

FIG. 2

rendezvous and dock with prepositioned spacecraft for servicing or retrieval. The Space Tug would normally return to the vicinity of the Shuttle Orbiter, with or without spacecraft, to be retrieved and returned to Earth in the Shuttle; there to be refurbished for another flight. Remote man control ability is also designed into the Space Tug to perform such intricate operations as remote spacecraft docking and servicing.

As a performance requirement reference, the current Space Tug can deploy about 8,000 lbs., retrieve about 3,400 lbs., and round-trip about 2,400 lbs. to geosynchronous orbit. The Components that were selected, including those that are modified, to make up this current Baseline Space Tug were evolved from a history of extensive and comprehensive trades analyses. From this background evolved a system that provides the required operational capability with the least risk for the lowest cost. In examining some of the subsystems, it is notable that many are, for the most part, already operationally proven. For example, the main engine is the current RL-10 modified for increased specific impulse (Isp) and idle-mode ability. Space proven 2219 Aluminum Alloy is used for the main tanks. The operational hydrazine Auxiliary Control System (ACS) is employed. Aviation demonstrated composites are planned for the shell. The newly introduced items will be primarily avionics and the spacecraft deployment/docking/retrieval system. Many components for these systems are available and the development job is primarily one of engineering design. Because of the continuing rapid advances in avionics technology, it is always prudent to wait to commit to the final components at the time of final design and development. Of course, the technological advances in the other subsystems areas will be utilized accordingly.

The previous analyses of many different subsystems, components, propellant types and capacities, from which the current Baseline concept evolved, were all affected by the mission requirements and available technologies. The basic Space Tug, as part of the Space Transportation System, was first identified in NASA's Integrated Manned Space Flight Program plan of 1969; the development history has been continuous since that time.

PAST SPACE TUG DEFINITIONS

The space operational stage, named Space Tug, from the Integrated Manned Space Flight Program was an all purpose system. Carried aloft by the Space Shuttle, it would deliver satellites; be a manned vehicle for Earth orbit operations; transported to a lunar orbit space station, it would make manned and unmanned sorties to and from the lunar surface. It was the system that interfaced with almost every item to be operated in space.

The initial definition studies on the Space Tug led to a concept having a number of elements and kit-on equipment to meet the many requirements. Examples of these elements and kits are illustrated in Figure 2. The propellant capacity for the primary propulsion module, which is the element analogous to the current Space Tug, optimized out at about 45,000 lbs. This was based on the earth orbital and lunar sortie operations.

The secondary propulsion module which could be cryogenic or storable, is sized for low ΔV long payloads. The reusable elements could be space based and returned to Earth occasionally for major refurbishment.

As time passed and the Shuttle mission model for the 1980s became more clear, interest was concentrated on unmanned spacecraft delivery and retrieval, although manned operations were considered up until the Systems Studies of 1973-1974. This change of requirement with time is outlined in Figure 3. In the Integrated Manned Space Flight Program era the requirement for 10,000 lbs. round-trip to geosynchronous was

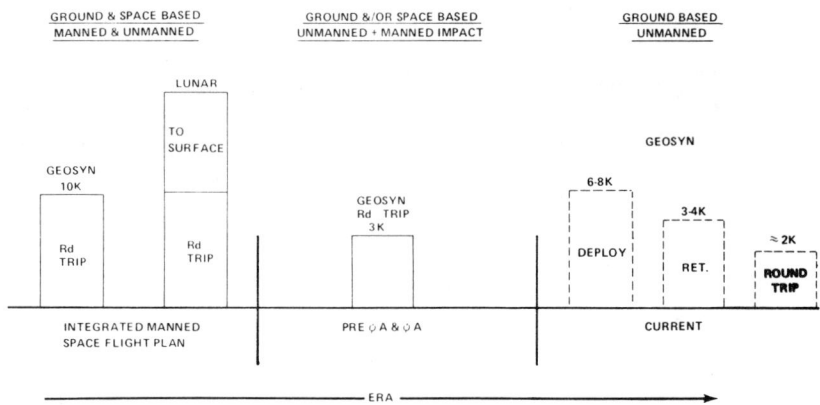

FIG. 3

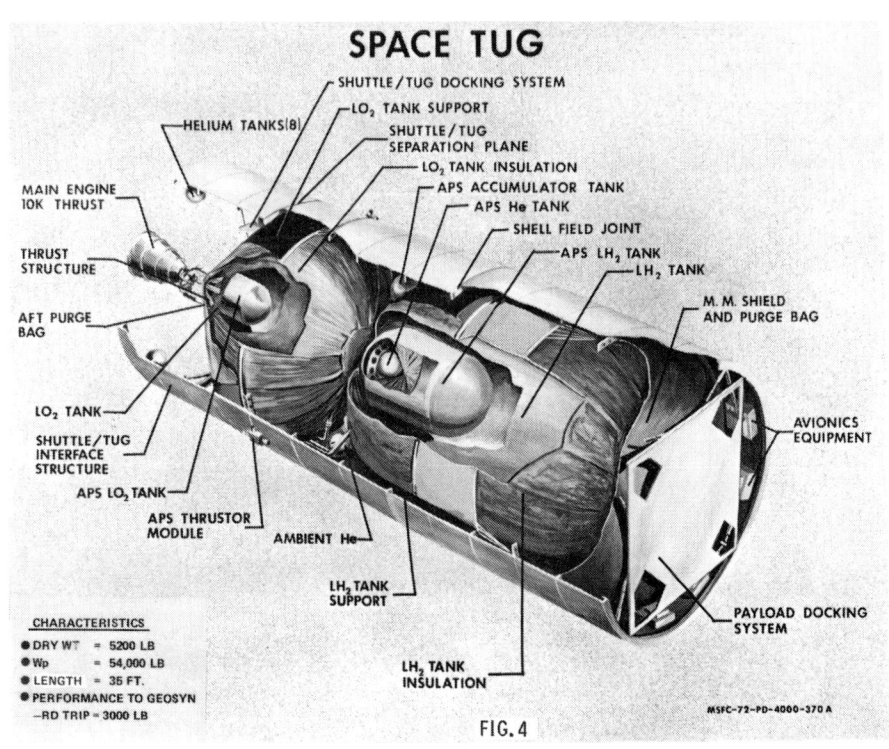

FIG. 4

accomplished by tandem staging; by means of on-orbit assembly via two Shuttle launches.

Following the Integrated Manned Space Flight Program Space Tug Baseline concept was the era of pre-Phase A and the Phase A efforts. During this period a Space Tug concept was defined that could round trip 3,000 lbs. to geosynchronous orbit via one 68K lbs. Shuttle. This concept was known as the "Point Design" Space Tug. To achieve this ambitious performance level, it required the utmost in technological demand. Some of the characteristics are described in Figure 4. This stage would need light weight, high performing components throughout. It was optimumly designed based on the Shuttle's capability and the implied performance needs. Some of the components needed to accomplish this performance level included a new high performance LOX/LH_2 engine, LOX/LH_2 APS, thin material tanks and structures, light weight efficient insulation and an autonomous light weight avionics. As before, all subsystems were to have long life and have economical operations which included ease in refurbishment. Such high technological requirements led to high development costs. The optimization to achieve this capability led to a stage length of 35 ft., thereby leaving only 25 ft. for spacecraft.

During the pre-Phase A and Phase A era, various conceptual design analyses were made, both by the United States and ELDO. Impacts were analyzed for space based and ground based Space Tugs of different lengths and capacities. Cost comparisons were also made for customizing Space Tug sizes for mission requirements. A concept was examined using a short Space Tug to handle the long low ΔV spacecraft and then using them in tandem for the high ΔV spacecraft, that were nominally shorter. An example of this arrangement is shown in Figure 5. Although the short Space Tug concept facilitated the long

FIG. 5

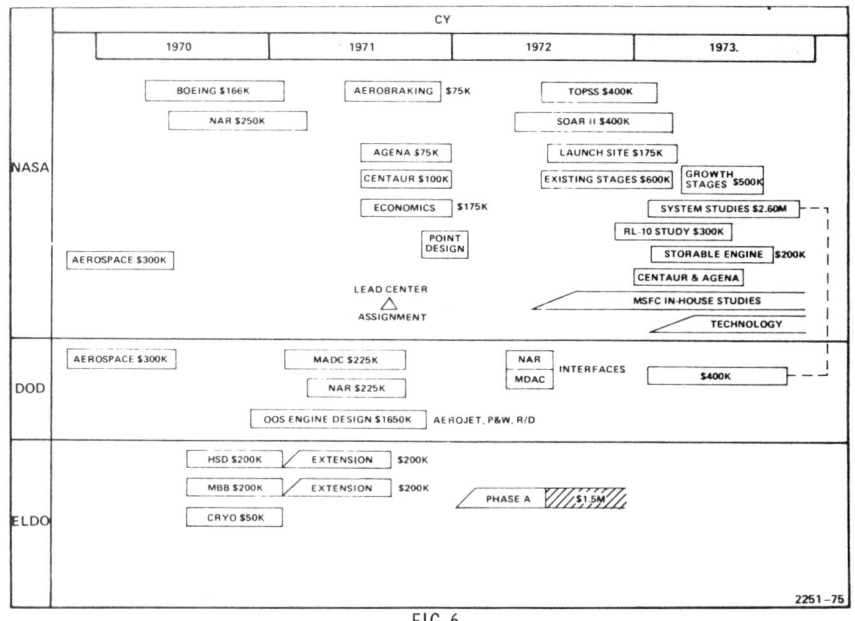

FIG. 6

spacecraft, the performance loss for geosynchronous missions was large. Consequently, the Space Tug design approach to meet such a high capability would be aimed toward a single larger stage. However, it was clear that if the requirements were reduced, a smaller stage approach would offer some flexibility. This could be done by either utilizing a primary smaller Space Tug concept or applying an auxiliary or kick stage. This kick stage would be effective in conjunction with the tandem assembly, or larger Space Tug for that matter, for very high energy missions as well as handling long low ΔV spacecraft.

By the beginning of the Systems Studies in 1973, a considerable amount of analyses had been made on Space Tug options and related activities. These studies are outlined in Figure 6. They included the efforts performed and sponsored by NASA as well as those by DOD and ELDO. Because of the development cost of a Space Tug, some effort was also initiated considering the use of existing stages as an interim to Space Tug operations. Also a novel concept was the application of aerobraking in returning from high orbits to a low Earth orbit. This aerobraking technique could increase the payload carrying capability considerably, as much as three times the round trip mass to geosynchronous orbit. Evaluations were also made on the consideration and impacts of various mission models that would effect the Space Tug. At this point in time three major points affecting the situation became prevelant. First, the full capability Space Tug could not be developed in time for its IOC to be commensurate with the Shuttle in 1980. Second, the high development cost of this type Space Tug would undesirably peak simultaneously with the Shuttle development costs. Third, the type of mission model envisioned for the 1980s had a reduced requirement on a space stage, particularly in the early 1980s. Although it was evident then as now that a high performing reusable space stage will always prove to be the most economical approach in the long term, development funding requirements as related to time is always a deciding factor. Consequently, the approach at that time was to examine the impacts of an interim stage followed by a Space Tug designed to meet the later more stringent requirements. The interim stage could be either an existing stage or a low development cost, limited capability, Space Tug type stage available early that would be later phased-developed into the desired higher performing Space Tug. Also under consideration was the impact of having the Space Tug use storable propellants as well as cryogenic. As mentioned in the earlier studies, it was shown to be beneficial to have a short stage (storable or cryogenic, prime or auxiliary), particularly for the long spacecraft. A storable propellant Space Tug would offer this shorter length; coupled with the high density propellant and resulting high mass fraction may deliver the required capability.

To evaluate the overall situation, the Systems Studies and other analyses and efforts were initiated to assess the overall problem. This assessment arrangement is described in Figure 7. The controlling parameter was cost. This included initial investments, time of peak funding, and total operations costs through the 1980s involving amortizations, transportation and payload benefits costs. The affecting parameters were concepts, IOC dates, Shuttle constraints and the Mission Model. Identified in these activities were a number of Space Tug concepts and interim stage options. The net result of the assessment was the need for a cryogenic Space Tug as early as practical but with a lesser performance capability need than previously identified. The application of a low cost Initial Upper Stage (IUS) was planned to resolve the early funding concern.

In an attempt to optimize the Space Tug length vs. performance, an analysis was made, based on the current Mission Model. The results of this analysis is illustrated in Figure 8. It was found that a 35 ft.

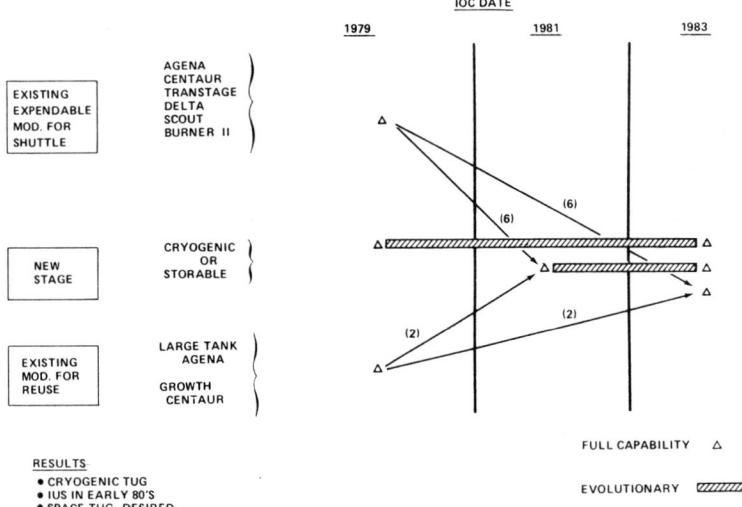

FIG. 7

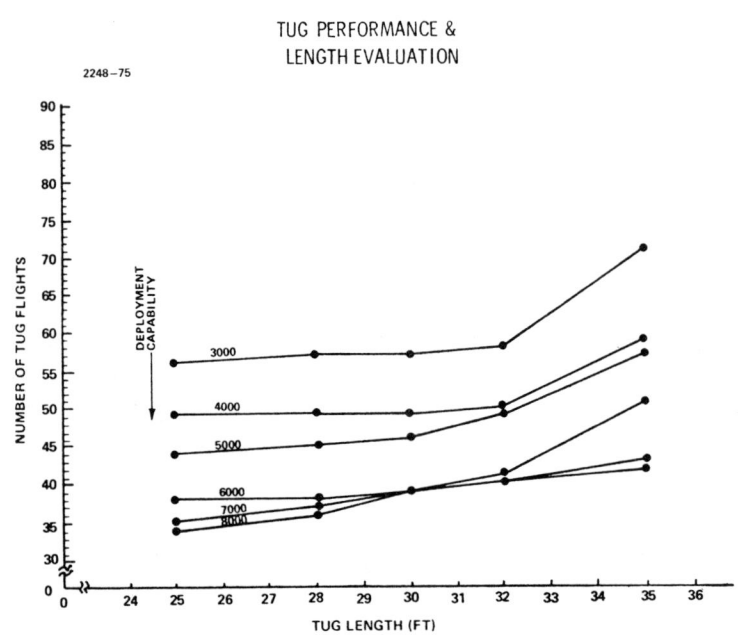

FIG. 8

length Space Tug would require more flights, and hence, greater costs than one shorter. A break in this flight versus length plot was found at about 32 ft. From there to a 25 ft. length was little reduction in the required number of flights to meet the then reference Mission Model. A Baseline Space Tug length of 30 ft. was selected. This gave adequate capability for efficient multiple spacecraft deployment with adequate retrieval capability and also allowed a margin for growth.

The output of all these efforts was the current Baseline Tug Definition and is directly related and dependent on the required Mission Model, and of course, by the Shuttle capability. The comparisons of cryogenic and storable propellant Space Tug systems is shown in Figure 9. The key to cost effectiveness gained from the current Mission Model is the heterogeneous (round trip) requirement. A Space Tug that can deliver 5,000 to 6,000 lbs. and retrieve a spacecraft of about 1,500 lbs. after deploying spacecraft(s) to geosynchronous orbit is the most overall cost effective system; all items considered. Hence, the single stage cryogenic Space Tug is the most effective

CURRENT BASELINE SPACE TUG CHARACTERISTICS

The current Baseline Space Tug is that resulting from all the aforementioned studies, analyses and requirements. The characteristics of this system are summarized in Figure 10. It is a point of departure for the continuing efforts in support of the Space Tug. As applicable technological advances and/or better component definitions become available or the requirements change, this Baseline will change. The development cost of this current Baseline Space Tug is estimated to be about 60% of the previously described Full Capability Space Tug. It is interesting to note, however, that this current Baseline Space Tug employs some structures, materials and avionics components considered to be high technology and, hence, high risk/high cost, a few years ago. The aforementioned steady timely advancements in technology

CHARACTERISTICS SUMMARY

A summary of the characteristics of the presently defined baseline Space Tug is presented below. The intent of this baseline is to provide a point of departure for future studies and analyses. These characteristics are expected to change and become firmer as the baseline evolves as a result of future Tug activities.

VEHICLE DESCRIPTION

ENGINE Pratt & Whitney RL-10-IIB (Retractable Nozzle)
ACTUATOR Hydraulic
APS SYSTEM 24 Hydrazine thrusters (25#)
STRUCTURE
　Skirts - Graphite Epoxy/Aluminum Composite
　Tanks - Aluminum Alloy/Elliptical Bulkheads
　Tank Supports - Fiber Glass Struts
　Thrust Structure - Fiber Glass Strut Truss
THERMAL CONTROL SYSTEM
　Tank Insulation - Goldized Super floc
　Active System for Fuel Cell
　Heat Pipes for other Avionics
PAYLOAD CAPABILITY TO GEOSYN-
CHRONOUS ORBIT
　Deploy   7926 lbs
　Retrieve   3396 lbs
　Round trip 2070 lbs -- 2400 lbs
AVIONICS SYSTEM
　Antenna - Electronically steerable phased array
　Platform - Strapdown
　Power - Fuel Cell (2) plus Battery
　Data Management - Data Bus
　SC Retrieval - Laser Radar
　SC Deployment Inspect - TV

MAIN ENGINE PERFORMANCE

	THRUST(LBS)	I_{sp}(SEC)
Full	15000	456.5
Pumped Idle	3750	434.7
Tank Head Idle	157	377

VEHICLE CHARACTERISTICS

Length 30 ft
Diameter 14.67 ft.
Dry Weight 5150 lbs
Burnout Weight 5755 lbs
First Ignition Weight 56,779 lbs
Deployment Adapter & Shuttle Systems 1900 lbs
Ground Liftoff Weight 58,679 lbs

PAYLOAD SENSITIVITIES*

	DEPLOY ONLY	RETRIEVAL ONLY
$\frac{\partial PL}{\partial W_S}$	-2.62	-1.38
$\frac{\partial PL}{\partial I_{sp}}$	0	0.23
$\frac{\partial PL}{\partial O}$	-0.38	0
$\frac{\partial PL}{\partial I_{SP}}$	83 LB/SEC.	59 LB/SEC.

ROUND TRIP
$\frac{\partial PL}{\partial I_{sp}}$ 43 lb/SEC

*Partials are explained in Section 2.2.1

FIG. 10

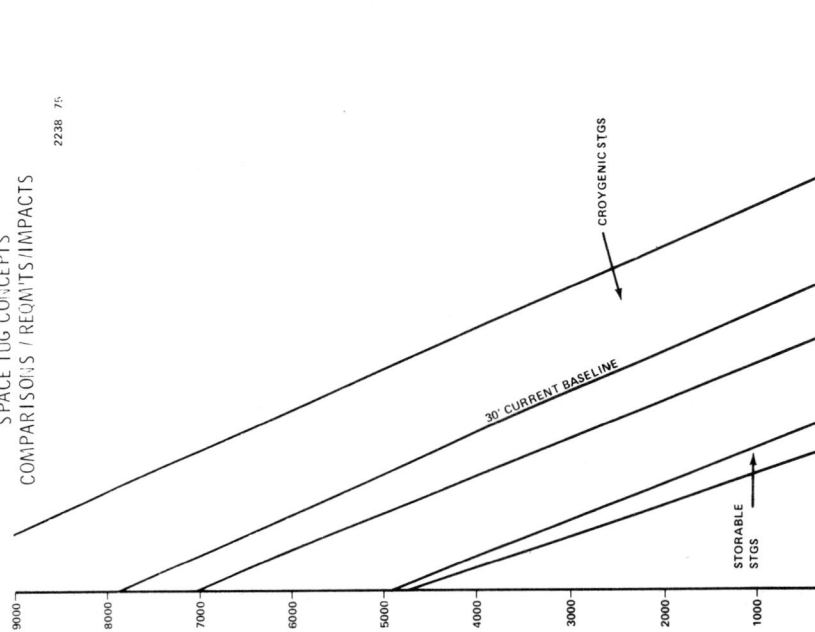

SPACE TUG CONCEPTS
COMPARISONS / REQMTS/IMPACTS

FIG. 9

have brought these elements within acceptable risk. The technology commitment for this current Baseline Space Tug is that available in 1978.

As mentioned, the effects of technological gains and applications have led to weight reductions and increased performance in some subsystems. A summarization of some of these are described in Figure 11. Noteworthy of these is in the avionics subsystem. The general functions of the avionics has been about the same from the initial Space Tug definition to the current Baseline; however, the weight has been cut in half. This is attributable to technological advances leading to lighter computers, guidance and control, power generation and other components. For example, the utilization of the Light Weight Advanced Fuel Cell over a Modified Shuttle Fuel Cell will reduce the component weights over 200 lbs. The structure weights were reduced about one-third by the application of thinner tapered tank walls and light weight composites. Propulsion subsystem weights, as would be expected, have remained about constant. The resultant gains in technology over the years has allowed the utilization of lighter weight developed components and materials with few changes giving a low cost Space Tug that can meet the current requirements.

Some of the specific design characteristics of this current Baseline Space Tug are worth noting. These are outlined in Figure 12. It is, of course, designed to meet the standards required of the Space Shuttle. Being reusable, the number of missions before scraping is of interest. Applying a component and subsystem interchange philosophy, it would be difficult to establish a true "life limit" to an initial Space Tug. Hence, for evaluation purposes, an overall average life of all of the components is estimated between 20 and 100 missions. It is to have

SUBSYSTEMS IMPACTS VS. TIME

ITEM	INITIAL	PRE-ØA & ØA	POINT DESIGN	CURRENT
ASTRIONICS	2000#	500-800#	960#	920#
PROPULSION (RL-10)	1400#	-	1350#	1300#
STRUCTURE (α wp)	2900# (wp=45K)	-	2000# (wp=54K)	1970# (wp=50K)
(α wp)	0.876	0.884 - 0.869	0.885	0.8837
MAIN ENGINE	RL-10 (Isp=444 sec.)	Advanced (Isp=460 sec.)	Advanced (Isp=470 sec.)	Mod RL-10 (Isp=456 sec.)

FIG. 11

CURRENT BASELINE SPACE TUG
SPECIFIC DESIGN CHARATERISTICS

o FUNCTIONAL AND PHYSICAL COMPATIBILITY WITH SHUTTLE PER "SPACE SHUTTLE SYSTEM PAYLOAD ACCOMMODATIONS, LEVEL II, PROGRAM DEFINITIONS AND REQUIREMENTS, JSC 07700, VOLUME XIV."

o REUSABLE - 20 MISSIONS MINIMUM, 100 GOAL.

o EARTH ORBIT DEPLOY 1 TO 3 SPACECRAFT, RETRIEVE 1.

o AUTOMATIC FLIGHT OPERATION WITH OPTIONAL GROUND COMMAND AUGMENTATION/OVERRIDE.

o AVIONICS - FAIL OPS/FAIL SAFE (MINIMUM)

o APS - FAIL OPS/FAIL SAFE

o PROPELLANT SYSTEM - REDUNDANT VALVING THROUGHOUT

o SAFETY FACTORS

  - STRUCTURES:

	ULTIMATE	YIELD
o IN SHUTTLE	1.4	1.1
o IN SPACE	1.25	1.05

  - RELIABILITY

    - MISSION SUCCESS = 0.97 (6 DAY MISSION)

    - SAFE RETURN (STS) = 0.999

FIG. 12

multiple spacecraft deployment ability. Being able to dock with on-orbit spacecraft, the Space Tug can perform either retrieval or servicing operations. Flight operations can be with different levels of autonomy. Subsystems having many components, i.e., avionics, APS, propellant, are designed with redundancy. This meets with the safe operating requirements and gives a mission success reliability of 0.97. The structures were designed primarily to the Shuttle induced environments.

The earliest possible schedule, shown in Figure 13, for Space Tug development that would result in an IOC of December 1983, is to conduct Phase B in about two years from now followed by Phase C/D two years later. This creates a situation where interface inputs must be provided to the spacecraft community, ground supporting elements and to the Space Shuttle all now under development. This is being resolved by the application of three inter-related major activities. These are: 1. Supporting Research and Technology (SRT) efforts oriented in direct support of the Space Tug. The favorable results of these continuing activities have been described earlier. 2. Supporting Studies being conducted to analyze the various Space Tug operations, both ground and flight, the subsystems designs, the interfaces - to spacecraft, to Shuttle, to ground, and the impacts of Mission Model perturbations. 3. Coupling these two major efforts are the unique application of some early test activities normally performed during later Phase B and C/D, called Simulation/Demonstration. By taking the components and operation procedures developed under the SRT effort and utilizing the approaches, subsystems designs and procedures defined in the Supporting Studies, laboratory controlled simulation tests are to be conducted involving the inter-relationships and interfaces, to demonstrate the abilities, or determine the abilities, of the subsystems. These physical tests, that include both hardware and computers, help determine the specific parameters that need to be put into the other developing programs. In essence, it is a method to verify the "paper analyses" and provide a basis for optimum design and operations selection, resulting in reduced risks, and leading to reduced development costs. As an example of potential cost savings, estimates for the avionics subsystems are that a 4% to 6% investment of the total required, in technology efforts, both SRT and Simulation/Demonstration, now would yield a 25% development savings later. This same cost ratio would probaby apply to many other systems of the Space Tug. In addition, by getting hardware and software test operations started early with Simulation/Demonstration the foundation will be set for the needed interface and operations simulation systems when the Space Tug is phased into the Space Transportation System (STS).

GROWTH POTENTIAL

Over the years at each stage of Space Tug definition or "Baseline" point, the concern at the time is the next progressive step. Historically this has been to both increase size and capability as well as to decrease size and capability. As described in this paper, the Space Tug definition is directly dependent upon its requirement, viz., the current projected Mission Model, and lift capability of the Space Shuttle to low earth orbit. As it is generally desirable to extend man's operations in space, it is desirable to consider the Space Tug extension to a manned carrier as it was defined originally. For low earth orbit operations, a single Space Tug via the current Space Shuttle could do some ΔV limited manned activities. For high ΔV flights, e.g., large plane changes or geosynchronous operations, tandem staging via two Shuttles, as described in the initial Space Tug concepts would be most useful. Since the current Baseline Space Tug definition evolved from manned Space Tug concepts with few detailed design demands changed, it would be applicable as such with little impact. This is

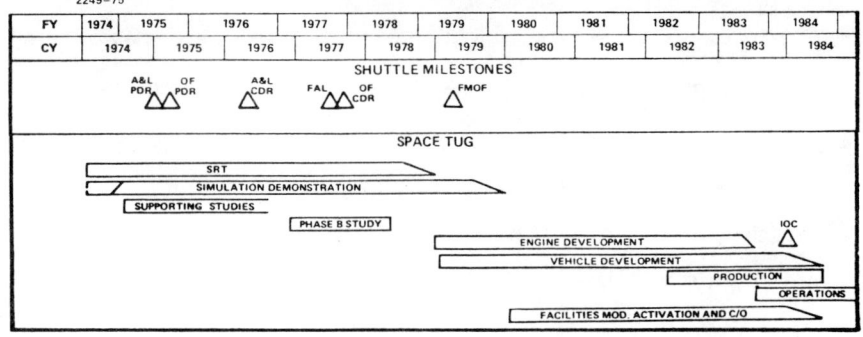

FIG. 13

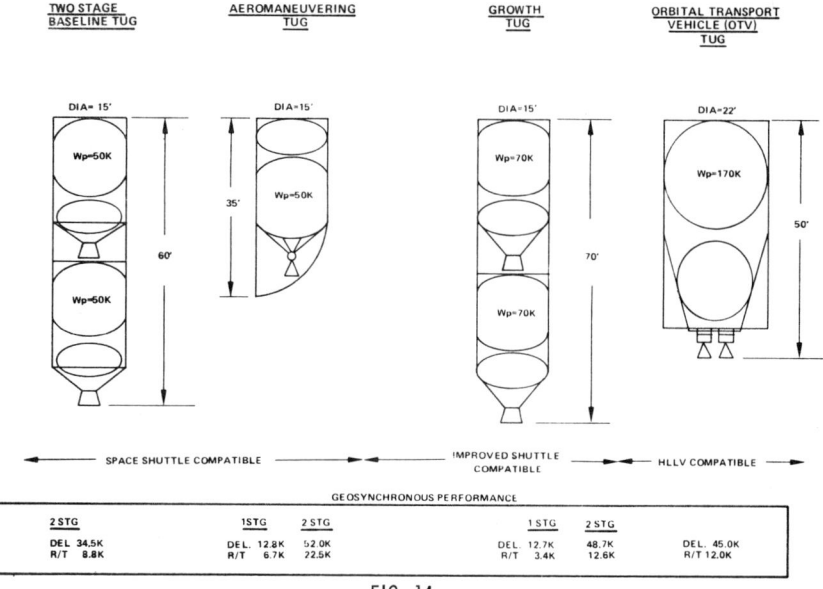

FIG. 14

apparent when reviewing the design along with the established reliabilities and redundancy levels of many of the subsystems. Presently it is estimated the additional development cost to bring the current Baseline Space Tug to an acceptable manned system would be a very small percentage of the total. This investment would be to make a few modifications, to provide data and override capability to the crew, and for manned operations qualification.

In exploring the growth concept of the Space Tug there are a number of options open. Some of these are presented in Figure 14. Based on the current Shuttle capabilities there are three options to increase performance using the current Baseline Space Tug as a point of departure. One, a higher performance (Isp = 470 sec.) engine could be substituted (as in the earlier high performance version) which would increase the round trip ability to geosynchronous orbit about 12%. Second, the aforementioned tandem staging could be employed raising the round trip to geosynchronous about 3 1/3 times. Third, the previously discussed aeromaneuvering (or aerobraking) could be used in single or tandem modes with considerable performance gains. If the Shuttle's capability to low earth orbit were increased, then the Space Tug could proportionately grow. The degree of growth of our current concept is in many ways unlimited since either on-orbit assembly or on-orbit fueling modes can give much higher gross weight in low earth orbit.

The national space program of the future will be one of expanded operations, both manned and unmanned. A potential program scenario is shown in Figure 15. In parallel with comprehensive earth orbit satellite, lunar satellite and planetary spacecraft programs, a successive series of extended manned operations may be pursued. Such a potential program may first involve low earth orbit space stations, followed by geosynchronous orbit space stations, then eventually a lunar orbit space

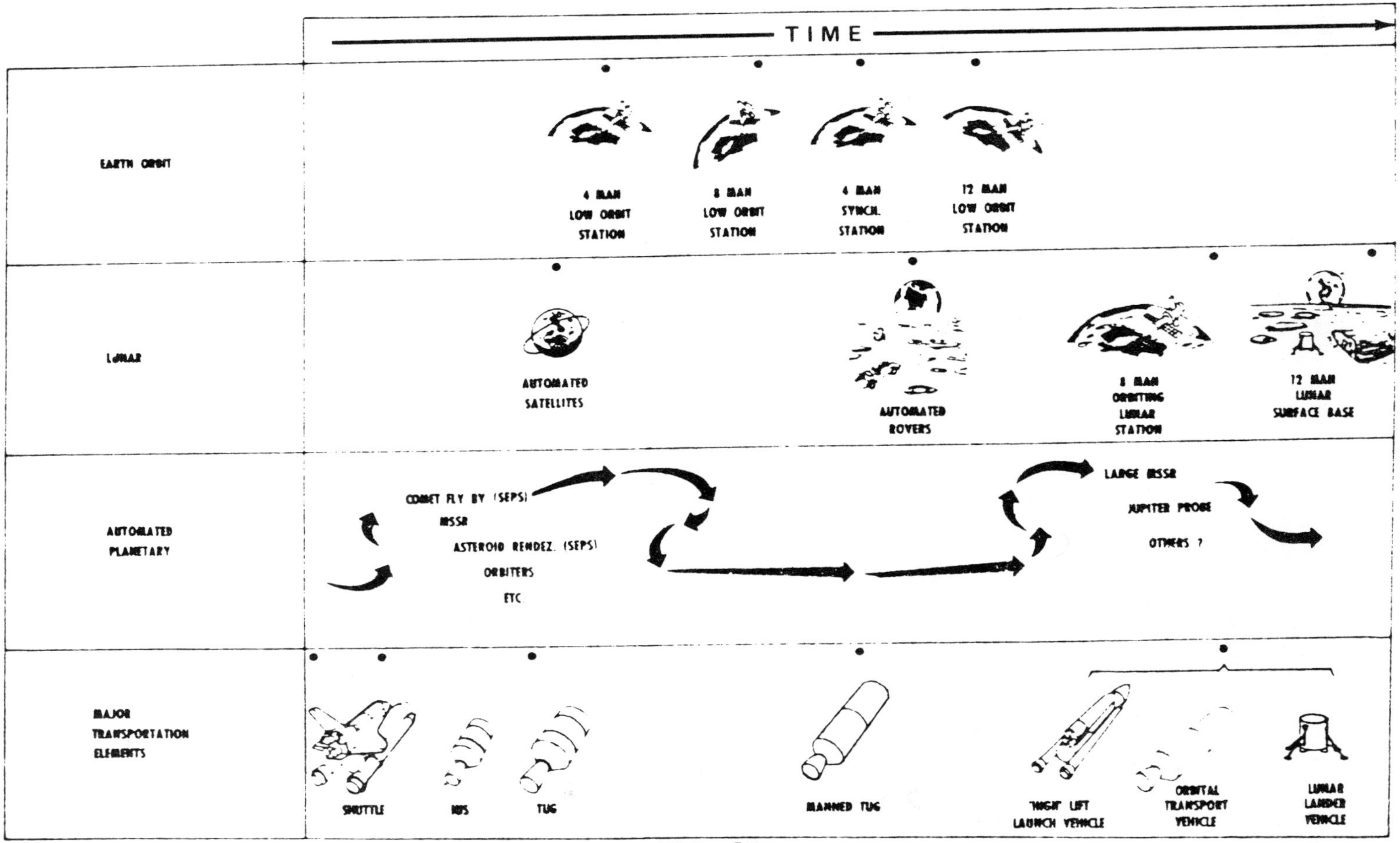

FIG. 15

station leading to manned lunar operations again. Versions of the Space Tug would, of course, be prime links in all these type operations. As one views this type of potential program scenario, the similarity to the 1969 Integrated Manned Space Flight Plan is apparent.

CONCLUSIONS

Beyond the low earth orbit operation regime of the Space Shuttle, there is a definite need for a cost effective space operational stage. To be cost effective it must be reusable and have operational versatility. Its size, type of propellant, configuration and degree of sophistication is dependent upon its expectations. Currently this needed space stage is known as the Space Tug. The currently defined Baseline Space Tug is the concept that best meets the current requirements. The Baseline Space Tug of the future will be defined by its needs.

ACKNOWLEDGEMENT

There are numerous contributors to the background data for this paper. All of those who over the years presented their ideas and worked diligently to formulate them into the various concepts. These were from the United States governmental agencies and industries as well as those from the European Launch Development Organization consorita. For this particular compilation there are two specific contributors that should be recognized. Mr. Thomas W. Barrett for his help in extensive historical research and Mr. Charles F. Huffaker for identifying the effects and applications of this current Baseline Space Tug to the current Mission Model.

AAS 75-140

ALTERNATE TUG/IUS APPROACHES

A. O. Tischler*

Operating, economic, and schedule objectives for the
Shuttle upper stage are reviewed, and the performance
and cost characteristics of upper stage candidates are
discussed. It is concluded that a new two-stage, solid-
motor IUS, with a mass of about 26,500 pounds is the
most cost effective approach to IUS requirements. In
addition, a hydrogen-oxygen stage, with a mass of about
22,500 pounds and used with a booster stage comprising
two perigee burn motors of the IUS, completes a payload
deployment/recovery/replacement system which is compatible with schedule constraints imposed by space program
funding limitations.

INTRODUCTION

Since most of the people attending this session are predisposed
by occupational affiliation to favor one of the designated IUS or Tug
candidates my invitation to discuss alternative solutions cast me in
a role analogous to that of the turtle in the punch bowl. To be in-
vited to suggest any but the preconceived solutions is, I must presume,
evidence that democratic principles prevail in this forum.

Nearly forty years ago one of my college professors, Dr. Max
Morris, long since expired, used to ask his class repeatedly "What are
we trying to do, men, go to the moon?" My simple-minded answer to the
question of what we're trying to do about providing a propulsive
adjunct to the Shuttle-to carry payloads to orbits beyond the Shuttles
reach- is displayed in this first chart.

I believe the objectives of an upper stage system for use with
Shuttle are encompassed by the following:

1) To lift a series of semi-identified payloads from a nominal
Shuttle orbit to some much higher orbit, with the geosynchronous
orbit being one that is necessary to enclose.

2) To return some of these payloads - but only those for which
the economic justification is clear and sufficient - to the

_____
* Consultant - Bethesda, Maryland

2) continued

   Shuttle orbit, for capture, refurbishment and ultimate reuse.

3) To refurbish, or, more realistically speaking, to resupply some payloads in their orbits.

4) To satisfy the economic restraint that no large investment in developing new upper stages be made during the period of Shuttle development.

5) To do all of the foregoing at the least total cost, based on the present value of future investments.

It is my contention that these five objectives, when coupled with a relatively few additional substantiated facts, are sufficient to lead to a selection of one of the five IUS candidates, provided that the selection is to be based on engineering and economic, rather than less tangible considerations. Furthermore, one of several payload recovery/replacement systems appears to have inherent cost advantages over others. The purpose of this presentation is to expose the rationale of this selection process for a payload deployment system and to identify the candidate system for the recovery/replacement system.

At first glance the problem of making an appropriate selection of an upper stage or stages at this time, with a rather inexact definition of the specific payloads to be carried, their numbers, their orbital destinations, their exact schedules, their (real) development costs, their intrinsic hardware value, their associated launch and other transport costs, their operational costs, their maintenance and/or refurbishment cost, and a number of other parameters defined by Jim Murphy of MSFC last Tuesday appears hopeless. To paraphrase one of America's great musical composers "That ain't necessarily so." A process of inductive reasoning can, under such circumstances, disclose that under any reasonable set of circumstances one approach to a solution consistently yields results that are as good as, or in some instances better than, the results using other approaches.

To illustrate that method of selection let me ask if any of the unfettered men in this audience would have difficulty in choosing a companion for this evening from either a 25 year old 110-pound 5 foot 4 inch college graduate with an animated fascination about technological progress or a 60 year old 190-pound 5 foot 2 inch grandmother whose principal life interest was her own self-preservation. Note that it was not necessary to identify your objectives in detail to permit you to make this choice. I'd be the first to admit, however, that a consistent selection is less likely when all the options are over age or overweight unless it happens to be your own gradmother.

A study of the list of objectives leads me to conclude that in the near time frame objective No. 4 which restricts upper stage development funds until after the Shuttle development is completed is now a dominant consideration. It is, in fact, so dominant that NASA has acceded the development of an interim upper stage to DOD. DOD, in turn has announced, in testimony presented to Congress, that its objectives are to satisfy the particular requirements of Air Force missions with this IUS. That seems to me to be an unfortunate and short-sighted position, since a practical solution that might satisfy both NASA and Air Force requirements is possible, but it does generate two consequential conclusions.

First, since none of the Air Force payloads planned for use during the period in which the IUS is expected to be operational is laid out to be recovered and reused, and since recovery of even the propulsion stage alone is costly in terms of payload deployment capability, the IUS can be expected to be an expendable-stage system. Clearly my objectives No. 2 and No. 3, which call for payload recovery and resupply, are displaced by the Air Force policy to another time frame, corresponding in initial operational capability to the latter part of the 80's decade.

Second, DOD is not likely to jeopardize its ability to lift military payloads on schedule because of possible delays in developing the Shuttle. Therefore it is logical and likely that the IUS should be sized so that it could, during an interim period preceding the planned interim operation with Shuttle, be launched on the existing Titan III launch vehicle. Titan III, as most of you know, can carry a little less than half the mass that Shuttle can carry. Those facts lead me to a conviction that the IUS must be sized so that the Shuttle can carry two IUS/payload combinations in each flight, if need be, and that consideration knocks the hydrogen-oxygen propellant combination out of contention for the IUS because of its bulk.

Let me now add that an IUS, properly designed to be flown two at a time in Shuttle, will have a payload capability in the two-ton range for geosynchronous orbit, that is, four tons per Shuttle flight. None of the payloads identified by NASA for geosynchronous placement exceeds this mass; therefore it is adequate to satisfy these published requirements. The majority of the geosynchronous orbit payloads listed are, in fact, in the one-ton class so that the small IUS would already be able to carry two payloads at a time to their destination. A larger IUS appears to be without justification.

Now let's look at the aging overweight existing stages. Table I spells out the mass, performance and dimensional characteristics of the four existing stages studied by the contractor-manufacturers under SAMSO sponsorship.

In view of the fact that geosynchronous orbit payloads tend to size the stage system the logical stage size for a half-shuttle IUS is roughly 26,000 pounds and for a Titan III, about 25,000 pounds. Among these stages only Trans is therefore of the right size.

Table I

EXISTING STAGE CHARACTERISTICS

	AGENA	DELTA	TRANS	CENTAUR
INITIAL MASS, LBS	14,900	12,100	26,900	35,600
BURNOUT MASS, LBS	1,450	1,890	3,780	5,410
λ	0.90	0.84	0.86	0.85
LENGTH, FEET	20.2	29.3	14.8	30.0
DIAMETER, FEET	5.0	4.8	10.0	10.0
Is, SEC	291	304	302	444

Table II

IUS COSTS ($ x 10^6)

	DDT&E	UNIT PRODUCTION	UNIT OPERATIONS
AGENA	50	7	0.5
TRANS	30	2.7	0.5
CENTAUR	40	9	0.5
SRM'S	42	1.25	0.25

Three of them use Earth storable liquid bipropellants with engine specific impulse in the 300 second range. That limits them to one-way trips whenever the destination is geosynchronous orbit.

Centaur carries the world's highest performing operational chemical rocket engines, but it's too big in both mass and bulk to be carried, two at a time, in the Shuttle bay, and too small to do the best job with only one system.

The system that I believe best fits the IUS requirements is a new one which did not appear on the list of existing stages but which has been pursued by Boeing under SAMSO sponsorship. That system uses two tailor-made stages of solid propellant motors to configure an IUS capability. The reason for its candidacy is displayed in Table II, which lists estimated costs of converting three of the existing stages to Shuttle use, the development costs of the new solid-motor configuration, as well as unit production and unit operations costs for each system.

Delta is omitted for my lack of published cost data. The other liquid propellant system development and unit production costs are based on data supplied by their contractors in reports now admittedly more than a year old. These stages, however, are up to 17 years old and I see no sound reason for rejecting these one-year old data.

Boeing estimated the development costs for the solid motor stages on the basis of four new solid motors. Factually only two new motor stages are required for IUS fulfillment. I use these old estimates to cost these motors (that's an accidental pun) into their most pessimistic cost outlook because they will nonetheless be much less expensive overall than any of the other systems. The SRM stage production cost number is a more recent Boeing estimate. The operational cost estimates for the liquid stages come from NASA. The operational cost for the solid motor stages come from me, and I'll willingly define the ratio of these operational costs as compared to liquid propellant systems if challanged to do so.

Note particularly that the development costs of the solid motor stage system are of the same order as the conversion costs of the liquid propellant stages. The unit production costs, on the other hand, are less than half those of Trans, and not in the same stadium as those of the other contenders.

Further study of the table shows that for as little as five payloads carried to high orbit the solid motor configuration is the least expensive in overall costs.

As to the ability to carry the spectrum of payload traffic all arguments can be swept aside with a single statement. A two stage solid motor IUS will consistently out perform Agena, Delta, and Trans. Remember that I've previously set Centaur aside for other reasons.

The deployment capability of the solid motor IUS to geosynchronous orbit, which will be used hereafter as a reference orbit for comparison purposes, lies somewhere between 4200 and 4700 pounds, the variation depending on the attainable solid propellant specific impulse and the mass of the motor cases, interstage structure and that of the guidance and communications and attitude control systems. I'll not burden this audience with discussion of these details. Some of the IUS properties are listed in Table III, in which you see that my own estimates of payload capability are based on extremely conservative values of both propellant specific impulse and motor mass fraction.

Thus, we've identified the winning IUS on straightforward engineering and economic principles. This decision would be pronounced academic, since the real decision may have been settled last week, will certainly be made next week, and is to be announced shortly after Congress reconvenes.

Let us now return to objectives 2 and 3, which were displaced, you'll recall, until the development costs of a high orbit recovery system can be accommodated by the space program budget.

Recovery of space payloads from high orbit is technologically difficult. It is possible to recover, per Shuttle flight, only about half the mass that can be deployed to high orbit per shuttle flight. That statement provides the clinching reason why NASA, as well as DOD, should accept a deployment system sized for half a Shuttle load. There

Table III

SOLID MOTOR IUS CONFIGURATION

SYSTEM		
	TOTAL MASS, LESS P/L	26,800 POUNDS
	TOTAL LENGTH, LESS P/L	18 FEET
	MAX. DIAMETER	7 FEET
PERIGEE KICK STAGE		
	MASS	20,250 POUNDS
	SPECIFIC IMPULSE	290 SECONDS
	PROPELLANT MASS FRACTION	0.90
	LENGTH	7 FEET
	DIAMETER, MOTOR	7 FEET
APOGEE KICK STAGE		
	MASS, MOTOR	5,760 POUNDS
	MASS, INCL. G&C, ACS	6,540 POUNDS
	SPECIFIC IMPULSE	290 SECONDS
	PROPELLANT MASS FRACTION	0.90
	LENGTH, INCL. G&C, ACS	11 FEET
	DIAMETER, MOTOR	5 FEET
PAYLOAD		
	USEFUL P/L	4,200 POUNDS

is no reasonable reason to develop a system for deployment of recoverable payloads that has a mass capacity in excess of that which can be recovered.

Because recovery is difficult it is also expensive. Each recovery of a 2-ton class payload from high orbit will demand a dedicated Shuttle flight and, of course, the operational costs of a suitable recovery stage. These costs will total about $15M, based on present operational cost estimates. Thus we can say that no 2-ton class payload having an intrinsic value of less than $15M is worth recovering from high orbit. That consideration will certainly limit the number of payloads to be recovered to some fraction of the number deployed, since many of the payloads rapidly become technologically obsolete, and no longer have high intrinsic value. These facts must be borne in mind when assessing the return on investment of the recovery, since they will limit further the investment that can be afforded. I'll also plug in at this point that any cost analysis based on recovery and reuse of all or even the majority of payloads deployed yields results that are economic nonsense.

The recovery system that really makes some sense in terms of economic value functions, which I won't have time to discuss here, is one that matches the payload deployment capability, and vice versa. I'll repeat that. The recovery system mass capability should be the same as the deployment system mass capability. While I won't derive a proof of that statement I think it is nearly self-evident.

A recovery/replacement system operating from Shuttle orbit to geosynchronous orbits must have a delta V capability of somewhat more than 28,000 fps. It must carry useful payload back from orbit, or for the replacement mission, which presumes carrying the same equipment mass to orbit as well as return it, must carry payload both ways. Limiting the recovery systems to those propellant combinations for which operational experience has been attained this steep delta V requirement necessitates the use of hydrogen-oxygen propellants in at least one stage of the recovery/replacement system. The use of basic performing propellants in this system would penalize the payload capability by a factor of two or more, and, even more important, would put it out of the range of matching the IUS deployment capability.

There are several ways in which a recovery capability can be devised to match the deployment capability provided by a two-per-Shuttle IUS. Note that what I just said is a bald **argument** that the IUS deployment capability should not be abandoned. Once existing, it represents the most expedient, the most reliable, and the most cost-effective method of placing payloads in high orbit.

One of the recovery schemes that can match the IUS deployment capability is the so called Tug. This one-stage hydrogen-oxygen system suffers from exceptional sensitivity to degradation of its mass fraction and/or its specific impulse. Seventeen years of exposure to developing such high-performance hydrogen-oxygen systems have taught me that such sensitivity is an open door to development cost escalation. It was a costly lesson that unfortunately some of my former colleagues have chosen not to learn. I don't believe that this approach would really satisfy my 5th objective, which necessitates keeping the overall cost at a minimum.

Another approach is that of a two-stage hydrogen-oxygen system, with both stages recoverable. These stages can be almost identical. Such a system was studied extensively in Europe by consortia of contractors about five years ago. The studies showed the two-stage recoverable hydrogen oxygen system to be feasible. The use of two stages reduces the sensitivity of the system to degradation of performance parameters, but it also necessitates recovery of the stages by the Shuttle at different times. The required phasing maneuvers and other operations are not onorous but they do take doing. I am not hooked by this answer. However, since there are a number of European representatives present at this conference it is quite likely that there is someone here far more qualified to speak about this approach than I am.

My favorite solution to the payload recovery system, for which I want to acknowledge valuable contributions by Mr. John Humphries, now working with General Electric in the field of nuclear power generation, capitalizes on the continued existance of the IUS as a least costly adjunct to the recovery system. Since the solid propellant IUS comprises a perigee kick stage and an apogee kick stage, the perigee kick stage, doubled up to accommodate double the total mass

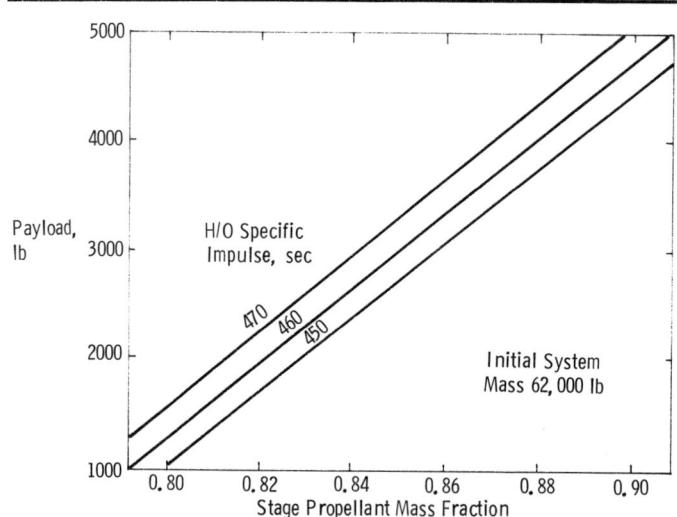

Figure 1

at departure from Shuttle, makes a near-perfect booster for a new
hydrogen-oxygen recovery stage. This approach roughly halves the size
of the hydrogen-oxygen stage, as compared to the single stage Tug, and
that should have a beneficial effect on the development cost. More
important the use of a booster significantly reduces the sensitivity
of the hydrogen-oxygen stage to its mass fraction and specific impulse
parameters, and this should have the effect of reducing its development cost substantially. This is illustrated by the next plot (Figure 1),
which shows the payload that can be recovered from geosynchronous orbit
by a boosted hydrogen-oxygen stage that I'm about to describe in greater
detail as a function of the performance parameters. Note that to obtain
a 4000 pound recovery capability demands a mass fraction of about 0.88
and an engine specific impulse of about 460 seconds.

The next sketch (Figure 2) shows crudely what this solid booster
hydrogen-oxygen stage might look like. Note first the two solid perigee
boost motors, each virtually identical to the perigee kick motors of
the IUS. Because the IUS is a 31,000 pound total mass system, including
payload, and the recovery system is a 62,000 pound total mass system,
without payload, the recovery system must use two IUS booster motors.
These motors, however, can derive their guidance from the upper stage,
then they are essentially just motors. Each of the motors in the IUS,
incidentally, will cost less than the G&C package.

The mass of the hydrogen-oxygen stage is about 22,500 pounds. The
layout of the stage, which is remarkably like that detailed by the
European study group, shows the oblated hydrogen tank forward, with four
oxygen tanks clustered around the engine. The engine size depicted here
corresponds to the existing RL-10 (Centaur) engine. Note that for a
stage of 14 feet diameter the stage length is also 14 feet; bore and
stroke are identical.

The RL-10 thrust level is much higher than it needs to be in this
application, and its performance needs to be upgraded. Within the
envelop shown a smaller thrust engine with a higher expansion ratio
can be made by squeezing the throat of the existing engine. The higher
expansion ratio thus obtained without the mechanical complexity of a
trombone nozzle will upgrade the engine performance.

The next slide shows a plot of the performance to geosynchronous
orbit that can be obtained with a boosted (Figure 3) hydrogen-oxygen
stage. This plot seems to confuse some people so pay attention. What
is plotted here is the payload that can be recovered from geosynchronous
orbit as a function of the delta V supplied by a booster, based on an
initial mass at Shuttle orbit of 62,000 pounds. The booster is
assumed to be a pair of solid motor, as I've previously described, with
propellant perigee impulse of 290 seconds and a propellant mass faction
of 0.90; both values are well within the state-of-the-art existing now.

As one moves to the right on this chart the delta V produced by
the booster increases obviously the intercept of the curves with the
ordinate, corresponding to 0 delta V produced by the boosters, represent the recoverable payload for a single hydrogen-oxygen stage, that

RETRIEVAL/REPLACE SYSTEM

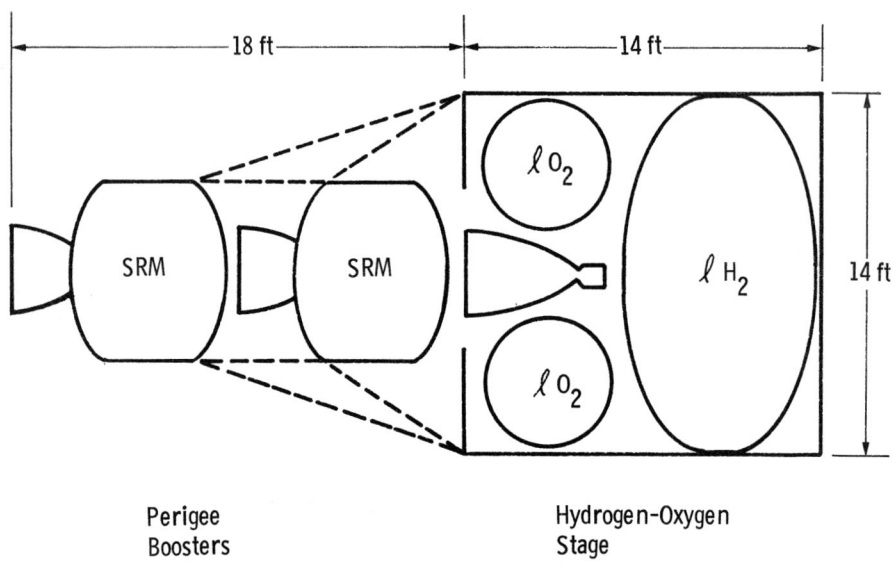

Figure 2

GEOSYNCHRONOUS REPLACEMENT CAPABILITY SOLID-BOOSTER RECOVERABLE H/O STAGE

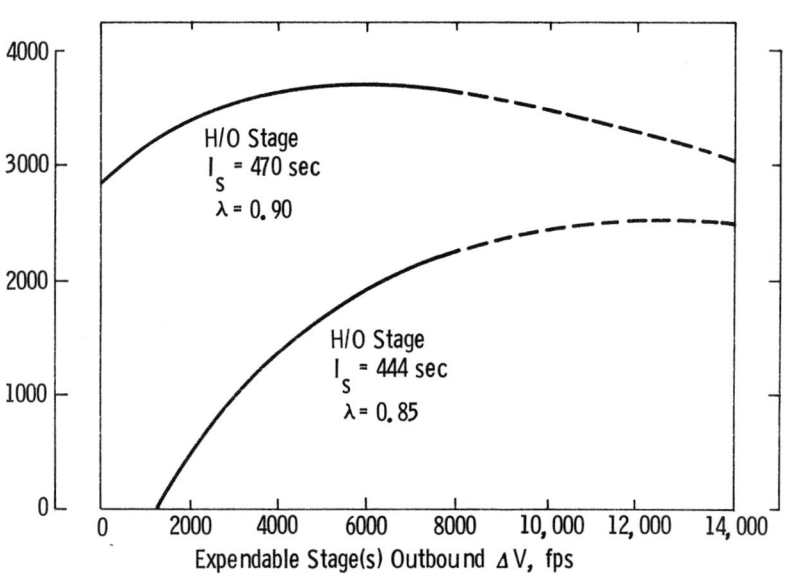

Figure 3

is, the Tug. The intercept with a line projected up from 14,200 fps represents a system where the boost supplies the entire outbound trip, and the hydrogen-oxygen stage is used only for payload retrieval. The curves are dotted beyond a boost of just over 8,000 fps because that delta V corresponds to the perigee burn.

The capabilities of two hydrogen-oxygen stages are shown (Figure 4). Although these systems are rubberized for purposes of making this plot the performance parameters of the stages remain fixed. The lower curve corresponds in performance parameters, Isp and mass fraction, to the present Centaur stage. The upper curve corresponds to a hydrogen-oxygen stage having a mass fraction of 0.90 and specific impulse of 470 seconds. Since the lower curve represents very conservative performance values and the upper stage represents a stage with performance values too optimistic to be realized at low cost, the payload performance of a _real_ boosted hydrogen-oxygen stage lies between these curves, and I'm not going to try to tell the world of space system experts where.

But note a few very salient points. For any set of performance parameters the boosted hydrogen-oxygen stage will always out perform the single stage system, and supplying boost to the extent corresponding to perigee burn on the outbound trip is just about where the best payload performance will be realized. Beyond that, depending to some extent on the stage performance values the retrieval payload can actually decline.

One last point, with conservative but realistic performance parameter values, the retrieval capability falls into the 4,000 pound range; that is, it falls into the same range of masses as the deployment capability of the simple and inexpensive IUS system previously described. The replacement capability for the same system is plotted here. Replacement, that is a two way trip carrying payload both ways, is limited to about 3,000 pounds. Note that these recovery and retrieval figures match or exceed those given by Mr. Tier for a much more sensitive single stage system.

Let me note one point on a closely related subject. The solid motors described for the IUS, coupled with an existing solid motor stage, can meet the requirements of all planetary missions. Missions now being planned and when I say all I mean all without any exception. Thus, there is no need to throw away the recoverable hydrogen-oxygen stage at any time.

Let me now review the bidding. A simple two-stage solid-motor IUS, with a mass of about 26,500 pounds, capable of putting more than two tons to geosynchronous orbit, is the most cost effective approach to meeting the IUS requirements. Two such systems can be carried on each Shuttle flight.

A hydrogen-oxygen stage with mass of about 22,500 pounds, used with a booster stage comprising two perigee burn motors of the IUS, can recover roughly two tons from geosynchronous orbit, thus essentially

GEOSYNCHRONOUS PAYLOAD RETRIEVAL CAPABILITY SOLID-BOOSTER RECOVERABLE H/O STAGE SYSTEM

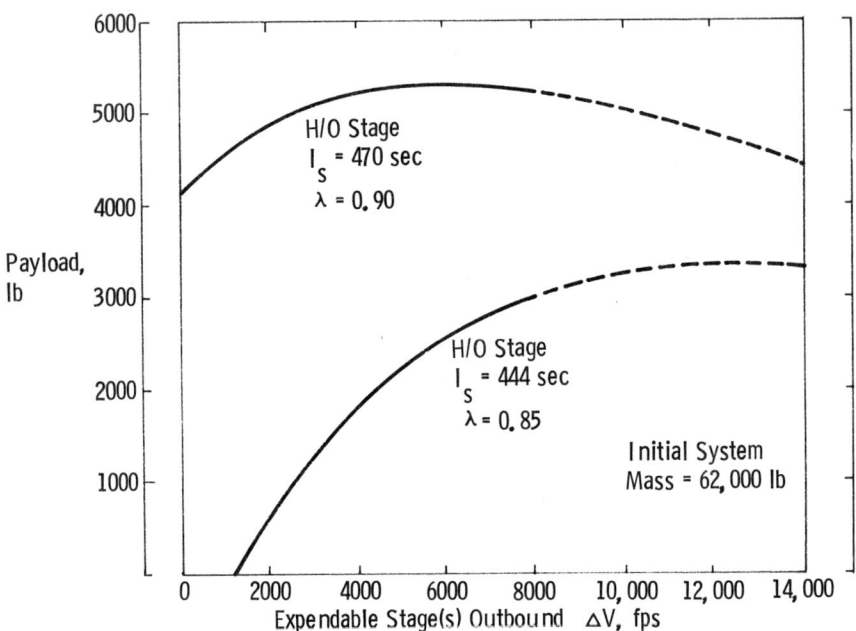

Figure 4

matching the IUS deployment capability. Although other systems using hydrogen-oxygen propellants could also match this capability the proposed recovery system <u>appears to be</u> the most cost effective approach because the boosted system is less sensitive to stage performance parameter degradation than other systems and the stage itself is small compared to the proposed single-stage Tug, to use a specific example.

The deployment/recovery/replacement system described is compatible with the schedule constraints imposed by the limitations in funding presently being experienced by the space program.

AAS 75-141

ORBITAL TRANSPORTATION IN THE 1980'S
AND BEYOND

Hubert P. Davis+

Orbital transportation beyond the low earth orbit operating regime of the Space Shuttle will be required for the 1980's and beyond. The characteristics and first order requirements of the mission arenas are discussed in context with a broad spectrum of future space transportation systems. Several concepts are highlighted and identify the distinctly different requirements imposed by manned vehicles versus unmanned vehicles. Considerable analytic and design activities are necessary prior to selection of orbital transportation systems to be developed after the Interim Upper Stage (IUS).

INTRODUCTION

Missions in the 1980's and beyond having destination orbits beyond Space Shuttle operational orbits will require space transportation in the form of orbital transfer vehicles. These vehicles will proceed from a low earth orbit of 200 to 500 kilometers for payload delivery missions to higher energy orbits; retrieval missions of payloads being returned to low earth orbit; round trip missions for payloads being deployed and retrieved; or, servicing missions to several orbital locations. The various mission arenas and system approaches are shown on Fig. 1.[1] This paper provides an overview of the characteristics and first order requirements of the arenas and highlights several diverse concept approaches.

TRANSPORTATION REGIMES

The orbit-to-orbit transportation in the 1980's will be supported by initial launch from the Space Shuttle. The near term capabilities of the Shuttle system will be 65,000 lbs. or approximately 30 metric tons to a

---
+ Manager, Future Programs Office, Lyndon B. Johnson Space Center, Houston, Texas 77058

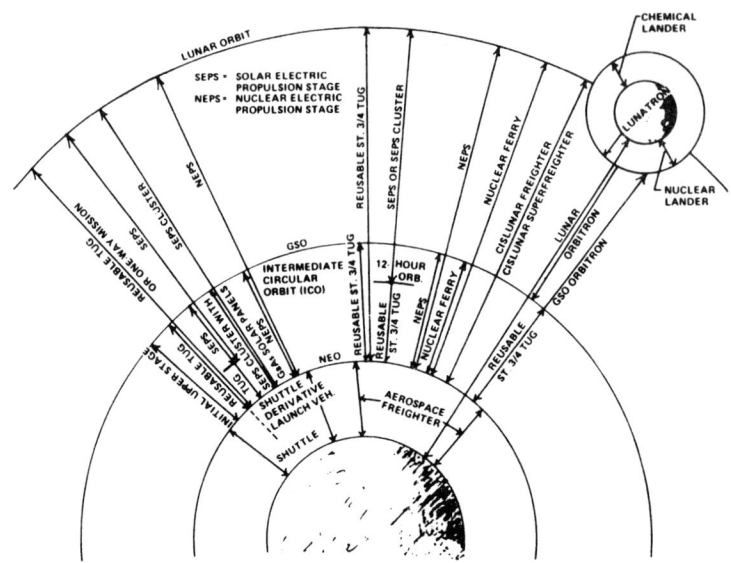

Fig. 1 Geolunar System - Transportation Options

350 kilometer altitude orbit at an inclination of 28.5°. The inclination is consequent to a due east launch from the Kennedy Space Center. To date, studies performed have given indications the basic Shuttle vehicle will evolve during the mid to the late 1980's and will achieve a significantly enhanced payload delivery capability. I believe we can confidently anticipate payload capability of the current Space Shuttle well beyond 80,000 lbs. in the late 1980's. This matter of course growth of the Shuttle system deserves careful attention in the design of the orbit-to-orbit systems for use in the mature operational period of the Shuttle system.

To accompany the Space Shuttle, a large launch vehicle termed Heavy Lift Launch Vehicle (HLLV) may be required for the late 1980's and beyond. Two of the various possibilities are illustrated on Fig. 2.[2] A straightforward application of the Shuttle main engines, solid rocket boosters, and external tank or its tooling is illustrated by the middle figure. A launch vehicle of this class can be expected to deploy a payload ranging from 70 to 110 metric tons. This represents a two to three-fold increase over the evolved Shuttle payload capabilities and, primarily,

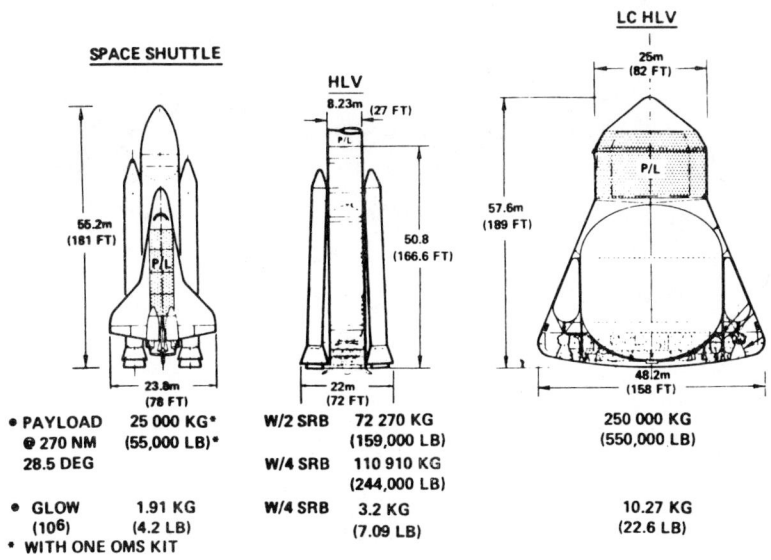

Fig. 2 Earth Launch Vehicles

may be motivated by the need to launch payloads of either a physical size or mass inconsistent with the confines of a Shuttle payload. Should a large scope program such as the Satellite Solar Power Station as suggested by Peter Glaser[3] and Gregory and Woodcock[4] be initiated, it will be economically better to develop a far more capable launch vehicle as indicated by one possible configuration shown at the extreme right of Fig. 2. Such a vehicle may have a launch capability of 250 metric tons or more and involve a vertical landing mode of recovery. Future state-of-the-art structural and propulsive technology may not permit achieving the design of the single stage version with a reasonable relationship between the gross liftoff weight and payload in low earth orbit. Single stage versions of this vehicle are attractive from an operations point of view because no transport of booster elements from the downrange landing is necessary.

An orbit of great interest to space activities supporting terrestrial affairs is the geosynchronous equatorial orbit. This orbit requires a velocity change from low orbit of approximately 4.3 kilometers per second and offers the opportunity once the orbit has been achieved to maneuver in longitude with very low energies. Lunar orbit was of great interest during the Apollo program and undoubtedly will be revisited by unmanned and manned spacecraft. The primary distinction between the geosynchronous equatorial and lunar orbits is not the energy required for the transfer but rather the coast time necessary between the departure burn to the transfer ellipse and the circularization burn at the destination orbit. Coast time for chemical rocket vehicles traversing to geosynchronous orbit is approximately 5-1/2 hours between burns. A typical traverse for similar high thrust vehicles from near earth orbit to lunar orbit requires approximately 3 days. The energy requirements necessary to achieve a 100 kilometer orbit about the lunar poles is approximately 4.1 kilometers per second from a 300 kilometer, 28° inclined low earth orbit.

Landing on the surface of the moon from the lunar orbit requires a velocity change allowance of approximately 2.2 kilometers per second, which includes hovering time for precision landing. The velocity change for return from the lunar surface to lunar orbit depends upon the orbital phasing of the destination orbit and is time dependent with a 14 day cycle. The minimum energy traverse for an in-plane launch requires a velocity change of 2.0 kilometers per second for rendezvous with a companion craft remaining in orbit about the lunar poles. Mission modes for return from either lunar orbit or geosynchronous orbit will be discussed later.

TRANSPORTATION REQUIREMENTS

Deployments to low earth orbits beyond the altitude capability of the Shuttle, to geosynchronous orbit for many of the applications missions, and to planetary trajectories to support the space science activities are a part of the current Shuttle mission requirements that the IUS candidate vehicle is anticipated to fulfill in the early 1980's. Depending upon the mission characteristics, extent of use, and relative costs of the stages and their payloads, reusable upper stage systems with payload retrieval capabilities may become imperative beyond 1984. Much work remains in the identification of missions and of the system characteristics to achieve those missions before final decisions can be made that are relative to the cost effective extent of reuse of systems.

Payload retrieval and round trip in the late 1980's is, however, an entirely separate issue from the relative economics of expendable and reusable vehicle systems. Even if further analyses of stage reuse results in unfavorable economics, bonafide mission requirements may require the development of the payload retrieval and round trip capability. This requirement may stem from the desire to achieve missions not otherwise obtainable or from the secondary economic criteria that retrieval of a malfunctioning payload in geosynchronous orbit becomes essential to permit a large program to proceed with evidence of the cause of equipment malfunction, which cannot be acquired without return of the payload to Earth.

Manned requirements are not now in the Shuttle mission model beyond the altitude achievable by the Shuttle. However, there are many potential missions, including the geosynchronous sortie, which incurs a residence within or near the transfer vehicle for a period of 2 weeks. Beyond 1985, permanent space facilities such as the manned geosynchronous space station may require placement, resupply, and rotation of personnel.

TRANSFER VEHICLE REQUIREMENTS

There are several requirements issues present in the design of the orbit-to-orbit vehicle that will be referred by the generic term orbital transfer vehicle (OTV). The most prominent is the determination of the payload performance capability and consequently the size of the OTV. This determination will be made by the end use program requirements, which will emerge between now and the time the design for such a system must be selected. The extent of reuse and the logistics resupply techniques influence the OTV size. These issues will be determined by economic considerations. During the Space Tug work of 1971, consideration was given to leaving the vehicle in orbit about the Earth and providing resupply of propellants and other consumables in a low earth orbit rendezvous and docking sequence. This plan was shelved for the subsequent Space Tug studies in favor of Shuttle return of the vehicle to the launch site after each mission for inspection and refurbishment. With the deferral of the Space Tug to the mid 1980's, it is appropriate to review the relative merits of basing the OTV on the surface of Earth or basing it in a parking orbit about the Earth. The choice of basing locale is relevant to the design characteristics of the OTV and its launch vehicle.

The OTV mode of return from high energy orbits is another design requirement issue being reviewed. The work done in the early 1970's on the Space Tug utilized a propulsive return to low earth orbit from geosynchronous orbit for rendezvous and docking with a Shuttle Orbiter. This mission mode selection requires the expenditure of approximately 2.4 kilometers per second to reduce the perigee velocity of the transfer ellipse from geosynchronous altitude to the low earth orbit circular velocity. The weight necessary to apply this velocity increment may be utilized in a different way. The Earth's atmosphere is available immediately below this low earth orbit destination and can be applied to reduce velocity requirement in two ways. Atmospheric braking may be utilized to reduce the velocity to the low orbit circular velocity in steps by repetitively entering the tenuous upper portions of the Earth's atmosphere. This mission mode has been studied by the Marshall Space Flight Center in a concept called "Aeromaneuvering Orbit-to-Orbit Shuttle" (AMOOS).[5] The more familiar Apollo spacecraft mode for atmospheric braking is the one pass entry into the atmosphere of the Earth with a sufficient reduction of velocity to return directly to the surface of the Earth. The trade-offs between the weight and operating costs of vehicles designed for these differing mission modes for the return journey are not complete. Work remains to be done before the selection of the Earth return mode.

When considering the OTV configuration for the outbound journey and the propulsive portion of the return journey, the number of OTV stages to be employed and degree that they are modularized are major design issues. Later, a concept will be shown of a highly modularized two stage vehicle for transfer of payloads of the 60 metric ton class, which was the mission requirement in the earlier studies of the reusable nuclear stage. Orbital docking is now expected to be required for OTV modular assembly as well as for the accomplishment of work with its payloads, either manned or unmanned, in the high energy destination orbits.

Operational control of OTV may be directed from an Earth based mission control center, a low earth orbit vehicle such as a Shuttle or manned space station, or within the OTV in an autonomous mode. A time phased mix of these three mission control schemes is a likely option for the orbital transportation vehicle of the late 1980's. Again, a significant number of analyses will be necessary to arrive at a minimum cost approach to operations control for the mission model that will evolve during the next several years.

GEOSYNCHRONOUS MISSIONS CONSIDERATIONS

The geosynchronous mission has been repeatedly identified as the end-use driver mission in the determination of OTV design requirements. This situation will almost certainly remain true in reviewing the future mission models.

The return mission mode offers several options as illustrated on Fig. 3.[2] The all propulsive, aeromaneuvering, and Apollo-type return modes are

	RETURN TO EARTH ORBIT		DIRECT RETURN TO EARTH	
	PROPULSION ONLY	PROPUL + AEROBRAKING	BALLISTIC	GLIDER
RETURN OPTIONS	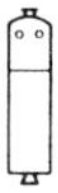	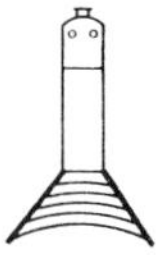		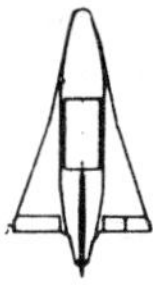
OTV OPTIONS	• SINGLE STAGE —LO2/LH2 • COMMON STAGE —LO2/LH2 • COMMON STAGE —LO2/MMH	• SINGLE STAGE —LO2/LH2		
PAYLOADS CREW/EQUIP MAINTENANCE PROVISIONS	KG (LB) 5000 (11,000) 1000 (2,200)		KG (LB) 6360 (14,000) 1000 (2,200)	KG (LB) 11360 (25,000) 1000 (2,200)

Fig. 3 Geosynchronous Mission Modes and Requirements

described earlier. An aerobraking, entry and landing glider is identified as the fourth return mode where integral propulsion provides the deorbit burn. As many as ten aerobraking passes may be required before the final entry and runway landing. The relative merits of these four approaches are to be determined.

Recently, a brief study was done at JSC and MSFC of manned missions to the geosynchronous orbit. A sortie mission velocity budget defined by the JSC study is given in Table 1. This mission total is approximately

Table 1

MISSION DELTA-VELOCITY BUDGET
GEOSYNCHRONOUS SORTIE

EVENT	MISSION DELTA-V FPS	STAGE 1 DELTA-V FPS	STAGE 2 DELTA-V FPS
LEO FIRST BURN (2.2° PLANE CHANGE)	8039	≈7000	≈1039
MIDCOURSE CORRECTION	100		100
CIRCULARIZE AT GEO (26.3° PLANE CHANGE)	5836		5836
RENDEZVOUS (INITIAL DOCKING, 4 SATELLITES)	540		540
45° LONGITUDE SHIFTING	900		900
DE-ORBIT BURN - GEO	5836		5836
MIDCOURSE CORRECTION	100		100
CIRCULARIZE AT LEO	8039	≈7000	8039
INITIAL RENDEZVOUS AT LEO	135	135	135
CONTINGENCY	100	50	50
TOTAL DELTA-V	29,625	≈14,185	≈22,575

NOTE: LEO PARKING ORBIT OF 160 NM CIRC., 28.5°

9 kilometers/second and represents a more severe mission requirement than the payload deployment missions studied for the Space Tug. The additional velocity is required for multiple visits to satellites at the high orbit location. The budget is based upon a two stage transportation vehicle that utilizes essentially identical propulsive stages. The first stage imparts a velocity of approximately 2.1 kilometers per second to the second stage and payload. The second stage in an all propulsive mission mode completes the mission and provides a total velocity of approximately 7 kilometers per second.

The crew rotation and resupply of a permanent facility in geosynchronous orbit constitutes a slightly less severe velocity budget, see Table 2. The total mission delta-V is reduced to 8.6 kilometers per second with the first stage imparting approximately 2.1 kilometers per second to the second stage and payload.

CONCEPTS OF OTV

Several OTV concepts have been brought forward in various space transportation studies and will now be briefly discussed. Marshall Space Flight

Table 2

MISSION DELTA-VELOCITY BUDGET
GEOSYNCHRONOUS CREW ROTATION/RESUPPLY

EVENT	MISSION DELTA-V FPS	STAGE 1 DELTA-V FPS	STAGE 2 DELTA-V FPS
LEO FIRST BURN (2.2° PLANE CHANGE)	8039	≈7000	≈1039
MIDCOURSE CORRECTION	100		100
CIRCULARIZE AT GEO (26.3° PLANE CHANGE)	5836		5836
RENDEZVOUS WITH SPACE STATION	135		135
DE-ORBIT BURN - GEO	5836		5836
MIDCOURSE CORRECTION	100		100
CIRCULARIZE AT LEO	8039	≈7000	8039
INITIAL RENDEZVOUS AT LEO	135	135	135
CONTINGENCY	100	50	50
TOTAL DELTA-V	28,320	≈14,185	≈21,270

NOTE: LEO PARKING ORBIT OF 160 NM CIRC, 28.5°

Center proposed a two stage variant of the baseline Space Tug for the manned geosynchronous sortie mission of the recent NASA study.[6] The second stage and crew module are illustrated on Fig. 4.

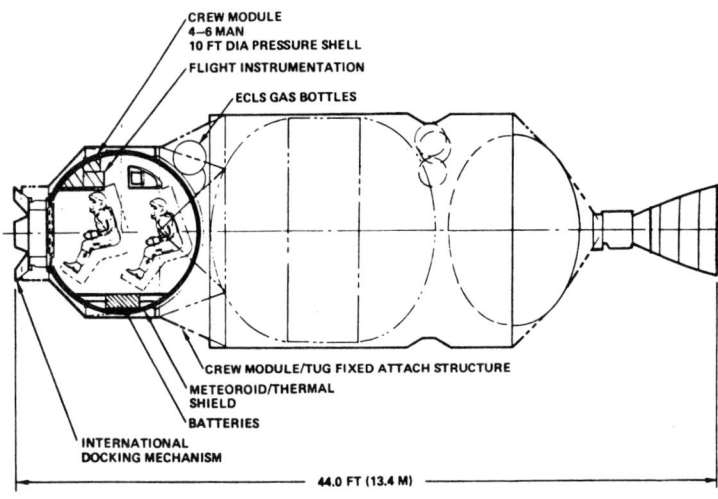

Fig. 4 Manned Tug

The crew module is a 4 to 6 man vehicle approximately 10 feet in diameter
and capable of providing support for a relatively short duration Sortie
mission. The MSFC analysis results shown in Table 3 indicate that the
two stage reusable vehicle can accommodate an approximate 4 metric ton
crew module for the round trip mission. Improvements in the Shuttle to
a higher launch capability discussed earlier and utilization of a larger tug
vehicle than the current baseline Tug can improve this payload capability
to approximately 6 metric tons. In addition, expending the first stage
rather than retrieving it increases the payload performance of the two
stage system to 6.3 metric tons.

Table 3

BASELINE/GROWTH TUG PERFORMANCE

2010-75	65K SHUTTLE	IMPROVED SHUTTLE
TWO STAGE BASELINE TUG (FULLY REUSABLE)	DEL 34.5K R/T 8.8K	DEL = 38.5K R/T = 10.3K
TWO STAGE BASELINE TUG (EXPEND FIRST STAGE)	DEL 36.0K T/T 9.7K	DEL = 41.0K R/T 11.2K
SINGLE STAGE GROWTH TUG (70K PROP.)	---	DEL = 12.7K/16.2K* R/T = 3.4K/4.5K*
TWO STAGE GROWTH TUG (FULLY REUSABLE)	---	DEL = 48.7K/55.0K* R/T = 12.6K/14.6K*
TWO STAGE GROWTH TUG (EXPEND FIRST STAGE)	---	DEL = 52.2K/58.0K* R/T 13.8/15.9K*

* 160 X 160 NM/100 X 500 NM DEPART ORBITS

NOTE IMPROVED SHUTTLE INJECTED TO HIGHER ORBIT (100 N.M. X 500 N.M.) TO REDUCE ORBIT-TO-ORBIT STAGE ΔV REQUIREMENTS

An OTV concept is illustrated on Figs. 5 and 6.[7] This vehicle has a
larger propellant capacity than the baseline Tug (approximately 39 metric
tons per stage as compared with 23 metric tons of the baseline Tug).
This size is compatible with the Space Shuttle payload bay dimensions
but exceeds the ability of the Shuttle to place a fully loaded stage in
low earth orbit without on orbit OTV propellant top off. This larger
system is required to accommodate the approximately 5.7 metric tons crew
module illustrated on Fig. 7. The crew module illustrated has a set of

manipulators and discretionary payload of approximately 1 metric ton for deployment at the high orbit. This payload may consist of repair parts for satellites or replacement instruments.

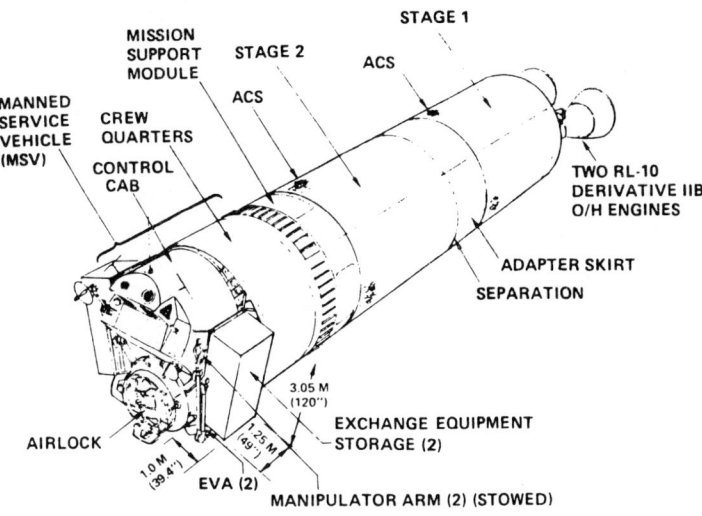

Fig. 5 GSMS Manned Orbital Transfer Vehicle - General Arrangement

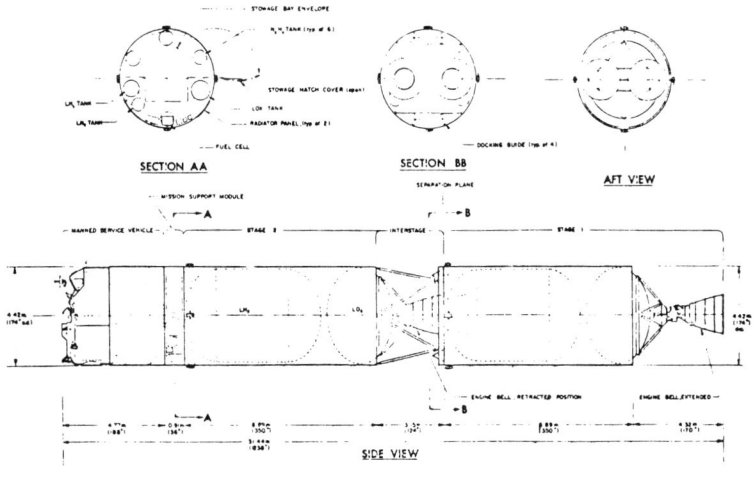

Fig. 6 GSMS Manned Orbital Transfer Vehicle - Inboard Profile

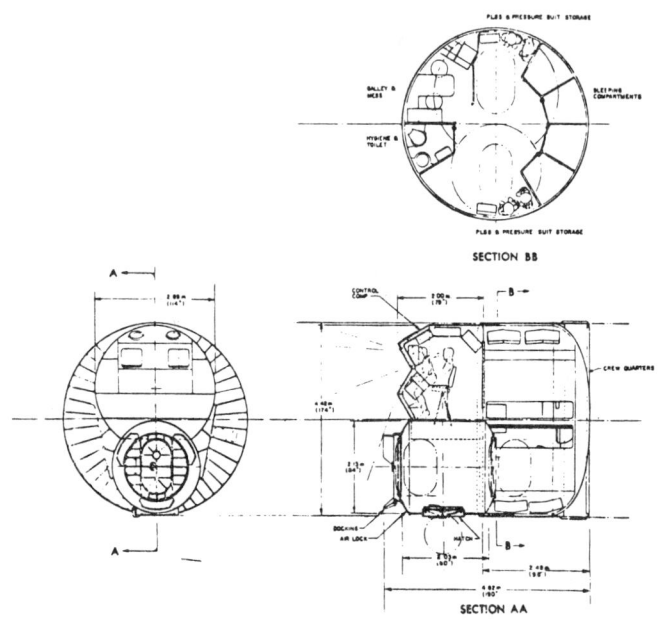

Fig. 7 GSMS Crew Module - Inboard Profile

A question is posed as what weight allowance should be made for the manned portion of the orbital transfer vehicle. The one outstanding past example of a manned OTV is the Apollo Lunar Module. The familiar view on Fig. 8 is of Lunar Module 5, Eagle, with this photo taken just prior to its rendezvous with Columbia for the return of Armstrong and Aldrin from the first lunar mission. The ascent stage had a separation weight of almost 5 metric tons, which included sufficient propulsion to achieve the launch from the surface of the moon and rendezvous. When the propellant weight and inert weight of the propulsion and attitude control systems are deducted from the weight of the later, lighter, Lunar Module 6 ascent stage, only 2 metric tons were required to build a crew module using 1960 technology that was capable of successfully housing two men for a period of 3 or 4 days. With this precedent, it is obvious the later technology now available would permit the construction of a two man crew module for geosynchronous or lunar orbit operations within a weight budget of approximately 3 metric tons. The questions become the weight allowance impact on mission capability and costs. The Apollo mission was an exploration mission. It was appropriate for this mission to increase fabrication costs to conserve weight. I well remember watching the machine operators

milling the corner fittings of the vehicle to shave off fractions of
ounces of weight in order to conserve the overall weight of the system.
I believe that our manned systems of the late 1980's should be designed
to be more tolerant of weight increase and propulsion stage performance
decrease than was the case for the relatively low number of missions
represented by the Apollo program. Of equal importance to manufacturing
costs is the probable requirement to extend crew stay time and mission
capability beyond that represented by the previous generation of manned
spacecraft so that a future crew module may fully achieve its utilitarian
purposes. I believe that these two factors will combine to indicate a
minimum crew module weight of four to five metric tons, and more attractive
vehicles may reach six metric tons inert weight.

Fig. 8 Lunar Module Ascent Stage

The variation in size of the liquid oxygen/liquid hydrogen OTV's considered
in the Future Space Transportation Systems Analysis Study is shown on
Fig. 9.[7] The small OTV's on the left portion of the chart are intended for
the sortie missions and future continuation of lunar exploration in the
Apollo mode, that is, infrequent visitation without permanent stations in
lunar orbit or on the surface. The loaded vehicle mass is approximately
50 metric tons for each of the two stages. The intermediate OTV's are capable

of supporting more ambitious missions with generous weight allowances for the payload modules and utilize two stage vehicles to support permanent geosynchronous or lunar orbit facilities. Alternatively, intermediate OTV's can support more capable sortie missions involving more than four people. The mass of these loaded vehicles approximates 100 metric tons per stage. If there is to be support of a large community of people in either geosynchronous orbit or on the lunar surface, even more capable vehicles will be required. The mass of these vehicles with oxygen/hydrogen propulsion systems can approach 300 metric tons.

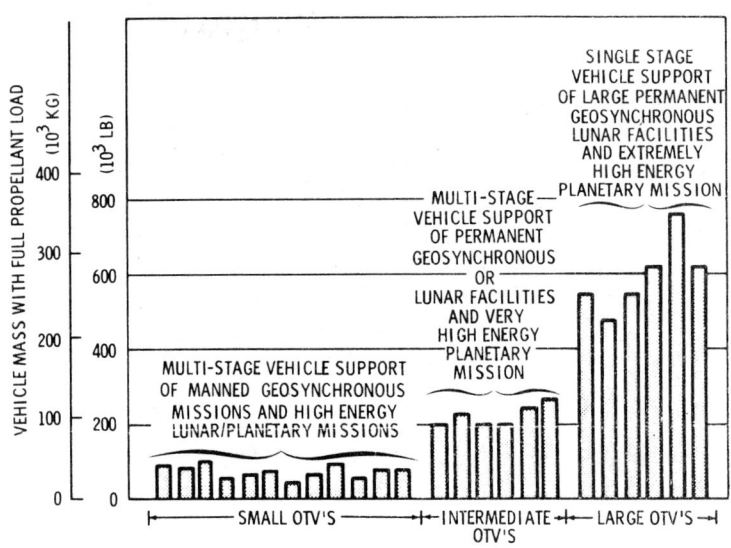

Fig. 9 Mass Comparison of LO_2/LH_2 Stage Options

If the intermediate class of OTV is to be required for infrequent missions, a heavy lift launch vehicle may not be available to provide the launch service from Earth. It may be advantageous to employ modularization to achieve the desired orbital mass with multiple Shuttle flights and orbital assembly. One brief analysis at JSC has indicated the possibility of a 200 metric ton OTV assembled in low earth orbit from eight Space Shuttle flights. This vehicle, illustrated on Fig. 10, is called the modular common stage OTV. It requires on orbit transfer of liquid oxygen into the tanks of the two stages. The Shuttle transports the liquid hydrogen to orbit already loaded into the OTV flight tanks, which are then attached to the two vehicles as the last two steps of orbital assembly prior to departure from low earth orbit.

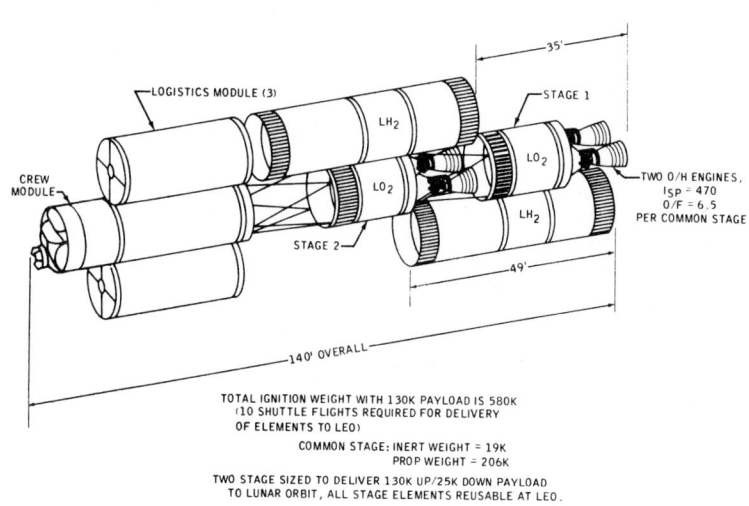

Fig. 10   Modular Common Stage OTV Concept

The lifting body type of direct entry vehicle is illustrated on Fig. 11 by the X-24C. This design is now being considered for a Mach 6-7 research aircraft role at the Flight Research Center in the mid 1980's.[8] This type

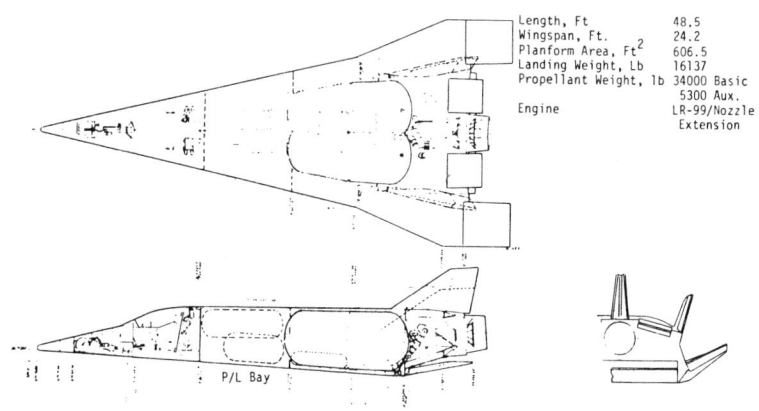

Fig. 11   X-24C-10C Configuration

of vehicle could be modified to serve the manned program needs from geosynchronous or lunar orbits. The performance capability of the vehicle will require a thermal protection system consistent with the entry heating produced by the more energetic mission. The thermal protection system weight, coupled with the amount of the propulsion capability integral to the vehicle, will determine its mass and size. This class of vehicle has not in recent years been studied. Updated weight and cost data are needed before the merits of such an approach can be assessed.

A conceptual design of a similar vehicle (Fig. 12) that looks very much like the Space Shuttle has been done.[9] This particular vehicle is designed for a multiple pass entry to reduce the heat loads to a value approximating the baseline Orbiter vehicle heat loads. The relative merits of single pass entry versus the multiple pass entry needs to be reviewed before systems selection can be made.

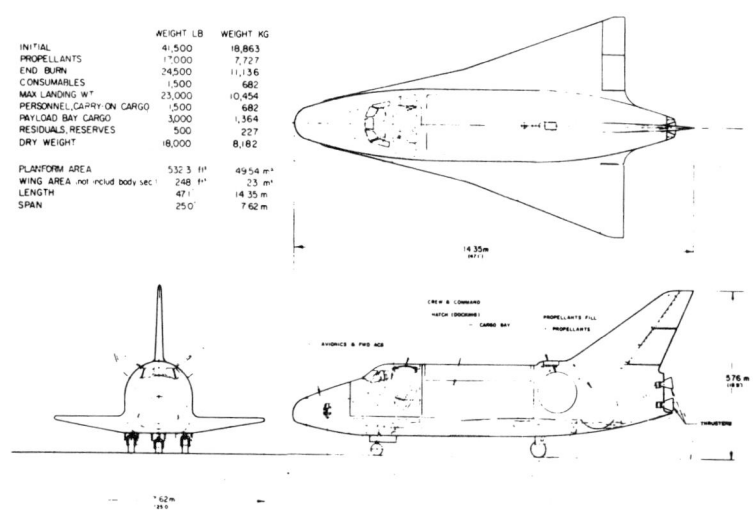

Fig. 12 Aerobraking Personnel/Cargo Glider

Unmanned transportation now appears to constitute a different class of problem than does the manned vehicle. The thrust level can be decreased and many departure burns utilized to achieve the outbound and inbound velocity increments. Additionally, the structural limitations of large space structures may determine the maximum permissible thrust levels instead of mission efficiency. These limitations may be necessary to avoid the weight of the structure being determined by the loads imparted by the transfer maneuver rather than by the loads of the satellite operational environment. A large space structure studied at JSC was found to have a single point acceleration limit of .010 g's and a four point acceleration limit of .036 g's.

The nuclear electric OTV shown on Fig. 13 is one concept of a low thrust system.[10] It utilizes a small reactor producing 25 megawatts of thermal

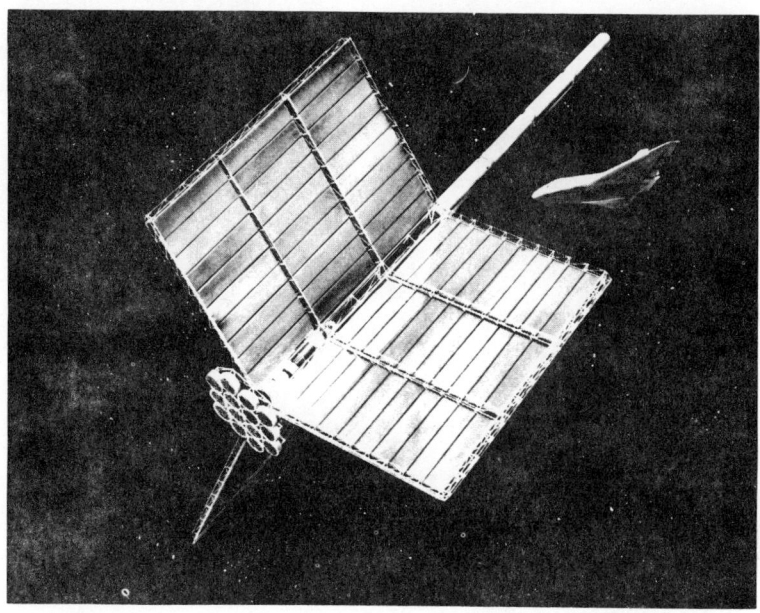

Fig. 13 5 MW Nuclear Electric Tug

power and operates with an overall efficiency of approximately 20 percent. The primary expanse of the vehicle is the space radiator necessary to achieve dissipation of the waste heat. This class of vehicle may be of interest if the usable life span and cost of the nuclear components can achieve more favorable values than past studies indicate. Another approach to achieving the electric propulsion energy supply is illustrated on Fig. 14.[10] This vehicle, called Solar Thermal Electric Propulsion Stage (STEPS), utilizes the energy of the sun rays to heat an operating fluid for operating turbomachinery, producing electric power for ion propulsion systems. The illustrated vehicle utilizes two 40 kilowatt generation modules, operates with a predicted overall efficiency of 14 percent, and is expected to produce approximately 50 kilowatts of jet power. A larger version of the same type machine could be built as shown on Fig. 15.[10]

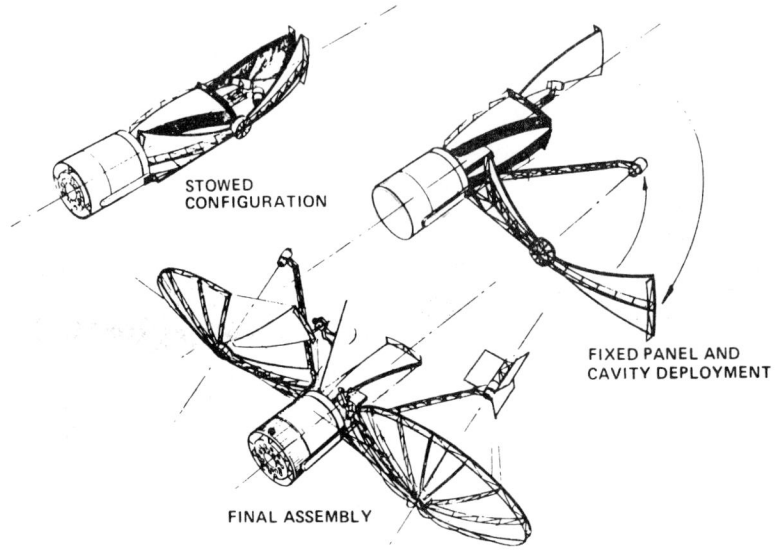

Fig. 14 Small Solar Thermal Electric Propulsion Stage

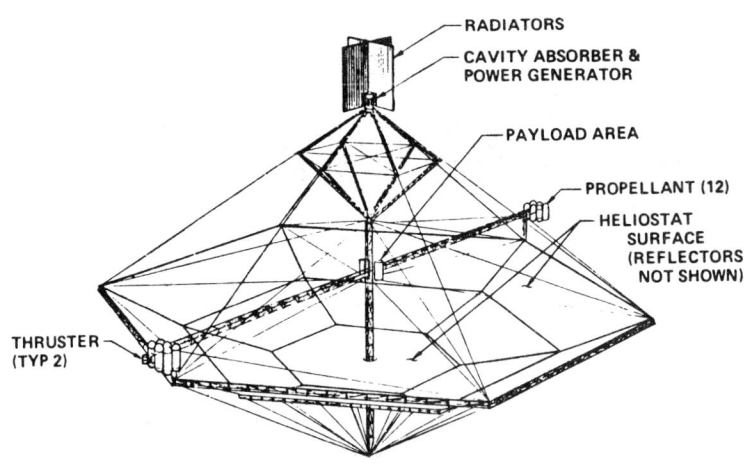

Fig. 15 Large Solar Thermal Electric Propulsion Stage

This vehicle uses a much larger heliostat type thermal collector with a power generation capability of approximately 37 megawatts and an overall efficiency of 14 percent yielding a jet power of approximately 20 megawatts. The 34 metric ton thrust level Nerva Engine required a reactor of approximately 1,600 megawatts thermal capacity to produce the 1,350 megawatts of jet power from the heated hydrogen working fluid. The same reactor applied to generation of electric power in a closed loop Brayton cycle for use in higher specific impulse ion thrusters might prove advantageous for a very ambitious set of missions.

A solar rocket powered concept was recently reviewed which the energy of the sun was utilized to directly heat the fuel.[11] This system illustrated on Fig. 16 has the unique distinction of beating the Carnot cycle losses. All of the energy, other than energy lost through thermal radiation, is imparted to the working fluid. It is expected the specific impulse levels of 700 to 900 seconds might be achieved by use of a solar heated hydrogen working fluid propulsion system.

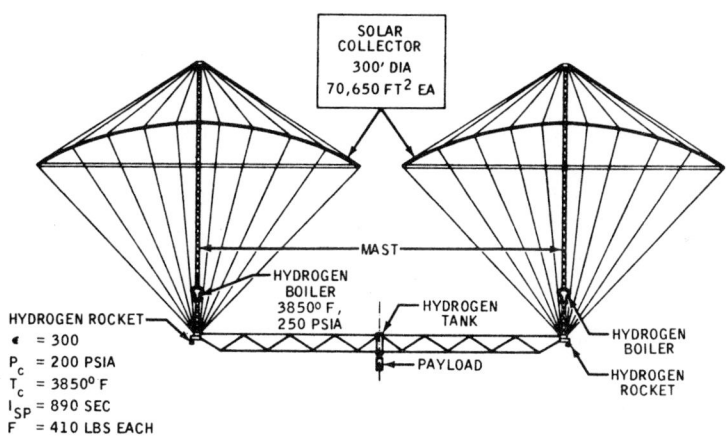

Fig. 16 Solar Hydrogen Rocket Powered Interorbital Shuttle

SYNOPSIS

These future possibilities, coupled with the mission capabilities they offer, constitute an exciting prospect for space flight in the late 1980's and beyond. The IUS is an important step toward achieving the orbit transfer capability the nation will require to accompany the Space Shuttle. As the IUS is now expected to be limited to deployment missions, early consideration must be given to the retrieval and round trip payload missions that may be required, not by economics considerations but rather by the requirement to achieve that mission capability. Space station support will be another important task for orbit-to-orbit transportation systems when a space station program is initiated. Chemical propulsion systems appear to have significant inherent growth potential to achieve the transfer of personnel and high priority payloads. The current and advanced technology chemical propulsion systems should be better defined so they may be utilized as the standard for comparison of the other more exotic propulsion schemes on the basis of technology risk and overall program cost. The high technology, low thrust systems appear to have a very attractive payoff for missions where mission elapsed time is not of central importance. Any predetermined cargo schedule may be productively fulfilled by vehicles of the very low thrust to weight ratio as illustrated by the solar electric propulsion systems. In conclusion, considerable effort remains in identifying the traffic to be anticipated through the remainder of the twentieth century and in systematically identifying the vehicle configuration possibilities for most efficiently accommodating that traffic.

REFERENCES

1. Dr. K. Ehricke, "Exoindustrial Productivity--The Extraterrestrial Imperative of Our Time," Rockwell International Corporation E75-5-1, May 1975.

2. Boeing Aerospace Company, Future Space Transportation Systems Analysis Study (NAS 9-14323), "Phase I Briefing" (Collection of Viewgraphs) D180-18759-1, April 29, 1975.

3. Dr. P. Glaser, "The Satellite Solar Power Station: An Option for Energy Production on Earth," Arthur D. Little, Inc., April 24, 1975.

4. G. Woodcock and D. Gregory, "Satellite Energy Systems," Boeing Aerospace Company, November 14, 1974.

5. Lockheed Missiles and Space Company, "Feasibility and Tradeoff Study of an Aeromaneuvering Orbit-to-Orbit Shuttle (AMOOS)" (NAS 8-28586), "Final Report," July 1974.

6. Marshall Space Flight Center, "Interim Report on Geosynchronous Space Station Program Options for the 1985-2000 Time Period," MSFC-PA-7-75, July 28, 1975.

7. Boeing Aerospace Company, Future Space Transportation Systems Analysis Study (NAS 9-14323), "Phase I Extension Orientation Briefing" (Collection of Viewgraphs) D180-18874, June 10, 1974.

8. Martin Marietta Corporation, "Summary Status of X-24 Program" (Collection of Viewgraphs), March 24, 1975.

9. Boeing Aerospace Company, Future Space Transportation Systems Analysis Study (NAS 9-14323), "Phase I Technical Report" D180-18768-1, May 9, 1975.

10. Boeing Aerospace Company, Future Space Transportation Systems Analysis Study (NAS 9-14323), "Phase I Technical Report Appendices" D180-18768-3, May 9, 1975.

11. Rockwell International Corporation, "Advanced Shuttle Projects Review Briefing" (Collection of Viewgraphs), July 1975.

AAS 75-152

AN INTEGRATED APPROACH TO TUG/IUS MISSION OPERATIONS
Billy S. King[*] and Roger A. Chamberlain[*]

Mission operations analyses performed for Tug and IUS studies were keyed to top-level functions which are necessary to meet upper stage mission requirements and are, therefore, independent of the detailed Tug/IUS design. A functional analysis at this level has inherent flexibility and is insensitive to design perturbations. Detailed "Building Blocks" of activity are developed in an "off-line" mode (in modularized form) which can be "plugged in" specific missions. This allows rapid response to new or changed mission requirements. Using mission profiles, support concepts are developed and support functions timelined for mission-operations crew sizing. Numerous functions (notably checkout) are performed pre-launch and on-orbit. Integration of software for ground and flight operations functions which are similar can reduce costly duplications and increase cost effectivity.

INTRODUCTION

The Space Tug Systems Studies (STSS) and the Interim Upper Stage (IUS) Systems Study recently completed have demonstrated the need for an operations approach which is insensitive to changing requirements and designs. During these studies, encompassing two years, numerous mission model changes occurred. These changes are to be expected as payload designs mature. Hopefully, user-community awareness of the full Shuttle Transportation System (STS) capability will generate new payloads and new uses for this national resource. The idealistic goal of desensitizing mission operations can be approached by operations analyses based on functions necessary to accomplish mission objectives. This paper will illustrate our method as applied to Tug/IUS mission operations which we call the "building block" approach. Obviously, the scope of this paper limits the discussion to concepts and illustrative examples abstracted from the total operations analysis.

Other approaches to this kind of analysis have been used but do not provide the ready visibility of completeness given by a "top-down" functional approach. Other approaches are inappropriate for configurations in

---
[*] Martin Marietta Aerospace, Denver Division, Denver, Colorado

preliminary design definition stages such as Space Tug. However, the functional "building block" approach is applicable to the IUS, which is a minimum modification of an existing design, as well as to the Space Tug.

METHODOLOGY

Use of the functional approach results in specification of functions necessary to accomplish mission objectives.

An example of the top-level functions (Figure 1) for expensable IUS, reusable IUS and a full-capability Tug flying a retrieval mission illustrate the flexibility of the functional approach. Those blocks in the diagram which are common at top level to all configurations (expendable IUS, reusable IUS and Space Tug) include launch through transfer orbit operations. At this point the IUS or Tug may be expended (e.g. Planetary Delivery) or returned for retrieval. Those blocks added for rendezvous, docking and retrieval of Spacecraft apply to the Space Tug only.

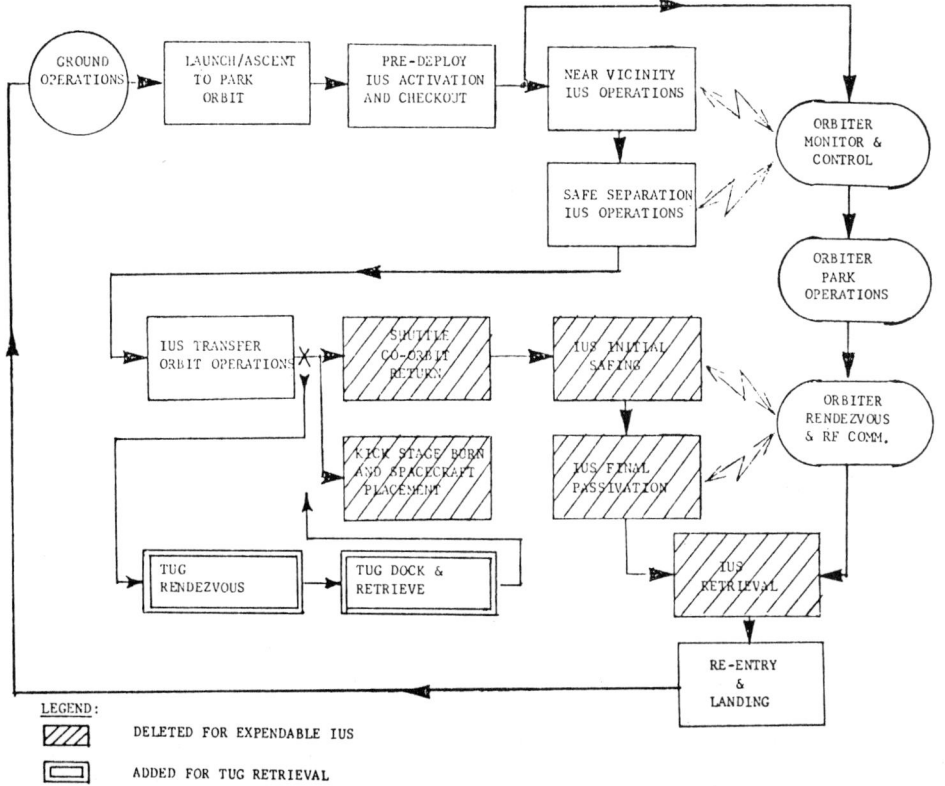

Figure 1 Functional Analysis Lends Itself to Building Block Structure

Many levels of supportive functions can be developed to fit the current state of design definition and subsequent analyses can delve deeper as the program progresses.

After development of top level functions, the traditional trajectory and mission analyses are performed for selected representative mission profiles which cover the range of mission variables and requirements.

Integration of the functional flows and mission profiles is accomplished in development of detailed mission sequences, which include ground station coverages. Detailed building blocks of events are developed for activities which are common to all missions or repetitive during a flight. This represents the initial entrance of subsystem design variables into the operations analysis.

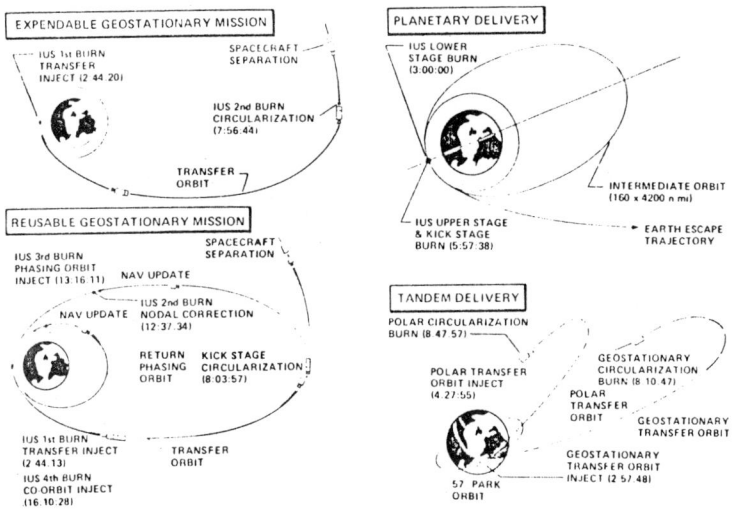

Figure 2 Reference Mission Bound Operations Requirements

The output of the operations analysis is used as input to numerous engineering trades (Figure 3) for communications link calculations, electrical power profiles, consumables budgets, thermal profiles, etc. However, the purpose of the detailed mission sequences discussed here is to permit resource allocations for operations support.

Both mission control center and flight crew activities are related to the detailed mission sequences. Functional allocation tradeoffs are conducted for onboard versus ground prime and backup performance. Ground

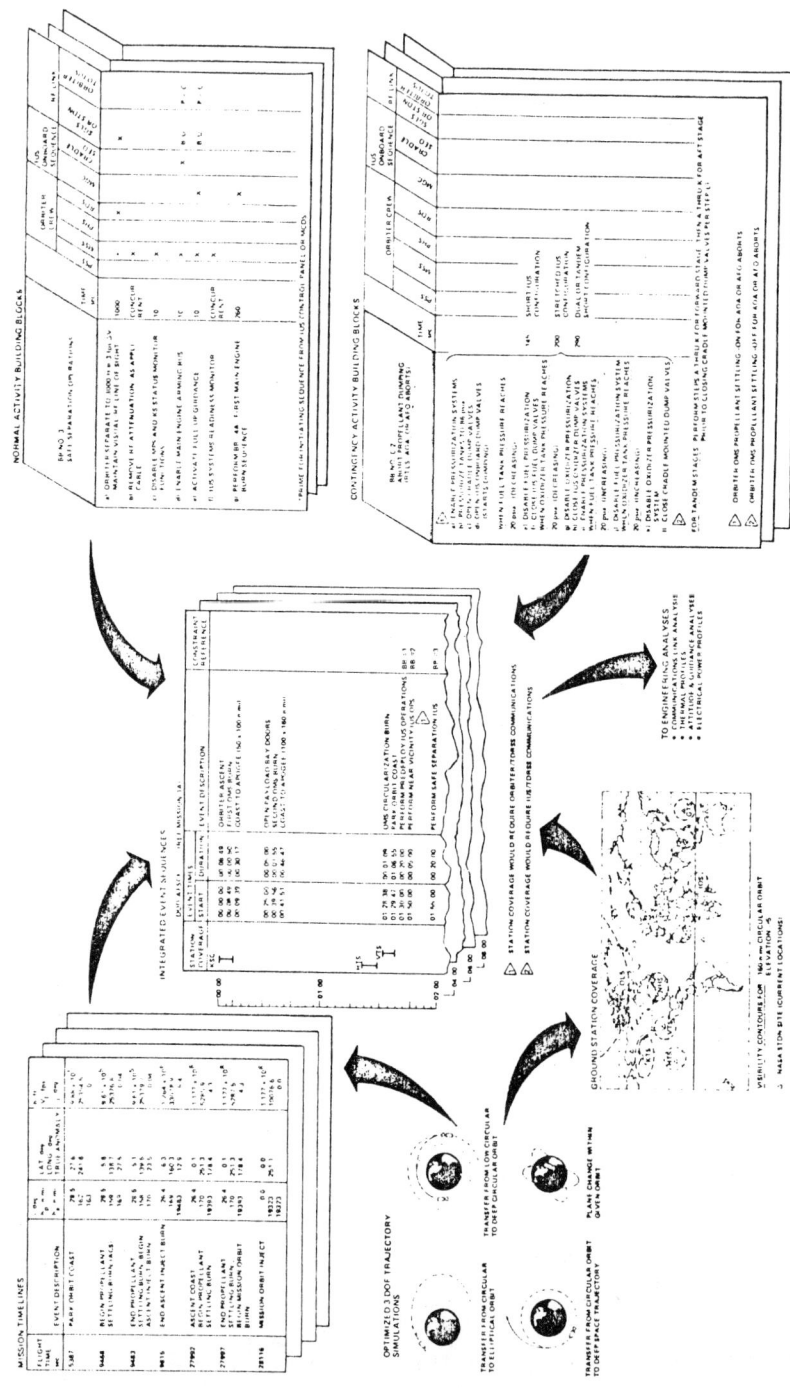

Figure 3   Inter-relationships of Operations Analysis Elements

station availability, level of autonomy desired, design capabilities, crew time, etc. are criteria used for the functional allocations. After a functional allocation is made, a hardware vs software implementation trade is conducted. This process is iterative in nature until an optimum mix is achieved. The degree of Mission Control Center involvement is determined from these trade-offs and decreases as autonomy increases. However, a requirement to monitor and back-up/override critical events of a totally autonomous vehicle necessitates having the same control center capability as would be required for prime control of these same events. This factor establishes the <u>minimum</u> control center resource requirement in conjunction with the discipline or skill requirements.

During previous programs one team of ground or launch operations specialists man consoles, provide input stimulii and monitor responses from flight hardware and instrumentation during prelaunch checkout and testing. In this process ground operations and checkout software is validated and exercised. Usually, during this same time frame, another team of flight operations specialists man consoles, provide input stimulii and monitor responses from simulators to prepare for flight control responsibilities and validate support software.

A consideration for reducing operations costs is the integration of the software support activities. Traditionally, ground checkout has been much more exhaustive than in-flight onboard checkout. Therefore, two different sets of support software have been developed. Since the shuttle program emphasizes rapid turnaround and cost effectivity, the use of common support software for both ground and flight operations functions would reduce software validation time and allow entering the flight operations phase with more confidence and less unique flight and ground crew training. Using the same software during ground operations and checkout monitoring for response to input stimulii provides the control center operator with signature responses from the actual flight instrumentation.

It is recognized that the Apollo/Skylab approach has resulted in the most successful space program in the world and that any new approach will have pitfalls and new learning curves to follow. However, we must learn from previous experiences and develop more cost-effective approaches to obtain public support for future space projects. It is with these goals in mind that this integrated approach to ground and flight operations is proposed.

RECOMMENDATIONS

A functional building-block approach to operations analysis can provide continuity throughout a program and should be given serious consideration. For this approach to be effective, it must be started early so that designs may be influenced to be easy to monitor and control. This allows easy verification by management that functions are being performed satisfactorily. The approach is carried forward from factory testing into ground operations and flight operations. This continuity from one phase to another will hasten the maturing of a program and reduce overall program costs.

AAS 75-153

TUG GROUND OPERATIONS - AN APPROACH TO
EARLY IDENTIFICATION OF OPERATIONAL IMPACTS*

John L. Best+

The Space Shuttle is being designed to provide economical transportation to and from low earth orbit. The mission model, however, also identifies missions to higher energy orbits and/or to the planets. In order to accomplish these high energy missions, additional propulsive stages are required.

The propulsive stages for performance of the high energy missions fall into three categories: the Interim Upper Stage (IUS), the Tug, and their associated kick stages. The IUS will be developed first, by DOD, with an operational data compatible with the operational date of Space Shuttle. The Tug will be developed by NASA for use during the 1983 to 1991 time frame. A transition period of at least one year is anticipated whereby both IUS and Tug will be used for accomplishment of high energy missions.

Previous Tug system studies basically provided ground operations requirements and concepts with limited information for the planning and fleet operations phases. No attempt had been made to analyze the interrelationships of these phases for optimizing overall program benefits or analyzing Tug fleet operational risk factors by studying the planning and operational phases as a "system". The preplanning and integration of the Tug with other elements of the STS and the Tug fleet operations phase had not been analyzed in sufficient detail for supporting midrange to long range program planning. An overall plan addressing both ground operational data and technical requirements that span the IUS/Tug planning and operations phases while narrowing options with emphases on more significant trade studies, was required.

---
*This work was sponsored by the NASA Marshall Space Flight Center under Contract NAS8-31011
*Martin Marietta Aerospace, Denver Division, Denver, Colorado

The recently completed Tug Fleet and Ground Operations Schedules and Controls Study[1] addressed both ground operational data and technical requirements that span the Tug planning and operations phases. This paper describes the approach taken to derive the requirements and to identify the impacts of operational requirements on Tug designs and interfaces. The more salient results of the study are also presented.

A companion study performed under another NASA contract and covering mission operations provided complimentary flight operations details. The two studies together provided operational planning data requirements, resource allocation, and control milestones for supporting the STS program.

The study tasks spanned three distinct phases. In phase 1, "strawman" processing flows, timelines, and resource requirements were developed. Numerous trades were performed to optimize the "strawman" processing flows. Where additional depth of analysis was required, special emphasis assessments were performed to compliment and expand the "greenlight" single-cycle processing flows.

Subsequently, the study operated on these optimized flows to develop requirements for other program phases. For example, the traffic impact was analyzed to establish the Tug fleet sizing. Contingency analysis was employed to realistically size the fleet under other than nominal conditions. Fleet management techniques were developed. Site activation requirements for the Tug were defined, based on the operational data developed early in the study. The period of transition from IUS to Tug was analyzed, giving special consideration to the period of time when concurrent IUS and Tug operations may be required. Requirements for Tug to spacecraft integration in the mission planning era was addressed considering such issues as Level I integration concepts and multiple spacecraft integration.

Finally, in phase 3, results were assessed to determine the impact on Tug design and development. An alternative concept for processing the Tug in an as-received condition in a factory clean environment was assessed. Each task resulted in a subplan that was integrated into an

overall plan. The subplan elements were subjected to a sensitivity
analysis before finalization. Supporting Research and Technology and
Recommended Additional Efforts were identified. By intent, the study
was fairly broad and shallow which is appropriate for operational
studies performed in early program phases. The objectives were to drive
out the "big swinger" requirements and to provide operational impacts
on the proposed designs. To that end several significant recommendations
were made.

For nominal Tug processing, factory clean environment in the VAB low
bay was recommended over 100K processing. Two processing cells are
required each with an LPS terminal. It was recommended that Level I
off-line integration should be performed in the TPF cell using selected
Orbiter simulation and that multiple spacecraft buildup should be performed off-Tug to reduce the turnaround times.

For contingency situations, the capability to perform spacecraft/Tug
mate and integration at the PCR should be provided. Payload changeout
provisions at the pad provide very valuable flexibility and that capability should be retained. Similarily, although vertical installation
at the pad was recommended, horizontal installation at the OPF should
remain open as an attractive option.

The study results indicated that the most cost effective approach to
WTR launches is to perform all maintenance and checkout at ETR. Tugs
would then be ferried to WTR where spacecraft integration would occur
in the PPR. It was recommended that payload integration, both analytical
and physical, be performed by the Tug project using techniques such as a
detailed user's guide, level I integration simulation built into the
checkout cells and software integration performed in an off-line simulation
lab.

An engineering model (Tug structural test article or propulsive test
vehicle) should be used for site activation. Activation planning should
recognize that the modification of facilities to accommodate tugs will
be in the high flight regime of the IUS program in 1980-1983. This will
require careful coordination to avoid operational and schedule incom-

patibilities. Activation schedules and milestones were developed as part of the study.

A mechanized computer assisted utilization planning system will be required. As part of that system the study performed some contingency analysis and defined provisions that should be incorporated in the facilities to support "real time" contingency capability. The study provided an analysis of the Tug logistics requirements, recommending transportation between ETR and WTR and between contingency landing sites to ETR by 747 piggyback cargo pod rather than rail, highway or marine transportation. Deferred spares procurement should be considered and the concept of Tug block build and delivery is attractive, especially for product improvement or technology upgrades.

Based on an impact analysis, it was recommended that minimum launch capabilities be provided at WTR for Tug missions. Basically, this concept provides for all maintenance, refurbishment, per-mission reconfiguration and post and pre-mission checkouts be performed at ETR. Tugs would be ferried to WTR, installed in the PPR and mated with the spacecraft. No maintenance or extensive checkout capability would be provided at WTR and no off-line facilities would be designated for Tug.

Although cost effective design concepts are necessary and provide one area for reducing costs, perhaps an even more fertile area lies in devising operational concepts that lend themselves to lower cost methods of doing business. Of course, these new methods can be implemented only if they are identified early and the capabilities are built into hardware, system designs, and management concepts.

This study has served that purpose by developing operations concepts and assessing the impact of those concepts on the baseline Tug design and the Orbiter interfaces. Where the baseline design did not support the most efficient method of operation, design changes have been recommended. Where the Tug-to-Orbiter interfaces did not adequately support the Tug operational requirements, the study provided recommendations for improvement. Perhaps one of the most significant contributions of this study, however, was the establishment of an "operational attitude" early in the Tug program. Appropriately this operational attitude is expected to solidify early Tug project planning with benefits already derived from the common contractor progress reviews and data exchanges. To be truly effective, the Tug project must continue to develop a maintainable and operationalized design while simultaneously developing appropriate fleet management and operations concepts.

AAS 75-154

SPACE TUG MISSION OPERATIONS

*Kenneth C. Nuss**

ABSTRACT

The Space Transportation System introduces a new era in the conquest of space with new man-machine interfaces and new missions for manned vehicles. Creating a detailed timeline by using the baseline Tug fitted to a realistic Tug mission provides a close approximation to the actual mission operations of the 1980's. The mission operations planned appear feasible. A Tug geosynchronous placement and retrieval mission is discussed.

INTRODUCTION

In the past, placement of unmanned earth satellites has been primarily dependent upon preprogrammed mechanical operations. Introduction of a manned vehicle into the unmanned-spacecraft delivery program creates a greater flexibility and allows alternative plans of action previously unavailable, but also involves greater operational complexity. Unmanned payload delivery via the Space Shuttle Transportation (STS) involves increased interaction between man and machine, thus requiring a flight operations plan compatible with both. In this paper, a flexible, standardized flight operations plan for Space Shuttle upper stages is presented. The geosynchronous mission presented represents a typical upper stage mission, based upon **NASA Reference Mission Number 1, Revision 1**, and serves as a reference for a Tug geosynchronous mission. A computer program was developed and structured so that, as key events of the mission change, activities tied to these events change accordingly to provide timelines for different specific missions. These timelines help the design engineer size the systems for upper-stage vehicles.

The baseline vehicle used for this paper is a cryogenic upper stage that fits into the Orbiter cargo bay. It is 14 feet in diameter by 30 feet long, weighs about 58,000 pounds, and is capable of delivering 6,000 pounds into geosynchronous orbit.

---

*Research Engineer, General Dynamics Convair Division, San Diego, California

The structure is comprised of LO_2 and LH_2 tanks, a Pratt & Whitney RL10 Category IIB main engine with extendable nozzle, and a body shell composed of a forward skirt, a main skirt, and an aft adapter. The whole structure conforms to the design of a rotating adapter pallet in the cargo bay of the Orbiter. An Orbiter manipulator arm aids in deployment and docking with the Orbiter.

A crew of four consists of a commander, a copilot, a mission specialist, and a payload specialist. The commander makes safety-related decisions, while the mission specialist handles all Tug-related operations, and the payload specialist handles payload operations. The pilot is responsible for deployment and retrieval operations.

Adequate umbilical connections are available for fueling, venting, and monitoring Tug subsystem parameters. Tug avionics subsystems are highly autonomous, of Level II autonomy. The agency for ground interface and control is the Tug Operations Center (TOC). Johnson Space Center has operational control when the Tug is mated with the Orbiter

MISSION OPERATIONS FOR A GEOSYNCHRONOUS MISSION

The flight-related mission events begin several minutes before launch with a final system check. In the final moments, about a minute before liftoff, propellant replenishment is discontinued, the vent valves are closed, and the tanks allowed to pressurize to provide added stability for launch loads. Tug fuel cells are activated, and power loads are switched from external sources to the fuel cells. Caution and warning data is monitored continuously by the Orbiter crew and ground. Shortly after liftoff, vent valves are re-opened and the Tug rides through the ascent profile and into orbit, requiring minimal attention from the Orbiter crew.

After Orbiter circulation in a 150-nmi parking orbit, a lengthy coast period follows — ranging with the intended position for the payload in geosynchronous orbit. The transfer from parking orbit to geosynchronous orbit is a Hohmann transfer. Thus, if our intended geosynchronous longitudinal position is 90°E, the transfer burn must begin at 90°W, or 180 degrees from target longitude minus earth's rotation. The likelihood that the Orbiter's ground trace will pass over the equator at a point exactly opposite the target longitude is very slim, so a phasing orbit is almost always required (an alternative would be to "walk" the payload at geosynchronous altitude). The phasing period also varies from mission to mission, depending upon the intended target position in geosynchronous orbit, and is usually larger than the parking orbit.

Tug-Payload Deployment

For the first time in space operations, a check of the upper stage and payload will be possible after ascent and in a zero-g environment. Thus, a defective upper stage or payload can be identified on-orbit and repaired before a release commitment is made. A thorough checkout before launch will be sufficient for most checks, but a status check of the vehicle and payload in space will prevent a commitment to deploy when the vehicle will not perform. Thus, deployment operations are planned so that commitment to deploy does not occur until there is reasonable assurance that the vehicle and payload are operational. Each event is based upon indications of previous events that all systems are "go."

The deployment sequence begins with the pilot giving the manipulator control station and remote manipulator system (RMS) arm an operational check. If it is not functional, there is no need to continue. Assuming that it checks out, the mission specialist gives the Tug and payload a thorough status check. Guidance and navigation systems are initialized, then verified operational. Meanwhile, the commander orients the Orbiter to a preferred deployment attitude. Forward umbilicals for the payload, if any, and fluid umbilicals are then detached from the Tug in preparation for rotation out of the cargo bay. The Tug is rotated out of the cargo bay to ensure clearances and to allow RF link establishment before separation, thereby providing for caution and warning monitor throughout. After rotation, the RF link is established, a final position and velocity update from the Orbiter is given, and the electrical umbilical disconnected.

The pilot then attaches the remote manipulator arm. Tug-adapter latches are released and the pilot, using the RMS, axially slips the Tug out of the aft adapter, lifting it clear of the cargo bay and positioning it above the Orbiter cabin. This release position allows minimum plume impingement during the Orbiter separation maneuver. The Tug is shown in the rotated and release positions in Figs. 1 and 2. The Orbiter releases the Tug, separates to a safe distance, then initiates the Tug attitude control system. A ground-Tug RF link is established for futher checkout operation from the ground if necessary. A final command enabling the main engine readies the Tug for its designated mission.

Figure 1. Tug in rotated position for deployment

Figure 2. Tug in extended position for release.

Orbital Operations

First burn for the Tug on a geosynchronous mission is a phasing burn. The Orbiter assumes an attitude to help monitor first burn. The computer provides the data for the burn attitude of the Tug, and the Tug assumes this attitude. Zero-g vents are disabled and the igniters turned on by automatic sequence. The engines operate in a tank head idle mode for 90 seconds, pressurizing the propellant tanks and settling them. The engine then transitions to full thrust for a 17-minute burn, placing the Tug into the phasing orbit. A second similar burn at perigee, one revolution later, sends the Tug and payload into a transfer orbit to geosynchronous altitude. This transfer lasts about five hours. During long coast periods, the Tug assumes an attitude that allows the interferometric landmark tracker to update the position and velocity vector.

At geosynchronoous altitude, the Tug performs a third burn, circularizing into geosynchronous altitude. The Tug then maneuvers to the required separation attitude, spins the payload if necessary, uncouples the umbilicals, and separates. The deployed payload is then given a TV inspection to help check proper operation. Payload status is reported to the ground.

The Tug then starts on the next phase of the mission, a pumped idle mode engine burn that places the Tug into a phasing orbit to the next payload for retrieval. Pump idle mode burns are less than full thrust, but the burn sequence is the same except for pressurization.

Upon reaching the target vehicle, the Tug circumnavigates the spacecraft, and inspects it with a TV camera while the docking tracker searches for the docking aid reflectors mounted on the spacecraft. The Tug aligns with the spacecraft, using the docking aids, and translates to the docking position. If TOC does not issue a hold command, the Tug continues to a hard docking.

At this point, the Tug transfers back to low altitude. At 160-nmi perigee, a main engine burn places the Tug and spacecraft into a phasing orbit in the plane of the Orbiter. The Tug circularizes into a 160-nmi orbit, positioning itself 10 nmi above and slightly ahead of the Orbiter. The Tug safes its main propulsion systems and becomes a passive vehicle for rendezvous and docking with the Orbiter.

Orbiter-Tug Rendezvous and Docking

Before proceeding with rendezvous and docking with the Tug, the Orbiter crew verifies that it can monitor Tug caution and warning data. It would be unsafe to proceed if it could not. The Orbiter RMS and rotating adapter systems are prepared and checked to ensure they are capable of docking. The Orbiter then determines range and range rate to the Tug and computes parameters for an intercept maneuver. The initial intercept maneuver is a Hohmann transfer to the vicinity of the Tug in the 160-nmi orbit. Several hours are allowed for final braking maneuvers and maneuvers to position the Orbiter and Tug into docking position. For docking, the Tug is positioned directly above the crew compartment ceiling window.

Once again, Tug safety status is verified before proceeding. The RMS end effector is attached to an RMS attachment fitting on the Tug. A contact switch in the attach point inhibits further attitude control thrusting at the instant of attachment. The remainder of the docking procedure is the reverse of the deployment procedure. The Tug is rotated and lowered to a position in line with the rotating adapter axis, tilted at 35 degrees. RMS movements are computer controlled with man-in-the-loop monitoring. The RMS inserts the Tug into the aft adapter with fine positioning alignment provided by adapter-mounted indexing devices. All Tug systems are safed and the cargo bay doors closed in final preparation for Orbiter re-entry.

OTHER MISSIONS

The geosynchronous mission is representative of a large number of missions, requiring the use of an upper stage. There are, of course, other mission types. Heavy-payload and high-inclination, low-earth orbits are very similar to the geosynchronous mission. The planetary mission, however, is significantly different, requiring an escape velocity. For most planetary missions, two burns are used to provide the escape velocity.

An attractive possibility for the Space Tug is delivery of multiple payloads. In some instances, it may be possible to carry as many as eight payloads to a given orbit, thus saving the cost of several launches. Another possible use of the future Space Tug is to use it as a servicing vehicle, either making repairs on a spacecraft, or retrieving the spacecraft for delivery to a repair station where repairs can be made and returning it to its functional orbit.

COMMUNICATIONS

An important part of Tug mission operations is the availability of ground tracking coverage. If the Tracking Data Relay Satellite System (TDRSS) is in operation, coverage for low earth orbits is excellent except for a small area of exclusion over the Indian Ocean. TDRSS coverage is diminished at altitudes approaching geosynchronous altitude, but at that altitude there is good ground coverage. The results of communications coverage analysis indicate a strong need for the TDRSS, especially for low earth orbit. It would be possible, but more difficult and requiring more autonomy, to operate without it.

CONCLUSIONS

Mission operations for the Space Tug have been investigated in considerable detail. Because of new man-machine interfaces and equipment sophistication, a standardized mission sequence is desirable and necessary. For a geosynchronous mission, representative of the Tug mission model, all planned operations seem feasible and within the capabilities of the Tug vehicle. Ground tracking coverage, an important element of the Tug operating concept, is limited without the aid of the TDRSS.

AAS 75-214

SPACEFLIGHT TRACKING AND DATA NETWORK SUPPORT
IN THE SPACE TUG ERA

L. R. Stelter*

and

Robert D. Godfrey[+]

This paper describes the Spaceflight Tracking and Data Network (STDN) which will be used to provide command, telemetry and tracking support to the Space Tug in the 1980's. A description of the mission support capabilities of the two major elements of the STDN - the ground based sites (GSTDN) and the Tracking and Data Relay Satellite System (TDRSS) - is provided. This paper also contains a summary description of the mission/STDN RF and Data Handling interface.

INTRODUCTION

The Spaceflight Tracking and Data Network (STDN) in the Space Tug Era will be composed of two major elements: a ground-based network (GSTDN), consisting of five orbital support and two launch support sites, and the Tracking and Data Relay Satellite System (TDRSS). The TDRSS element of the STDN will consist of two operational and one spare Tracking and Data Relay (TDR) spacecraft. The configuration of this network is shown in Fig. 1. This network will have the capability of providing at least 85% coverage to all mission Spacecraft above 200 km orbital altitude (100% coverage to all users above 1200 km orbital

---

* Associate Director for Engineering, Networks Directorate
[+] TDRSS Project Office
   NASA Goddard Space Flight Center

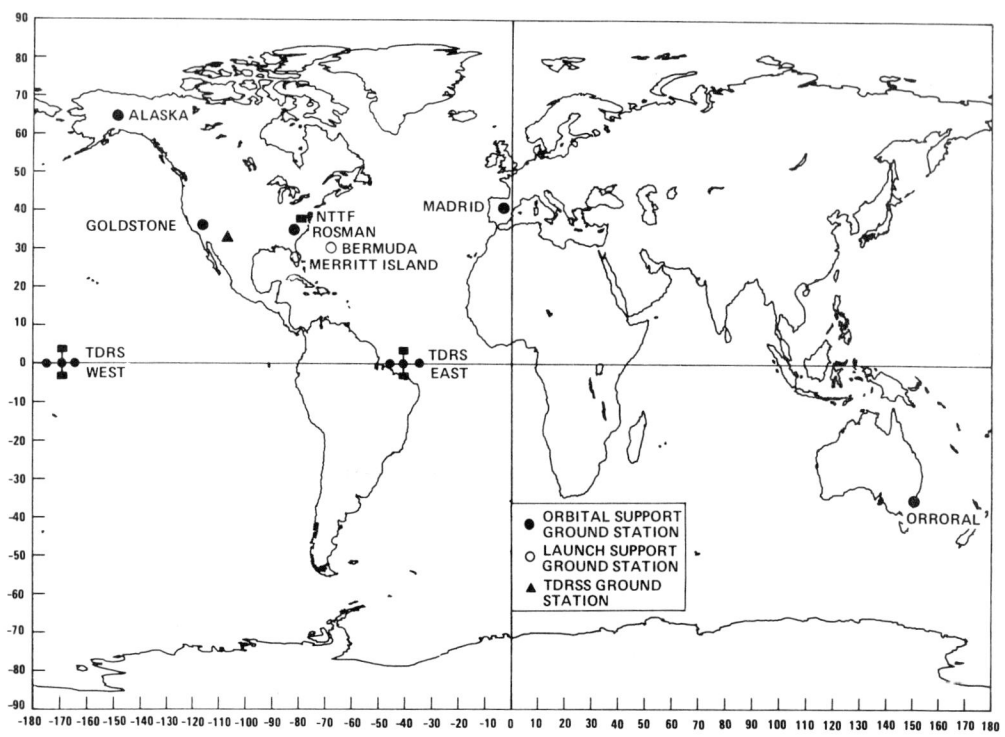

Figure 1. The Post 1980 STDN

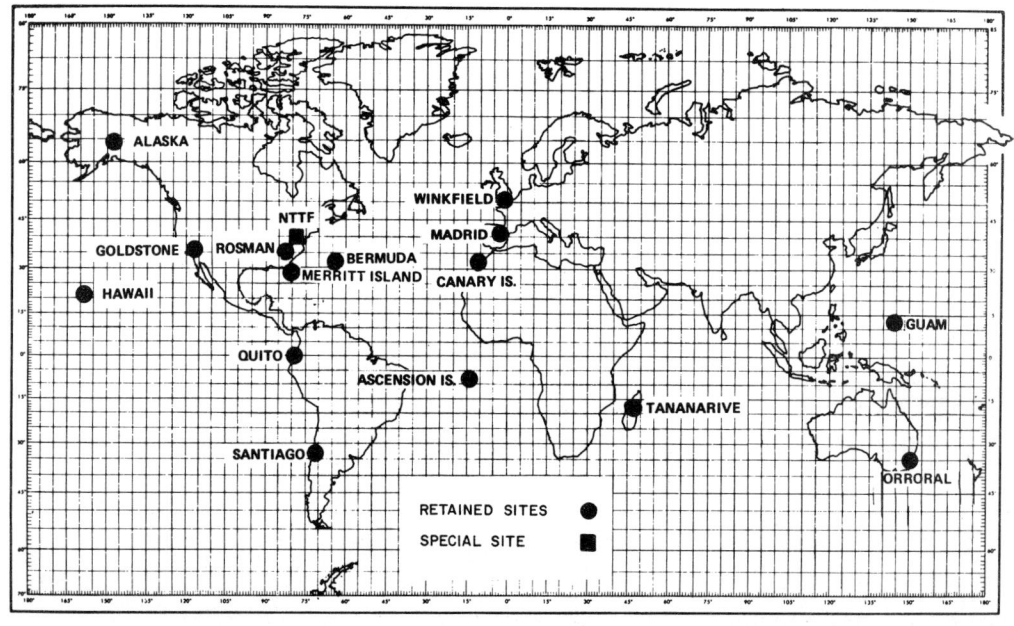

Figure 2. Current Spaceflight Tracking and Data Network

altitude). This configuration of the STDN will provide tracking, command and telemetry support services to the mission spacecraft in the most cost effective manner. In general, the ground-based network will provide support to mission spacecraft with orbital altitudes greater than around 5000 km while the TDRSS will support those missions with lower orbital altitudes.

The STDN is currently in the process of evolving from the totally ground-based network shown in Fig. 2 to the ground site/TDRSS network shown in Fig. 1. The mission spacecraft support services provided by the two elements will be compatible and complementary. The TDRSS elements of the Space Tug Era STDN will be operational by January 1, 1980. The ground-based elements of the network will achieve the Tug Era configuration by July 1980, although modifications required to support specific user requirements will be implemented.

The mission support sites and control centers of the STDN are tied together by the NASA communications system (NASCOM). This system distributes site operational instructions and mission spacecraft command data to the sites and site status information, mission spacecraft telemetry and tracking data from the support sites to the appropriate destination. The current configuration of the system is shown in Fig. 3. As the STDN evolves from its current configuration, as shown in figure 3, to the Tug Era network, the NASCOM system will change from its current narrow band transmission system, which is capable of transmitting only a part of the data received on site, to a wide band system which can transmit, in real time, most of the data received by each ground-based site and all of the data received by the TDRSS.

GROUND-BASED ELEMENT

The basic configuration of retained ground sites will remain unchanged. Many of the on-site systems will be upgraded but the operational philosophy will be the same as in the current STDN except for the handling of mission spacecraft data.

A typical STDN ground site configuration for the mid 1980's is shown in Figs. 4 and 5. The major systems contained within this site are:

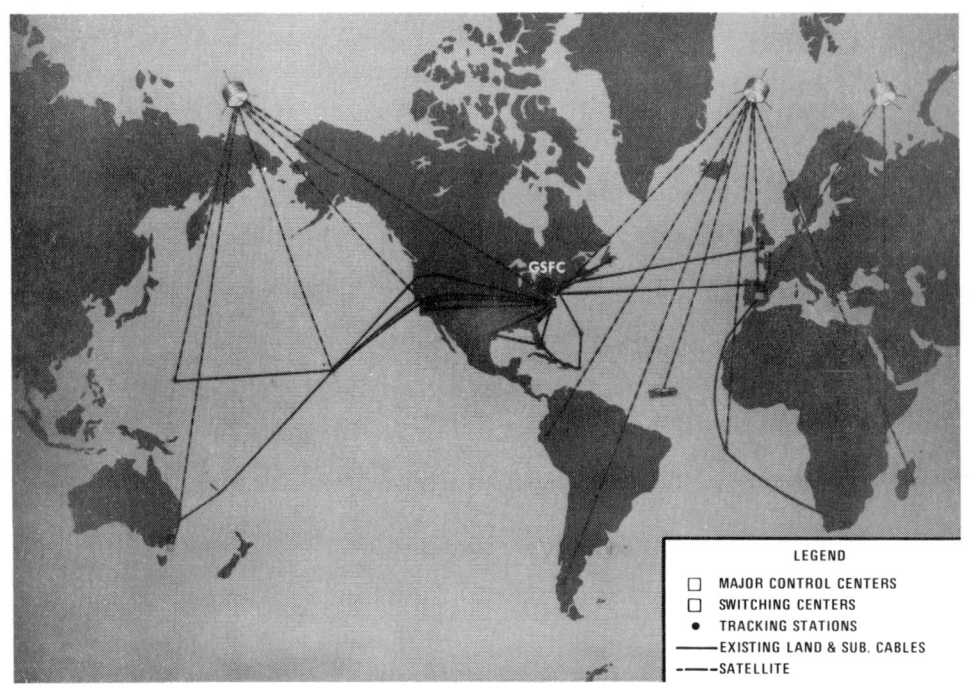

Figure 3. NASCOM Network Trunking Plan

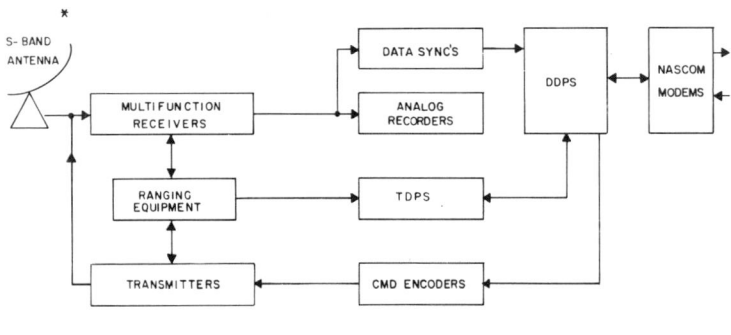

* KU-BAND MAY BE ADDED IF REQUIRED

Figure 4. Typical STDN Ground Site Configuration for the Mid 1980's

Antenna Systems

Each orbital site will be equipped with a 26-meter (diameter) antenna system. Some sites will also use a 9-meter antenna system, including all launch support only sites.

Operating Frequencies

These sites will be capable of providing support at S-band (2025 to 2120 MHz command link and 2200- to 2300-MHz telemetry link). Ku-band (13.4 to 14.05 GHz command and 14.6 to 15.25 GHz telemetry) capability will be provided when required.

Receiving System

These sites will be equipped with the STDN multifunction receivers (MFR), which have a maximum receive bandwidth of 20 MHz at S-band (2200 to 2300 MHz).

Transmitter System

These sites will be equipped with an SCE transmitter system which has a maximum bandwidth of 20 MHz. The maximum transmitter power will be 20 kW.

Range and Range Rate System

The ranging system will be the standard Goddard Range and Range Rate (GRARR) system, which is a coherent system utilizing ranging sidetones and a carrier doppler range rate system.

Telemetry System

The telemetry system at each site will be configured around the Digital Data Processing System (DDPS) and will be capable of supporting any data signal compatible with the GSFC aerospace data standards.

Command System

This system will be configured around the spacecraft command encoders and be capable of supporting any signal consistent with the GSFC aerospace data standards.

Figure 5. Typical STDN Ground Site

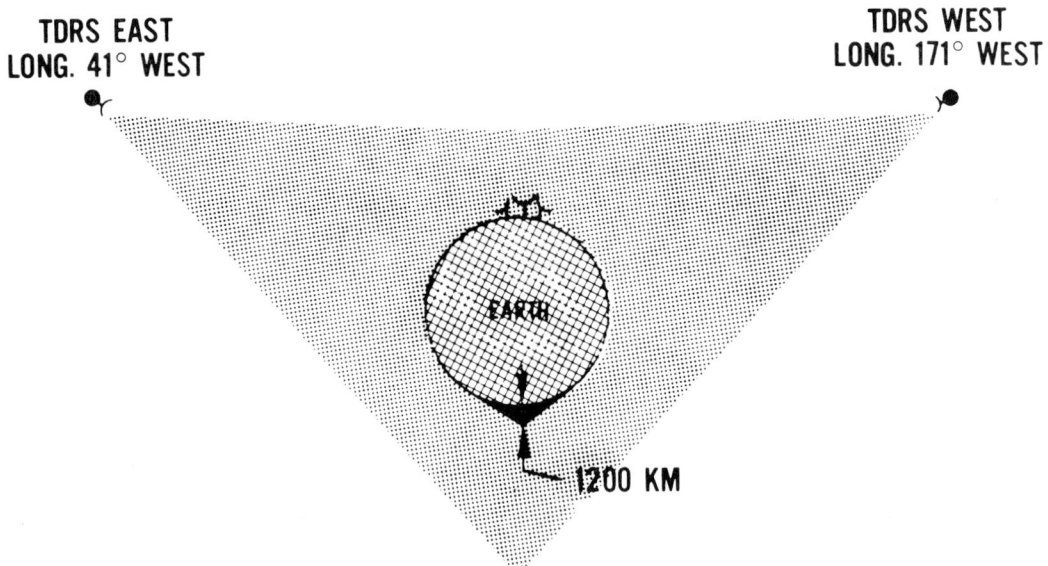

Figure 6. Tracking and Data Relay Satellite System

Site to User Data Distribution System

Wide band spade type NASCOM circuits will be provided for a normal "real time" operating mode. A minimum of one 56 kb/sec. circuit will be provided between each site and GSFC. When a mission spacecraft's data rate (e.g. when dumping tape recorders at a high rate) exceeds the NASCOM circuit capacity, the data will be recorded and transmitted when capacity becomes available.

The overall mission support capabilities of the GSTDN sites will remain essentially unchanged. However, due to the small number of sites remaining, the GSTDN alone cannot provide support to low earth orbiting users. The GSTDN will normally provide support only to users at high orbital altitude (>5000km) and during the early launch phase. The ground-based sites will also provide a limited backup support capability for the TDRSS supported users.

SPACE ELEMENT

The space based elements of the STDN, the Tracking and Data Relay Satellite System (TDRSS), will consist of two geosynchronous relay satellites, approximately 130 degrees apart in longitude (see Figs. 6 and 7), and a ground terminal located in the continental United States. Additionally, the system will include two spare satellites; one in orbit, and one in readiness for a rapid replacement launch. The purpose of the TDRSS is to provide telecommunications services which relay communications signals between low earth-orbiting user spacecraft and the mission spacecraft control and/or data processing facilities. A real-time, bent-pipe concept is utilized in the operation of the TDRSS telecommunications services. The system will be capable of transmitting data to, receiving data from, or tracking mission spacecraft over at least 85% of the user orbit.

The TDRSS will contain two types of telecommunications systems: a multiple-access system for the simultaneous telemetry support of up to 20 mission spacecraft, and two single-access systems for support of high data-rate mission spacecraft. These systems are depicted in Figs. 8 and 9. A detailed descrip-

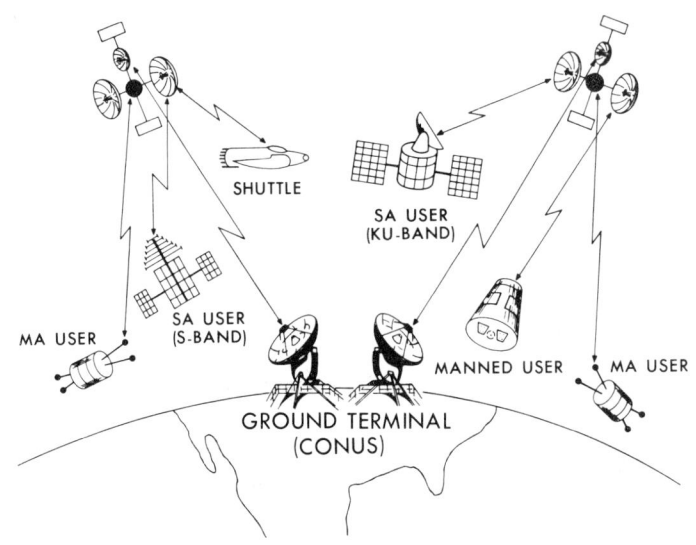

Figure 7. TDRSS Concept

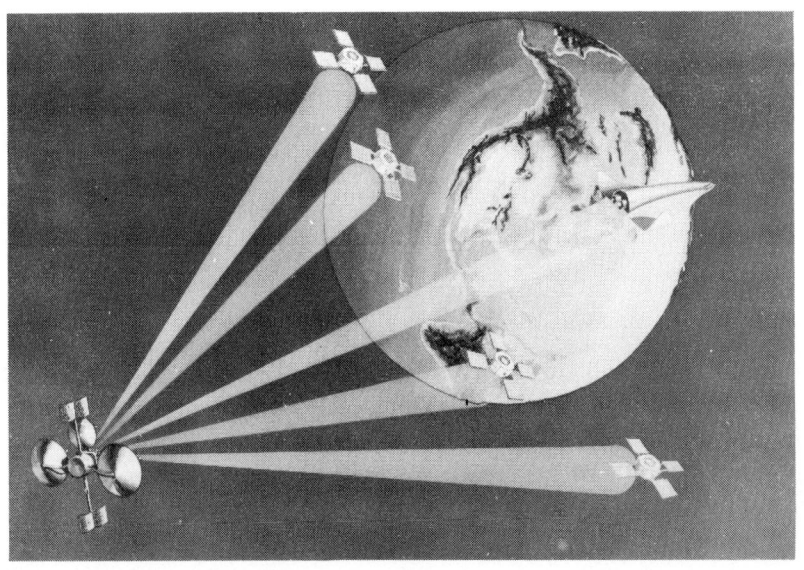

Figure 8. Multiple Access Service Illustration

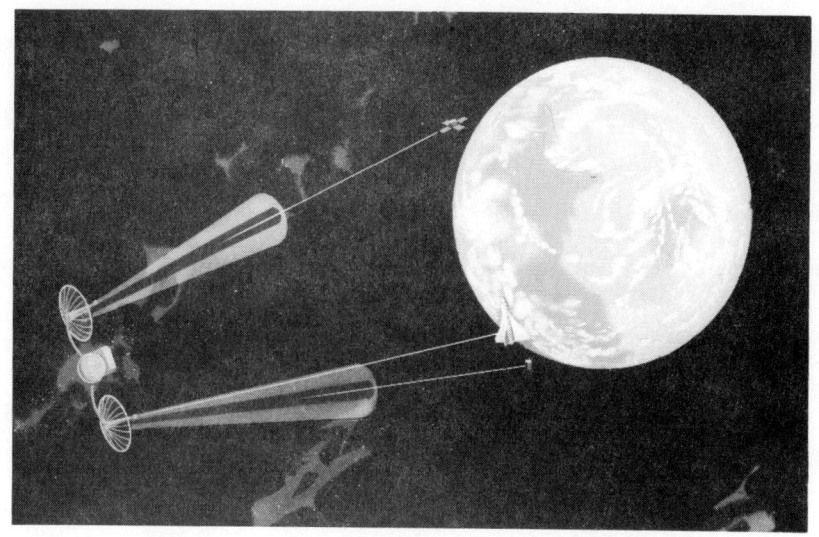

Figure 9. Single Access Service Illustration

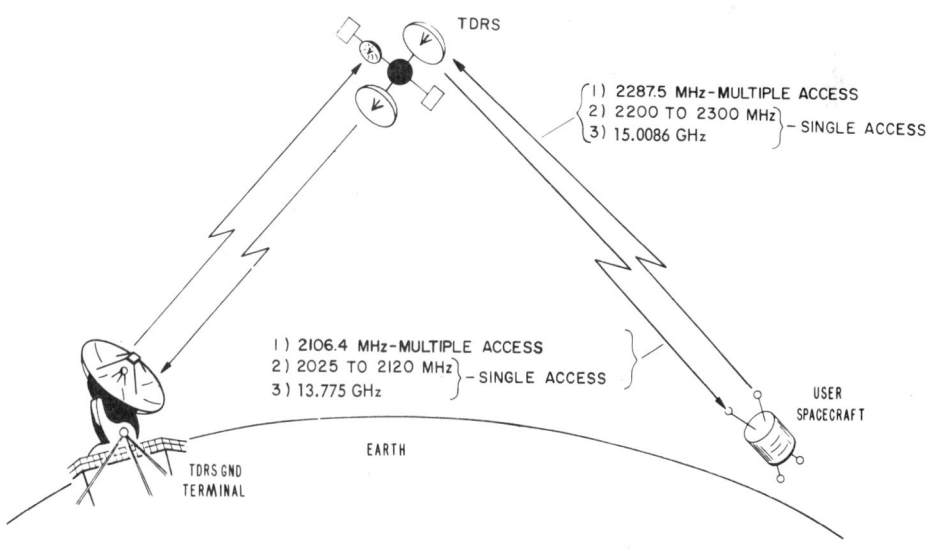

Figure 10. TDRSS Frequency Plan

tion of the mission spacecraft/TDRSS interface is given in the TDRSS Users Guide, Revision 2, STDN 101.2, Goddard Space Flight Center, May 1975.

TELECOMMUNICATION SERVICES

Multiple-access Communication Service System.

The Multiple-access (MA) communication service system is designed to provide simultaneous real-time and dedicated return link service to low earth-orbiting mission spacecraft with real-time data rates up to 50 kb/sec. Forward and return link support can be provided to all users of this system during the entire portion of their orbit visible to a TDRS, a minimum of 85% of the orbital period. Based upon current mission model projections, no scheduling restrictions should be encountered if this system is utilized for return link support. The forward link service provided by this system is time shared with a maximum bit rate of 10 kb/sec and supports one mission spacecraft at a time. This system will operate at S-band.

Single-access Communication Service System.

The Single-access (SA) communications service system is designed to provide a high data rate return link to mission spacecraft with real-time, playback, or science data requirements up to 300 Mb/sec and those requiring forward link data rates up to 25 Mb/sec. This system will be utilized only on a priority basis, and will not normally be available for dedicated support of a specific mission (with the exception of the Space Shuttle). Return link service is provided at S-band (up to 6 Mb/sec) and Ku-band (up to 300 Mb/sec).

Cross Support Communication Services.

Any mission spacecraft which is compatible with the MA system can receive forward or return link support from either the MA or SSA systems. Continuous MA support of a real-time return link or periodic SSA support of a high data rate experiment are appropriate support modes for these systems.

Tracking Services.

Each of these service systems can provide range and range rate tracking data for each mission spacecraft supported. These data will be provided by a PN

Range and Range Rate (R&RR) system and will provide tracking accuracies comparable to that currently available from STDN.

TDRSS FREQUENCY PLAN

The forward and return link frequency plan is summarized in Fig. 10 for the space-to-space links. Frequencies for TDRS-to-ground terminal and ground terminal-to-TDRS links are not shown since they do not affect the TDRSS/mission spacecraft interface.

SERVICE OPERATIONS AND DATA HANDLING

Fig. 11 provides an overview of the TDRSS user operation interfaces. The major operational usage features of the TDRSS are the availability of extended user spacecraft contact time and user spacecraft-to-user data facility telecommunications bandwidths in excess of user spacecraft data rates, thus providing real-time data transfer. Necessary real-time mission operations control and monitor data will be available in a standard form for all users. The real-time control of the TDRSS/user control center interface will be accomplished by TDRSS operations controllers in the Network Operations Control Center (NOCC). All forward and return link data between the Project Operations Control Center (POCC) (and/or other user data source or destination) and the user spacecraft will be transmitted in real time according to the user support schedule. No forward or return link data will be stored by the TDRSS.

MISSION SPACECRAFT SIGNAL DESIGN OPTIONS

The TDRSS incorporates a specific signal design, which is specified in detail in the TDRSS Users Guide. MA users must use the signal design specified. However, SA users can use unique signal designs provided they are compatible with the forward link Intermediate Frequency (IF) channel salient characteristics of the TDRSS. Therefore, alternate signal designs must be thoroughly coordinated with the Networks Directorate at GSFC to insure:

- The forward or return link spectral characteristics are compatible with TDRSS-planned utilization of available frequency bands.

- For forward link signals, the resulting transmitted spectrum is constrained by the applicable flux density considerations.
- The user provides equipment to handle any user-unique signal designs.

TDRSS/USER ORBITAL COVERAGE CAPABILITIES

The two-satellite TDRS system will provide the potential for near global real-time coverage (at least 85% averaged over 24 hours) for most users. This coverage can be provided to all MA users and to some high-priority SA users. The zone of exclusion (i.e., area in which coverage cannot be provided) is shown in Fig. 12 for a vehicle at a 200-kilometer altitude (outer figure) and a 1000-kilometer altitude (inner figure). Total coverage is provided for orbital altitudes greater than 1200-kilometers. The zone of exclusion represents the lower altitude coverage limits for the TDRSS users. The upper altitude coverage limits are 12,000 kilometers for the SA system and 2000 kilometers for the MA system, when viewed at the limb from the TDRS (worst case). In summary, the following coverage is provided:

- Minimum coverage at 200 kilometers of 85%.
- 100-percent coverage between 1200 and 2000 kilometers for MA service and 12,000 kilometers for SA service.
- Coverage decreases toward zero for synchronous altitudes.

The amount of coverage which can be provided to mission spacecraft is a function of the user's altitude and inclination. Users at low altitudes and low inclinations will pass through the zone of exclusion each orbit and receive the least coverage. Users at high altitudes and high inclinations will pass through the zone of exclusion only periodically (e.g., a user at 1000 kilometers and 99-degree inclination will pass through the zone of exclusion once per day or less although the duration of this passage will be greater than for a lower inclination mission). The coverage, as a function of altitude, is summarized in Fig. 13. This figure, as well as Fig. 12 shows the lower limit of coverage previously defined.

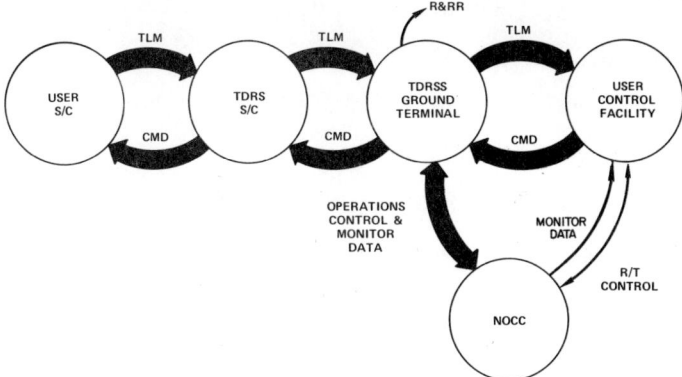

Figure 11. STDN (TDRSS) Operations

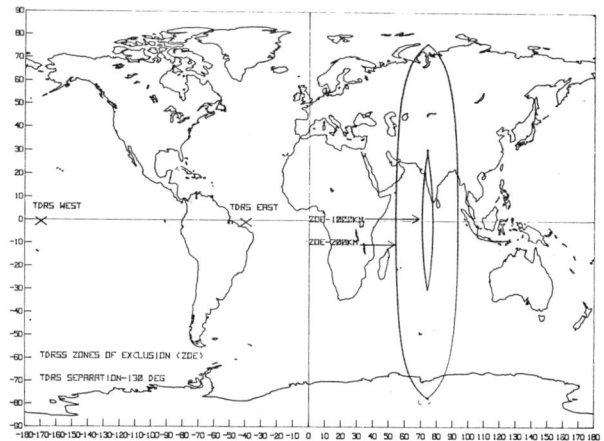

Figure 12. TDRSS Zones of Exclusion

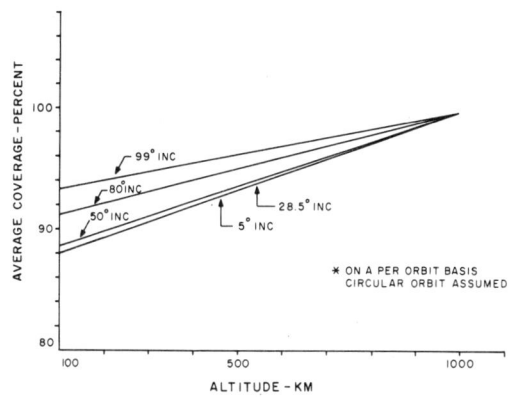

Figure 13. *Average Coverage vs User Altitude, Various Inclinations

FORWARD LINK

All forward links are SQPN modulated and consist of a Frequency Hop (FH) preamble for acquisition followed by the PN modulated data. The performance of the forward links for the various service systems is summarized in Fig. 14, for a representative set of mission spacecraft characteristics. This figure gives the performance without any error control coding. The performance of the forward link, when operating through an omni (0-dB gain) antenna, is of particular interest to a user and the following gives the achievable (uncoded) data rate for each service system, assuming the user employs efficient state-of-the-art receiving systems, and requires a 3-dB margin with a Bit Error Rate (BER) of 10^{-5}.

MA - 400 b/sec.
SSA - 3.4 kb/sec.
KSA - 102 b/sec.

There will be one data channel per forward link available with each of the three service systems (MA, SSA, and KSA). The data transmitted on the I and Q channels will be identical. Each SA system can support only one user at a time (except for the case where two users are close enough to be within the same antenna beam) although both the SSA and KSA forward links can be generated simultaneously by each SA system. There will be two SA systems per TDRS. The service characteristics of each system are given below.

MA System

This system is time shared by all MA users. All users operate at the same frequency and are discriminated by unique PN codes. There will be one MA forward link per TDRS.

SSA System

Users of this time-shared system are discriminated by frequency and unique PN codes. There will be two SSA forward links per TDRS.

KSA System

Users of this time-shared system are discriminated by polarization, unique PN codes, and beam pointing. There will be two KSA forward links per TDRS.

The forward link signal characteristics of the telecommunication system for all three services, are summarized below. This list represents the characteristics of the recommended signals design and is required for all MA forward links.

- All forward links will be spread spectrum.
- All normal forward links will be digital (SQPN modulated).
- All normal forward links will contain an FH preamble.
- Available RF bandwidths (3dB) (per link):

    MA - 5 MHz; SSA - 20 MHz, and KSA - 50 MHz.
- Maximum achievable data rates:

    MA - 10 kb/sec; SSA - 500 kb/sec, and KSA - 25 Mb/sec.
- All data are modulo-2 added asynchronously to the PN code.

Forward link service can be provided through any one, or any combination, of the three service systems, depending upon the mission spacecraft data requirements. Command data will only be handled in real time throughout the telecommunications path between the mission control center and the mission spacecraft. Command data destined for mission spacecraft are expected to be transmitted from user control centers in the exact formats required by the user spacecraft. Any command data acknowledgements may be derived by the user from real-time user spacecraft telemetry data. User spacecraft command support periods will be scheduled on the basis of required forward link coverage intervals, independent of return link and tracking requirements.

Under normal operating conditions, there will be no operational restriction on the use of the forward links in the normal power mode (assumed in Fig. 13), except for time-shared scheduling. A high power mode will be available on the SSA (2.5-dB increase over normal) and KSA (3.5-dB increase over normal), which will be restricted by TDRSS power considerations to a 50% duty cycle for

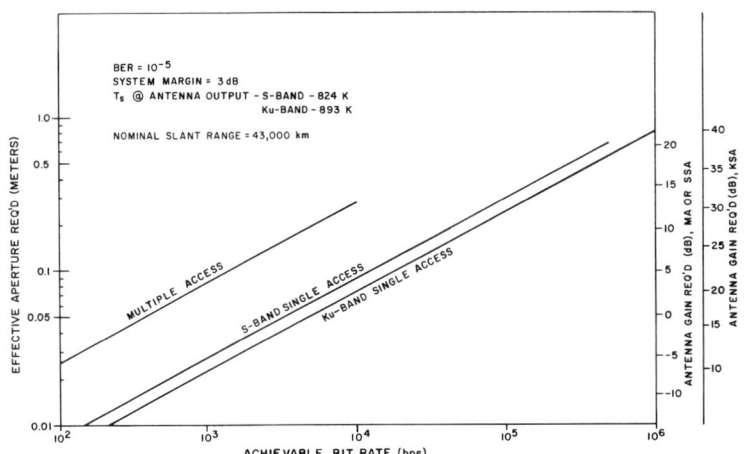

Figure 14. Forward Link Performance

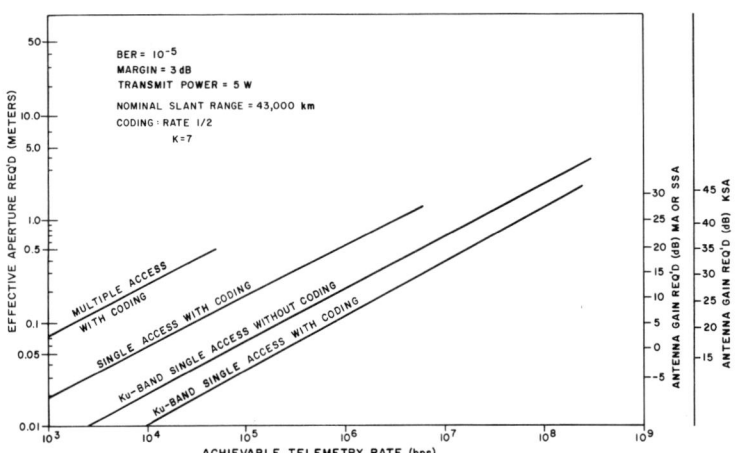

Figure 15. Return Link Performance

the SSA system. The KSA high power mode will not be restricted by TDRS power supply considerations. During periods when the TDRS is in solar eclipse, the use of the MA forward link may be restricted for short periods. These restrictions will be limited to one TDRS at a time and should not create a significant scheduling or support problem. The SA forward links, high power or normal, are not affected by solar eclipse. Due to flux density considerations, use of the high power mode may be restricted from a user coverage point-of-view when the user is located near the limb of the earth as seen from TDRS. This high power mode will normally be utilized for user spacecraft contingency support. A similar flux density limitation may occur when MA users are receiving forward link support from an SA system.

RETURN LINK

The detailed signal design and other TDRSS/user interface specifications are given in the TDRSS Users' Guide. All link performance and other characteristics specified assume that the return links are SQPN modulated digital signals.

The return link performance capability of the various service systems is summarized in Fig. 15 for a representative set of user characteristics. The TDRSS will supply, as a normal operating mode, a constraint length 7, rate 1/2 Viterbi decoder which will be available to all users. The MA, SSA, and KSA with coding curves assume a constraint length 7, rate 1/2 convolutional nonsystematic, transparent code for Forward Error Control (FEC).

Data Channel

The maximum number of independent, simultaneous data channels available with each service system of the TDRSS is:

- MA System. Two channels per system, one modulated on the I channel and the other on the Q channel.
- SSA System. Four channels per system, one spread data channel plus one nonspread channel modulated on the I channel and one spread plus one non-spread channel modulated on the Q channel.

- KSA System. Three channels per system, one spread channel and two non-spread channels.

This capability can be used, for example, to transmit simultaneously a real-time link and a playback link; or a real-time housekeeping and a science data link.

Return Link

Each SA system can support only one mission spacecraft at a time (except for the case where two users are within the same Ku/S-band combined beam) although both the SSA and KSA return links can be operated simultaneously by each SA system. There will be two SA systems per TDRS.

MA System

Each link will be dedicated to a specific user and can provide support for the entire visible part of the mission spacecraft orbit. All users will operate at the same frequency and are discriminated by unique PN codes. Sufficient MA capacity is planned to provide dedicated support to all MA users.

SSA System

This is a time-shared system and will not normally provide continuous support to any user. The users of this system are discriminated by frequency and beam pointing. There will be two SSA systems per TDRS.

KSA System

This is a time-shared system and will not normally provide continuous support to any user. The users of this system are discriminated by polarization and beam pointing. There will be two KSA systems per TDRS.

The return link signal characteristics of the TDRSS telecommunications system are summarized below. These characteristics are recommended for all return link signals and required for the multiple access return link.

- The MA return link must be spread spectrum, digital (SQPN modulation).
- The MA user's EIRP should be restricted to the minimum required for transmission of the desired data rate to minimize interference with other MA

users. Rate 1/2 convolutional coding must be employed on this link to help minimize the users' EIRP requirement.

- Spectrum spreading on the SA return link is not required for TDRSS operation. However, full utilization of the SA communications and tracking services requires spread spectrum modulation.
- The design of the SA return link signal is at the option of the user, subject to the restrictions previously specified.
- Available RF bandwidths (3 dB) (per link):

    MA - 5 MHz, SSA - 10 MHz, and KSA - 225 MHz.
- Maximum achievable data rates:

    MA - 50 kb/sec, with coding; SSA - 6 Mb/sec, and KSA - 300 Mb/sec.
- Data is modulo -2 added asynchronously to the PN code except for SA return link operation without added PN (no PN spectrum spreading), in which case the data modulation is QPSK.

Return link service can be provided through any one, or any combination, of the three service systems, depending upon the user's data requirements. Telemetry data will only be handled in real time throughout the TDRSS space-to-ground and terrestrial communications paths interfacing user spacecraft to user data facilities. No provisions for specialized project data formatting or data store and forward functions are provided. Only Viterbi decoding and bit synchronization functions will be performed at the TDRSS ground terminal. The possible need for time tagging of user spacecraft data at the TDRSS ground terminal is recognized. However, the provisioning of such capability will be based on user spacecraft project need. In addition, the measurement of return channel time delay (within the ground terminal) to one-microsecond accuracy is available. Return link user spacecraft support periods will be scheduled on the basis of required return link coverage intervals, independent of forward and tracking requirements.

USER TRACKING SERVICES

The TDRSS will have the capability to obtain user and TDRSS tracking data for orbit determination. Error analysis has shown that for orbit maintenance, one TDRS is capable of tracking user spacecraft to the same accuracy as the existing STDN. A comparison of the STDN with the TDRSS for long-arc tracking is presented in Table 1.

Table 1

TDRSS vs STDN for Long Data Arcs[1]

User	Position Uncertainty (Meters)	
	STDN	TDRSS[2]
EOS (910 x 910 km, i=99)	59[3]	58
Shuttle (435 x 435 km, i=50)	727[3]	703
SAS (560 x 510 km, i=3)	691[4]	757
Note		
1. Assumes 150-meter position uncertainty.		
2. Assumes bilateration tracking of TDRS.		
3. USB tracking.		
4. Minitrack.		

The TDRSS tracking system is a Pseudo Noise Range and Range Rate System. If the recommended TDRSS compatible signal design is utilized, the PN spectrum spreading code provides the range code.

Tracking service can be provided through any one, or any combination, of the three service systems, dependent upon the users' tracking requirements. Sampled range and doppler data for all users will be transmitted from the TDRSS ground terminal to the orbit determination facility at GSFC in real time. The tracking data message format will be identical to that used in the remainder of the STDN. For optimum utilization of services, support periods for tracking should be coincident with forward link support periods because both types of support require establishment of a forward link.

NASCOM TELECOMMUNICATIONS SYSTEM

The primary NASCOM telecommunication system facilities to be provided to support the TDRSS will utilize leased data communications services between GSFC, the TDRSS ground terminal, the Johnson Space Center (JSC) and other user locations as required. Domestic satellite as well as terrestrial digital facilities are the prime candidates to provide these leased services. Figures 16 (terrestrial data distribution system) and 17 (Domestic satellite data distribution system) illustrate the candidate systems.

NASCOM will provide terminal systems on these leased services as entry points for all ground-located TDRSS users. Time division multiplexing techniques on the leased services, will provide users with access for command, telemetry and tracking services. The individual user will have a discrete channel interface with NASCOM for the forward and return channels to and from his spacecraft. The telemetry user with a Shuttle/Orbiter multidisciplinary payload, may receive a Time Division Multiplexed (TDM) composite telemetry signal. The tracking service will be available to the user as a composite data signal containing all users' tracking data or as processed tracking information from a central orbit/attitude determination facility. All user data will be transmitted on the NASCOM system in real time. The NASCOM primary system facilities will provide diverse routing of a user service in the event of a leased service outage.

During communication outages, the full data load may not be supported with the real-time transmission. Users' real-time data will be supported on a pre-assigned priority basis. However, users' data will be protected at the TDRSS ground terminal by contingency storage which will be used only during NASCOM outages. Users must prearrange their physical and operating interfaces with NASCOM for recovery of such contingency-stored data. The NASCOM system will be designed to handle the projected data load. The scheduling system

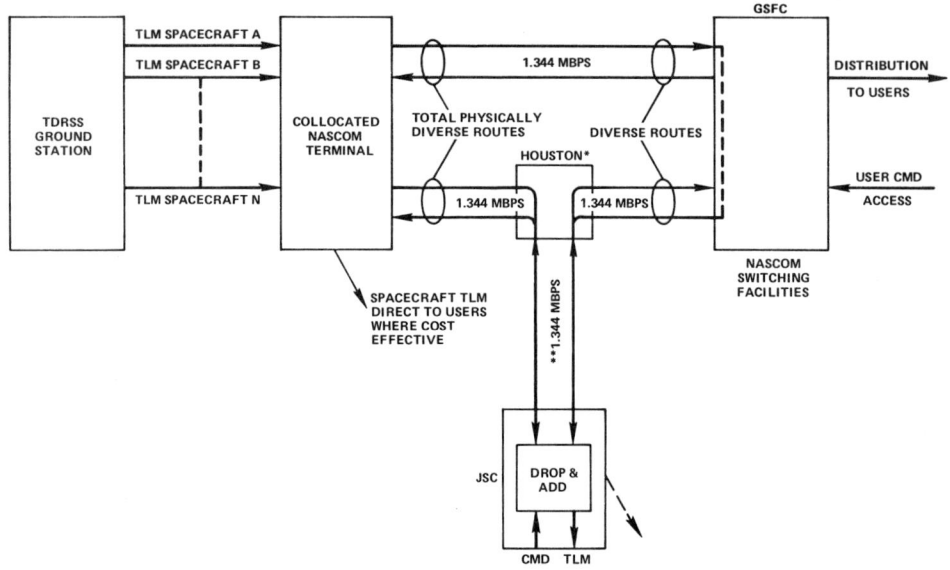

*FOR PROPOSED WSTF GROUND STATION LOCATION, SHUTTLE DATA CHANNELS TO JSC INCORPORATED INTO 1.344 MBPS LINES (ONE ROUTED VIA HOUSTON).

**DIVERSE FACILITIES ASSUMED.

***QUANTITY OR SIZE OF CHANNELS TENTATIVE, SUBJECT TO CHANGING LOAD PROJECTIONS.

Figure 16. Terrestrial Data Distribution System

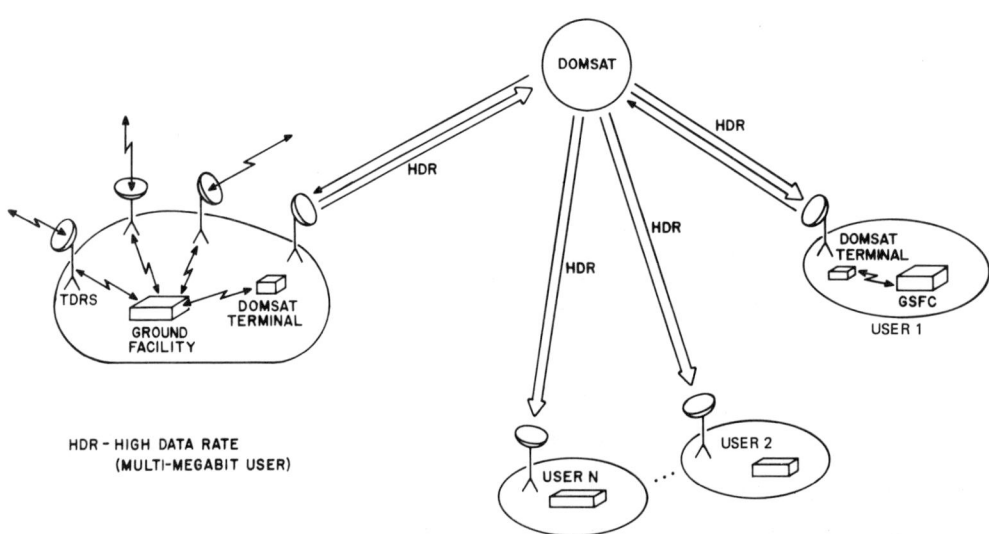

Figure 17. Domestic Satellite Data Distribution System

should prevent the occurrence of any overload. The NASCOM system will have a variable propagation delay. The propagation delay of a user channel is a function of the users' data rate and the users' composite data load on the leased service. An upper bound on user channel propagation is to be determined.

COMBINED SUPPORT

The GSTDN will provide the primary support for high orbital altitude missions while the TDRSS will provide the primary support for the low orbit altitude mission spacecraft. Some missions (such as the Space Tug), however, will operate in both the high and low altitude regions and may receive support from both elements of the STDN. STDN support of these missions can be accomplished through the use of an Office of Tracking and Data Acquisition (OTDA) funded transponder development, which will provide a mission spacecraft transponder compatible with both elements of the STDN.

GLOSSARY

CHANNEL. An independent stream of data contained within a link.

CHIP. Represents one bit of the PN sequence as opposed to data bits.

DDPS. Digital Data Processing System.

FORWARD LINK. The telecommunications link from the ground terminal through a TDRS to the mission spacecraft. It is used to relay user digital data, analog data, and tracking signals.

GSFC. Goddard Space Flight Center.

GSTDN. The ground based part of the STDN.

I DATA CHANNEL. Data stream transmitted by 0-degree and 180 degree phase modulation of the reference carrier.

KSA. Ku-Band Single Access system

LINK. Describes the entire composite stream of data transmitted by a single mission spacecraft to one of the TDRS service systems or from one of the service systems to a single mission spacecraft.

MA. Multiple Access System

NASCOM. NASA Communications System

OTDA. Office of Tracking and Data Acquisition, NASA Headquarters.

PN. Pseudorandom Noise.

Q DATA CHANNEL. Data stream transmitted by ±90 degree phase modulation the reference carrier.

RETURN LINK. The telecommunications link from the mission spacecraft through a TDRS to the ground terminal. It is used to relay user digital data, analog data, and tracking signals.

SA. Single Access system.

SCE. Spacecraft Command Encoder.

SQPN. Staggered quadriphase Pseudorandom Noise.

SQPN MODULATION. A modulation process in which the PN code modulating the I channel is delayed 1/2 chip relative to the independent PN code modulating the Q channel.

STDN. Spacflight Tracking and Data Network

SSA. S-Band Single Access system

T&DA. Tracking and Data Acquisition

TDRS. Tracking and Data Relay Satellite

TDRSS. Tracking and Data Relay Satellite System

AAS 75-147

THE SPACE TUG ECONOMIC ANALYSIS STUDY — WHAT WE LEARNED

Charles V. Hopkins*

INTRODUCTION

The Space Tug Economic Analysis study was performed in 1971-1972, by Lockheed Missiles & Space Company, Inc., (Prime contractor) and Mathematica, Inc., (subcontractor); it was funded by the NASA Marshall Space Flight Center under contract NAS8-27709. The objectives of this study were to place economic bounds on the selection of an upper stage for the Space Shuttle, and then to compare and evaluate vehicle concepts within these bounds.

As defined for this contract, the term "Space Tug" was broadly interpreted to cover a range of upper stages that might be flown with the Space Shuttle. The vehicle concepts that we evaluated covered the range from true Space Tug configurations (i.e., reusable propulsive stages with payload retrieval capability) to expendable orbit-injection stages (existing upper stages modified for the Shuttle). The expendable orbit-injection stage family has recently been redesignated under the name Interim Upper Stage (IUS). Vehicle configurations, sizes, and operational modes were selected to explore the full range of liquid chemical propulsion systems. Solid stages and reusable IUS configurations were omitted by direction. Advanced propulsion concepts (e.g., nuclear heat-transfer and electric propulsion) were likewise omitted but were evaluated under the subsequent Advanced Propulsion Comparison study (Contract SNSO-2).

The measures that we applied in the evaluation included system investment (upper-stage RDT&E costs), benefits (savings in operational costs), economic sensitivites to major system variables, and peak-year funding levels. The savings in operational program costs between alternative upper-stage concepts were the key to this analysis. Previous studies had shown that savings in payload costs (e.g., spacecraft reuse, relaxed weight and volume constraints) underlie the economic justification of the Space Shuttle. Consequently we selected as our economic measure the total program costs, i.e., the sum of payload costs, Shuttle user fees, and all upper-stage costs.

---
*Advanced Systems Engineering Specialist, Lockheed Missiles & Space Company, Inc., Sunnyvale, California

THE SYSTEM MODEL

To evaluate the large numbers of configurations and flight modes considered in this study we had to model the entire transportation system and its payloads. The system model, operated on a UNIVAC 1108 computer, synthesized Space Tug designs; calculated upper-stage performance, mission capture, transportation costs, and payload costs; matched the best mix of payloads and flight activity; and then summed the program costs into time-phased funding plans. Space Tug weights and sizes were varied parametrically to cover sizing effects. Payload costs and weights were likewise varied to account for varying degrees of Shuttle benefits.

THE ECONOMIC ANALYSIS

Mathematica evaluated the total costs of alternative upper-stage systems, as output by the Lockheed model, by establishing the recurring (production and operations) cost savings relative to some baseline system, discounting these savings, and then comparing the savings with the (discounted) investment that is made in each system by means of RDT&E cost. This gave a cost/benefit measure. The calculated savings were also transformed into another more easily understood format using the concept of allowable nonrecurring (or allowable RDT&E) costs. Allowable nonrecurring cost answers the question: "What are my recurring-cost savings worth, in terms of dollars I could spend during the RDT&E program?"

Another important output of the economic analysis was sensitivity studies of the major cost-driving variables. The objective of this effort was to see whether changes in these driving variables would reverse any of the trends that had emerged in the analysis. Mathematica analyzed the effect of excursions in Shuttle user fee; payload cost and payload refurbishment factor; upper-stage costs; and mission-model activity level, composition, and energy level. In general no significant trend reversals were observed, even for substantial changes in the number and type of missions flown.

STUDY FINDINGS: COST DRIVERS

In terms of the magnitude of total program costs, we found that payload costs are the largest single element, contributing over 70 percent of the total investment. Shuttle user fees contribute up to about 20 percent and Tug costs account for roughly 10 percent. Consequently those upper-stage factors that influence payload costs and Shuttle user costs are the most influential drivers. These include:

- Stage propellant capacity: this drives stage length and payload capability, as discussed below
- Stage length: this drives mission capture for long payloads
- Stage payload capability: this drives spacecraft delivery and servicing modes and also payload design margins. However, there is an economically optimum performance level beyond which diminishing returns occur.
- Multiple-payload capability: this drives the required numbers of Shuttle flights

STUDY FINDINGS: SPACE TUG AND IUS IMPLICATIONS

We found that the economic justification of a full-capability Space Tug (in comparison to an efficient IUS) depends on the economic criteria that you apply. If the criterion is total program cost, we concluded that a Space Tug is economically justified in the long run (i.e., gives favorable return on investment). The basis for this justification (Fig. 1) rests primarily on the potential savings in the cost of payloads that can be serviced to increase their lifetime. This conclusion assumes:

- A ten-year operational program with 300-500 spacecraft flight opportunitites
- Availability of Tug funding
- A properly sized Tug
- A discount rate of 10 percent

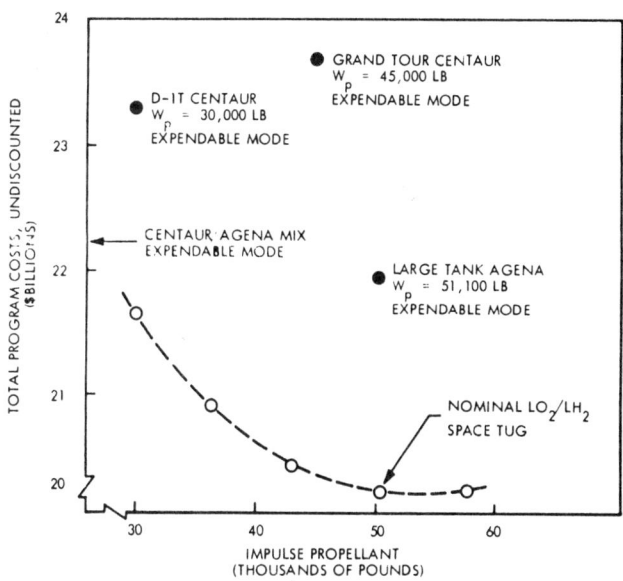

Fig. 1 Total Program Cost Comparison: Space Tug vs Expendable IUS Concepts

However, if the economic criterion is changed to low-peak funding, rather than return on investment, the expendable IUS becomes attractive. Such an evaluation implies, economically, that the scarcity of funds in the early Shuttle time period effectively drives the discount rate for Space Tug investment to very high values.

If an Interim Upper Stage is developed first, as now planned; we learned that the following features will characterize the most economical system:

- Large performance margins for payload benefits and/or multiple delivery
- Low cost per flight
- Short stage length for compatibility with large payloads
- Low RDT&E cost for minimum early peak funding
- Flexibility to fly with a range of backup expendable boosters during transition

If funding for the Space Tug program becomes available, the selected Tug concept should incorporate the following characteristics to achieve the most economical approach:

- Configuration: Single stage with minimum-length design
- Propellant Capacity: Roughly 50,000 pounds
- Flight Mode: Ground-based with multiple-payload capability

The selection of specific IUS or Tug concepts should be the subject of further economic analyses.

STUDY FINDINGS: PAYLOAD IMPLICATIONS

Although the Space Tug Economic Analysis was not a payload study per se some of the trends that we observed in this study have implications for spacecraft programs that will use the Shuttle/upper stage transportation system. These may be summarized as follows:

- Refurbishment and reuse of selected payloads (either through spacecraft retrieval or on-orbit servicing) returns large economic benefits; however, less than half of the payloads are worth such servicing.
- Data are not available to evaluate the tradeoff between payload maintenance and increased spacecraft lifetime through redundancy
- Large performance margins in the Shuttle/upper stage transportation system are of measurable economic benefit to the spacecraft.

AAS 75-149

A SURVEY OF THE ECONOMICS OF MATERIALS
PROCESSING IN SPACE [1]

B.P. Miller [2]

The technology of space materials processing is in the formative research stage. Although the results obtained to date in the Apollo and Skylab flights have been sufficietly encouraging and interesting to merit further experimentation, the demonstration of the technology needed to evolve an operational program or obtain commercial participation has not yet occurred. A survey of the economics of space materials processing has been performed with the objectives of identifying those areas of space materials processing that give preliminary indication of significant economic potential, and to identify possible approaches to quantify the economic potential. It is concluded that limited economic studies have been performed to date, primarily in the area of the processing of inorganic materials, but that the economics of space processing of biological material has not received adequate attention. Specific studies are recommended to evaluate the economic impact of human lymphocyte subgroup separation on organ transplantation, and on the separation and concentration of urokinase producing cells.

---

(1) This work described in this paper was performed for NASA under Contract NAS W 2558

(2) ECON, Inc. 419 N. Harrison St., Princeton, NJ

INTRODUCTION

The Historical Development of Materials Processing in Space

During the late 1960's several researchers in NASA began studies of the use of the space environment to perform material processing operations that are not technically feasible or not economically attractive on Earth. To a great extent, these studies were an outgrowth of the consideration of the problems of using welding techniques for assembling large objects in orbit from parts that had been prefabricated on Earth. The studies showed that the physical characteristics of the space environment, namely virtual weightlessness, a vacuum sink of unlimited capacity, and the availablility of solar energy, are potentially useful in materials processing. Of these, weightlessness is the most important as it cannot be duplicated on or near earth for more than a few seconds.

During the mid-1970's the interest in materials processing in space progressed from studies and ground based experimentation and technology development through a series of simple experiments in the Apollo 14, 16, and 17 missions. The Apollo experiments consisted of weightless solidification experiments and small scale electrophoresis tests.[1] The Apollo experiments were relatively encouraging and showed that worthwhile results could be obtained in space processing using relatively simple and inexpensive experiments.[2]

As a result of the success obtained in the Apollo Flights, a more extensive series of materials processing experiments was planned and implemented in the Skylab program. Figure 1 is a tabulation of the Skylab space processing experiments and

Figure 1. Skylab Experiments and Science Demonstrations

		SKYLAB MISSION		
		II	III	IV
MATERIAL PROCESSING FACILITY				
M551:	Metals Melting Experiment, Mr. R.M. Poorman, MSFC Astronautics Lab.	x		
M552:	Exothermic Brazing Experiment, Mr. J.R. Williams, MSFC Product Eng. Lab.	x		
M553:	Sphere Forming Experiment, Mr. E.A. Hasemeyer, MSFC Product Eng. Lab.	x		
* M555:	Gallium Arsenide Crystal Growth Experiment, Dr. R.E. Seidensticker, Westinghouse Res. Lab.			
MULTIPURPOSE FURNACE SYSTEM				
M556:	Vapor Growth of II-VI Compounds, Prof. H. Wedemeir, Rensselaer Poly. Inst.		x	x
M557:	Immiscible Alloy Compositions, Mr. J.L. Reger TRW Systems		x	x
M558:	Radioactive Tracer Diffusion, Dr. A.O. Ukanwa, MSFC Space Sciences Lab.		x	
M559:	Microsegration in Germanium, Dr. F.A. Padovani, Texas Instruments		x	
M560:	Growth of Spherical Crystals, Dr. H.U. Walter, University of Alabama		x	x
M561:	Whisker-Reinforced Composites, Dr. T. Kawada, Nat. Inst. for Metals Res., Japan		x	x
M562:	Indium Antimonide Crystals, Prof. H.C. Gatos, Mass. Institute of Tech.		x	x
M563:	Mixed III-V Crystal Growth, Prof. W.R. Wilcox, University of Southern California		x	x
M564:	Alkali Halide Eutectics, Prof. A.S. Yue, University of California, Los Angeles		x	
M565:	Silver Grids Melted in Space, Prof. A. Deruyherre, Katholieke University, Leuven, Belgium		x	
M566:	Copper-Aluminum Eutectic, Mr. E.A. Hasemeyer MSFC Product Eng. Lab.		x	x
SCIENCE DEMONSTRATIONS				
Diffusion in Liquids			x	
Ice Melting			x	
TV101 Liquid Floating Zone				x
TV102 Immiscible Liquids				x
TV103 Liquid Films				x
TV105 Rochelle Salt Growth				x
TV106 Deposition of Silver Crystals				x
TV107 Fluid Mechanics Series				x
TV117 Charged Particle Mobility				x
EXPERIMENTS PERFORMED ON EACH MISSION		3	13	14

* Not Flown-Storage Area Preempted by Skylab Repair Kit

NASA E375-15310 (1)
12-4-74

science demonstrations. In addition to the fifteen formally scheduled experiments, a group of minor ad hoc experiments called science demonstrations were added to the Skylab III and IV missions. These science demonstrations made use of hardware and materials available aboard the spacecraft. Although one (M555) of the fifteen scheduled experiments could not be performed as its stowage space was pre-empted by the Skylab Repair Kit, a high level of crew productivity was obtained during the Skylab Mission, and several of the space processing experiments were run twice. In general, the Skylab experiments dealt with melting and freezing processes and the behavior of fluids. Although the discussion of the specific results of the Skylab experiments is beyond the scope of this report, the results indicate the possibility of producing higher quality crystal materials in the space environment than can be produced by comparable processes at present on Earth. Specifically, Experiment M560 supported the possibility of making high quality crystals directly in wafer form in space, thus avoiding the wastage and degradation of material caused by the cutting, grinding and polishing processes used to produce such wafers on the ground. Moreover, Experiment 562 yielded crystal material with more uniform electrical properties than that achieved with comparable samples grown on Earth.[3]

The next step in experimentation in materials processing in space will occur in the Apollo-Soyuz Test Project (ASTP) planned for 1975. Figure 2 is a tabulation of the ASTP experiment program. The series of experiments involving the Multipurpose Furnace System (MA-010), and the Crystal Growth in Space experiment (MA-028), essentialy involve the continuation of the research with inorganic materials begun during the Skylab program. The other two ASTP experiments (MA-014 and MA-011) are concerned with a different area of space processing: the separation of biological materials to isolate specific materials that are important for medical research and applications. Both of these experiments will employ an electrical separation

Figure 2.

ASTP EXPERIMENTS

ELECTROPHORESIS EXPERIMENTS

MA-014: ELECTROPHORESIS - EPE
Dr. K. Hanning, Max Planck Inst.

MA-011: ELECTROPHORESIS TECHNOLOGY
Dr. R.S. Snyder, MSFC
Astronautics Lab.

Dr. P.E. Bigazzi,
State U. of New York

Mr. G.A. Barlow, Abbot
Laboratories

Dr. M. Bier, Veterans Admin.

MA-010 MULTIPURPOSE FURNACE SYSTEM

MA-041: SURFACE TENSION INDUCED CONVECTION
Dr. R.E. Reed, Oak Ridge Nat'l. Lab.

MA-070: ZERO-G PROCESSING OF MAGNETS
Dr. D.J. Larson, Grumman Corp.

MA-044: MONOTECTIC AND SYNTECTIC ALLOYS
Dr. C.Y. Ang, Northrop Corp.

MA-085: CRYSTAL GROWTH FROM THE VAPOR PHASE
Prof. H. Wiedmeier, Rensselaer Poly. Inst.

MA-060: INTERFACE MARKING IN CRYSTALS
Prof. H.C. Gatos, MIT

MA-131: SODIUM CHLORIDE-LITHIUM FLOURIDE EUTECTIC
Prof. A.S. Yue, UCLA

MA-150 MULTIPLE MATERIAL MELTING
USSR

COOPERATIVE EXPERIMENT

MA-028: CRYSTAL GROWTH IN PLACE
Dr. M.D. Lind, Rockwell international

NASA Hq. ES75-15563 (1)
1-16-75

technique known as electrophoresis to separate certain classes of human cells into groups that have different functions or properties. The weightlessness and quiet nature of the space environment are both important to the electrophoresis process. On Earth, electrophoresis and related techniques are successful only in arrangements where the separation medium is stabilized against convection and other mechanical disturbances either by containment in a porous supporting medium or by a stable flow regime. The electrophoresis experiments are of major importance to the biomedical community, and could possibly contribute to the development of new approaches to the treatment of disease, and the stabilization of transplanted organs.

The Relationship of the Space Shuttle and Spacelab to Materials Processing

The advent of the Space Shuttle and Spacelab in the 1980's will provide new opportunities for experimentation in the processing of materials in space. By its very nature, materials processing tends to involve repetitive use of the processing equipment. The Space Shuttle/Spacelab will combine the attractive features of the space environment with the capability for frequent, repetitive reuse of the processing equipment.[4] Using the Space Shuttle, an extensive program of applied research could result in demonstrations of technical feasibility that subsequently could lead to the development of operational or commercial processing of materials in space. Preliminary planning for the selection of space processing experiments for the Space Shuttle has been going on since 1972. In the interim between the flight of ASTP 1975 and the Space Shuttle/Spacelab in the 1980's, research and development in space processing of materials will continue with an active program of a ground based investigations and a series of suborbital rocket flights.

The Objectives of Economic Analysis

The purpose of this paper is to identify what areas of materials processing in space give preliminary indication of economic benefits, and to identify possible ways to estimate the size of the benefits. Since the field of materials processing in space is in the formative experimental stage, it is premature to attempt to justify operational or commercial processing on the basis of the results obtained to date. However, at this time economic analysis can be used as a management tool to identify potential benefit areas and guide research and development toward those areas of large potential economic payoff, as well as serving as a framework for developing the nature of prospective uses and users for materials processed in space. For example, alternative space materials processing research and development projects can be considered as alternative investment opportunities, and assessed from the viewpoint of net payoff on each of the projects. Using payoff as a guideline for project selection, projects with a high payoff should be undertaken, while those with a low payoff should not. For comparability between projects, payoff can be expressed as the net present value of benefits less the net present value of costs for a specified discount rate. With discount rate as a parameter, a project with an expected positive net present value at 15% discount rate is one that almost certainly should be undertaken. The lower the value of the discount rate at which the net present value of the benefits equals the net present value of the costs, the less the economic motivation to undertake the project.

Research and development in space materials processing may be considered to contain a relatively high degree of risk with respect to the economic payoff. In the private sector a risk component is often added to the base or normal discount rate. The magnitude of the risk factor is a measure of the risk aversness of the investor. The addition of the risk factor has the effect of favoring projects that are risk free. Some economists believe that the government does not need to believe in the same way as the private sector with respect to risk. Since the government undertakes many research and development projects with varying degrees of risk, it has the option of pooling risks in much the same manner as an insurance company and does not need to apply a differential risk penalty on its own projects. Thus, the government can undertake research and development projects that would be rejected by industry as being too speculative.

The Nature of Materials Processing in Space

There is a natural division between the applications of space processing to conventional materials technology and the biomedical applications. The biomedical applications involve organic materials, while the applications to conventional materials technology are concerned with inorganic materials.

The following paragraphs describe some of the potential products of space processing with emphasis on the potential uses and users of the products. Where possible, potential economic implications of the products are described, along with an approach that could be used to estimate the potential benefits. All of the applications described are speculative, some more so than others. In some cases the applications can be focused to a specific end objective, and an available data base exists to make possible a benefit model and an estimate of the benefits. Other applications are of a much broader research nature, and this represents a more difficult (but not impossible) data collection and modeling problem. For this reason, this review concentrates on these applications where the benefit model and data base are readily identifiable. The ability to quantify benefits at this time should not be construed as a measure of the worth of an application, but as a suggestion of the fact that the application is sufficiently focused to enable the identification of the uses, users, and the potential relationship between the technology and economic benefits. While the path to success may be more tenuous in some of the applications where no recommendation is made to estimate benefits at the present time, the implications of success in these areas may be as far reaching (from an economic viewpoint) as those areas that are presently quantifiable.

Inorganic Materials
Electronic and Electro-Optic Devices

The previously described and referenced Skylab space processing experiments demonstrated the possibility of producing crystals of higher quality, larger size, and more uniform electrical properties than can be produced at the present time on Earth. The most likely candidates for processing in space will probably be the semi-conductor and ceramic oxide crystals for electronic and electro-optical devices. Examples include silicon, gallium arsenide, and bismuth germanate. Other electronic devices such as rare earth iron garnet crystals for use in magnetic bubble memories also appear to be candidates for space processing. Studies performed in 1972 supported the technical and economic feasibility of space processing of sophisticated compound single crystals, and magnetic bubble memories.[5] However, the rapid innovation and advance of technology in electronic and electro-optic devices may introduce new or improved devices involving new technologies that could supersede these crystal applications by the mid-1980's. Thus, if economic benefits are realized through the production of electronic or electro-optic devices using materials processed in space, it is likely that the devices will not be simple improvements of devices that can now be made on Earth, but may be new compositions that result from experimentation with materials in the space environment. However, it should be borne in mind that the experiments with space grown crystals performed to date compare materials processed exactly the same way in space and on the ground, the only difference being the presence or absence of gravity. While this approach may be scientifically valid for isolating the effect of gravity, the practical significance of space processing will be apparent when space grown crystals are compared with the best crystals grown on Earth by any economically or technically viable method. This is particularly so, as according to some scientists, the results observed in space grown crystals could have been produced on Earth if different, experimental techniques had been used.[11] On the other hand, further experimentation in space may show that the absence of gravity does not produce spontaneous improvements in processes that also work satisfactorily on the surface of the Earth. Rather, it may be that the absence of gravity allows manipulations or process conditions that affect the properties of the materials, but will not work or cannot be controlled on the ground.

Since it is unlikely that space processing will be used for devices that are currently in production or development, it will be necessary to perform a technical forecast before the benefits of space processing can be estimated. A method of technical forecasting that could be used would be to draw upon the technical expertise of the Skylab and ASTP investigators in this area to describe the expected technology benefits or applications of their experiments. Using this information, industrial researchers could then be surveyed to determine usefulness of the expected results to specific research applications. A speculative estimate could then be made of possible product developments, and demand and cost estimates made for the products. While this approach appears to be straightforward, the results could be limited by the highly competetive nature of the industry, and the fact that the forecast will deal with products that have not yet been "invented".

High Temperature, High Strength Structural Materials

It has long been known that directional solidification of alloy systems can produce improvements in the mechanical properties of the material. Since 1958 studies have been performed on eutectic and near-eutectic melts of metals, oxides, and salts because of the unique microstructures that can be developed in these systems. These mixed phase composite materials have potential application in fields such as high temperature metallurgy, toughened ceramics, and super conducting systems. In the processing of eutectic materials the weightless space environment appears to suppress the random effects of convection, so that the heat and mass transport effects that govern solidification become highly predictable, yielding a convectionless directional solidification process.[6] A benefit/cost study of the use of space processed eutectic material in aircraft gas turbine blades performed during 1974 indicated that there would be an adequate demand to justify production of space processed blades both from a quantity and benefits derived standpoint. The study was based upon the technical assumptions that the space processed materials would yield an added 200° F temperature tolerance, reducing fuel consumption by 4% and doubling blade life for existing aircraft, with respect to blades processed on Earth. The results of the study indicate a cost savings of 4.744×10^9 (0% discount rate) for the 1980-91 time period, as opposed to a production cost (including an allocation of Space Shuttle operating costs) of 1.246×10^9 (0% discount rate). A review of this study by one of the leading producers of jet turbine blades raised questions concerning the technical and market assumptions, and, hence, the validity of the economic benefits. The review indicated that the use of eutectic alloy blades is presently limited by the absence of a reliable high temperature coating material which limits the possible increase in blade operating temperature to approximately 50° F. Moreover, based upon current marketing experience, it appears that aircraft gas turbine manufacturers are not willing to pay the increased price for eutectic alloy blades, even though eutectic alloy blades (now produced on Earth) exhibit life extension capabilities of four (4) to seven (7) times that of a conventional blade.[8] The reluctance to incorporate eutectic alloy blades may be due to the fact that the life of present blades is long when compared to the time between tear down of jet engines. Thus, it would appear at present that the need is not for increased performance or life, but perhaps for reduced initial cost, or for a corresponding improvement in the life of other jet engine components in order to make their life expectancies approximately equal to that of the eutectic blades.

Biological Materials

Processing Considerations

The primary direction of research in space processing of organic materials has been in the area of electrophoretic separation and purification of biological materials. Electrophoretic separation processes are based upon electrochemical effects that cause particles to take on electrical charges when suspended in aqueous solution. The nature of these charges is determined by the equilibrium between particle surfaces and ions in the solution, so that the charges are characteristic of the particles, but can also be manipulated to some extent by changing the composition of the solution. Forces can be applied to the charged particles by applying an electrical field to the solution. When this is done each particle will move along the direction of the field at a constant velocity such that the fluid drag forces are equal and opposite to the electric force on the particle. Thus, in general, each kind of particle suspended in a solution in this manner moves when an electric field is applied with a characteristic velocity determined partly by its chemical nature, and partly by its size and shape. Using this principle of electrophoretic separation, particles that move at different velocities can be physically separated and separately collected.

The use of electrophoretic separation techniques on Earth is limited by the effects of gravity. In space, the absence of a significant gravity field increases the suspension time of heavy particles in the solution, and supresses convective effects. On Earth, gels, paper, starches, and small dimensions are all used to minimize convection and sedimentation effects on the electrophoretic process. Since the heavy particles of interest include living human and animal cells, the increased dwell time of these heavy cells in the space environment is of major importance. Estimates indicate that it should be possible to increase the size of cell separated from approximately 1mm to approximately 6mm in space.

Freeze drying of biological materials may also be improved by the space environment. Freeze drying is often used as a means of preserving prepared biological material. The weightless environment may improve freeze drying by permitting evaporation of ice from all surfaces of a particle at the same time.

Two types of biological materials, namely molecular and cellular have been considered for processing in space. The common technique for the processing of both types of materials is electrophoresis. The areas discussed below are areas in which there is currently a high level of research activity, and these are probably areas in which additional progress will be made during the five year interval between the ASTP and early shuttle flights. As such, these are probably not the exact products that will be of interest in the 1980's, but are concrete examples of present needs for processes or products that could conceivalby be improved by materials processing in the space environment.

Cellular Material

The following paragraphs describe two potential applications of space based processing of cellular material that could have significant economic impact. It should be noted that other potential applications based upon the electrophoretic mobility of cells and their functional properties such as the separation of functionally defined lymphoid cells have been considered and may also be of major importance.[6]

Human Lymphocytic Group Separation

Human lymphocytes are a class of white blood cells that controls much of the human body's immune responses to disease and organ transplant rejection. For example, current kidney transplantation research is exploring the use of various lymphocyte culture response tests to enhance selection criteria for donor-recipients, a Mixed Leukocyte Culture (MLC) Response Test has been developed to yield an indirect measurement of the recipients immunological response to his donor. While limited primarily to parental and sibling donors, the MLC test has been successfully used to reject several potential donors who might, by prior criteria alone, be given a kidney with less than acceptable odds for survival. A modified version of the MLC test is also under development in an effort to detect cell-mediated immunization and produce a more biologically refined classification of cadaver transplant recipients. While mortality as a result of kidney transplantation has been reduced to approximately 10% during the past ten years, the problem of ultimate rejection of the transplanted organ remains an important consideration. Studies by the Rogosin Kidney Center indicate a one year predicted kidney survival of 44% for cadaveric transplants (for the period 1963-73), but indicate that this could be improved to greater than 80% by proper selection of cadaver graft recipients. While research tests for matching the donor and recipient have been developed that might lead to an 80% predicted survival for one year, these tests are not generally applicable in every day practice. The use of these research tests in every day practice would effectively reduce the number of transplants and increase the average time spent in dialysis by the potential recipient.

Thus, in many cases at the present time a transplant is effected even though the probability of predicted survival for one year (based upon matching experience and survival statistics) may be much less than 80%.

To date, two major classes of human lymphocytes have been identified. It has been hypothesized that there are several subgroups within these major groups, and that these subgroups play an important role in enhancing or suppressing the immunological reaction to the transplanted kidney. One of the hoped for results in the electrophoresis experimentation with cellular material in space is the identification of the lymphocyte subgroups. If this is successful, the donor/recipient matching process could be improved, and perhaps more importantly the factors that cause rejections could be specifically identified. If the latter is achieved, it may be possible to devise a serum to suppress or neutralize the rejection process without destroying the remainder of the patients immune systems. Thus, if successful, space materials processing experiments aimed at separating the lymphocyte subgroups could make possible a dramatic improvement in the survival of transplanted organs.

A review of the current literature indicates that an economic assessment of the potential benefits of improved human lymphocyte subgroup classification has not been performed. Two categories of economic benefits are associated with the survival of the transplanted kidney. The first category is the economic value of the human lives involved, while the second is the reduction of the costs of the federally supported dialysis program. The value of the lives can be estimated by determining the average age of death as a result of kidney disease, the average expected income of the kidney transplant recipients, and the expected increase in the life span as a result of the improved survival kidney transplants. Since 1972 the costs of dialysis have been paid for under the Social Security Act for those patients who have had a transplant or who receive dialysis more than once per week.

While estimates of the costs of dialysis range widely, it has been estimated that the annual cost for hospital performed dialysis is in the range of $15,000.00 to $47,000.00 per patient/year. It has further been estimated that out of the population of about 58,000 people in the US with serious and significant kidney disease approximately 10,000 are medically and psychologically equipped to handle long term dialysis.[10] Using these figures it can be seen that the direct cost of the federally supported dialysis program could reach $150,000,000.00 to $470,000,000.00 per year in the near term, and as medical techniques are improved and a larger part of the potential population is accepted for dialysis the cost could increase to nearly $1,000,000,000.00 per year. It should be noted that these figures represent direct costs of the dialysis program and do not include estimates of lost income and production effects. Discussion with Dr. Albert Rubin of Cornell University Medical College indicates that the mechanism for achieving economic benefits as a result of human lymphocyte subgroup separation would be the improved survival rate of the transplanted organs through better matching of the donor and recipient, and suppression of the rejection process. The latter could also result in an increase in the number of potential transplant candidates. Both of these factors could cause a reduction in the number of patients on dialysis. The data base for assessing the economic potential of this area exists as a result of the federally supported dialysis program. Data such as the incidence of kidney disease, the availability of transplants, the survival rates of the transplanted kidney, and the costs of the dialysis program are available from various federal and private sources. Since the technical feasibility of human lymphocyte subgroup separation has not yet been demonstrated, it is suggested that a probabalistic approach be used for the benefit assessment in order to indicate the range of uncertainty associated with the results.

It should be noted that other organ transplant operations where rejection of the transplanted organ is an important factor could also possibly benefit from successful human lymphocyte subgroup separation. In comparison to the evaluation of the economic benefits of improved kidney transplantation, the economic benefits of the more generalized problem of improved organ transplantations have not been examined in this study. The more generalized benefits may be more difficult to assess as a result of the lack of a comparable data base, however, hospital and insurance company records could possibly provide useful data. It is suggested that the generalized problem be explored in a detailed study of this subject.

Urokinase Research and Production

Urokinase is an enzyme, produced by human kidney cells, which has been found to be effective in removing blood clots from veins and arteries when administered in large doses.[3] Abbot Laboratories is attempting to develop processes to produce urokinase on a large enough scale for routine medical use, and have proposed a kidney cell separation experiment (MA-011) for the ASTP mission. Mr. Grant H. Barlow of Abbott Laboratories indicates that urokinase has been tried clinically for the prevention of pulminary embolism following surgery, and has been found to be medically effective. The original supply of urokinase was produced from human urine at a cost per dose of approximately $1500.00 Thus, the high cost and low availability of urokinase precludes its general use at the present time. Current efforts aimed at the production of urokinase in economically useful quantities are based upon the fact that human kidney cells grown in a culture produce the enzyme urokinase. Growth of the cells in a culture on Earth is limited by the available surface, and in the absence of gravity it may be possible to increase the surface in contact with the nutrient. The production of urokinase is further limited by the fact that only approximately 5% of the cells in the culture produce the enzyme. Various investigators, including Mr. Barlow, have hypothesized on the basis of ground tests using electrophoresis in a magnetic field that it may be possible to isolate the urokinase producing cells in a space based electrophoresis process. If the urokinase producing cells can be isolated it should then be possible to greatly increase the ratio of urokinase producing cells to non-producing cells in a culture. Thus, with a given quantity of kidney material, and a fixed time that might be established by limitations of the nutrient or the inability to remove toxic products from the culture, it should be possible to obtain a higher yield of urokinase from a concentrated mix of producing cells. This could conceivably lead to the capability to produce large quantities of urokinase at a reduced price.

A review of the available literature indicates that the economic aspects of the prospective capability to produce large quantities of urokinase at a reduced price have not been explored. In a manner similar to the recommended kidney transplant and dialysis study, the data base needed to support a study of the economic impact of space based materials processing on the production of urokinase should be available from various government and private sources. The incidence of death and disability induced by blood clots can be obtained from medical statistics, and the potential impact of urokinase can be estimated from the results of clinical studies. The demand for the product and its potential utilization can also be related to the price per dose. Since the ASTP experiment may produce rather specific insights into the feasibility of separating the urokinase producing cells from other human kidney cells, this area could rapidly assume increasing importance in the planning of Space Shuttle based space materials processing experiments. Consequently, it is recommended that a preliminary economic study of the prospective benefits of using space based facilities for the separation (and possibly growth) of urokinase producing kidney cells should be conducted.

Molecular Material[6]

Several molecular biological products, all with the common base of the potential use of electrophoresis for purification, have been considered as typical candidates for space materials processing. Although the spectrum of potential medical application of these materials could be extremely broad, some experts believe that space based electrophoresis may not play as decisive a role in the separation of molecular species as in the separation of cellular species. Some of these molecular products and potential applications are described in the following sections:

Erythropoietin

Erythropoietin is a hormone capable of increasing the total number and volume of circulating red blood cells in a normal animal. Its potential value lies in the treatment of renal failure or anemia, and the ability to avoid complications of the present form of treatment involving repeated transfusions of whole blood or red cells. Relatively high purity erythropoietin is required for the treatment of humans, but has not yet been obtained in sufficient quantity for clinical testing. There is some expectation that space based electrophoresis may prove to be a practical separation and purification method. No estimates have been made of the demand for clinically effective erythropoietin; however, it may be possible to estimate the economic impact of the development of clinically effective erythropoietin by examining the incidence of anemia and the cost of present treatment using transfusions of whole blood or red blood cells.

Other Research Applications

Several other research applications for space based electrophoresis have been suggested, including the production of highly purified quantities of Factor VIII (antihemophilia factor), purification of subunit viral vaccines, separation

of antibodies for cancer research, and the identification of
sub classes of lippoproteins for use in the study of ar-
teriosclerosis. All of these potential applications of
electrophoresis in a gravity free environment are primarily
of a more speculative research nature than the other organic
materials applications discussed in this review. Several
of these research applications could possibly lead to major
breakthroughs in the field of medical science with attendent
large potential economic impact. However, in each case the
technology is highly speculative, and considerable further
research and development should be performed before specific
economic implications can be considered.

Conclusions and Recommendations

An ultimate objective of the space materials processing
programs is to bring the technology to the point where industry
will want to participate on a commercial basis. The demonstrations of technical feasibility in space needed to obtain
commercial participation in the program have not yet occurred,
although the results obtained to date have been sufficiently
encouraging and interesting to merit further experimentation.
In the absence of demonstrated feasibility, it is premature
to attempt to justify operational space materials processing
on the basis of the results of experimentation to date. However, the results to date have been encouraging and supportive
of the conduct of further experimentation. Given the constraint of a limited budget, and the opportunity for many
experiments, economic analysis can be used as a management
tool in the ranking and selection of experiments. In this
context, economic analysis can be used to identify areas of
large potential benefits, and to serve as a framework for
developing the nature of prospective uses and users for
materials processed in space.

Limited economic studies have been performed to date, primarily in the area of space processing of inorganic materials. While these studies indicate the prospect of economic benefits in the area of electronic and electro-optic devices, the technology in this area has competitively and rapidly advanced in the past years, and the devices considered desirable today may be superseded by the time that the Space Shuttle is available to support further large scale experimentation. Further economic studies in this area should concentrate in a speculative fashion on the potential impact of space processing of inorganic materials, based upon technology forecasts into the mid-1980's.

The economics of space processing of cellular and molecular materials for biological and medical purposes has not received adequate attention and should be the subject of further study. An examination of the economics of this area could help focus research objectives, and provide impetus to a broader program. While the success of specific processes remains to be demonstrated, the economic impact of a successful space research program can be demonstrated in at least two applications. Specifically, it is recommended that economic studies be performed on the potential of the impact of human lymphocyte subgroup separation on organ transplantations, and on the separation and concentration of urokinase producing cells leading to the capability to produce large quantities of urokinase at a reduced price. In both of these areas the data base exists to support a substantive, quantitative economic assessment of the impact of a successful research and development program.

REFERENCES

1. Final Report of the Space Shuttle Payload Planning Working Groups. Materials Processing and Space Manufacturing. May 1973. NASA Goddard Space Flight Center.

2. Extract From The Space Applications Program, 1974. Overview of NASA Application Programs Prepared for the 1974 NASA/NAE Applications Summer Study. May 1974.

3. Testimony to: by:
   Special Programs   Space Processing Applications
   NASA Headquarters   1975

4. Study for Identification of Beneficial Users of Space (Phase I). Contract NAS8-28179. Final Report, Volume I. December 10, 1972.

5. Economic Analysis of Crystal Growth in space. Final Report, Contract NAS8-27842. July 1972

6. Space Processing - Status, Prospects, and Problems - 1974. Steg and McCreight. General Electric Company, Space Sciences Laboratory. September 1974.

7. An Analysis of Jet Turbine Blades as A Space Processing Candidate (Preliminary) Monthly Report, 15 August - 15 September 1974. Contract NAS8-29881. Engineering Experiment Station. Auburn University.

8. Personal Correspondence. R.L. Harnel, Materials Technology Department, TRW Systems Group. 27 February 1975.

9. Rogosin Kidney Center Annual Report - 1973. The New York Hospital. Cornell Medical Center.

10. Appropriations Hearings for 1974. Department of Health, Education and Welfare.

11. Crystal Growing in Space: Significance Still Up in the Air. Science, Volume 187, 14 February 1975.

AAS 75-150

THE ECONOMICS OF SATELLITE MAINTENANCE[*]

Wilfred L. DeRocher, Jr.[+] and Richard G. Sosnay[‡]

The primary goal of the space transportation system--
to reduce the cost of space programs while satisfying
their mission requirements--can be enhanced by the
proper choice of a satellite maintenance concept.
This paper develops life-cycle costs of performing the
automated satellite program in three competitive modes--
expendable, ground-refurbishable, and on-orbit maintainable. It is shown that on-orbit maintenance is
the most economic maintenance mode for both low- and
high-earth orbits.

INTRODUCTION

The quantity and diversity of space shuttle missions that may occur in the 80's depend on their being economically viable while satisfying mission objectives. This paper identifies a technique for establishing comparative life-cycle costs for three alternative satellite maintenance modes--expendable, ground-refurbishable, and on-orbit maintainable. The expendable mode consists of launching the desired number of satellites to obtain the desired on-orbit fleet size and then replacing each satellite as it fails until the program is complete. The ground-refurbishable maintenance mode is like expendable except that when a satellite fails it is retrieved from space, repaired, and checked out on the ground and then again placed in orbit. The on-orbit maintainable mode also starts like the expendable mode except that when a satellite fails, an on-orbit servicer carries replacement modules to the failed satellite, replaces the failed modules as well as any degraded modules, and then the on-orbit servicer returns to earth.

[*] This work was performed under NASA Contract NAS8-30820.
[+] Departmental Staff Engineer Martin Marietta Corporation,
[‡] Senior Engineer   Denver Division, Denver, Colorado, 80201

APPROACH

We suggest the use of an analytic method of calculating the costing parameters because of the available level of input data, the ease of performing the sensitivity studies of the costing parameters input, and the proper sensitivity, with regard to the input data/output costs, of the decision as to which maintenance mode can provide the most economic benefits. The analytic method involved the ground rule of maintaining a constant availability and a constant reliability across the three maintenance modes for the selected mission model. The mission model consisted of 47 different satellite programs with a total of 340 operating cycles. This could be accomplished in the expendable mode by 340 expendable satellites, in the ground refurbishable mode by 93 original satellites, 36 replacement satellites and up to 247 ground refurbishments, and in the on-orbit maintainable mode by 93 original satellites and 247 servicings.

ANALYSIS

The most important parameter in determining the satellite maintenance cost is the parts factor, which is the ratio of the cost of the replaced modules (or refurbished parts) to the total satellite unit cost. Parts factors represent the portion of the satellite that is repaired or replaced. They are based on a failure analysis of the satellite and represent two categories of failure--random failures and wearout failures. A delta between on-orbit maintainable parts factors and ground refurbishable parts factors was used to take into account external appendages, additional painting, cleaning, and decontaminating, and a final acceptance test. Parts factors for all of the satellites varied from 0.06 to 0.38, with an average value of 0.16 for the on-orbit maintainable mode, and varied from 0.13 to 0.52 with an average value of 0.28 for the ground-refurbishable mode.

The results of the baseline cost analysis are shown in Fig 1. The launch costs increase from the on-orbit maintainable mode through the expendable to the ground-refurbishable mode and represent approximately 10% of total satellite program costs. The on-orbit maintainable mode saves approximately 11 billion dollars in satellite costs over the expendable mode.

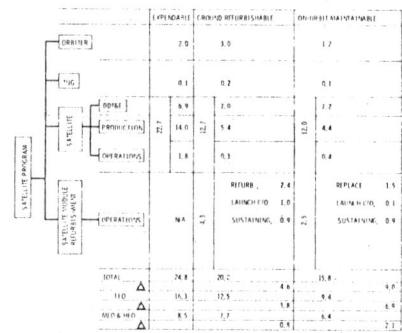

Fig. 1   Baseline Cost Summary in Billions of Dollars

This is partially offset by 2.5 billion dollars of module refurbishment costs. The total effect is that 9 billion dollars, or 36%, can be saved by on-orbit maintenance over expendable, or 4.4 billion dollars, 22%, over ground refurbishment. The maintenance concept costs are shown below to represent only 0.1 billion dollars and thus do not much affect the savings. These savings can be used in the form of higher quality or more space missions.

The life-cycle costs of an on-orbit maintenance concept in the form of a pivoting arm on-orbit servicer were generated using the considerations shown in Table 1. These considerations were developed from definition of the airborne hardware, mission schedules, loss rates, launch site requirements, and operational needs. The resulting costs are: (1) DDT&E - $29M, (2) production - $17M, (3) operations - $57M, and (4) total life-cycle cost - $103M. Thus the maintenance concept costs are seen to be a small part of the total satellite program costs and savings.

A sensitivity study was performed to determine (1) accuracy and validity of input data and effects of data inputs on results, and (2) effects of future changes in data on study results. The "influence coefficient" method was used. Table 2 presents a summary of the maximum expected variations in savings for the changes in cost parameters investigated. Fig 2 presents a profile of possible variations in savings as a function of mission model size for the most important input cost parameters.

Results of the cost sensitivity study show that mission model changes and parts factor variations have the greatest effect on study results, but that on-orbit maintenance is still feasible with reductions of up

TABLE 1 PIVOTING ARM ON-ORBIT SERVICER COST CONSIDERATIONS

ABS ELEMENT	CONSIDERATIONS
STRUCTURE	450 LB
MECHANISM	160 LB
CONTROL ELECTRONICS	30 LB
AIRBORNE SPARES	FIVE FULL UNITS, SUBSYSTEMS - PARTIAL
AIRBORNE SUPPORT EQUIPMENT	450 LB
LOGISTICS	LOGISTICS MANAGEMENT INVENTORY CONTROL OMI MANUALS TRAINERS TRAINING
GROUND SUPPORT EQUIPMENT	MECHANICAL - 40 UNITS ELECTRICAL - 15 UNITS
FACILITIES	REARRANGEMENT
OPERATIONAL SITE SERVICES	LAUNCH OPERATIONS FLIGHT OPERATIONS MAINTENANCE

TABLE 2 COST SENSITIVITY VARIATIONS

COST PARAMETER	VARIATION	AVERAGE OR NOMINAL VALUE	CHANGE IN SAVINGS*
NUMBER OF OPERATING CYCLES	CUT BY 25% AND BY 50%	340 FLIGHTS	SEE FIG. 2
PARTS FACTOR	± 0.08	0.16	±1,005
PRODUCTION SERVICE FACTOR	± 0.06	0.09	± 353
NONRECURRING SERVICE FACTOR	± 0.03	0.04	± 296
LAUNCH CHECKOUT RATIO	± 0.03	0.09	± 244
ORBITER LAUNCH COST	± $3.0M (± 25%)	$12.0M	± 208
TUG LAUNCH COST	± $0.3M (± 30%)	$ 1.1M	± 2
LOAD FACTOR - ORBITER	± 0.3	0.70	- 242 / + 579
LOAD FACTOR - TUG	± 0.15	0.85	- 1 / + 1
DIFFERENTIAL LAUNCH COST-SHARING FACTOR ORBITER AND TUG	± 0.15	0.21 / 0.09	± 662

*MILLIONS OF DOLLARS

to 50% in the number of missions currently in the mission model.

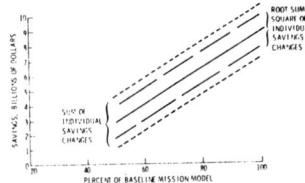

Fig. 2 Cost Sensitivity Analysis

CONCLUSIONS

A technique for comparing the life-cycle costs for three alternative satellite maintenance modes--expendable, ground-refurbishable, and on-orbit maintainable--has been described and illustrated. The technique is useful when the level of satellite definition is low and when direct identification of cost sensitivity effects is desired. The on-orbit maintainable mode has the greatest economic benefit when applied over a wide range of automated satellite mission models for the 1980's. This result is insensitive to wide variations in the cost parameters. However, the magnitude of the savings do vary. The cost of the on-orbit maintenance system is small ($\approx 1\%$) when compared to the savings in the 47 satellite programs.

AAS 75-170

BURNER II INTERIM UPPER STAGE
AAS TECHNICAL PAPER*
Henry Kudish[+]

INTRODUCTION

In order to define candidate configurations for the interim upper stage (IUS) task, the Air Force awarded study contracts to contractors of existing upper stages. This paper summarizes the results of the study contract awarded to the Boeing Aerospace Company to study the Burner II Interim Upper Stage.

BURNER II IUS VEHICLE CONFIGURATIONS

Using Burner II design philosophy, Boeing Aerospace Company, Space Systems Division, has developed a family of solid rocket motor stages that meets the requirements for a low-cost, expendable IUS. This vehicle family is shown in Fig. 1.

The two-stage configuration is the basic vehicle from which the family is derived. The vehicle consists of a forward stage with a 4700 pound solid-propellant motor and an aft stage with an interstage, a 20,000 pound solid-propellant motor, and an aft skirt structure. The forward stage also consists of an avionics bay, motor support structure, an integral conically shaped spacecraft support and staging adapter, hydrazine-fueled reaction control system (RCS), electrical umbilical interfaces and staging provisions for the interstage and aft motor.

---

[*] Prepared for technical papers that will later be published in the Proceedings of the American Astronautical Society

[+] Burner II Interim Upper Stage Study Manager. Mr. Kudish is affiliated with Boeing Aerospace Company, Seattle, Washington (Organization 2-2910 MS 84-04, (206) 773-2739).

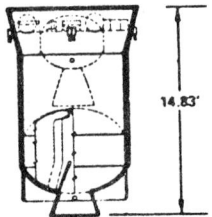

Fig. 1   Burner II IUS Vehicle Family

The avionics bay contains the guidance and control (G&C) system, the telemetry, tracking and command (TT&C) system, the data management system, the electrical power and distribution system, and an optional celestial sensor. The G&C employs a strap-down inertial sensor unit for primary guidance supplemented by radio and celestial update when desired for higher-accuracy missions. A conventional SGLS/STDN S-band 20-watt transponder provides for TT&C. The data management subsystem employs a high-speed general purpose digital computer, with 32,000-word capacity, which performs all data formatting functions as well as all G&C computations. The control outputs are processed by a TVC/RCS power amplifier. This unit also contains the caution and warning logic. Electrical power is supplied by 165-ampere-hour primary batteries, supplemented by a separate pyrotechnic battery for transient loads. The number of primary batteries can be tailored to the mission.

The interstage consists of a cylindrical section and a short conical section that mates with the aft end of the forward stage. Two trunnion shafts are located on the conical part of the interstage. These mate with the forward airborne support equipment (ASE) structure. A fitting is bolted on the cylindrical section that will provide an attachment interface for the Orbiter remote manipulator arm. The forward stage is joined to the interstage with explosively separated staging nuts and bolts. These staging devices are also used at the spacecraft/IUS interface.

A separate spider-beam structural adapter is used to provide a support and staging interface for NASA spacecraft. This adapter attaches to the standard DoD interface contained in the conical forward-stage adapter.

The aft skirt is a skin stringer structure that bolts onto the aft end flange on the aft motor. The skirt includes three trunnion shafts that mate with the aft Airborne Support Equipment structure.

The two solid rocket motors are very similar in design. Both employ Class II propellants, filament-wound cases, and movable nozzles. The movable nozzles are powered by electromechanical actuators, two of which are mounted on each motor. The large motor has an overall L/D nearly equal to 1.0 with a total length of 92 inches, and a maximum loaded weight of 21,200 pounds, including 20,000 pounds of propellant. This motor can be offloaded to 17,300 pounds of propellant. The small motor has an overall L/D of 1.33, length of 78 inches and loaded weight of 5,100 pounds, including 4700 pounds of propellant. Both motors will use safe and arm devices that will be mounted on the IUS structure remote from the SRM igniters, and connected by an explosive transfer assembly.

The three-stage vehicle configuration is the two-stage vehicle with an added interstage and large motor.

Thrust Vector Control

Thrust vector control is provided by an electromechanical actuator. The actuator is a ball screw, directly driven by a permanent magnet brush type motor through a simple gear train. A single controller box provides amplification and feedback electronics for both motors.

Reaction Control System (RCS) Description

The RCS system is a blowdown pressurized hydrazine monopropellant system. The system consists of the thrusters assemblies, tanks and associated plumbing. The plumbing includes the tubing to transfer the propellant from the tanks to the thrusters, burst discs at each tank outlet, fill valves for the hydrazine and the pressurant and a leak check fitting. The system has no fluid interfaces with the Shuttle Orbiter.

Guidance, Navigation and Control

Guidance and targeting techniques have been derived which compute the ΔV at first and second solid motor burns to accomplish a Hohmann or non-Hohmann transfer between a low-earth orbit and a geosynchronous orbit. The non-Hohmann orbit transfer can be selected from a large family of possible choices.

Inertial Navigation System

The guidance hardware chosen for the Burner II IUS is a modification of the Teledyne TDS-3C IMU containing three two-degree of freedom elastically suspended mechanically tuned strapdown gyros which act also as accelerometers.

Electrical Power System

The electrical power system is divided into two sections. One or more 165 ampere-hour silver-zinc batteries supply power to the guidance, navigation and control system; the telemetry, tracking and command system; the data management system; and the signal conditioning equipment associated with the data management function. Spacecraft power would also be supplied from this source.

A separate silver-zinc battery, with 13 ampere-hour capacity supplies redundant caution, warning, and safety devices; thruster solenoids; and solid rocket motor ordnance devices and TVC actuators.

AAS 75-171

TRANSTAGE INTERIM UPPER STAGE

Peter B. Teets[*]

The Transtage provides the basic "building block"
module for the IUS. It satisfies all requirements
with minimum modification and investment. It also
provides the maximum upper-stage flexibility that
will minimize IUS program risks and costs. This
paper discusses how Transtage IUS provides minimum
development risk, 100% mission model capture, significant performance margin, and low-risk Transition
for Spacecraft from today's Expendable Launch
Vehicles to the Space Transportation System.

INTRODUCTION

The Transtage IUS offers minimum technical risk because it requires minimum modification to meet IUS requirements. The current Transtage (Figure 1) was originally designed for orbit-to-orbit missions, especially geostationary delivery. All subsystems are space-proven. The guidance and navigation system has set industry standards for orbit placement accuracy. It carries sufficient propellant to meet all DOD IUS requirements. Because it performs the basic IUS mission today, the only modifications required for IUS are those needed for STS compatibility. Transtage size, weight, and performance capability are the same for Titan or IUS use (Fig. 2).

---

[*] Program Manager, Transtage IUS, Martin Marietta Corporation, Denver, Colorado

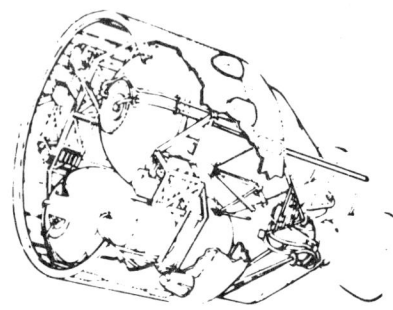
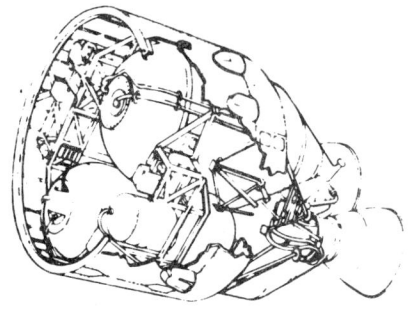

TITAN TRANSTAGE

- FOURTH STAGE OF TITAN IIIC
- DESIGNED FOR ORBIT-TO-ORBIT TRANSFER
- PERFORMS
  - SPACECRAFT THERMAL MANEUVERS
  - ORBIT CIRCULARIZATION AND PLANE CHANGE
  - SPACECRAFT SEPARATION
  - POSTSEPARATION MANEUVERS
- UP TO EIGHT PAYLOADS DEPLOYMENT ON SINGLE FLIGHT
- MATURE, FLIGHT PROVIDED PAYLOAD DELIVERY SYSTEM
- 17 OF 18 SUCCESSFUL MISSIONS SINCE 1966

LENGTH	14.7 ft
DIAMETER	10.0 ft
DRY WEIGHT	3,517 lb
LOADED WEIGHT	27,177 lb

Figure 1 Current Transtage The Original Orbit-To-Orbit Transfer Stage of the Industry

MODIFICATION FOR IUS

- STRUCTURE
  - MINOR RING FRAME AND STRINGER MODES
  - ATTACH FITTINGS
- AVIONICS
  - NEW TT&C SYSTEM
  - C&W PROVISIONS
- PROPULSION SUBSYSTEM
  - PROPELLANT DUMP
  - PREVALVES
- AIRBORNE SOFTWARE
  - 70% OF MODULES ARE UNCHANGED FROM TITAN

LENGTH	14.7 ft
DIAMETER	10.0 ft
DRY WEIGHT	3,498 lb
LOADED WEIGHT	27,077 lb

Figure 2 Only STS Compatibility Modifications Are Required for IUS Use

TRANSTAGE IUS PROVIDES NEEDED FLEXIBILITY

Our Transtage IUS program is based on the "building block" concept (Fig. 3). It consists of three family elements: (1) the basic, existing length Transtage IUS; (2) the dual Transtage and existing TE-364-4 kick stage; and (3) the stretched Transtage. The basic Transtage IUS can accommodate the entire DOD mission model and 83% of the NASA mission model. Because it is only 15-ft long, it can carry spacecraft 45 ft long in the Shuttle cargo bay. Because it is short and weighs only 27,000 lb, two can be carried in tandem with separate spacecraft in the same Orbiter flight.

The basic IUS can be easily adapted to form the second family element of our program because of a unique design feature. The Transtage was originally designed to permit on-orbit separation of the control module from the propulsion module. It is built today with the control module and propulsion modules assembled separately. It is a simple assembly sequence to add an additional propulsion module to Transtage IUS and create the dual Transtage. The only new design required is a simple assembly sequence to add an additional propulsion module to Transtage

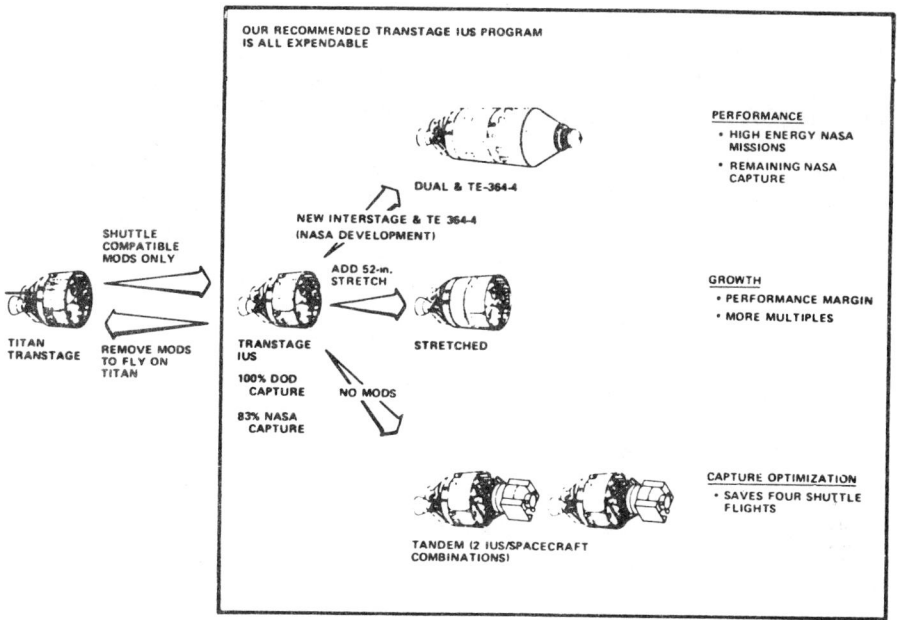

Figure 3 Our Building Block Concept Offers Maximum Flexibility To Minimize IUS Program Risk and Cost

IUS and create the dual Transtage. The only new design required is a simple interstage structure adapter almost identical dimensionally and simpler in design to the one that currently attaches Transtage to Stage II of the Titan IIIC vehicle. By using these same Transtage IUS elements, the dual Transtage can offer the same minimum modification/minimum risk advantages of the basic Transtage IUS. The dual Transtage will be used to capture the high-energy NASA planetary missions. The existing TE-364-4 kick stage added to the dual Transtage, will permit capture of all NASA missions.

The third family element of our Transtage IUS program is the stretched Transtage. This vehicle offers additional performance margin, performance growth, and mission capture advantages to both NASA and DOD programs. Geostationary performance capability is increased from 4200 lb for the short Transtage to 6400 lb. The performance improvement permits more multiple spacecraft to be carried and reduces the number of Shuttle flights required to capture the mission model. The stretch involves adding propellant by increasing tank length 52 in. The tank domes and cones are left unchanged while the barrel sections and skirt are simply manufactured 52 in. longer on the same production line.

TRANSTAGE IUS PROVIDES 100% MISSION CAPTURE

Our Transtage IUS program captures the entire mission model. The existing length Transtage IUS can deliver 100% of the DOD mission model and 83% of the NASA mission model. The remaining 17% of the NASA mission model is captured by using the dual Transtage configuration and the existing TE-364-4 kick stage. Note that the dual Transtage and TE-364-4 kick stage combination offer planetary programs greater capability than any currently existing expendable booster combination.

OPTIMUM PLANNING WITH TRANSTAGE WILL REDUCE TRANSITION RISK

Some risk is always present in the transition from one major program to another. Optimum transition planning, however, can reduce these risks. The Transtage offers the opportunity for reduced risk transition planning. The Transtage is unique in that it has the same physical characteristics and delivery capability, whether launched on a Titan or in an Orbiter. This unique feature guarantees minimum IUS program risk during the transition period from expendable launch vehicle to Shuttle. The integrated Transtage-spacecraft combinations could be flown on the Shuttle or, as backup, on Titan III.

CONCLUSION: TRANSTAGE IS THE IDEAL INTERIM UPPER STAGE

The Transtage has optimum size. It requires minimum modification. It can optimize total program costs. It is already interfaced with many of the DOD spacecraft. It can fly on either Titan or STS with a minimum of transition problems. It is a mature stage. It is simple, safe, and reliable. It is already designed for the primary mission of the IUS program, geostationary delivery. It will minimize Orbiter interface risk. It will accommodate both DOD and NASA requirements. It offers maximum program flexibility to perform the IUS mission with minimum program risk regardless of changes to STS or user requirements. The Transtage is the ideal Interim Upper Stage.

Acknowledgement

The author gratefully acknowledges the contributions of all members of the Transtage IUS System Study Team at Martin Marietta Corporation, Aerojet Liquid Rocket Company, and Delco Electronics. Particular appreciation must go to Messrs. T. Goyette and J. Nelson of Martin Marietta Corporation.

AAS 75-172

CENTAUR AS AN INITIAL UPPER STAGE FOR THE SPACE SHUTTLE

D.J. Jones* and D.A. Heald**

ABSTRACT

Centaur is one of the candidate stages for the Space Shuttle Interim Upper Stage (IUS).

This paper describes Centaur IUS versions and reviews the modifications from the existing D-1T Centaur. Performance capabilities are presented including the results of mission capture analyses. Flight and ground operations highlights are explained. Orbiter and payload interface requirements and capabilities are covered. Relative cost data showing the economic benefits of reusability is presented.

In comparing life cycle costs with expendable stages it is found that the operational cost saving for Reusable Centaur offsets the slightly higher development costs after one or two years of operations. Reusable Centaur is a modification of the operational D-1T Centaur created by increasing the diameter of the LH_2 tank and lengthening the LO_2 tank to increase propellant capacity. The Centaur is inherently reusable so that minimum subsystem modifications are required. The stage can deliver 5,000 pounds of payload to synchronous equatorial orbit in the reusable mode without the use of kickstages. The stage captures the complete National Mission Model. The fact that the stage is reusable means a saving to the government of several million dollars per flight using Reusable Centaur instead of an expendable IUS.

The Space Shuttle will deliver large payloads to low earth orbit. A large number of planned missions require an upper stage to reach higher orbits, such as synchronous, or to escape for

*IUS/Tug Program Manager, General Dynamics Convair Division, San Diego, California
**IUS Study Manager, General Dynamics Convair Division, San Diego, California

planetary probes. Minimum development funds will be available for a new upper stage (Tug) until Shuttle development funding requirements begin to decline. Therefore, existing upper stages, modified for Shuttle compatibility, are being considered as an Interim Upper Stage (IUS) to be available at or near Shuttle IOC. Centaur D-1T, shown in Fig. 1 is the basis for our IUS concepts. The Helios flight in December 1974 demonstrated Centaur's ability to perform IUS type missions including a three-hour coast and four starts.

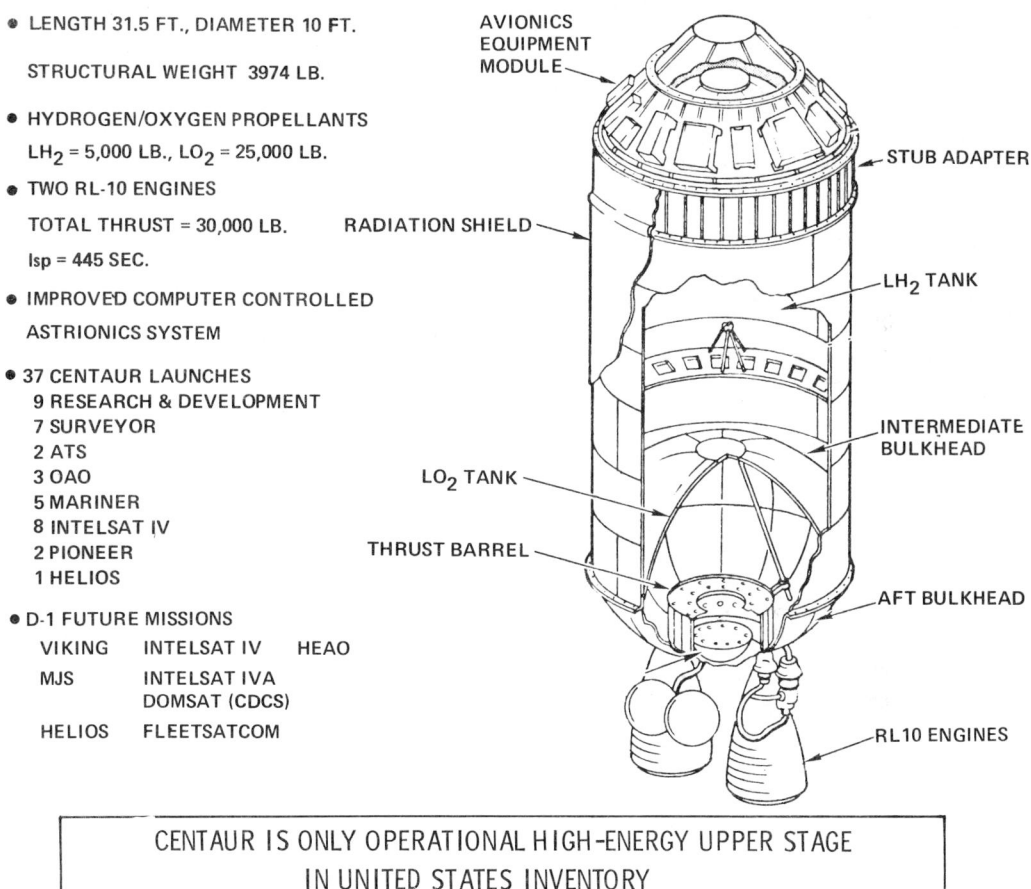

Fig. 1 D-1T Centaur.

There are two alternative operational modes: an expendable mode, wherein the IUS is expended while delivering the payload; and a reusable mode, wherein after payload deployment, the IUS returns to rendezvous with the Orbiter for return to earth and reuse. Expendable IUS requires the least early-year development dollars. Reusable IUS results in lower total life-cycle costs due to its recovery and reuse. The General Dynamics Convair approach in all programs has been to provide a basic capability to accommodate all DOD requirements at a minimum cost. A delta cost will then provide the additional capability to accomplish all NASA missions. The ability to accomplish the National Mission Model with small changes from the DoD program and the potential for reusability are a direct result of the high performance capability inherent in the Centaur vehicle.

The expendable Centaur (EC) programs provide very high capability with minimum program risk. Fig. 2 shows that the basic program consists of a 25-foot-long stage designed for geosynchronous and high-altitude missions and a "stretched" version (identical to the D-1T length), which is designed for planetary and escape missions. The 31-foot EC is identical in design to the 25-foot EC except for the addition of a cylindrical section to the LH_2 tank. The highest velocity Pioneer mission requires a small spin-stabilized kick stage, modified from the existing design for Helios.

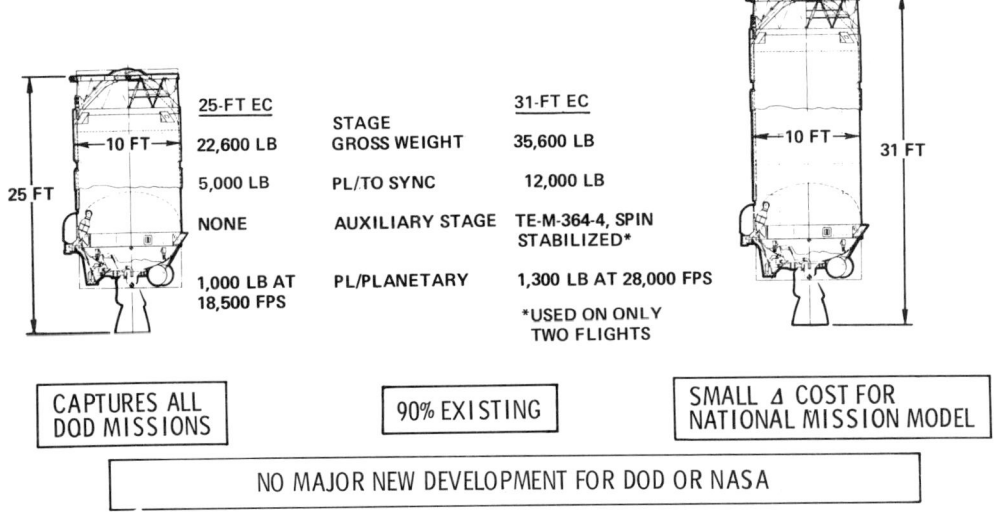

Fig. 2 Expendable Centaur IUS program.

These configurations use approximately 90% (by value) of existing Centaur equipment, thereby providing high program confidence. The 25-foot EC can deliver almost 5,000 pounds to geosynchronous and the 31-foot EC, although designed for earth escape missions, can deliver more than 12,000 pounds to geosynchronous. The high performance of these stages provide significant payload margin above the anticipated single payload weights. This allows multiple-payload deployment capability, which greatly reduces total transportation costs.

An extremely attractive option is the "short" expendable version pictured in Fig. 3. This stage at slightly greater development cost, delivers almost 12,000 pounds to orbit and is less than 22 feet long. It combines the advantages of short length, high performance, and single configuration accommodation of the entire National Mission Model. This stage, with the use of a simple-spin stabilized kick stage for very high energy missions, has sufficient performance and payload length availability to satisfy all users. The short length/high performance combination makes it particularly attractive for multiple-payload deployment, such as eight Navigation Satellites.

The reusable Centaur (RC) programs have low total program cost at slightly higher initial development cost. The inherent high performance of the Centaur provides the basis for reusability with approximately 70% (by value) of existing Centaur equipment used on the reusable configurations. The basic reusable program consists of a 28-foot-long vehicle and a 22-foot-long vehicle (Fig. 3), which are identical except for different cylindrical sections of the propellant tanks.

Fig. 3 Centaur IUS with multiple payloads.

The 28-foot RC shown in Fig. 4 can deliver more than 5,000 pounds to geosynchronous orbit in a reusable mode and can perform the most demanding high-energy mission in an expendable mode without the use of an auxiliary propulsion stage. The 22-foot version or an alternative 19.5-foot version is designated for long payloads. The high performance of these stages provides the users significant flexibility, in either the reusable or expendable mode, and can be used for both single and multiple payload deployment. For instance, the 28-foot Reusable Centaur can place nearly 16,000 pounds into synchronous equatorial orbit, or could provide the plane change between different orbits.

All Centaur IUS designs are fail-safe and provide for inflight propellant dump in case of Shuttle abort. All Centaur IUS vehicles use the existing Pratt & Whitney RL10 engine, which has demonstrated reliability and reusability.

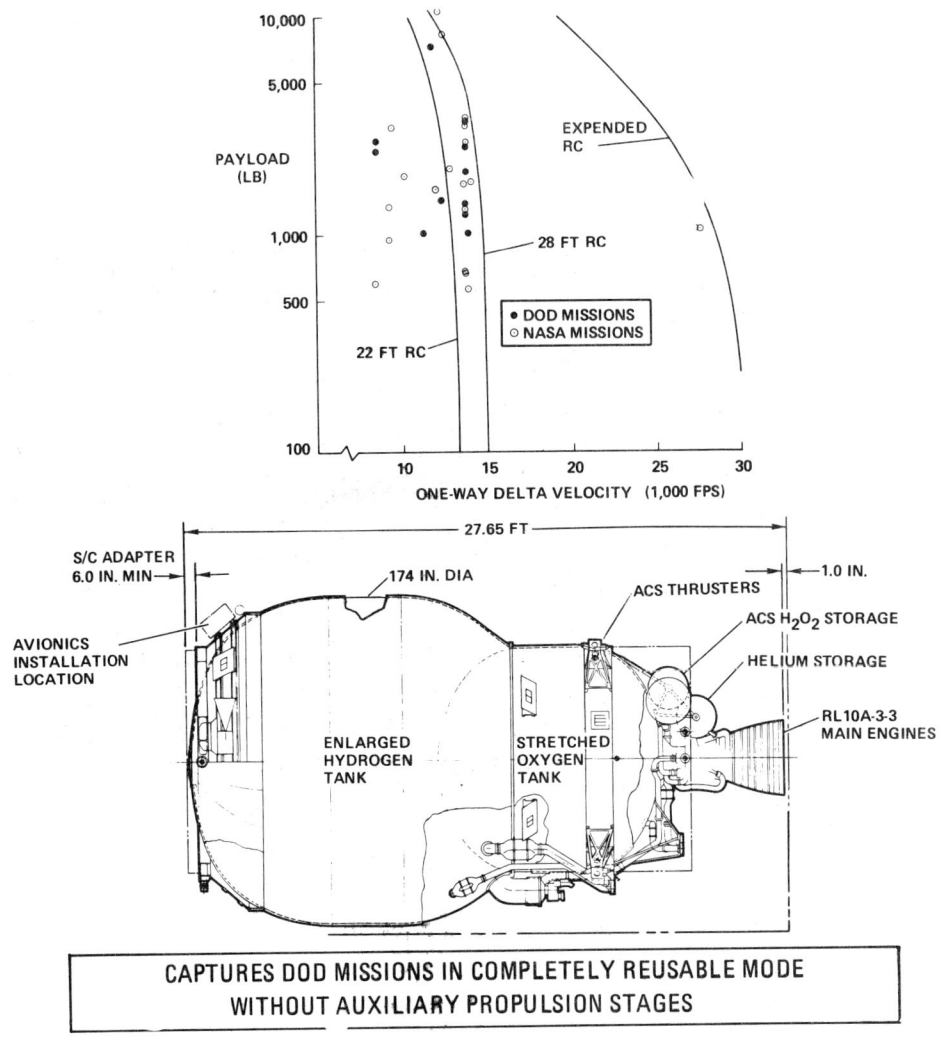

Fig. 4 Reusable Centaur IUS general configuration and performance capability.

The inherent high performance of the Centaur is the basis for assuring a low-risk IUS program while providing significant benefits. Centaur performance enables candidates to be defined without resorting to high-risk weight reductions and/or major engine modifications, or the need for large kick stages and allows the incorporation of all necessary features to provide a safe man-rated vehicle. This performance margin will also permit increased redundancy to further enhance reliability, spacecraft growth, and mission flexibility. Centaur high performance, coupled with the use of clean cryogenic propellants, provides the basis for reusability with the associated reduction in life cycle cost. Recurring cost becomes a very important criterion if the Space Tug is delayed and the IUS operational span becomes greater than three to four years.

AAS 75-173

DELTA AS AN INTERIM UPPER STAGE (IUS)

R. P. Dawson and J. F. Meyers[*]

The Space Transportation System will require an upper stage — Space Tug or Orbit-to-Orbit Shuttle (OOS) — to achieve maximum effectiveness. A reusable upper stage, although highly desirable, requires a substantial initial cash outlay, which would compete for early Shuttle development program funds. Hence, a less expensive expendable upper stage, to be used in the interim, has been the subject of various Government studies in recent years.

This paper describes the results of one of these studies, the Delta IUS System Study, which addressed the use of Delta launch vehicle hardware to satisfy this requirement. Basic Delta hardware is described, along with the changes that would be required to make it compatible with Shuttle.

The study has shown that with the addition of two inexpensive propellant tanks and other minor changes, the basic Delta stage can be easily converted into a tri-tank IUS that can capture all proposed DOD and NASA earth orbital missions with ample performance margin. Furthermore, with the addition of simple, low-cost auxiliary stages based on existing solid propellant motors, the Delta IUS can capture all of the NASA planetary missions proposed. The combinations of IUS elements proposed — the tri-tank family approach — provides the flexibility to select the least expensive IUS suitable for the mission, thus enhancing cost effectiveness.

---

[*] McDonnell Douglas Astronautics Company

This paper reports the results of a study conducted by McDonnell Douglas Astronautics Company for the Space and Missile Systems Organization (SAMSO) of the DOD. The objective of the study was to provide preliminary designs, interface definitions, cost estimates, and program definitions for a Delta-derived interim upper stage (IUS) system that can be used with the Shuttle in the 1980-1984 time period to extend the STS operating regime.

The Delta IUS is based on the second stage of the basic Delta launch vehicle. This basic stage is slightly more than 19 feet long, has a maximum diameter of 5 feet at the forward interface, and contains approximately 10,000 pounds of storable propellants in stainless steel tanks. The single pressure-fed TRW Apollo lunar module descent-derived engine provides approximately 10,000 pounds of thrust and has multistart capability.

The baseline expendable Delta IUS concept shown on Fig. 1 (referred to as the Delta Tri-Tank IUS) consists of three standard Delta tanks attached together with manifolded propellant lines and a single engine mounted on the center tank. Existing Delta gas bottles and attitude control modules are used. The avionics equipment is mounted within a new forward skirt in a manner similar to that of the basic Delta stage.

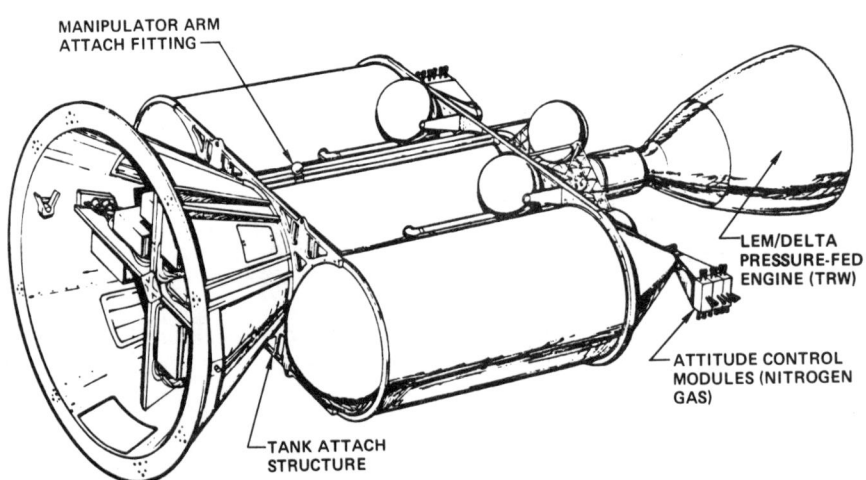

- CAN PERFORM 100 PERCENT OF BOTH DOD AND NASA EARTH ORBITAL MISSIONS WITH MORE THAN 1,000-LB PAYLOAD MARGIN

Fig. 1 Baseline Delta Tri-Tank IUS Configuration

This baseline tri-tank configuration serves as the basic building block for an entire family of Delta IUS configurations, as shown on Fig. 2. The expendable family consists of a basic tri-tank IUS, planetary versions with auxiliary stages, and a single-tank version. The reusable family consists of a reusable tri-tank configuration with expendable auxiliary stages as required. A short-length alternative was also studied.

The performance capabilities and principal characteristics for the expendable Delta IUS configuration family are summarized on Fig. 3. The baseline Delta tri-tank version can deliver 4,820 pounds to geosynchronous orbit. Thus, the Delta IUS can capture all proposed DOD and NASA earth-orbital missions with ample performance margins. For the NASA planetary missions, auxiliary stages are used.

Other subjects addressed in the study were Orbiter interfaces, operations, and program aspects, including required inventories and life-cycle costs. Major conclusions of the Delta IUS System Study are summarized on Table 1.

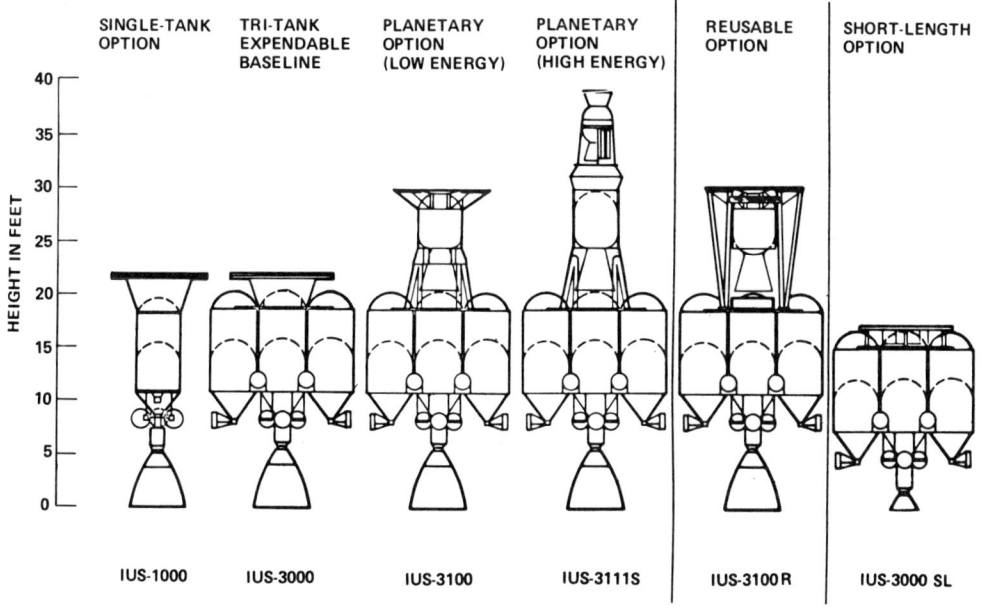

Fig. 2 Delta IUS Family

	BASELINE TRI-TANK (IUS-3000)	SINGLE-TANK OPTION (IUS-1000)	PLANETARY OPTIONS			SHORT-LENGTH OPTION (IUS-3000SL)
			(IUS-3100)	(IUS-3111)	(IUS-3111S)	
GEOMETRY						
• OVERALL LENGTH	22 FT 5 IN	22 FT 5 IN	29 FT 6 IN	37 FT	37 FT 8 IN	15 FT 10 IN
• DIAMETER MAXIMUM	14 FT 10 IN	55 IN - TANK / 120 IN - ADAPTER	14 FT 10 IN	14 FT 10 IN	14 FT 10 IN	14 FT 10 IN
WEIGHTS						
• PROPELLANT LOADING						
DELTA STAGE (LB)	30,172	10,047	30,172	30,172	30,172	30,172
AUX STAGE (LB)	–	–	7,298	7,298/2,333	7,298/2,333	–
• GROSS STAGE WEIGHT (LB)	34,422	12,294	42,740	45,439	45,539	34,342
ENGINE TYPE						
• DELTA STAGE	TRW LEM-D ($\epsilon = 54$)	TRW LEM-D ($\epsilon = 54$)	TRW LEM-D ($\epsilon = 54$)	TRW LEM-D ($\epsilon = 54$)	TRW LEM-D ($\epsilon = 54$)	TRW LEM-D ($\epsilon = 16$)
• AUX STAGE	–	–	MM STG III ($\epsilon = 55$)	MM/TE364-11 ($\epsilon = 55/33.9$)	MM/TE364-11 ($\epsilon = 55/33.9$)	–
PERFORMANCE CAPABILITIES						
• GEOSYNCH (LB) (3,376) (ΔV = 13,961 FPS)	4,820	830	8,530	NA	NA	3,380
• PLANETARY (LB) (ΔV = 27,943 FPS)(1,052)	0	0	PL-01 10,573 REQ.	698	1,105	0
(ΔV = 17,375 FPS)(5,270)	0	0	11,073 AVAIL.	5,270	5,436	0
• HIGH ALTITUDE (LB)(3,000) (ΔV = 14,683 FPS)	3,982	549	NA	NA	NA	2,620

Fig. 3 Expendable Delta IUS Tri-Tank Family

Table 1

DELTA IUS SYSTEM STUDY CONCLUSIONS

Resource Commitments

- Low RDT&E costs
- Low unit costs
- Minimum cost impact on Orbiter or spacecraft
- Expendable IUS program more economical than reusable program

Transition Requirements

- Parallel ground launch vehicle program into 1980's at ETR and WTR
- Orderly spacecraft/launch vehicle transition
- Delta maturity and commonality a valuable asset

Design Approach/System Evaluation

- Low technical risk; required safety features added
- Minimum modifications to existing Delta guidance system, engine, tanks, ACS, etc.
- No new SRM's and subsystems needed for auxiliary stages

Operational Characteristics

- Delta IUS meets all DOD and NASA mission requirements
- 1500-pound payload margin available for geosynchronous missions
- Tri-tank family matches IUS to mission; no expensive overkill
- Dual side-by-side installation possible with single-tank IUS

Management

- Experienced Delta team to be used
- Government working interfaces well defined and effective
- Program approach provides ample development time

AAS 75-174

SPINNING SOLID PERIGEE STAGE
H. A. Rosen,* C. R. Jones,+ and L. M. Bronstein†

> If the Space Transportation System is to serve in the 1980's for launching geostationary satellites that presently use the Delta and Centaur boosters, the cost of STS launches must be competitive. This can be accomplished through the provision of a multiple-launch capability in the orbiter, combined with the design of a low cost spinning solid perigee kick motor (PKM). The PKM will be attached to the spacecraft and function essentially as the Delta third stage does today.

Geostationary space system designers may be able to select from among six boosters in the 1980's. The presently-used Delta, Centaur, and Titan IIIC will be joined by the Japanese N rocket with a capability equivalent to that of the Delta, and the European Ariane with a Centaur class capability. And finally, the Space Transportation System (STS) will be available with its dramatically increased capacity.

The increase from the current three boosters to six may be illusory because the STS capacity and launch rate are designed to launch all potential payloads and the U.S. Government plans to phase out the U.S. expendable boosters in the 1980's. At the same time, the ESA Ariane and NASDA N rocket will introduce the first serious competition to the U.S. monopoly in the booster business. If the U.S. is to

---

* Vice President, Engineering, Hughes Aircraft Company Space and Communications Group

+ Associate Division Manager, NASA Systems Division, Hughes Aircraft Company Space and Communications Group

† Project Manager, Systems Laboratories, Hughes Aircraft Company Space and Communications Group

maintain its preeminance in this service, the STS designers must consider the geostationary space system designer needs.

A review of current geostationary systems shows the predominate use of the smallest, lowest cost vehicle, the Delta. The geostationary systems can be categorized as experimental/scientific or operational systems. The operational systems, which provide on-going communication and meteorological services, will require replenishment in the 1980's. The STS designers should provide launch capabilities for replenishment satellites in the Delta and Centaur class. The operational space system designer has a very important constraint, however — cost.

The expected reimbursable cost for the Delta and Centaur, if the STS were not available and a 5% inflation rate is the only cause of cost increase, is shown in Fig. 1. The NASA established objective is to operate the STS orbiter at a direct cost of $10.5M per orbiter flight (1971 dollars). The reimbursable cost of the orbiter flight alone, assuming 60% of the direct cost for indirect cost and 5% for inflation, is nearly twice the Delta cost and only slightly less than the Centaur cost. Furthermore, the STS requires an upper stage in addition to the orbiter to launch a geostationary system. The Delta users could not

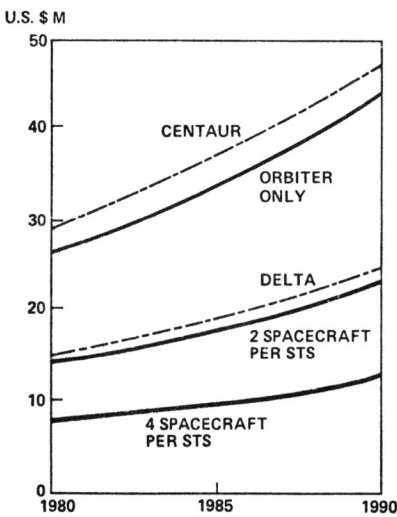

Fig. 1 Multiple Launch STS Projected Cost (Upper Stage $1M Total Cost Per Spacecraft)

survive at this cost level and the Centaur users would clearly prefer an opportunity for a lower cost.

The weight and volume capacity of the orbiter offers the opportunity for a significant reduction in launch cost, namely multiple launches. If the upper stage is assumed to have a full cost equal to $1M for each satellite, a two-satellite STS launch would be one-half the projected cost of a Centaur launch; a four-satellite STS launch would be one-half the projected cost of a Delta launch. The objective should thus be to develop a two- and four-satellite launching capability from the STS orbiter with an upper stage whose full cost is equal to or less than $1M per satellite.

The STS upper stage must raise the satellite from the orbiter altitude of approximately 295 km (160 miles) to a geostationary circular orbit of 35,800 km (22,500 miles). The Delta third stage solid rocket motor fires at perigee raising the satellite from a 185 km parking orbit to the 35,800 km geostationary altitude and the satellite is spin stabilized during the solid motor firing; hence, the expression "spinning solid perigee stage." The STS could employ the same launch sequence concept through the use of an unguided spinning solid rocket motor upper stage as a perigee kick motor (PKM). The apogee boost required for orbit circularization and inclination removal would be supplied by a spacecraft-mounted apogee kick motor (AKM) as all Delta and Centaur launched spacecraft use today. This approach has two major advantages:

1) The unguided, spin stabilized solid rocket motor perigee stage plus the satellite-mounted apogee kick motor is the lowest possible cost solution (less than $1M is a reasonable goal).

2) The spinning perigee stage would take less volume, weight, and power from the orbiter than any known alternative.

A spinning solid perigee stage appears both feasible and achievable at a cost less than $1M, but two factors need to be addressed (see Fig. 2). First, the current interface definition between the spacecraft and booster, as illustrated by the Delta, may not be the most effective in the STS era. A more appropriate spacecraft interface definition may be between the Delta second and third stage or, in the STS case, between

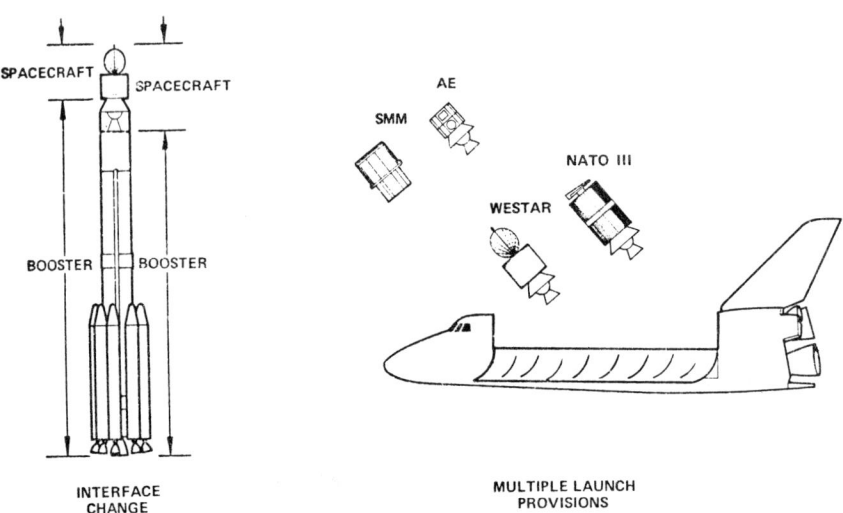

Fig. 2 U.S. Government Considerations For Cost Competitive STS

the orbiter and whatever is required after orbiter separation. The second issue, and perhaps the most difficult for the Government, is the multiple launch provisions. Given that the new interface definition is accepted, low orbit satellite launches, high orbit launches, and even spacelab type missions could be combined in a single orbiter flight. Most satellite system operators prefer launching their satellites one at a time, and this means the Government should strive to allocate orbiter bay space, weight, and power; establish standard mechanical and electrical attachments; schedule and plan grouping of payloads; and price the launch service equitably.

In summary, the technical problems connected with multiple launches using STS appear solvable using a low cost spinning solid perigee stage. The Government must solve the operational scheduling and grouping problems of multiple launches in order to have a cost competitive STS.

AAS 75-292

AGENA INTERIM UPPER STAGE

J. H. Guill*

The Space Shuttle is a very ambitious project, challenging the historical utilization and exploitation of space. The development of space technology over the past 20 years has fostered the proliferation of spacecraft concepts and a concordant proliferation of expendable launch vehicles to carry these spacecraft into space. The Space Shuttle, through application of advanced technologies and operational concepts, will eliminate this entire launch vehicle stable, forcing all users into one system — the Space Shuttle.

For those missions which require propulsive energy above the capability of the Space Shuttle, an upper stage is needed. The needs of transitioning from expendable launch vehicles into the Shuttle indicates the desirability of an existing upper stage to fill this role. The Agena has been selected by the USAF as a candidate for this Interim Upper Stage (IUS) assignment.

The Agena is unique in the breadth and scope of its space accomplishments. It has flown over three hundred and thirty Air Force and NASA missions. It has completed 52 percent of all successful United States space missions. The record of this very successful and versatile vehicle covers a span of applications including a multitude of different military and scientific earth-orbital, lunar, planetary, and manned programs. Continuing development has constantly improved the Agena performance capability, reliability, and applicability to anticipate and meet the needs of a rapidly changing national and international space program.

The basic Agena was analyzed in depth to determine the minimum changes necessary to meet IUS interfaces, safety, and performance requirements, and was evaluated against the mission models provided during the course of the Agena IUS study. The Agena can exceed all defined IUS requirements, including the highly critical transition from today's boosters and payloads to the Shuttle environment and its payloads.

---

*Manager — Shuttle Upper Stage Programs, Lockheed Missiles & Space Company, Inc., Sunnyvale, California

BACKGROUND

The President announced in January 1972 that the National Aeronautics and Space Administration (NASA) would develop the Space Shuttle with the goal of providing routine reliable transportation to and from space. This announcement established the policy that the Space Shuttle should eventually replace all of the current expendable launch vehicles. This national goal, as expressed by the President in 1972, contained as an integral part the requirement that DoD explore the Shuttle system to determine how its use could lead to better and more effective utilization of space for military operations.

Because the Space Shuttle will, under the stated policy, replace the current expendable launch vehicles, the Air Force was required to ensure that critical national defense missions could be supported in an effective manner during transition to the Space Shuttle and after the advent of routine Space Shuttle utilization. These national defense requirements included the need to support polar-orbit operations, and specific missions which required an upper stage to achieve orbits beyond the capability of the Space Shuttle only.

To accommodate high energy missions beyond the basic capability of the Space Shuttle, DoD and NASA achieved a working arrangement in which the Space Shuttle upper stage augmentation could be achieved in two steps. Initially, the upper stage capability could be achieved by selection of an interim upper stage. The responsibility for the management of the IUS program was allocated to the USAF. The second phase of the Space Shuttle upper stage augmentation program would be achieved by the Space Tug, which would be defined by NASA.

OBJECTIVE

The Air Force Systems Command has been assigned as the responsible command for the DoD Space Shuttle program. The development of an IUS has been included within this responsibility. The objective of the IUS program is to develop a minimum-RDT&E-cost, minimum-risk IUS suitable for transitioning payloads into the Space Shuttle during the years 1980 to 1984. This IUS would be used for deploying NASA and other non-DoD payloads during the period prior to the operational availability of the Space Tug. The IUS has been defined as a modification of an existing upper stage. The Agena, whose history and assignments are summarized in Fig. 1, has been selected by the USAF as a candidate for the IUS assignments.

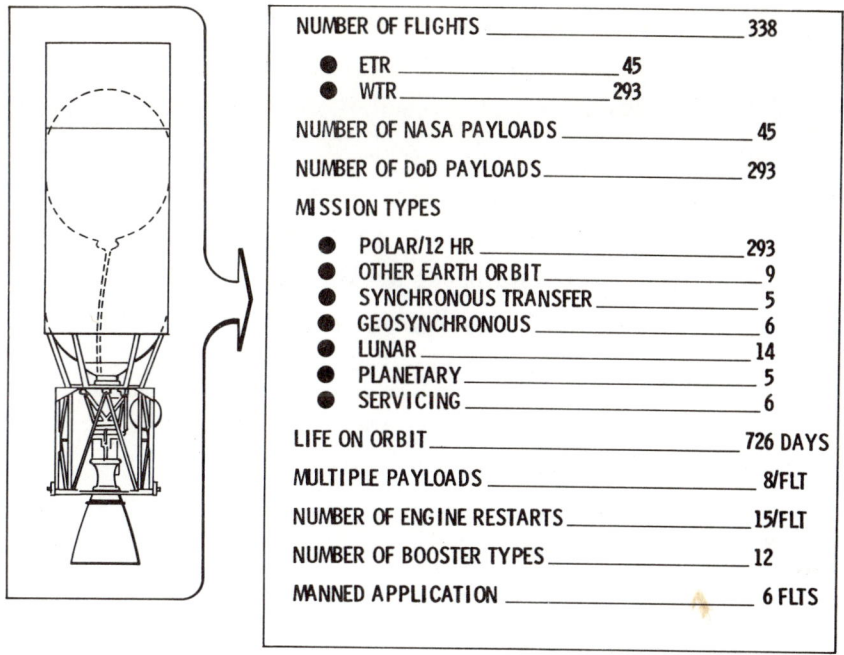

Fig. 1 The History and Assignments of the Agena

SCOPE

The purpose of the 9-month Agena IUS contract was to conduct a system study of the Agena upper stage as a candidate for the IUS assignments. The scope of the study was to provide SAMSO with preliminary designs of four Agena concepts: (1) an expendable Agena IUS system; (2) a short-length expendable Agena IUS system; (3) a reusable Agena IUS system; and (4) a short-length reusable Agena IUS system.

Since the Space Shuttle is fixed in design and operational concept, it will generate standardization – standard performance to Earth orbit, standard environments, standard interfaces, standard procedures, and standard services. This will mean that spacecraft users must tailor their spacecraft designs to meet these standards. The only flexibility available to these spacecraft users will be that limited flexibility which is included within the standardized envelopes.

Since the Space Shuttle will deliver spacecraft into Earth orbit only, any additional propulsive systems to other orbits must be included along with the respective spacecraft.

PHILOSOPHICAL APPROACH

The philosophy of the Shuttle system challenges the historical utilization and exploitation of space. The development of space technology over the past 20 years has required the users of space to specialize. There is now a multitude of users, whose spacecraft vary widely in concept, design, size, function, and position in space. Launch vehicles were developed to match this proliferation of users. Yet at the very time when spacecraft vary widely in size, function, and placement, the nation is in the process of eliminating the entire launch vehicle stable, small, medium, and large boosters with their multifold upper stages, and forcing all users into one launch system — the Space Shuttle.

How will spacecraft users who need to place their spacecraft at positions in space other than the Shuttle orbit provide propulsive capability? One answer is to tailor a propulsive system to each individual spacecraft. Small spacecraft placed in low energy orbits could be equipped with small propulsive devices. Small spacecraft with high energy orbit requirements would have a higher energy propulsive device to carry it to that orbit. Heavy spacecraft which might be placed in high energy orbits would have still greater propulsive requirements. The constant tendency of spacecraft designers has been to increase the size and weight of spacecraft (e.g., see Fig. 2 which catalogs the growth of communication satellites). One might conclude that each spacecraft user, (if he should require an orbit or trajectory greater than that which could be provided by the Space Shuttle) should provide his own propulsion system. If this should occur, then the Space Shuttle might replace all the existing boosters, but in turn would foster a new proliferation of upper stages, each designed and adapted to specific spacecraft and specific energy requirements.

Because this concept would be counter-productive to the entire Space Shuttle concept, an alternative propulsion system concept is to design a single upper stage which would perform all the projected space missions, regardless of spacecraft size or orbit placement. However, such an upper stage would necessarily be designed for the most difficult mission, i.e., it would need to place the largest spacecraft into the most highly energetic orbits or trajectories, and would be an overkill for all the remaining spacecraft missions. This upper stage would be costly to build and costly to operate, thus imposing excess charges on smaller spacecraft users and consequently defeating the low cost concept of the Space Shuttle system. This concept is represented by the Space Tug, which has now been postponed into the 1980s.

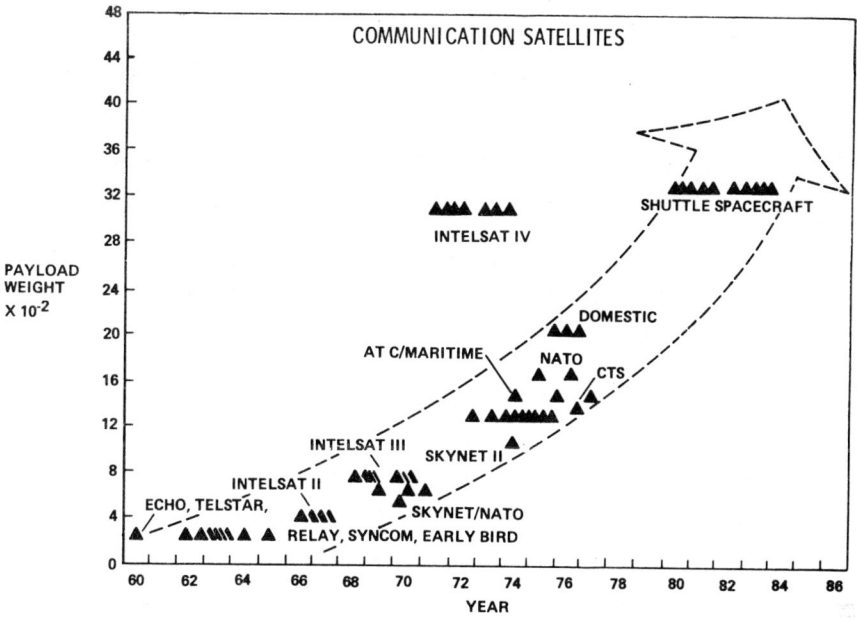

Fig. 2 Spacecraft Weight Growth for Communication Satellite

An alternative upper stage concept, perhaps more compatible with the varying sizes and requirements of the multifold spacecraft which will be delivered into Earth orbit by the Space Shuttle, is a modular upper stage. In this concept propulsive modules would be assigned to varying size spacecraft and propulsive energy requirements. This type of modular upper stage could be standardized, in a manner similar to the Space Shuttle, but the propulsive modules and their respective costs could be roughly tailored to variable spacecraft requirements (Fig. 3). The modular upper stage is a compromise between the tailoring of propulsive systems to each spacecraft and the design of a single upper stage which could perform the most energetic mission.

A modular upper stage could consist of all-solid modules or all-liquid modules. But there are disadvantages in each of these concepts. The all-solid modular system is attractive because of the relatively lower cost of developing and producing solid motors. But to maintain this low cost feature, solid motors must be simple in concept. This, in turn, requires relatively high thrust in motors, or larger size, which impose high g loads on small spacecraft near the termination of motor burn. Due to tail-off and/or thrust termination inaccuracies, solid motors are relatively inaccurate in positioning spacecraft into highly precise

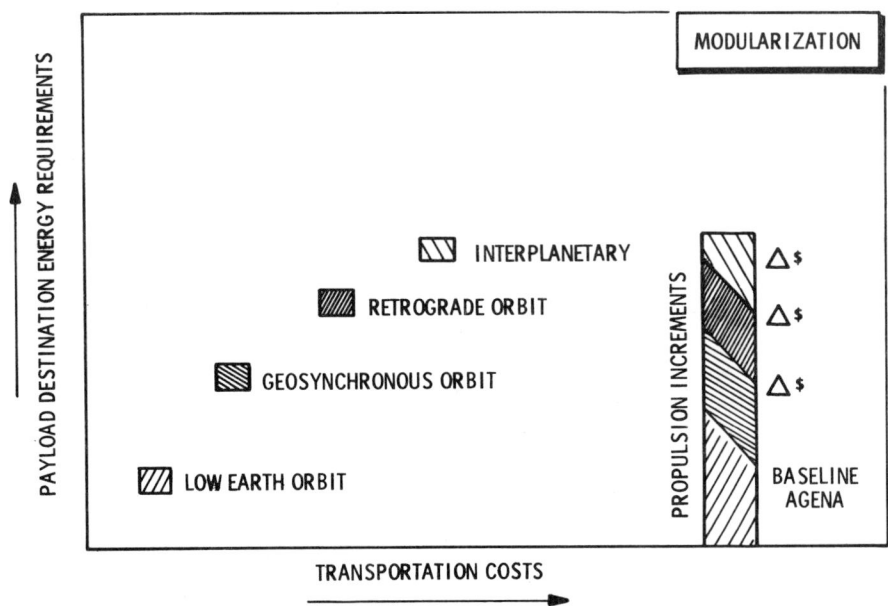

Fig. 3 Modular Upper Stage User Cost Consideration

positions. Solid motors are point design systems, loaded well in advance of application, and are not flexible if the spacecraft requirements change or changeout to alternative spacecraft is required. These and other less important disadvantages such as difficulty of restart capability and low specific impulse mitigate against an all-solid modular stage.

The higher cost liquid system, on the other hand, provides greater flexibility in propellant loading, in restart, in providing lower thrust levels, and in more accurate thrust vector control.

By combining solid and liquids in a modular concept, the best of both concepts may be obtained, or alternatively, the worst of both systems may be eliminated. If one places the avionics in a small upper liquid stage, then the solid motor elements become simply that — solid motors. The liquid stage can perform low weight and low energy spacecraft placements without the use of solid motors. As the mission energy requirements increase, solid modules may be added at the bottom to provide added thrust (Fig. 4). If even more energy is required, other solid modules may be added. This hybrid modular concept makes best use of the characteristics of higher thrust, less accurate solid motors and best use of lower

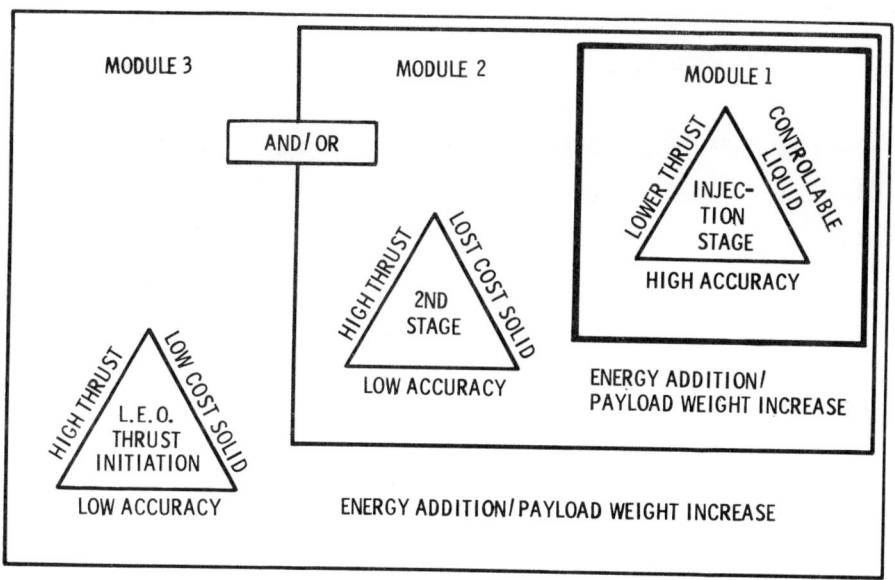

Fig. 4 Hybrid Modularization Concept

thrust, highly accurate liquid systems. Furthermore, scaled costs can be assigned for the delivery of various types of spacecraft to various orbits or trajectories. It is important, however, that the accurate liquid system always remain at the top of the stack to remove the inherent inaccuracies and lack of flexibility of the solid motors.

THE AGENA IUS CONCEPT

The Expendable Agena IUS developed by LMSC during the course of the Agena IUS systems study used the above concept as a model. However, two key additional factors also influenced the Expendable Agena IUS concept. The first was the groundrule that the IUS would be a modification of an existing upper stage. It is interesting to note that the Agena is about the optimum size to match this requirement. The maximum use of existing hardware was specified by SAMSO. Thus, the Agena in its current form was selected as the liquid upper stage. The Minuteman III third stage motor was selected as the solid motor propulsive module. This motor was selected since it met the required geometry, thrust level, thrust vector control, reliability, and proven performance.

There was another key factor inherent in the Expendable Agena IUS concept selection. This was transition. Transition is the process of reducing by consolidation the number and types of expendable launch vehicles in the pre-Shuttle era and providing expendable launch vehicle backup during the early Shuttle years. This backup capability would be maintained until the Shuttle has successfully passed through the validation and certification phases.

It is the purpose of the remainder of this paper to address the Agena IUS configuration selected, emphasizing the influence of the hybrid modular upper stage philosophy and emphasizing the requirements of the transition from expendable launch vehicles into the Shuttle era.

A current Agena upper stage can serve as baseline for adaptation to an Agena IUS configuration. This baseline Agena, which is compatible with Thorad, Atlas, and Titan III boosters, is modified slightly for compatibility with the Shuttle. The Agena IUS configuration is shown in Fig. 5. The basic configuration can deliver approximately 70 percent of the missions in the latest DoD/NASA/nonNASA mission models. With the addition of Minuteman III third stage motors mounted on the aft rack, 100 percent of the entire DoD and NASA mission models can be accomplished.

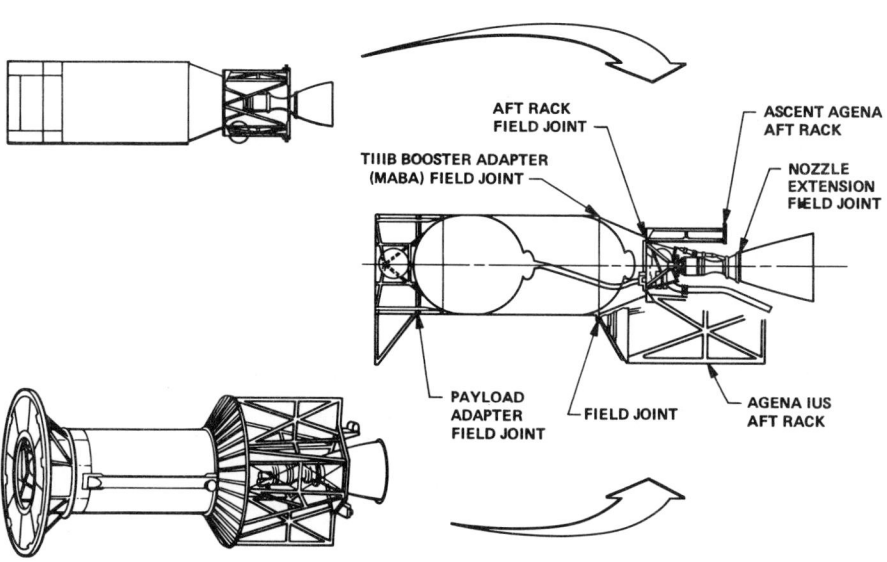

Fig. 5 The Agena IUS configuration is a minor modification of the transition Agena and can be adapted to either application in the field.

The Agena as described serves as the liquid upper stage of the solid/liquid hybrid modular stage discussed above. In applications where added energy is required, solid Minuteman III third stage motors are attached to the Agena aft rack. Either two or four motors can be added, depending on the payload weight and the trajectory requirements. This configuration also meets all Orbiter requirements — loads, center of gravity, length, safety, interface compatibility, and deployment. For example, the Orbiter design is such that heavy payloads must be placed aft in the cargo bay to meet structural load requirements (Fig. 6). Therefore, the Agena IUS concept, where weight is concentrated in the aft position in the cargo bay, meets these requirements precisely. The Agena IUS concept meets all Orbiter deployment requirements, including potential single-point failure of the Remote Manipulation System (Figs. 7, 8, and 9).

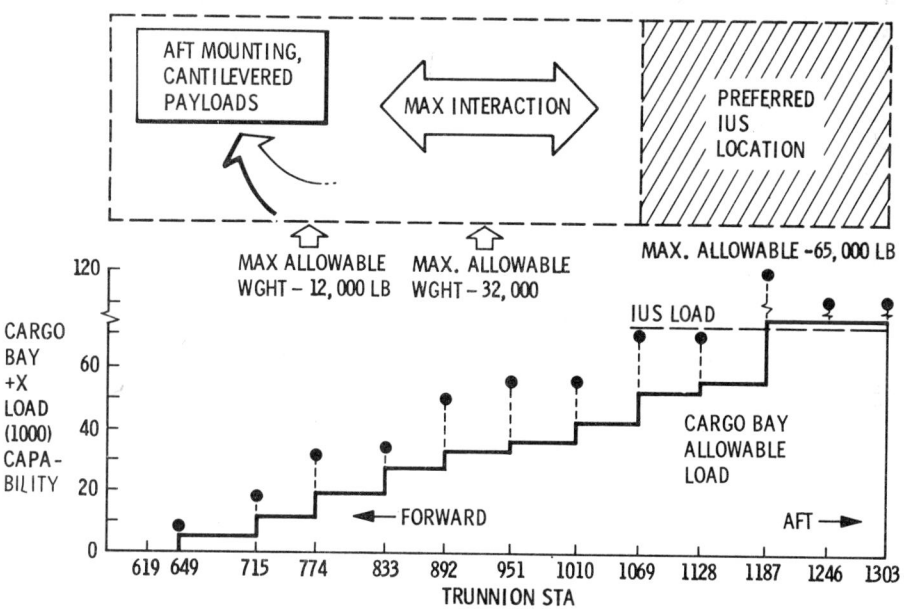

Fig. 6 The design of the Shuttle requires that the IUS and other heavy payloads be mounted in the aft section of the cargo compartment

Although the Agena IUS meets all Space Shuttle and mission requirements, it also meets all transition requirements. These will be discussed in detail below.

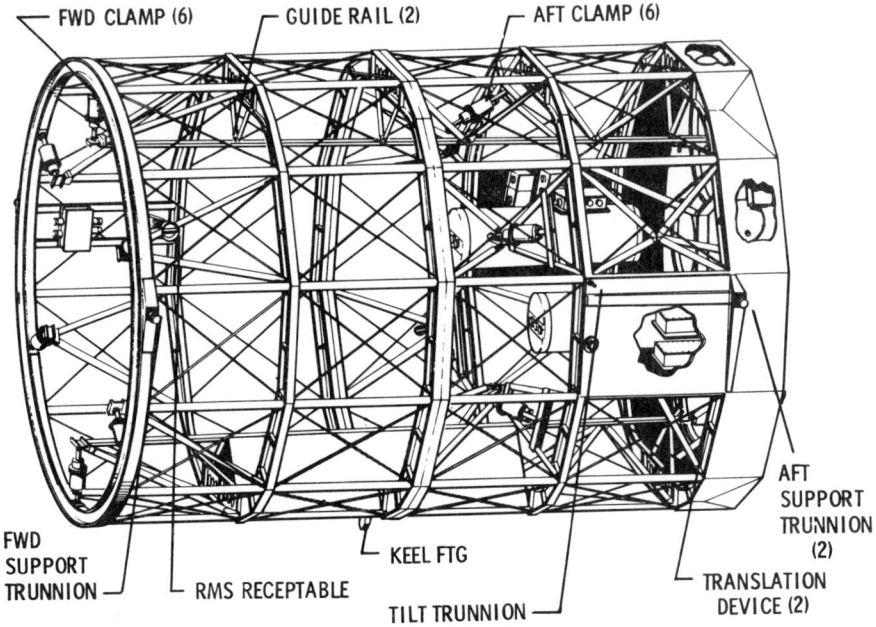

Fig. 7 Agena IUS Pallet Concept Equipped With Backup Deployment System

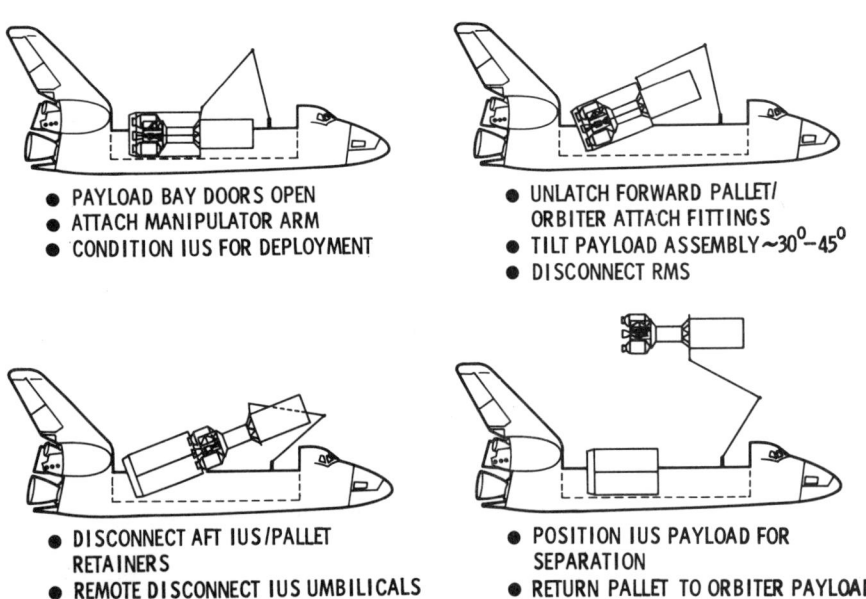

Fig. 8 Agena IUS Normal Deployment Sequences

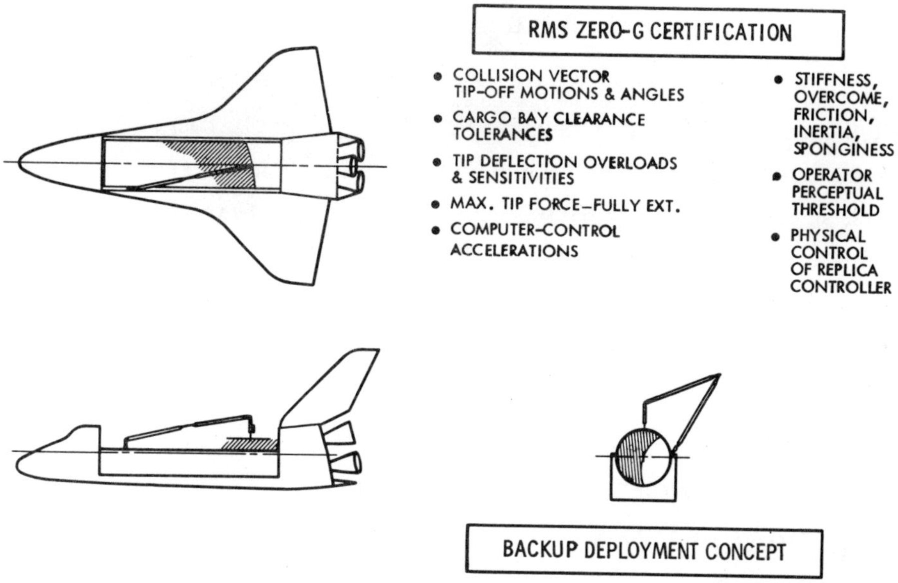

Fig. 9 Certification of key components of the Shuttle is planned. The RMS is the key to deployment; during its early certification period a backup deployment technique must be available

TRANSITION

The current space program began with a relatively small number of launch vehicles and a few unsophisticated satellites. But as the knowledge of the potential uses of space broadened, the number of satellites and the number of users of space-gathered data increased. Concurrently, there was a proliferation of launch vehicles and launch complexes located both at the Eastern Test Range and the Western Test Range. Many of these launch systems competed in launch capability with others, thus driving down the number of launches using any one specific launch vehicle and, at the same time, driving up the cost of each launch vehicle as the number of launches became fewer. Concurrent with this, the sophistication of satellite technology was greatly improved, thus permitting a satellite to remain on orbit for a much longer time than was achievable during the early years of the space program. This, in turn, reduced the number of launches necessary to obtain the same amount of data that was previously obtained. Thus, by 1975, the nation had a launch stable much larger than was required; at the same time the costs for retaining this total launch stable and their related launch complexes have risen to burdensome levels.

The Space Shuttle is intended to reverse this trend. The Space Shuttle is intended to replace the entire existing expendable launch stable and to reduce launch costs through reduction of the number of launch vehicles, launch complexes, and launch crews and also through the ability to reuse expensive hardware. But the transition from the existing expendable launch vehicles to the Shuttle cannot be done abruptly. Existing spacecraft must continue to fly; new spacecraft must be designed for the Shuttle but be able to fly on backup expendable launch vehicles in case of technical, operational, or schedule delay of the Shuttle (Fig. 10).

Because this transition from the existing expendable launch vehicles to the Shuttle cannot be performed abruptly — immediately accruing the cost savings promised by the Shuttle — it would appear wise to consolidate the current proliferation of launch vehicles both to effect savings and to lead efficiently into Shuttle transition. In addition, the Shuttle may not be able to fly all existing and new payloads immediately. The Shuttle incorporates much new technology and a high degree of new operational techniques. It may, consequently, be subject to technical, operational, or schedule perturbations to a greater degree than are expendable vehicles. Thus, there must be a backup capability of the consolidated expendable launch vehicle stable during the early years of Shuttle flight validation. Because Shuttle payloads will inherently be heavier than expendable launch vehicle payloads, it will be necessary for backup launch vehicles to match these increased weight requirements, even if the added weight is only nominal. Furthermore, the backup capability will probably have to be available for several years to coincide with the full validation and certification of the Shuttle system.

It is, of course, highly desirable that the IUS retain compatibility with expendable launch vehicles during the initial phase of the Shuttle era. Compatibility requires minimum changes to convert the IUS into an upper stage to fly on an expendable launch vehicle such as the Thorad or Titan. The Agena IUS design, based on the existing Agena, is already compatible with the Thorad and Titan boosters. In fact, the Agena has been used as an upper stage on the Thor and Titan boosters more than any other upper stage. Consequently, the Agena IUS design lends itself to easy field conversion to a transition Agena that is compatible with the existing expendable launch vehicles.

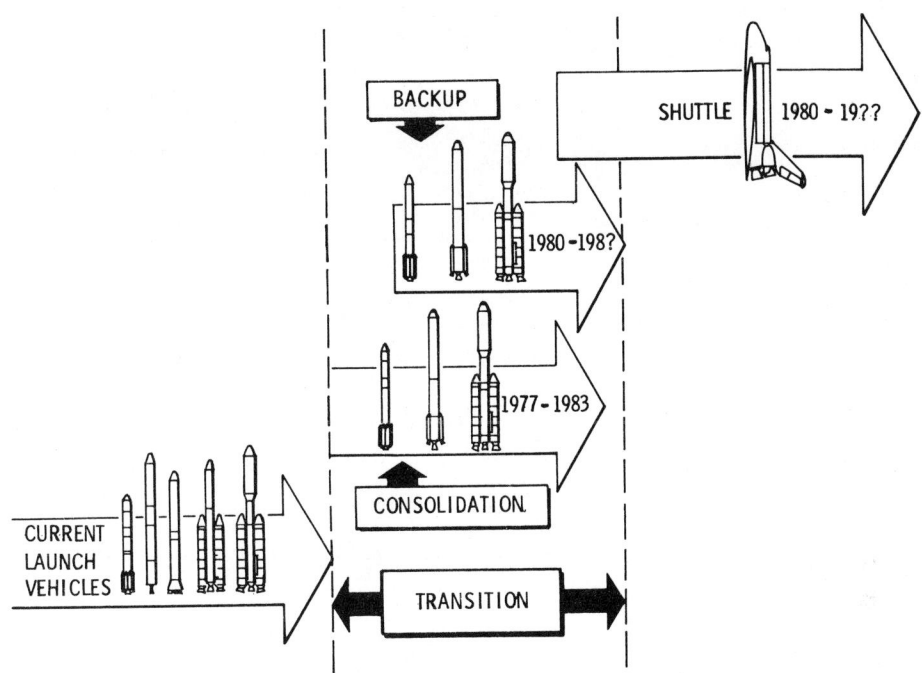

Fig. 10 The transition launch vehicles can be divided into two categories: (1) consolidated launch stable; and (2) backup launch stable. It would be optimum if these two could contain the same launch vehicles

The compatibility of the selected IUS with the Thor, Titan IIIB, and Titan IIID should be examined. The Agena is compatible with all versions of these vehicles. The Agena has flown over 150 flights on different versions of the Thor booster. The Agena can be directly substituted for the Delta second stage as shown in Fig. 11. Consequently, the Agena can be directly used on the current 29XX or 39XX versions of the Thor/Delta booster. The Agena has had over 100 assignments on the Atlas booster. More significantly perhaps, the Agena has also flown extensively on the Titan III booster. Numerous studies have been performed relating to the addition of the existing Titan IIIB/Agena launch vehicle to Complex 40 at ETR. These studies have been recently updated and the transfer implications are summarized in Fig. 12.

The Agena is compatible with Titan IIID. Its performance is compared with the Transtage, Delta, and Centaur in Fig. 13.

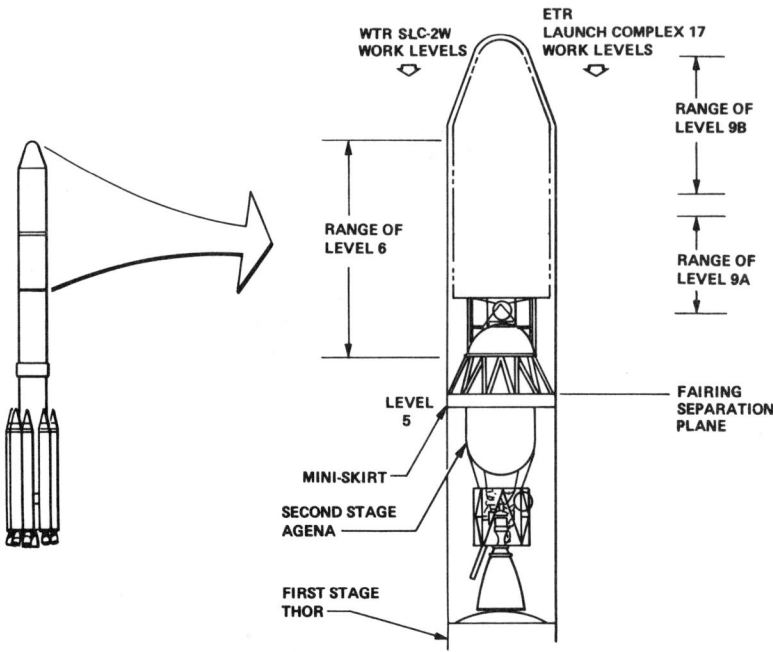

Fig. 11 The Thor/Agena has flown over 150 times. The Agena is completely compatible with the 29XX series and the 39XX series

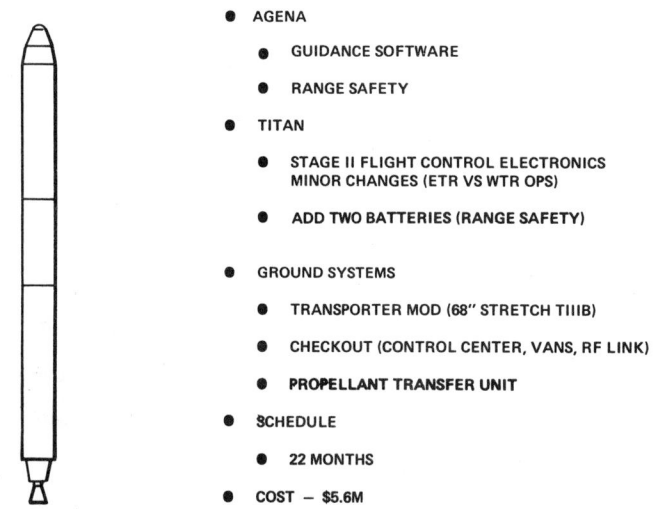

Fig. 12 The Titan IIIB/Agena exists at WTR. The addition of this existing launch vehicle at ETR is simple.

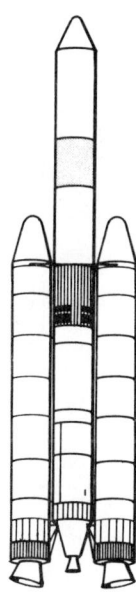

UPPER STAGE	INERT WT (LB)	PROP. WT. (LB)	I_{SP} (SEC)	SYN EQ PERF. (LB)*
TRANSTAGE	3,542	23,472	302	3,230
DELTA	1,666	10,292	306	3,283
AGENA	1,250	13,950	300	4,450
CENTAUR	4,480	30,760	444	7,400

*PAYLOAD CAPABILITY TO GEOSYNCHRONOUS ORBIT.

Fig. 13 The Titan III can be equipped with various upper stages. A performance comparison of the Titan IIIC/E with the Transtage, Delta, Agena or Centaur as the upper stage is shown.

The consolidation of expendable launch vehicle should be based on an assessment of their mission capture potential as well as cost. This has been done for published DoD and NASA mission models (Fig. 14). No consideration has been made in this analysis for multiple payload opportunities, because the recent flight failures involving multiple payloads mitigates against this approach unless absolutely necessary. It has been assumed that one payload only is flown on each launch. Weight allowance has been included for the spacecraft adapter for mounting the payload onto the upper stage. Allowance has been made for aerodynamic shrouds. The combined total of missions in the published models is 150, of which 74 are DoD and 76 are NASA.

The expendable launch vehicles have been grouped as a function of their energy level capability and their cost: low, medium, and high. Analysis shows that the 39XX Agena (Thor/Agena) configuration can capture both the low and medium energy NASA missions and can capture almost twice as many DoD missions as can the Delta. The TIIIB/Agena configuration, with the Algol option, outperforms the Atlas/Centaur configuration and is substantially lower in pre-flight

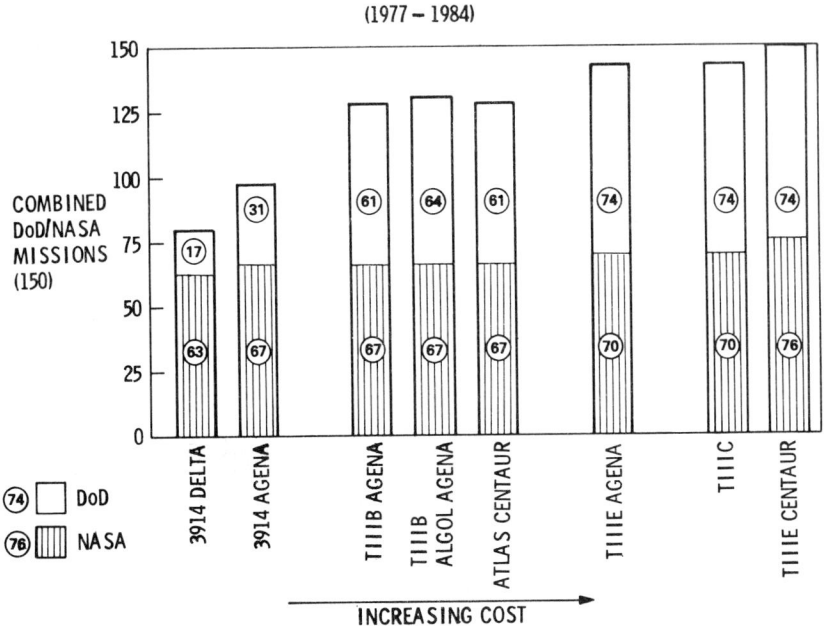

Fig. 14 Capture analysis of pre-Shuttle missions (those designed specifically for expendable launch vehicles).

cost. The highest energy interplanetary missions must clearly be performed by the TIIIE/Centaur, but not by the Titan IIIC or the Titan IIID/Agena, unless the Agena is uprated. Uprating the Transtage is of no benefit because its inert weight offsets performance improvements (Fig. 15).

It is apparent from this mission capture analysis that the optimum choice based on performance is clearly the 39XX Thor/Agena for low energy missions, TIIIB/Agena with the Algol option for medium energy missions, and the TIIIE/Centaur for the high energy missions.

Because the Space Shuttle is intended to replace the current expendable launch vehicles, it is required that critical missions be supported in an effective manner both during the transition era and after the advent of routine Space Shuttle operations. It follows that the consolidation of the expendable launch vehicles and the selection of the IUS vehicle must be accomplished in a coordinated manner. In our analyses of the Agena as an IUS and as a transition upper stage on expendable launch vehicles, we have considered changeout to be a very important factor.

	W_{INERT} (LB)	W_{PROP} (LB)	I_{SP} (SEC)	PAYLOAD CAPABILITY (LB)
TITAN IIIC/ IMPROVED TRANSTAGE	3,542	23,472	312	3,510
TITAN IIID/ IMPROVED AGENA	1,297	15,302	311	5,020

Fig. 15 A comparison of the performance of the Titan IIIC with uprated Transtage and a Titan IIID with uprated Agena

Our Agena IUS design is suitable to be field-modified, replacing a minimum number of components at field joints and thus allowing the Agena IUS to become a transition Agena to be flown as an upper stage on expendable launch vehicles. This concept permits the production of a minimum number of IUS and transition stages, since they are freely interchangeable. Thus, the Agena IUS reduces the total cost and the risk during the transition era.

The mission model used in the backup assessment has been based on the DoD and NASA mission model that was furnished by the Government for use on the Agena IUS system study. No consideration has been made for multiple payload opportunities on expendable launch vehicles for the reasons given above. Instead, it was assumed that one payload only would be flown on each launch. Weight allowance has been included for the payload adapter for mounting the payload onto the upper stage. Allowance has been made for an aerodynamic shroud.

The expendable launch vehicles were grouped as a function of energy level capability and cost, and, as previously stated, analysis shows that the 39XX Agena (Thor-Agena) mission capture greatly exceeds that of the Delta (Fig. 16). In fact, the Delta can capture only a small percentage of the Shuttle-era low energy missions, indicating a lack of potential on the part of the Delta for application as a Shuttle-era backup vehicle. The TIIIB/Agena with the Algol option clearly outperforms the Atlas/Centaur and essentially equals the capture capability of the more costly TIIIC. Once again the most energetic missions can only be captured by the Titan IIIE/Centaur; some projected interplanetary missions cannot be captured even by this vehicle. It is clear from this capture analysis that the optimum choice for backup applications during the Shuttle era are the same three launch vehicles that were found to be most attractive for consolidation. This choice is the 39XX Thor/Agena, TIIIB/Agena with the Algol option, and the TIIIE/Centaur.

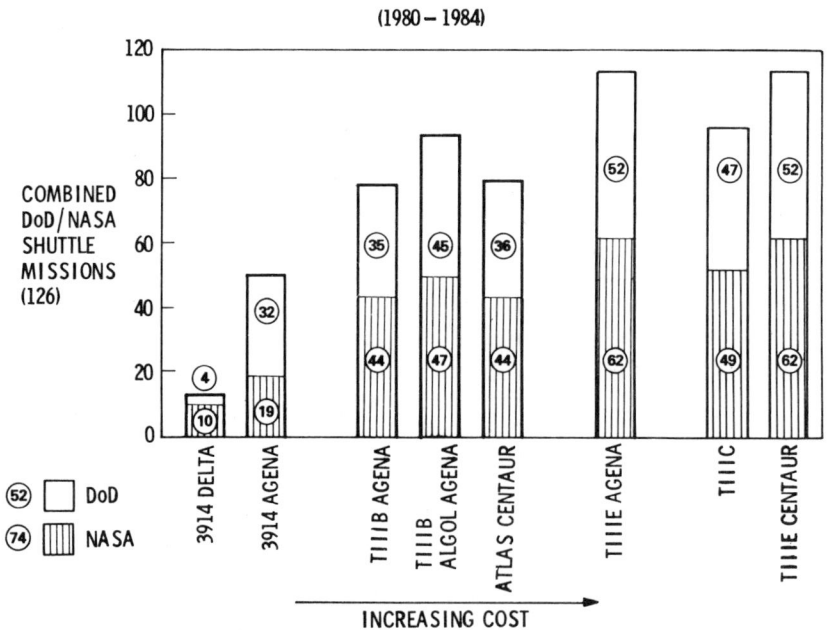

Fig. 16 Capture analysis of early Shuttle missions when flown on backup expendable launch vehicles.

One other factor of importance emerges from the analysis of the introduction of the Space Shuttle into the nation's space program. The Shuttle will become operational at ETR from 2 to 4 years before it will become operational at WTR. There will be an overlap, consequently, between the need for backup expendable launch vehicles at ETR and the normal utilization of expendable launch vehicles at WTR. This permits a planned utilization of expendable launch vehicle hardware purchased as backup vehicles for ETR. By purchasing a few low energy launch vehicles (Thor/Agena) a few medium energy vehicles (Titan IIIB/Agena with Algol option), and a few high energy vehicles (Titan IIIE/Centaur) for backup to the Space Shuttle at ETR, these vehicles would be immediately available for payload changeout and launch in case of Shuttle delay. If there is no Shuttle delay, then these vehicles could be transferred to WTR and launched as normal expendable launch vehicles prior to the introduction of the Space Shuttle at WTR. This plan, which calls for an essentially identical consolidated launch vehicle family at ETR and WTR, would minimize the excess hardware remaining when the Space Shuttle becomes fully operational at ETR and WTR.

The recommended launch vehicle stable at ETR and at WTR during the transition phase is shown in Figs. 17 and 18.

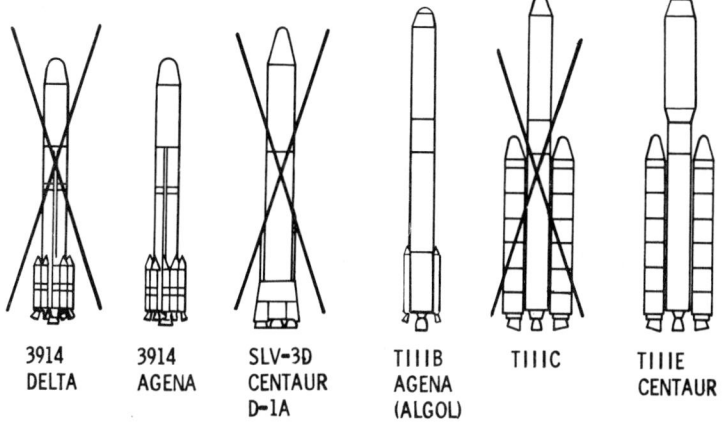

Fig. 17 Recommended Expendable Launch Vehicle Consolidation at ETR

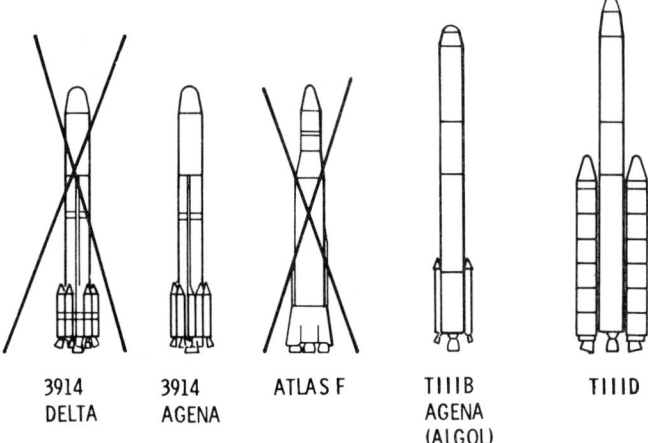

Fig. 18 Recommended Expendable Launch Vehicle Consolidation at WTR

Consolidation of the expendable launch vehicles and launch complexes at both WTR and ETR results in a savings to the Government of more than $700 million. This savings must be off-set by cost additions resulting from integration of the Agena into the ITL and into SLC-2 at WTR and LC 17 at ETR. The net savings to the Government is still at least $700 million.

The major source for cost savings comes from the phase out and deactivation of surplus launch complexes. Complex phase-out results in a reduction in support personnel, maintenance and refurbishment labor, spares, and material. The increase in launch rate at the remaining complexes results in relatively small increases in these areas.

CONCLUSIONS

Assignment of the Agena as the Shuttle IUS and the phased development of both the transition Agena and the Agena IUS system can be accomplished in combination with transition. The combined cost package, if Agena is selected as the IUS and also for transition assignment on expendable launch vehicles, would result in a savings to the Government of over $600 million. This includes full implementation of the transition consolidation and total RDT&E for the Agena IUS. By selecting the Agena, the Government will save money, not be exposed to new funding outlays. At the same time, the Government will be utilizing an upper stage that has high reliability, proven in more than 330 flights, and has great flexibility to meet the demanding requirements of transition to the Shuttle. Consolidation using the Agena imposes on the user and operator communities the identical disciplines that will characterize the Shuttle.

AAS 75-155

RL10 DERIVATIVES FOR IUS/TUG*

J. P. B. Cuffe[+]

THE RL10 ENGINE

The RL10 is an oxygen/hydrogen space engine with the highest performance of any operational rocket engine. One hundred and four RL10 engines have flown on lunar, planetary and earth orbit missions without a single failure or malfunction. On a recent Titan/Centaur flight, the RL10 engines were each fired 4 times, a record for any pump fed rocket engine. Over 9000 firings have been made on RL10 engines since it was first run in 1959. The engine is currently in production for the Centaur upper stage. There are missions planned for this stage until the shuttle is operational. The RL10 will remain in production through this decade to supply the engines for these vehicles.

THE SPACE SHUTTLE UPPER STAGES

The Space Shuttle, as an element of the Space Transportation Systems, is designed to operate from earth to low (160 miles) orbit. At least half of the planned Space Shuttle missions require an additional stage to deploy and retrieve payloads in higher orbits (for example, 22,000 mile geosynchronous orbit). Primarily because of funding restraints, it is not presently planned to develop a new upper stage concurrently with the Space Shuttle. Instead, a phased approach will probably be followed, where a modified existing vehicle will be used as an interim upper stage (IUS planned to be developed by the Air Force) for at least the first three years of Shuttle operation, when it will be replaced by a new full capability cryogenic Space Tug, which will be the responsibility of NASA. However, an IUS with growth capability which could be evolved into a vehicle of Space Tug performance would result in considerable savings in development risk and cost.

---

[*] The work described in this paper was performed under NASA Marshall Space Flight Center contracts: NASA-28989, NAS8-31151, and NAS8-31008.
[+] Pratt & Whitney Aircraft, Florida Research and Development Center, West Palm Beach, Florida

INTERIM UPPER STAGE

Candidates for this stage are modified versions of the existing storable Agena, Delta and Transtage and cryogenic Centaur vehicles, together with a new multi-stage solid vehicle. These expendable vehicles, which have completed their concept definition studies under contract to AFSC/SAMSO, can all be developed for less than $100 million and are only required to deploy payloads. The cryogenic Centaur is the only candidate which is practical to use as a reusable vehicle. This Reusable Centaur IUS, which can still be developed for less than $100 million, would be expected to have lower total program costs than the various expendable vehicles, particularly if the period for IUS operation was greater than three years. The transition to the Space Tug would also be easier, since there would be no change in propellants.

SPACE TUG

This is a new high performance, reusable vehicle, able to deploy and retrieve payloads from geosynchronous orbit. The performance and reuse requirements make oxygen and hydrogen the only practical propellant choice. By using an engine derived from the RL10, rather than a new advanced engine, the cost of developing this vehicle has been greatly reduced.

GROWTH INTERIM UPPER STAGE

The Centaur IUS may be "grown" into a vehicle with the performance capability of the Space Tug, including the ability to retrieve payloads. The required performance increase is obtained by replacing the two RL10's presently used with a single RL10 derivative engine.

RL10 FOR CENTAUR IUS

The production RL10 engine would be used in the expendable Centaur IUS with no change. This engine has already demonstrated a life in excess of that required for ten Reusable Centaur missions. The long life capability of this engine is due to a combination of factors, including the use of noncorrosive propellants, conservative design margins, development maturity and its expander power cycle. Only trim adjustments to the engine's mixture ratio setting and cooldown sequence are needed for the engine to be ready for the Reusable Centaur IUS. With RL10 production and product support continuing through this decade, the qualification and production programs for Reusable Centaur engines can be phased in without any interruptions.

RL10 DERIVED ENGINES FOR SPACE TUG AND GROWTH CENTAUR IUS

The specific impulse of the RL10 can be dramatically improved by using a large expansion ratio nozzle. By making this nozzle retract over the engine, there is no increase in length when the tug is installed in the Space Shuttle's payload bay. By modifying the engine so that it can run at low thrust (pressure fed without its pumps rotating) and intermediate (maneuver) thrust as well as at full thrust, the efficiency and flexibility of Tug operation is improved. An engine of this type, the RL10 Derivative II, was selected by McDonnell Douglas and General Dynamics – Convair for their 1973 Space Tug designs and also by NASA MSFC for their 1974 baseline Space Tug. This engine enables the Tug to achieve its performance requirements (retrieve 3500 lb from geosynchronous orbit) with adequate margins: a new advanced engine is not needed under the current groundrules. Should a very short ($\simeq$25 ft) Space Tug be required in order to accommodate long payloads, then the 55 in. RL10 Derivative II engine can be integrated with a toroidal oxidizer tank and safely run at high mixture ratios. If the two RL10 engines in the Reusable Centaur IUS are removed and replaced by a single RL10 Derivative II, the resulting increase in specific impulse, and reduction in inert weight and fluid losses, allows the performance of this vehicle to be increased so that it meets Space Tug requirements.

Should Tug performance requirements be increased so that a new engine would be required, then the expander power cycle is a "natural" for this size of Space engine. The RL10 Category IV engine, which is a new design using the RL10 power cycle, enables Tug payload to be increased by up to 20%, though engine development cost would be doubled. However, such an engine will be much cheaper to develop than a high pressure engine using the SSME staged combustion cycle, with its need for an active closed loop control system and high operating pressures and heat transfer rates.

SIMULATION DEMONSTRATION PROGRAMS

These programs are planned to establish requirements and demonstrate concepts prior to the initiation of the Space Tug development. The objective of one of these programs, presently underway, is to measure the performance of an RL10 with a nozzle extension which increases its area ratio to that of the Derivative II ($\epsilon = 205:1$). Work on another program which will lead to testing a pressure fed

RL10 under Space Tug conditions has recently been initiated. These low funding level programs, together with the maturity of the basic RL10 engine, will ensure that the development and qualification of the RL10 Derivative II engine for the Space Shuttle Upper Stage will proceed as planned.

SUMMARY

(1) The RL10 will continue as an ongoing program through this decade.

(2) The RL10 is immediately available for installation in an expendable Centaur IUS and can be requalified for use in a Reusable Centaur IUS for as little as $1 million.

(3) The RL10 Derivative II engine, with a development cost of $35 to $45 million, is the cost effective selection for the full capability Space Tug.

(4) A short (25 ft) Space Tug which meets NASA performance requirements may be obtained with a toroidal oxidizer tank and RL10 Derivative II engine.

(5) By replacing its two RL10A-3-3 engines with a single RL10 Derivative II, the Reusable Centaur IUS can be evolved into a vehicle which meets Space Tug performance requirements.

(6) A new advanced engine is not presently needed for the Space Tug. Should Tug performance requirements eventually be increased so that a new engine is required, then a new expander cycle engine with a two position nozzle is the best choice.

(7) The simulation demonstration programs which are either underway or planned will lead to smooth development of the Space Shuttle Upper Stage engine.

AAS 75-156

ADVANCED SPACE ENGINE
COMPONENT TECHNOLOGY STATUS

A. T. Zachary[+]

Key to the achievement of an "all up" Space Tug, and
thus realize maximum benefit from the space transportation system, is the availability of a low-cost,
high-performance engine system. To define this engine
system, studies have been conducted under Air Force
and NASA direction, and pertinent hardware technology
programs have been initiated. From among the several
candidate systems studied, the cryogenic, stage-combustion cycle, high chamber pressure system in conjunction with a large-area-ratio nozzle was selected
for component and subsystem demonstration. Main turbopump, high-area-ratio thrust chamber, preburner,
and igniter components are being evaluated by NASA
under contract to Rocketdyne.

INTRODUCTION

The Space Transportation System consists of the Space Shuttle now under development and a final propulsion stage. The final propulsion stage development has been divided into two phases with the near-term system to be selected from existing vehicles to fulfill interim payload requirements and a final "all up" system to achieve maximum benefit from the Space Transportation System.

To define a high-performance engine system suitable for an all up Space Tug application, studies have been conducted under Air Force and NASA

[+]Advanced Space Engine Project Manager, Rockwell International/Rocketdyne Division, Canoga Park, California (Dept. 578, AA69, (213)884-3374).

direction. From among several candidate systems studied, the cryogenic staged combustion cycle, high chamber pressure system in conjunction with a large-area-ratio nozzle was selected as superior in the 8000- to 50,000-pound-thrust range under consideration for the Space Tug[1].

SUMMARY

The Advanced Space Engine effort was initiated with the NASA-LeRC system studied[2] based on a 20,000-pound-thrust and a 400:1 area-ratio nozzle design point. The study conducted at Rocketdyne (Contract NAS3-16751) resulted in selection of an engine system that features compact design, light weight, and high performance. The overall configuration, with the nozzle in the retracted and extended positions, respectively, is shown in Fig. 1.

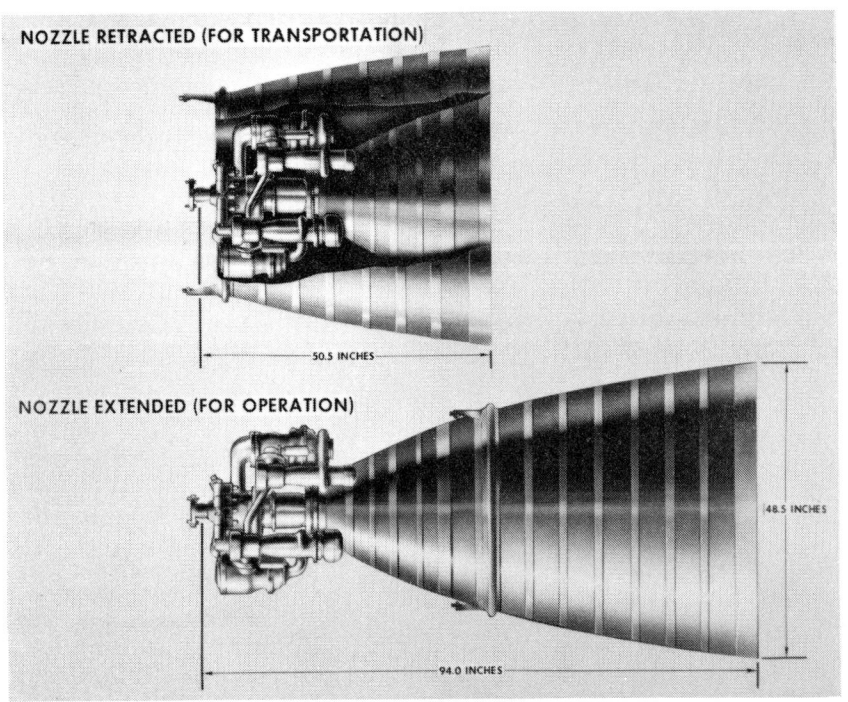

Fig. 1. Advanced Space Engine

Subsequent to completion of the Engine Preliminary Design Program, a series of component and subsystem technology contracts have been initiated by the NASA-LeRC that are directed toward the overall goals of developing small, high-performance component technology; establishing the feasibility of a small, high-performance engine system, and establishing a baseline for predicting Space Tug performance.

The first of these contracts is the Small, High-Pressure Hydrogen Turbopump Program (NAS3-17754) to design, fabricate, and test two turbopump assemblies (Fig. 2) with workhorse preburners to provide the turbine drive hot gas.

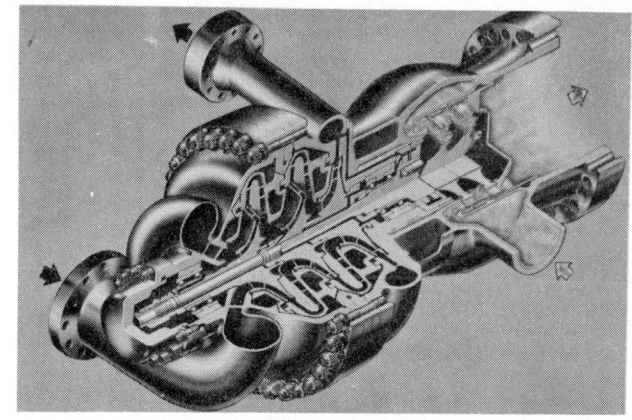

Concurrent with the turbopump effort, a preburner unit was designed, based on the engine preburner design (Fig. 3) prepared under the Engine Preliminary Design Study (NAS3-16751) and fabricated to provide the turbine drive fluid during turbopump test operations. Testing is now in progress.

The main LO_2 turbopump (Fig. 4) is being designed, fabricated, and tested

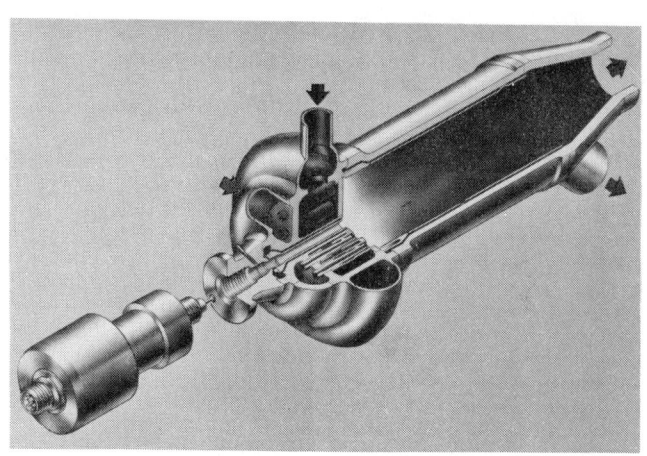

Fig. 3. Preburner Assembly

under Contract (NAS3-17800) to the NASA-LeRC. Two units are being fabricated along with two workhorse preburners similar to those previously described in conjunction with the Hydrogen Turbopump Program.

The advanced Thrust Chamber Technology Program (NAS3-17825) is directed toward design, fabrication, and test of two thrust chamber assemblies. Each assembly consists of an injector, igniter, combustion chamber (ϵ = 8:1), a regeneratively cooled nozzle (ϵ = 8:1 to 175:1), and a dump-cooled nozzle (ϵ of 175:1 to 400:1). The complete thrust chamber assembly is shown in Fig. 5. Testing of the main injector is now in progress (Fig. 6).

In a significant step toward the integration of the components now being fabricated and tested, a program is being initiated by the NASA-LeRC for the design, fabrication, and test of combustion system assemblies. The assemblies will consist of thrust chamber hardware made available from the Advanced Thrust Chamber Program (NAS3-17825) coupled with flight-type preburners to be designed, fabricated, and tested under the new program. The system logically represents the first step toward achieving a full engine system. The integration of the components will be accomplished in a manner that will provide maximum simulation of actual engine conditions and will include the use of primary control valves to facilitate system transient control.

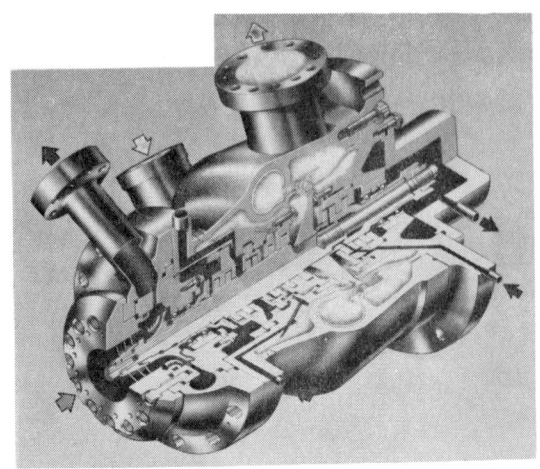

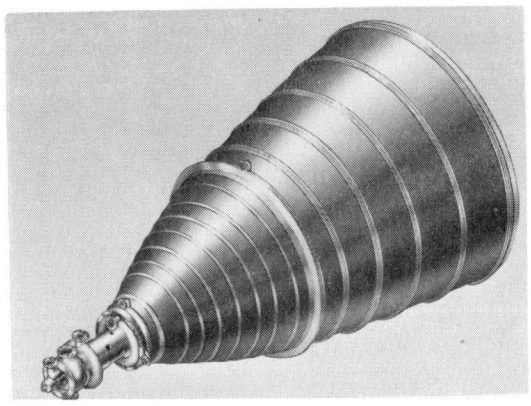

Fig. 5. Thrust Chamber Assembly

CONCLUSIONS

Significant effort in analysis, design, fabrication, and testing of key components of the Advanced Space Engine is currently in progress. The effort and achievements to date strongly indicate that a viable high-performance system can be achieved within the constraints established to ensure meeting the requirements of high-energy, upper-stage missions.

Fig. 6. Main Injector Test

However, the approach pursued will not only provide baseline technology for the Advanced Space Engine as now envisioned, but will also provide a mix of key components that will allow the investigation of alternative concepts. These alternatives encompass not only perturbations within the baseline design, but also the utilization of key components in conjunction with other concurrent experimental efforts (e.g., utilizing the high-pressure turbopumps with the aerospike thrust chamber effort now in process), also directed toward achieving the maximum overall engine effectiveness for space missions.

REFERENCES

1. R-8807, O_2/H_2 Advanced Maneuvering Propulsion Technology Program Engine System Studies Final Report, Vol. II; 25,000-Pound Thrust Bell Engine Configuration Design and Analysis, AFRPL-TR-72-4, Rockwell International/Rocketdyne Division, Canoga Park, California.

2. R-9269, Advanced Space Engine Preliminary Design, NASA Lewis Research Center, Contract NAS3-16751, Rockwell International/Rocketdyne Division, Canoga Park, California.

AAS 75-157

AEROSPIKE DEVELOPMENT STATUS*

J. Campbell[+] and H. G. Diem[++]

The aerospike system has unique design and operating features that make it an attractive concept for many future aerospike vehicle applications. For volume or length limited applications, such as Space Tug, the aerospike is attractive because of its short length; 20 to 25 percent the length of comparable bell nozzle engines.

AEROSPIKE CONCEPT

The aerospike concept is a truncated annular spike nozzle (radial in-flow type), which utilizes a small amount of secondary flow introduced into the nozzle base region. This nozzle has been successfully tested in a number of Air Force-, NASA-, and company-sponsored programs.

The primary flow (high-pressure gases), which produces the major portion of the engine thrust, is exhausted from an annular-type combustion chamber and expands against the metal surface of the center truncated-spike nozzle (Fig. 1). The characteristics of the primary flowfield upstream of the base (shown as region 1 in Fig. 1) are determined by the annular throat geometry, the nozzle wall contour, and the ambient pressure. The annular primary flow continues to expand beyond the nozzle surface and encloses a subsonic, recirculating flowfield in the base region (region 2

[+]Manager, Combustion Processes, Rockwell International/Rocketdyne Division, Canoga Park, California (Dept. 588-194, AA67, (213)884-3315).

[++]Program Manager, Advanced Upper Stage Engine Programs, Rockwell International/Rocketdyne Division, Canoga Park, California (Dept. 578, AA28, (213)884-3050).

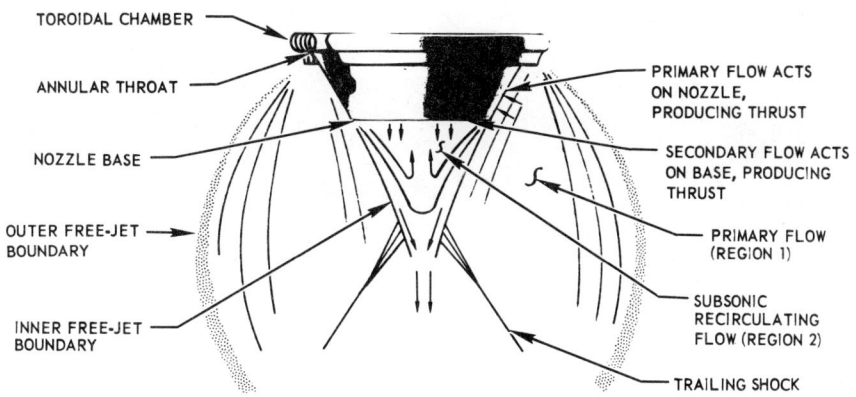

Fig. 1. Aerodynamic Spike Flowfield Illustrated Under Altitude Conditions

of Fig. 1). The pressure acting on the nozzle base contributes additional thrust to the nozzle. When a small amount of secondary flow is introduced into the base (added to the recirculating flow), the base pressure is further increased with a resulting overall I_s increase.

The effect of ambient pressure on the free-jet boundary and the flowfields causes the performance to be near an "ideal" variable-area-ratio nozzle at all P_c/P_a values; this effect is called altitude compensation.

The aerospike concept can be adapted to a wide range of vehicles and configurations. Design flexibility makes possible engine configurations ideally tailored to the vehicle afterbody shape. Typical systems include round configurations for Space Tug to oval or linear configurations for single stage-to-orbit or hypersonic flyback vehicles.

A design feature of the aerospike concept is that it lends itself to segment and modular construction. The approach of module and segment development achieves development cost savings through reduced hardware, smaller test facilities, and reduced test propellant consumption.

AEROSPIKE TESTING

Aerospike designs ranging in thrust from 400 to 250,000 pounds have undergone test firings (Fig. 2). The work has encompassed thrust chamber as well as complete turbopump-fed propulsion system demonstrations. Both storable and cryogenic propellants have been utilized.

The O_2/H_2 Linear Engine program was of a particular significance. Under NASA sponsorship, the feasibility of

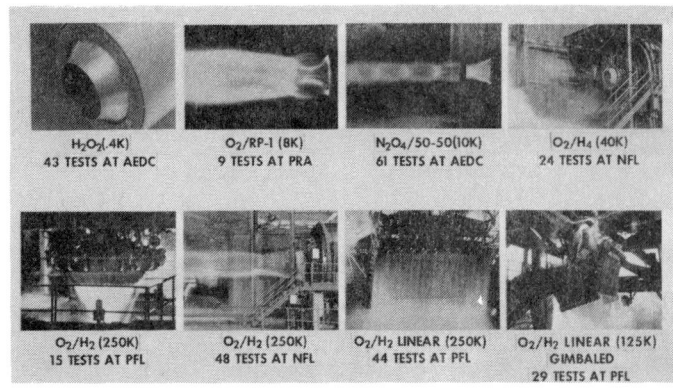

Fig. 2. Hot Firing Aerospike Tests

fabricating and assembling a full-size linear aerospike rocket engine complete with all necessary power systems components and control devices was demonstrated (Fig. 2). The engine was designed to deliver 200,000 pounds of test-site thrust (250,000 pounds vacuum thrust) at a chamber pressure of 1200 psia, an engine mixture ratio of 5.5, and an expansion ratio of 119:1.

Sixty-seven tests were conducted for an accumulated mainstage of over 4000 seconds on two engine assemblies. Three tests with durations in excess of 500 seconds were demonstrated. The overall performance of the linear engines was excellent and good correlation of data for both engines was obtained.

SPACE TUG APPLICATION

During the past several years a program has been in progress under Air Force Rocket Propulsion Laboratory sponsorship, Air Force Contract F04611-67-C-0116, to evaluate a 25,000-pound-thrust lightweight, high-performance, O_2/H_2 aerospike engine system for application to possible high-energy, space maneuvering missions. The thrust chamber assembly uses 24 identical combustor segments retained by two continuous rings. The operating chamber pressure is 1000 psi; the expansion area ratio is 200:1. The nominal operating engine mixture ratio is 5.5:1 with an off-design operational capability of ± 0.5 mixture ratio units. The thrust chamber and nozzle are regeneratively cooled with both hydrogen and oxygen.

Based on the results of segment testing, the design of the complete thrust chamber was finalized and fabrication has now been completed. The completed assembly is shown in Fig. 3. Final measured weight of the chamber assembly was 240 pounds--29 pounds under the contract specification of 269 pounds.

Fig. 3. 25,000-Pound-Thrust Aerospike Thrust Chamber Assembly (Aft View)

Testing of the thrust chamber assembly has been initiated at the Rocketdyne test facilities at the Santa Susana Field Laboratory. Blowdowns and ignition tests were successfully accomplished. Mainstage testing has been interrupted due to propellant leakage from feed lines on the back side of the chamber, which caused fire damage during a test attempt. Repair of the chamber is presently in progress and testing is expected to resume under NASA sponsorship. The planned test program will include tests at simulated altitude conditions using a facility diffuser system.

CONCLUSIONS

The extensive analytical and experimental work conducted on the aerospike concept over the past 15 years provides a broad base of technology for application to a variety of missions and applications. The program currently in progress on the 25,000-pound-thrust O_2/H_2 aerospike engine is providing the ground work for propulsion systems that can be utilized in many applications. The continuation of this program provides engine technology to meet future upper-stage and Space Tug system requirements.

AAS 75-159

SOLID ROCKET TECHNOLOGY
ADVANCEMENTS FOR SPACE
TUG AND IUS
APPLICATIONS[*]

W. Ascher, R. L. Bailey,
J. W. Behm, and W. Gin [+]

Condensation

It presently appears that in order for the Space Shuttle Tug or Interim Upper Stage (IUS) to capture all the missions in the current mission model for the Tug and the IUS, an auxiliary or kick stage, probably using a solid propellant rocket motor, is required. The purpose of this paper is to present and review the two solid propellant rocket motor technology concepts being sponsored and pursued by the NASA Jet Propulsion Laboratory to meet the general requirements of the motor designs applicable to that auxiliary or kick stage. One concept, called the Advanced Propulsion Module (APM) motor, is an 1800-kg, high-mass-fraction motor, which is single-burn and contains Class 2 propellant. The other concept, called the High Energy Upper Stage Restartable Solid (HEUS-RS), is a two-burn (stop-restartable on command) motor, which presently contains 1400 kg of Class 7 propellant. The details and status of the motor design and component and motor test results to date are presented, along with the schedule for future work.

---

[*] This paper presents the results of one phase of research carried out at the Jet Propulsion Laboratory, California Institute of Technology, under Contract NAS 7-100, sponsored by the National Aeronautics and Space Administration.

[+] Jet Propulsion Laboratory, Pasadena, California

INTRODUCTION AND SUMMARY

The motor technology work is oriented to the auxiliary or kick stage of the Space Tug and the IUS of the Shuttle envisioned for mission applications in the 1980 decade. Planetary missions such as the Pioneer and Mariner class of outer planet (Jupiter, Saturn, Uranus, and Neptune regions) spacecraft typically require large velocity increments. Earth orbit missions can range from low earth orbiters to geosynchronous and some highly elliptical earth orbiters. Both the planetary missions and earth orbiters would start from Shuttle Orbiter altitudes.

The paper offers a description of the current state of the art of solid propellant rocketry applicable to kick stage implementation in the Shuttle Tug or IUS and then presents the technical advancements being developed under NASA Office of Aeronautics and Space Technology sponsorship.

The current technology of kick stage solid propellant motors is represented by the TE-M-364-4 motor used in the injection propulsion unit of the Mariner Jupiter/Saturn 1977 (MJS'77) spacecraft. Future outer planet spacecraft are anticipated to require a kick motor of about 1800-kg size. Designs of this size using current and advanced technologies are compared as to their performance and design characteristics. The features of the resulting advanced technology motor, called the APM motor, include a 1.27-m-diameter, Kevlar-49, filament-wound chamber, a castable silicone elastomeric insulation filled with phenolic microballoons, and a re-entrant nozzle with a carbon-carbon exit cone with an expansion ratio of 80. The propellant, a hydroxyl-terminated polybutadiene/ammonium perchlorate/aluminized composite formulation, was selected to minimize cost and risk as well as to retain a Class 2 hazard rating.

The present techniques for providing multi-burn capability for solid propellant rockets are cited, and the advantages of on-command thrust termination by liquid quench for an upper stage motor are noted. The motor developed to demonstrate this technology is a test-weight, 1.06-m-diameter, composite glass chamber motor using a composite plastic nozzle with an expansion ratio of 45. The motor, which was originally loaded with a berylliumized double base propellant, is currently loaded with Trident-type, cross-linked, double base propellant. The quench subassembly is an externally pressurized, internally housed, two-compartment bottle containing a water-glycol solution, which can be commanded to perform two quench operations. Two full-scale motors have been test fired and successfully terminated on command under ground, open air test conditions.

The technology readiness dates for the high-mass-fraction motor and the restartable motor are approximately mid-1977 and mid-1976, respectively, when the ongoing technology efforts are expected to be completed. At that time, the development of correctly sized and configured motors can be initiated by the using agencies to be applied to the IUS and/or the Space Tug auxiliary stage.

AAS 75-160

REQUIREMENTS AND CONSIDERATIONS IN SELECTING SPACE TUG PROPULSION SYSTEMS

Christopher J. Cohan*

A high-capability Space Shuttle upper stage — i.e., Space Tug — will be required for high-earth-orbit missions. One major issue associated with the Tug will be selection of a propulsion system. This selection must be based not upon propulsion characteristics alone, but upon these characteristics evaluated with respect to the Tug requirements. This paper discusses those requirements and how they relate to the propulsion system selection. Comparisons of the candidate chemical propulsion systems are also made.

INTRODUCTION

The Space Shuttle, which has the goal of providing low-cost transportation to and from low earth orbit, is the primary element of a new National Space Transportation System. For missions beyond the low-earth-orbit capability of the Shuttle, an additional propulsive stage is required. Such missions account for approximately 40% of the projected Shuttle traffic. This paper will be directed at the Space Tug rather than the Interim Upper Stage (IUS), which will be the Shuttle upper stage during the first few years of Shuttle operations.

The propulsion system is probably the single most important system in the Space Tug vehicle since, by definition, the vehicle is a propulsive stage. This paper addresses the requirements and considerations involved in selecting a Space Tug propulsion system, with emphasis on these factors as they relate to chemical propulsion systems.

SUMMARY

The Space Tug and its payload are transported to low earth orbit in the Space Shuttle payload bay. In orbit, the Tug is deployed from the Orbiter and initiates its mission. If the Tug is reusable it returns and rendezvous with the Orbiter. The potential operational modes for the Tug are payload delivery with the stage expended, payload delivery with the stage recovered for reuse, payload retrieval and payload round-trip missions.

---

*Design Specialist, General Dynamics, Convair Division

The candidate propulsion systems for the Tug fall into the broad categories of chemical, electrical, and nuclear. Chemical propulsion systems, because of their operational status, will probably be the main candidates for a basic Space Tug during the 1980's. The major propulsion system characteristics that must be evaluated are performance, weight, size, safety, reliability, and cost. These parameters must be considered not only for the engine but also for the propellant storage. Selection of a Tug propulsion system requires that these characteristics be evaluated within the framework of the Tug requirements.

The major areas of Tug requirements are mission characteristics, STS requirements, and payload requirements plus the ever-present constraints of funding and schedule. The mission characteristics relate to such things as velocity requirements, payload weight and geometry, and number and scheduling of missions. The energy requirements directly impact the propulsion system. The energy can be provided by a single stage having the required energy level or by a multiple-stage system of lower energy level stages. The tradeoff involves cost, accuracy, and reliability. The parameter of payload length is related to the propulsion system in that the length of the stage is to a large extent dictated by the propulsion system. The stage length limits the maximum payload length and the multiple payload capability, either as multiples on a single stage or as tandem stages. Multiple payload capability is significant in terms of reducing transportation costs to the user.

Since the Tug is transported in the Orbiter, it is subject to constraints imposed by the STS requirements. The major physical requirements are the weight and geometry limitations. Given these limitations, high propulsion performance will yield the highest stage performance, while high-density propellants have a size — i.e., stage length-advantage. Orbiter center of gravity constraints can impact propulsion selection through consideration of stage location and abort propellant dump. Safety will be a major factor since the Tug is carried in the manned Orbiter.

As part of the STS, the Tug should contribute to the goal of providing low-cost space transportation. In general, the key to low-cost operations is reusability of the stage. With the large number of upper stage flights anticipated for the Shuttle, the lower operations costs of a reusable system should offset the higher development cost, resulting in lower total program cost. Reusability is related to the Tug propulsion system through the factors of propulsion system reusability and performance. With the velocity requirement for a reusable system being essentially twice that of an expendable system, a high-energy prouplsion system has the best potential for an effective fully reusable system.

Payload requirements for injection accuracy and reliability could have an impact upon propulsion selection when evaluating a three- or four-stage system made up of low-energy stages. A requirement for payload retrieval is similar to a reusability requirement, in that the stage must return to the Orbiter. The energy requirements are higher than the reusable delivery mission since the payload is carried on the return leg.

The advantages and disadvantages of the conventional chemical propulsion systems are shown in Table 1. In comparing the performance potential of these systems it is necessary to consider both the propulsion performance, I_{sp}, and the propellant mass fraction characteristics. Typical characteristics are shown in Table 2. Cryogenic propellants — i.e., LO_2 and LH_2 — have the highest I_{sp}. Storable propellant stages have a mass fraction advantage as a result of the higher-density propellants and, hence, less weight for propellant containment. Solid propellant stages generally have the lowest I_{sp} and propellant mass fraction. In addition, the requirement for a two-burn mission, (perigee burn and apogee circularization burn) requires a two-stage solid system.

Table 1
TUG PROPELLANT COMPARISON

Propellant	Advantages	Disadvantages
Solid	Simple Low Cost High Density Storability	Low Performance Single Burn
Storable	High Density Storability	Low Performance Corrosive Toxic Hypergolic
Cryogenic	High Performance Nontoxic	Low Density Thermally Active

Table 2
TYPICAL TUG PERFORMANCE PARAMETERS

Propulsion System	I_{sp}	Propellant Mass Fraction
Solid	295	0.88/0.8*
Earth Storable	338	0.93
Cryogenic	465	0.90

*First Stage/Second Stage

Theoretical (no losses such as gravity losses, venting, or engine start) payload delivery performance has been calculated for each of the three basic chemical propulsion systems. This performance data is shown in Figure 1 for both the expendable and reusable system modes. The systems were sized to maximize geosynchronous payload for a stage plus payload weight of 60,000 pounds. Expendable stage performance comparisons show that the cryogenic stage has significantly more performance than either solid- or storable-propellant stages. Considering the reusable mode, only the cryogenic system has the potential for a fully reusable single-stage system with an effective payload performance capability.

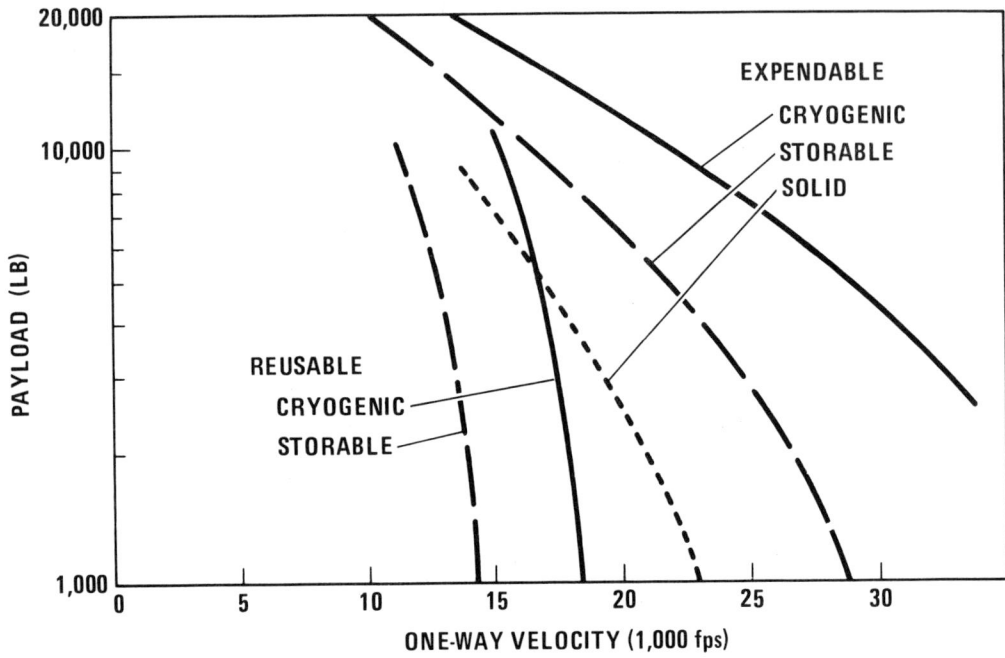

Fig 1. Theoretical Tug Performance.

The propulsion related requirements summarized here would also have to be considered in evaluating any "hybrid" systems — i.e., systems that attempt to combine the better features of alternative approaches.

In summary, the dominant requirements are the mission energy requirements, reliability, and cost considerations. The Space Shuttle requirements and payload accommodations act as constraints on the Tug, limiting such factors as weight, geometry, and center of gravity. With the manned Orbiter, safety considerations become very important. The upper stage must contribute to the STS goal of low space transportation costs, which is a strong incentive for Tug reusability, which impacts the propulsion system selection through the relatively high performance requirement.

AAS 75-161

SPACE TUG PROPULSION SYSTEMS - STORABLE VERSUS CRYOGENIC

W. E. Pipes[*]

To extend the capability of Space Shuttle beyond near-earth orbit for unmanned missions, an upper stage or Space Tug is required. The Space Tug will be used to deliver spacecraft to planetary, geostationary, and other earth orbits beyond the Shuttle Orbiter capability. To obtain the maximum cost benefits from the Space Transportation System (STS), it will also be capable of spacecraft retrieval, servicing and inspection. The Space Tug will be a sophisticated high-performance, reusable vehicle capable of autonomous spacecraft rendezvous and docking with a high degree of reliability and safety. The selection of storable or cryogenic propellants for the main propulsion system on the full capability Space Tug can significantly effect program cost, performance, and risk.

In March 1973, four independent Space Tug System Studies were initiated under the joint funding of both NASA and DOD. The four studies were performed concurrently--two on cryogenic Space Tugs and two on storable propellant Space Tugs. After completing one of the storable Space Tug studies Martin Marietta Corporation (MMC) performed a company funded assessment to establish if cryogenic or storable Tugs best meet the STS requirements.

The cryogenic Tug has the advantage of significantly higher specific impulse while the storable Tug has the advantages of high density (shorter stage and better mass fraction), less complexity, and long on-orbit life.

The assessment was initiated by reviewing all Space Tug System Studies and selecting a composite point design for the storable and cryogenic Tugs based on consideration of all contractors' data. Vehicle performance was then determined, a mission capture analysis performed, programmatics established and costs developed. Based on these data, key issues such as operations, safety, complexity, and risk were assessed. Following Tug point design evaluation, sensitivity studies were performed on Tug as well

---

[*] Lead Propulsion Engineer, Space Tug, Martin Marietta Corporation, Denver, Colorado.

as the mission model to evaluate the impact on Tug selection. Tug nominal, maximum, and minimum performance was established. Cryogenic Tug length was reduced by changes in the main propulsion system. These variations in Tug performance and length impact the number of Shuttle flights required to perform the mission model. The spacecraft and mission model requirements were also varied to determine impact on Tug selection. From these data, conclusions were drawn relative to selection of a storable or cryogenic Space Tug.

TUG DESCRIPTION AND PROGRAM SUMMARY

The point design for the storable Space Tug is 28 ft 8 in. including the docking mechanism for retrieval. The vehicle carries 59,540 lbs of propellant, has a dry weight of 3033 lbs, and a mass fraction of 0.948. The vehicle performance to geostationary is delivery 4630 lb (5650 lb deliver only), retrieve 1630 lb, and round trip 1190 lb. The propellants used are monomethylhydrazine (MMH) fuel and nitrogen tetroxide (N_2O_4) oxidizer. The main engine is pump fed at 12,000 lbs thrust and delivers 338 seconds specific impulse at a chamber pressure of 800 psia, mixture ratio of 1.9, and area ratio of 300:1.

The cryogenic Space Tug point design is 34 ft 4 in. long and carries 57,369 lbs of propellant. The dry weight and mass fraction are 4853 and 0.914 with retrieve mechanism and rendezvous avionics. The vehicle performance to geostationary is delivery 9900 lbs (10,350 deliver only) retrieve 6000 lbs and round trip 3670 lbs. The main engine is a derivative of the Pratt Whitney RL-10 delivering 15,000 lb thrust and 461.6 seconds specifc impulse at a chamber pressure of 400 psia, mixture ratio of 5.4, and area ratio of 180:1.

The programmatics and cost shown in Table 1 are based on the payload mission model and capture ground rules used in the Space Tug System Studies. The model contained a total of 437 Tug boosted spacecraft from December 1983 to 1990 for both NASA and DOD. With multiple deliveries the storable Tug accommodated the model with 307 Shuttle/Tug flights while the cryogenic Tug required only 243 flights due to its higher retrieve performance. The Design Development Test and Engineering (DDT&E) cost for the cryogenic Tug is higher than the storable Tug due to its increased complexity. However, the reduced flights result in lower operations cost and lower production cost due to the reduced fleet size. Adding the Shuttle cost of 10.5 million dollars per launch results in a lower total transportation cost for the cryogenic Tug.

Table 1

POINT DESIGN PROGRAMMATIC AND COST COMPARISON

Item	Storable Tug	Cryogenic Tug
Programmatics		
Total Flights (100%)	307	243
(Buildup)	274	224
Average Crew Size	174	186
Fleet Size		
Main Stage	15	10
Kick Stage(s)	3-10K	2-10K; 5-4K
Costs (M of $)		
DDT&E	236	281
Production	191	179
Operations	224	194
Total Tug	651	654
Shuttle	2877	2352
Total Transportation	3528	3006

RESULTS AND CONCLUSIONS

Performance

The storable Tug performance is adequate in the delivery mode but has marginal geostationary retrieval capability. The length advantage of the storable Tug offers greater flexibility in payload length and results in fewer delivery flights than the cryogenic Tug. By use of delayed retrieval as shown in Figure 1, the storable Tug can retrieve any payload from geostationary orbit; however, this will result in increased flights if the geostationary retrieval traffic is high.

The principle of delayed retrieval is to convert the residual propellants available after spacecraft delivery into reducing the energy of the orbit of another spacecraft that is to be retrieved. Residual propellants are generally available because the Tug delivery capability is significantly greater than the largest spacecraft in the mission model. After spacecraft delivery, the Tug rendezvous with the spacecraft to be retrieved in a fashion similar to a round-trip mission. The Tug then burns the excess propellant from the delivery mission and deorbits the spacecraft to a lesser-energy orbit. It then releases the spacecraft and returns to the

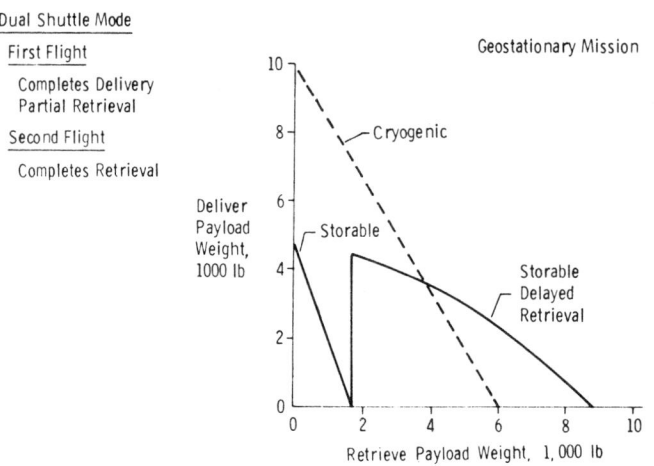

Fig. 1 Delayed Retrieval Achieves Required Performance

Orbiter. The spacecraft, now in a lesser-energy orbit, is then recovered at some later date in exactly the same fashion as for a normal retrieve-only mission. The net impact of the entire delayed retrieval operation is to increase the effective spacecraft retrieval capability of the storable Tug from 1630 lb to more than 6000 lb, at the expense of only one additional rendezvous and docking operation.

The cryogenic Tug has surplus performance that can be used to shorten the Tug length and reduce Shuttle flights. Figure 2 is a carpet plot of length and performance for the cryogenic Tug. It is significant that increasing performance beyond 9000 lbs delivery capability does not reduce Shuttle flights. However, converting the surplus performance to reduced Tug length does save Shuttle flights.

An interesting observation which became apparent during the assessment is the ratio of cryogenic to storable geostationary performance capability. For the point design Tugs, the cryogenic performance in the expendable mode is 1.36 times that of the storable Tug and 3.68 for retrieval. This data is summarized in Table 2 and it is apparent that for the same delivery performance the cryogenic retrieval capability is greater. The primary factors affecting the performance ratio between the two is specific impulse and delta velocity required.

Safety/Complexity/Risk

The cryogenic Tug main propulsion system is more complex than the storable main propulsion system and has more development risk items. These items

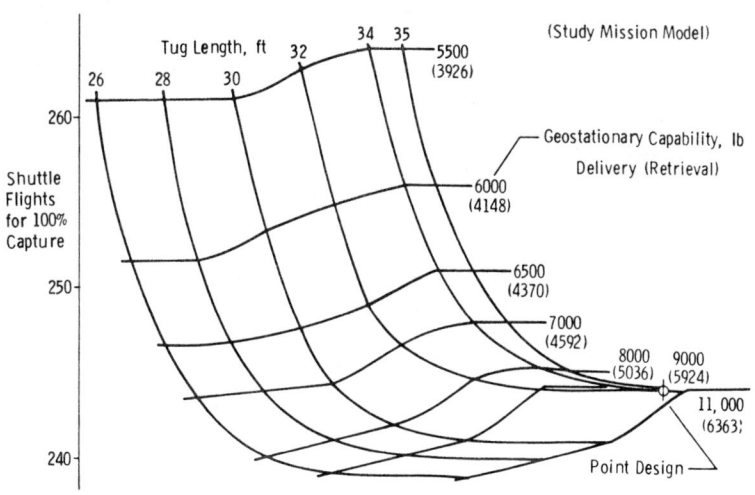

Fig. 2 Cryogenic Tug Length and Performance vs Shuttle Flights

Table 2
GEOSTATIONARY PERFORMANCE CAPABILITY COMPARISON - POINT DESIGN

	Cryogenic, lb ($Isp = 462$, $\lambda = 0.914$)	Storable, lb ($Isp = 338$, $\lambda = 0.948$)	Capability Comparison Cryo/Storable
Expend Tug	19,010	13,970	1.36
Return Tug	10,350	5,650	1.83
Round Trip	3,670	1,190	3.08
Retrieve	6,000	1,630	3.68

could impact development cost risk; however, the surplus performance minimizes performance risk.

Although both Tugs are safety-manageable, the physical stability, predictability, and simpler propulsion system of storables results in a potentially safer Tug with storables than with cryogenics.

Mission Model Sensitivity

Changes to the mission model and its requirements have a significant impact on Tug selection and configuration. The cryogenic Tug is sensitive to payload length, while the storable Tug is sensitive to payload retrieval weight. The most significant mission model driver is the percent retrieval from geostationary orbit. Since mission models change and requirements change, a sensitivity study was performed to determine the impact of varying the retrieval requirements. The change in requirements included:

(1) retrieve all payloads; (2) retrieve only the expensive sensor type payloads; (3) retrieve only low earth orbit payloads (no geostationary); and (4) no retrieval. The results of the analysis show that the cryogenic Tug has clearly fewer flights for models with a high percentage of retrieval missions. However, as retrieval requirements are reduced, the storable Tug becomes very competitive on the basis of total number of flights, as shown in Figure 3.

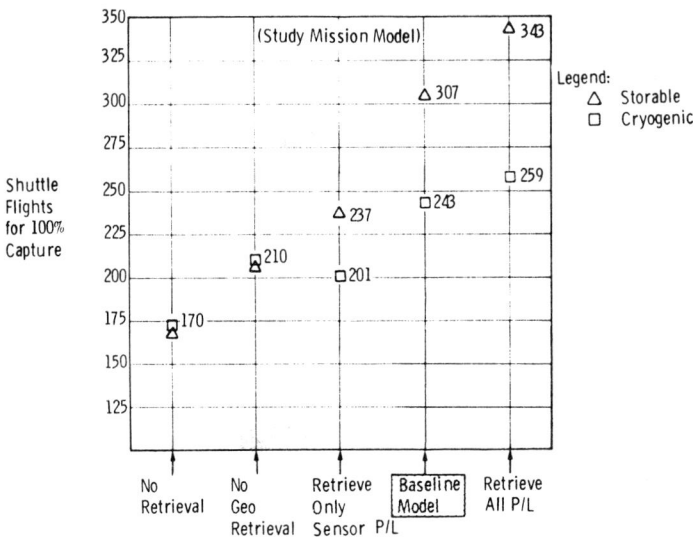

Fig. 3 Selective Payload Retrieval vs Shuttle Flights

Conclusions

For high traffic geostationary retrieval, the cryogenic Tug is clearly superior.

If routine geostationary retrieval is eliminated, the storable Tug is competitive.

AAS 75-162

MIXED-MODE PROPULSION SYSTEMS FOR FULL CAPABILITY SPACE TUGS

R. Salkeld* and R. Beichel**
Aerojet Liquid Rocket Company

INTRODUCTION

Missions exceeding the range of Earth-to-orbit shuttles require space tugs, the effectiveness of which will be a key determinant of the effectiveness of space transportation systems as a whole. For best economy, tugs should be fully reusable and for most efficient servicing and checkout, early tugs should be ground-based. Thus, they must ride to and from the Earth in shuttle payload bays, and hence be constrained by rigid weight and size limits. The purpose of this paper is to show how application of mixed-mode propulsion system optimization, and new high pressure oxygen-cooled engine cycles, can give significant increases in payload weight capability and sharp decreases in vehicle size for full capability tugs.

SUMMARY

Previous work has shown that appropriate application of the mixed-mode propulsion principle can produce performance gains and vehicle size reductions for single-stage-to-orbit (SSTO) shuttles.[1,2] Using an improved propulsion system, reusable mixed-mode tugs are designed to be 60% shorter than reference single-mode O_2/H_2 designs, at no penalty in payload weight, or alternatively, to be 43% shorter with a geosynchronous payload weight <u>increase</u> of 21%.

The cited improvements can be accomplished by: 1) incorporation of mixed-mode propulsion using O_2, MMH, and H_2 as propellants; 2) use of a compact high-pressure O_2-cooled dual-fuel engine embodying a

---
*Consultant
**Senior Scientist

new cycle; 3) use of a columbium rolling diaphragm nozzle extension as a compact, lightweight means of obtaining large nozzle expansion ratios; 4) utilization of an O_2/H_2 mixture ratio of 7:1; and 5) storage of the O_2 in a toroidal tank of spherical segments in the aft peripheral space available around the engine.

Reference tug designs using O_2/H_2 mixture ratios of 6:1 and standard RL10-IIB engines are given in Fig. 1 where the configurational benefits resulting from the mixed-mode approach, the compact engine, and toroidal tanks of twelve spherical segments are illustrated. Note that the payload: dry-weight ratios of the maximum payload mixed-mode case is 40% higher than for the tandem-tank O_2/H_2 concept using the oxygen-cooled engine.

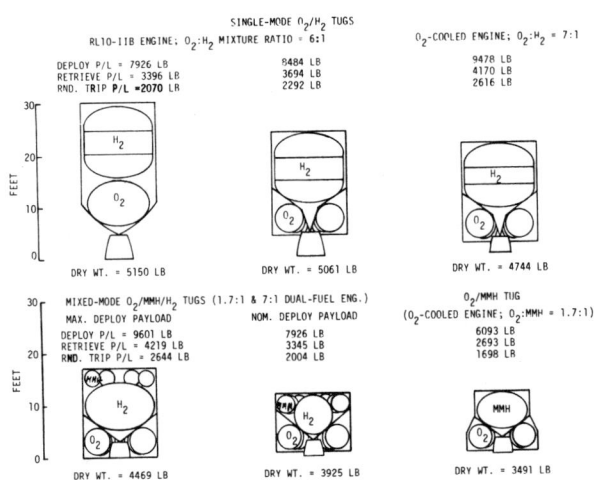

Fig. 1. Comparison of Single-Mode and Mixed-Mode Tugs First Ignition Wt. With Max. Payload 64,000 Lbs

PROPULSION SYSTEM CONSIDERATIONS

The high performance propellant combination O_2/H_2 was utilized in all design analyses of mode 2 operation. Applicable candidate propellants were evaluated for mode 1 operation and alkylhydrazine fuels appeared the most promising.

A typical mode 1 engine staged-combustion cycle using O_2/MMH propellants and a mode 2 engine cycle of O_2/H_2 are discussed. The basic features of the cycles are the same as those presented previously for SSTO engines.[2] The cycles can be either assembled as independent engines (O_2/MMH or O_2/H_2) or combined into a single dual-fuel engine and, therefore, are completely independent. The dual-fuel version operates as simply as two separate engines operated in sequence. Common componentry are utilized where possible to minimize engine weight, two preburners (oxidizer- and fuel-rich) are incorporated in the design to minimize preburner temperatures (increase life and reduce weight) and to eliminate interpropellant turbomachinery seals, and a gas-gas combustion cycle is adopted to allow stable dual-fuel operation with a common injector and combustion chamber.

In the case of the O_2/MMH mode of operation, the MMH flows through a mechanically driven integrated boost pump through the main fuel pump to the fuel-rich preburner. In the preburner it burns with a small amount of oxygen producing an approximately $2000°F$ fuel-rich gas, which is cooled in a heat exchanger prior to being used to drive the turbine of the fuel turbopump assembly.

The major change in the cycle, to accommodate the high fuel-rich preburner temperature, is seen to be the addition of a heat exchanger with the elimination of an oxidizer-rich preburner. The heat exchanger is approximately the same weight as a preburner, and eliminates the need for cooling the fuel-rich gas turbine and the main injector. The selection of 2700 and 1800 psia chamber pressure, respectively, for the two modes of operation was based upon a throat heat flux of 71 btu/in^2-sec.

The preliminary design of the full capability space tug engine is a scaled-down version of the mixed-mode SSTO engine.[2] The engine configuration includes a deployable nozzle. It is regeneratively cooled to an area ratio (ϵ) of 130:1, is radiation cooled to ϵ = 280:1 using columbium, and to $\epsilon = 400:1$ using titanium.

Three nozzle options were examined: a rolling bladder nozzle, a telescoping nozzle, and a permanently installed nozzle. The selected rolling bladder-type nozzle has been thoroughly demonstrated in hot firings, where it is rolled out and rolled back perfectly while hot. It is nested when retracted against an aluminum honeycomb support structure, requires simple gas actuation, is self-stabilized by gas pressure, requires no hot gas seal at the attachment point, and is reusable.

A comparison of space tug candidates is made in Fig. 2. It is seen that the dual-fuel engine with rolling bladder or telescoping nozzle is the lightest and smallest engine despite having the high expansion ratio of 400:1.

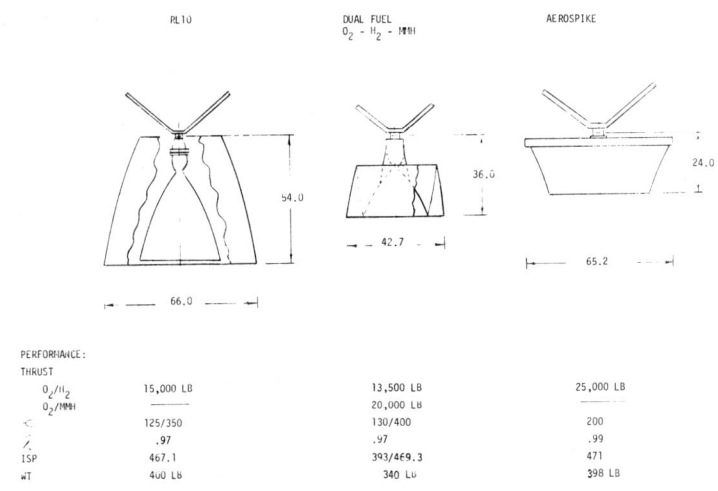

Fig. 2. Comparison of Candidate Space Tug Engines

REFERENCES

1. R. Salkeld, "Mixed-Mode Propulsion for the Space Shuttle," Astronautics & Aeronautics, Vol. 9, No. 8, Aug. 1971, pp. 52-58.

2. R. Beichel, "Propulsion Systems for Single-Stage Shuttles," Astronautics & Aeronautics, Vol. 12, No. 11, Nov. 1974, pp. 32-39.

AAS 75-163

A CANDIDATE MISSION USING THE SHUTTLE AND SOLAR ELECTRIC PROPULSION

John H. Duxbury* and Robert C. Finke**

The NASA planetary missions place the highest performance demands on the Space Transportation System. The options available for the most demanding planetary missions will depend on the Interim Upper Stage (IUS) selected and the programmatic phasing of the IUS and the NASA Tug. A possible approach for minimizing the impact of this limitation is to take advantage of the high-performance characteristics of electric propulsion. An initial flight having high scientific interest and a 15 kW solar electric propulsion unit for launch in the early 1980's is described in this paper. The mission would serve as a precursor flight for demonstrating the technology maturity of electric propulsion and its readiness for applications such as augmenting Shuttle performance for planetary missions.

INTRODUCTION

Solar electric propulsion technology has demonstrated the potential for exceeding the performance of chemical propulsion vehicles for high-energy missions. Ion thruster flight tests, such as the SERT II, have been carried out and thruster technology has advanced significantly in terms of performance and life. In addition, developments in solar array technology offer practical designs for lightweight, high-power solar arrays for electrically propelled vehicles.

One of the more basic problems facing mission planners for the Space Shuttle Transportation System is the performance limitation placed upon

---

 * Supervisor, Spacecraft Studies Group, Project Engineering Division, Jet Propulsion Laboratory, Pasadena, California
 ** Chief, Electric Propulsion Branch, Spacecraft Technology Division, National Aeronautics & Space Administration, Lewis Research Center, Cleveland, Ohio

the system by the relatively low performance for planetary missions attainable from currently envisioned upper-stage chemical rockets (Ref. 1).

Present high-performance chemical stages, such as the Centaur D-1A, has a specific impulse of 444 sec, weighs 1,950 kg dry, and carries 13,500 kg of propellant. By comparison, a 3,000 sec I_{SP} solar electric stage with the same total impulse (13 million lb_f-s) would weight 1,290 kg dry and carry 2,000 kg of propellant, a differential of approximately 12,000 kg that does not require launching into orbit.

Also, several studies have shown that a solar electric stage combined with the Shuttle can increase payloads placed into geosynchronous orbit up to 50% (Ref. 2). Future reductions in power plant specific weight would permit missions to be performed with higher specific impulses (up to 20,000 sec) with a consequent increase in performance.

An early solar electric propulsion mission, such as an out-of-the-ecliptic mission briefly described in this paper, offers a potential two-fold benefit. First, it will accomplish an interplanetary mission in the NASA Planetary Mission Model that has high scientific interest. Second, it will demonstrate the capability and the maturity of the technology for future electric propulsion applications such as a solar electric upper stage for Shuttle planetary missions.

SCIENCE OBJECTIVES

The Sun continuously emits plasma and other radiation from its surface and sporadically releases intense radiation from isolated spots. The plasma emitted contains super-heated gases that can conduct electricity, interact in interesting ways with magnetic fields, and travel at supersonic speeds.

The scientific objectives of the SEP Out-of-the-Ecliptic mission are: (1) to observe the three-dimensional structure of solar features at radio and x-ray wavelengths, (2) to determine the properties of the solar wind as a function of latitude and time, (3) to determine the configuration of the interplanetary field and its relation to solar fields, and (4) to study the modulations of cosmic rays to determine the spectrum of intersellar cosmic rays. A representative 30 kg instrument package capable of supporting these science objectives is described in Ref. 3.

MISSION DESCRIPTION

Optimal launch dates for the Out-of-the-Ecliptic (OOE) mission occur twice a year (June and November) and correspond to the points where the Earth passes through the node of the solar equatorial plane and the ecliptic plane. Launching at these favorable dates gains 7.2^o of heliographic inclination for the spacecraft. Excess launch vehicle injection capability is used to add to this initial 7.2^o orbital inclination. The SEP capability is then used to further incline the spacecraft orbit in an incremental fashion by thrusting at the nodes of the resultant orbit.

The objective is to maximize solar inclination to attain an orbit near the Sun's poles and mapping the intermediate space. For the SEP delivery mode, final solar inclination will depend upon many factors including mission duration, thruster lifetime, thrust duty cycle (ratio of thrust time to orbital period), launch vehicle capability, and spacecraft mass. A plot of solar inclination for a spectrum of missions versus mission duration is shown in Fig. 1 (Ref. 5).

SEP SPACECRAFT DESIGN

The SEP spacecraft consists of two modules: the solar electric propulsion module and the mission module (Fig. 2). This concept has been given considerable design attention at JPL (Ref. 4, 5).

The mission module includes the science payload and supporting subsystems, telecommunications, command computer, data management electronics, power conditioning equipment, and articulation/attitude control. The engineering subsystems use Mariner or Viking derived designs packaged in a form compatible with the Mariner standard chassis. The overall configuration is a rectangular shape more suitable for SEP spacecraft. Electrical power is furnished from the propulsion module solar array to the

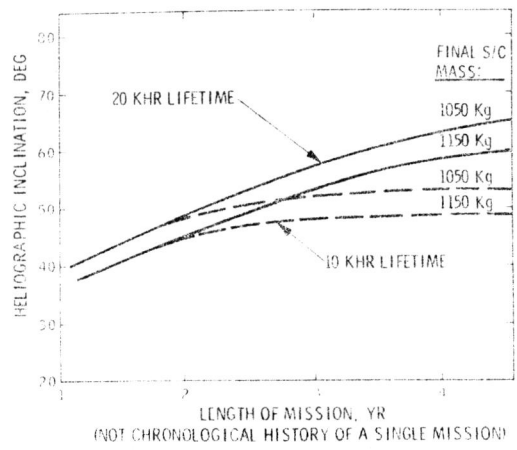

Figure 1. Heliographic Inclination vs. Trip Time for 15 kW SEP, Circular 1.0 AU Orbit and Shuttle/IUS (performance equivalent to Titan 3E/Centaur D1-T)

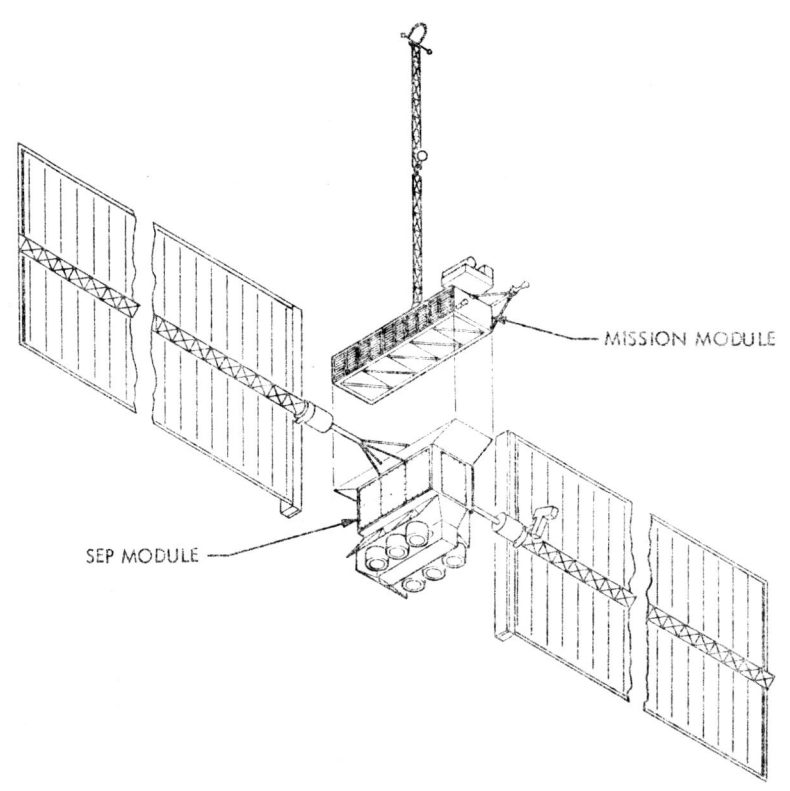

Figure 2. SEP Spacecraft Configuration for OOE Mission

mission module. Attitude control torques are provided by a gas expulsion system during coast. The mission module mass, as shown in Table I, is 345 kg for the OOE mission.

Table I. Mission Module Mass Summary

Subsystem	Design Base	Mass (kg)
Science	Mixed	30
Structure	Mariner	61
Telecom	MJS	37
Power	Viking/MJS	60
Computer Command	Viking	18
Flight Data	MJS	13
Attitude and Articulation Control	MJS	61
Pyrotechnics	Viking	4
Engineering Support (T/C, Cabling, Devices)	Viking/Mariner	61
	TOTAL	345 kg

Solar Electric Propulsion Module

The SEP module contains the primary propulsion unit and power source (solar arrays) of the SEP spacecraft. During the thrust phases of the mission, vehicle attitude stabilization torques are generated by the SEP module ion thrusters. The SEP module design takes advantage of capabilities existing in the mission module for such support functions as data handling and transmission, attitude-control electronics, and command computer rather than duplicating these functions at increased mass, volume, and cost. Unlike the mission module, many of the hardware elements in the SEP module represent new technology currently under development by NASA. A "building block" architecture has been selected for the SEP module which allows the base design to grow or shrink in incremental hardware steps to match the varying mission-to-mission propulsion performance requirements. The intent of this particular design approach is to minimize SEP module redesign costs for applications beyond the initial mission.

The SEP module for the OOE mission includes six 30-cm mercury-ion bombardment thrusters, each on a two-axis gimbal mounted to a body-fixed frame, six power processors, propellant tank and distribution lines. Each thruster is hardwired to a power processor. A spare thruster/power processor combination is provided to protect against a single failure. The solar array is comprised of two identical solar wings which produce 8 kW each at launch. Each power processor draws approximately 3 kW when its companion thruster is operating at full thrust. Thus, five thrusters operating simultaneously produce a total thrust subsystem demand of 15 kW on the solar array leaving 1 kW of power for the mission module. The mercury propellant requirement is 1,600 kg. The SEP module dry mass, using the scaling equation developed in Ref. 5, is 705 kg.

System Mass and Performance Study

The SEP spacecraft mass summary for the OOE mission is tabulated below:

Mission Module	345
SEP Module (dry)	705
S/C Final Mass	1,050
Propellant	1,600
S/C-L/V Adapter	50
S/C Liftoff Mass	2,700 kg

For the above 1,050 kg final mass vehicle, Fig. 2 indicates that heliocentric latitudes exceeding 60° can be reached with 4-year flight times and a thrust subsystem lifetime of 20,000 hours.

TECHNOLOGY PROGRAM STATUS

A NASA program goal is to bring primary electric propulsion to technology readiness by 1979. Major activities are continuing at LeRC to define thrust subsystem requirements, develop a thrust subsystem engineering model, conduct life tests and demonstrate multiple thruster operations. Power processors are being redesigned and developed to functional model status, integrated with thrusters, and verified by system level and lifetime testing. Other critical system elements such as gimbals, propellant storage and distribution, and thrust subsystem-spacecraft interface hardware and software will be simulated and integrated into a functional subsystem. Thrust subsystem interactions and integration problems are being investigated to the extent necessary to clearly define critical interfaces and system requirements and reduce risk in future applications.

The engineering model will be used to bridge the gap between technology development and flight application readiness. It is a device comprised of elements which are believed to be qualifiable, and which have characteristics believed to be suited to anticipated flight application requirements. This device is designed for ground tests in which the full range of inter-element integration problems are explored and functional capabilities of the system demonstrated.

Engineering Model Thruster (EMT)

The 30-cm Engineering Model Thruster is being developed in-house at the NASA Lewis Research Center and coordinated with a contract effort at the Hughes Research Laboratories (Ref. 6).

The major goal of this development program is a thrust of 0.135 N (0.0305 lb) at a specific impulse of 3,000 sec and a maximum power input of 2.75 kW (not including power conditioning losses). Recent test data, including direct thrust measurements, indicate that these goals have been met (Ref. 7).

To demonstrate adequate thruster life, a 10,000-hour endurance test of the first EMT was carried out at Hughes Aircraft Company (HAC) in a large vacuum facility (Ref. 8). This test determined the potential life-time of the EMT, evaluated long-term control stability, and ascertained life-limiting phenomena not observable in short-term tests (Ref. 9). Ten-thousand hours of operating were attained. Potential life-limiting mechanisms were ascertained and retrofits to eliminate these were implemented. A 15,000-hour life test of the upgraded thruster will be started in late 1975.

Power Processor (PP)

The PP interfaces between the thruster and the source of raw power. It must accept voltage variations from 200 to 400 V and transform this raw power to voltages and currents compatible with the thruster needs.

In addition, the PP provides the closed-loop control required to maintain the thruster at optimum operating conditions, senses overloads, and provides automatic protection against any supply output shorted to itself or to another power supply. The PP accepts command inputs from the mission module command computer to select thruster operating points and provides data to the mission module telemetry subsystem to indicate thruster operating status and to provide diagnostic information in case of system malfunction. The efficiency goal for the PP is 91% at full power output with a 0.96% reliability for 15,000 hours. It is designed to dissipate its waste heat by radiation to deep space.

To attain these goals, the LeRC is presently under contract with TRW to design final PP circuitry and to fabricate and delivery five functional models for thermal vac testing and for combined power processor/thruster system and life testing.

Multiple Thruster Array (MTA)

The thrust subsystem components, EMT's, PP's, gimbals, propellant tank, etc., are being integrated as they become available for system interaction tests with all components mounted in a simulated spacecraft body in the 7.6 m dia by 22.8 m long vacuum facility at LeRC. System level tests are to be performed to define the thrust subsystem characteristics and interfaces and to minimize operational risk.

CONCLUDING REMARKS

High specific impulse propulsion systems should be considered as a necessary adjunct to the Shuttle Space Transportation System to increase its capabilities and potential uses. Solar electric propulsion represents the most realistic near-term approach of attaining specific impulses greater than 3,000 sec. Status of the program leading to SEP thrust subsystem technology readiness by 1979 has been explained.

One possible early flight, the Solar Electric Propulsion Out-of-the-Ecliptic mission, has been briefly described. This mission would adequately demonstrate technology maturity and provide meaningful performance data for future planetary SEP spacecraft and Space Transportation System electric propulsion upper-stage applications. In addition, significant science return would also be obtained from heliocentric latitudes never before explored.

REFERENCES

1. Cork, M.J., Driver, J.M., Wright, J.L., "Shuttle/IUS Performance for Planetary Missions," Paper No. AAS 75-273, AAS 21st Annual Meeting, Denver, Colorado, August 26-28, 1975.

2. "Extended Definition Feasibility Study for a Solar Electric Propulsion Stage Concept Selection," Midterm Briefing, Contract NAS8-27360, Exhibit E, June 20, 1973.

3. "Out of Ecliptic and Solar Stereoscopic Mission," MS(74)34, European Space Research Organization Report on Mission Definition Study, December 16, 1974.

4. Duxbury, John H., Paul, Gary M., "Interplanetary Spacecraft Design Using Solar Electric Propulsion," Paper 74-1084, AIAA, San Diego, California, October 1974.

5. Duxbury, J.H., "A Circular 1.0 AU Out-of-the-Ecliptic Mission Using Solar Electric Propulsion," III European Electric Propulsion Conference, Hinterzarten, Federal Republic of Germany, Paper 74-242, October 1974.

6. Poeschel, R.L., "An Engineering Model 30-cm Ion Thruster," AIAA, New York, NY, Paper 73-1084, November 1973.

7. Banks, B.A., et al, "Direct Thrust Measurements of a 30-cm Ion Thruster," AIAA, New Orleans, La., Paper 75-340, March 1975.

8. Caldwell, J.J., "Test Facility for 6,000 Hour Life Test of 30-cm Mercury Ion Thruster," Hughes Aircraft Company, Presented at AIAA/ASTM/IES/NASA 7th Space Simulation Symposium, Los Angeles, California, November 12-14, 1973.

9. Contract NAS3-15523, Thruster Endurance Tests, Hughes Research Laboratories, 1972-1974.

AAS 75-164

STATUS REPORT ON NUCLEAR ELECTRIC PROPULSION SYSTEMS (NEP)*
J. W. Stearns, Jr. [+]

Condensation

INTRODUCTION

Nuclear electric propulsion and power systems planning is emerging today as a low-cost capability for high-total-energy systems to be launched by Shuttle. Missions may include planetary exploration, Earth orbit transportation, nuclear waste disposal, and power for large space stations and lunar bases.

During the early 1960s, NEP was based on the fast spectrum reactor technology under the SNAP-50/SPUR program at CANEL[5][++], at potential power levels of 300 to 500 kWe and specific mass less than 20 kg/kWe. By the mid-1960s, thermionic power conversion became feasible with efficiency approaching 15 percent, a significant contribution to potentially reliable systems design when coupled with heat pipe radiators. Until 1973, when all nuclear space programs were terminated, the NASA-AEC Space Nuclear Systems Office was moving toward the development of a NEP system which would enable a space flight rendezvous with Comet Halley in 1986.

Future NASA space missions require increased electrical power and increased propulsion energy. Thermionic power conversion promises 20 to 30 percent efficiency, and electric propulsion subsystems development is near completion under solar power programs. A small scale R&AD program in nuclear power and propulsion was reinitiated in FY 1974, to investigate new options for system development.

Shuttle-launched missions involving nuclear power systems will require a large initial investment in weight and cost. It is essential, therefore, that full

---

*This paper presents the results of one phase of research carried out at the Jet Propulsion Laboratory, California Institute of Technology, under Contract NAS7-100, sponsored by the National Aeronautics and Space Administration.

[+] Mr. Stearns is Assistant Manager, Energy Systems Section, Jet Propulsion Laboratory, Caltech, Pasadena, California (Bldg. 122-123, [213] 354-6156).

[++] The numbers in superior positions refer to references listed at the end of the complete paper.

use be made of the high energy available for cost effectiveness. Extended lifetime, high-energy missions with large, multiple payloads are typical of this capability. Planetary missions with NEP presently emphasize the extended exploration of the outer plants, one of many examples of which is shown by the mission sketches of Table 1[10]. Of the NASA geocentric missions, a great deal of interest is presently centered on the Satellite Solar Power Station[11], for which a summary comparison between reusable NEP and chemical propulsion tug operations is discussed in the full length paper available from the AAS or the author. Major cost savings with NEP occur because of the reduced number of HLLV launches.

SPACECRAFT CONCEPTS

During the Advanced Propulsion Comparison (APC) Studies conducted by SNSO[13], a wide range of payload sizes and masses was considered for both geocentric and planetary missions. Heavy emphasis was placed on economic comparisons and cost benefits. A multipayload, multimission spacecraft was shown by APC to be economically necessary for our next generation of spacecraft development, requiring a totally new design approach[14]. This concept is illustrated in Fig. 1. During FY 1976, studies are also being conducted into the feasibility of replacing the ion engines with a high-density plasma thruster placed behind the reactor. For geocentric missions, the requirement for high power (1 to 5 MWe) at exhaust velocity < 30 km/s appears to match the operating characteristics of the MPD arc jet.

Table 1

PLANETARY EXTENDED MISSION SKETCH FOR JUPITER AND ITS SATELLITES, USING A 400-kWe NEP SPACECRAFT LAUNCHED BY SHUTTLE TO LEO.

(1300 days)	21,800 kg in 26.0/0.0 orbit	(68 days)	Orbit Io and drop 500-kg lander; sample return
(25 days)	Orbit Callisto and drop 500-kg lander; sample return	(14 days)	Drop a 500-kg Jupiter atmosphere probe
(102 days)	Orbit Ganymede and drop 500-kg lander; sample return	(~1500 days)	Return surface samples to Earth orbit
(73 days)	Orbit Europa and drop 500-kg lander; sample return	(3082 days total)	

Fig. 1. Conical End-Thrust Multimission Spacecraft Configuration (120 to 400 kWe)

SPACE NUCLEAR POWER TECHNOLOGY

Prior to 1973, development work was directed toward in-core nuclear thermionic power conversion (Fig. 2)[9]. The thermionic fuel element (TFE) consisted of a series string of uranium-fueled thermionic converters. Emitter temperature was 1850 K and the collectors were cooled to 1050 K by the liquid metal (NaK) reactor coolant. Individual thermionic converter efficiency was approximately 15 percent, but interconnection and other system losses brought overall power subsystem efficiency to 10 percent.

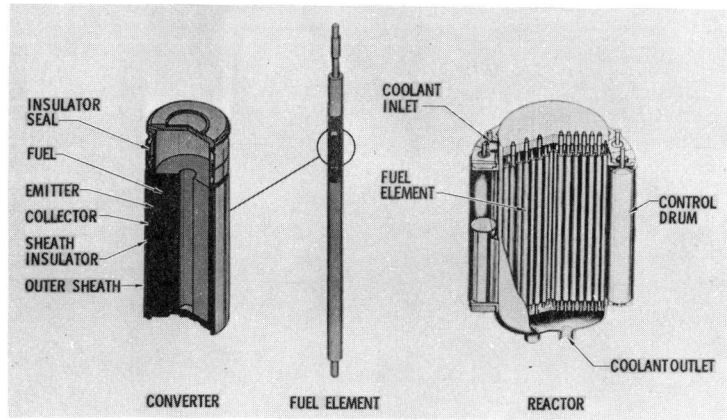

Fig. 2. In-Core Nuclear Thermionic Reactor Design

Since 1974, research in thermionic converters promises higher efficiency at lower temperature. In addition, a heat-pipe-cooled reactor with out-of-core thermionic power conversion is beginning to appear feasible. The concept is shown in Fig. 3. In addition to development of a fuel/heat pipe thermal interface and a heat pipe/thermionic converter electrical isolator development, scheduled for demonstration in FY 1976, there is a very distinct relationship between converter efficiency and temperature which must be taken into account (Fig. 4). Current state of the art requires a tungsten heat pipe, both expensive and difficult to fabricate. At a slightly lower temperature (~1600 K), molybdenum heat pipes and a Mo/UO_2 cermet fuel can be used at greatly reduced cost. However, heat rejection system mass will double at this lower temperature, requiring approximately a 30 percent increase of thermionic converter efficiency to break even. Thus there is a performance incentive to keep converter operating temperatures high, consistent with low-cost materials and manufacturing.

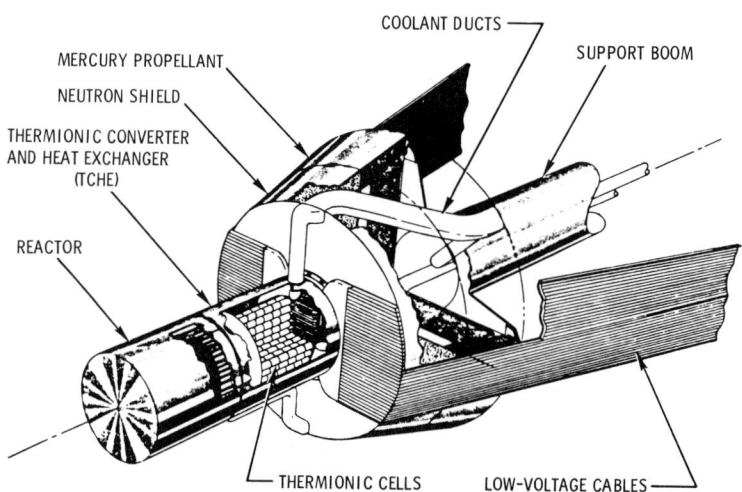

Fig. 3. Heat-Pipe-Cooled Reactor Concept with Out-of-Core Thermionic Power Conversion

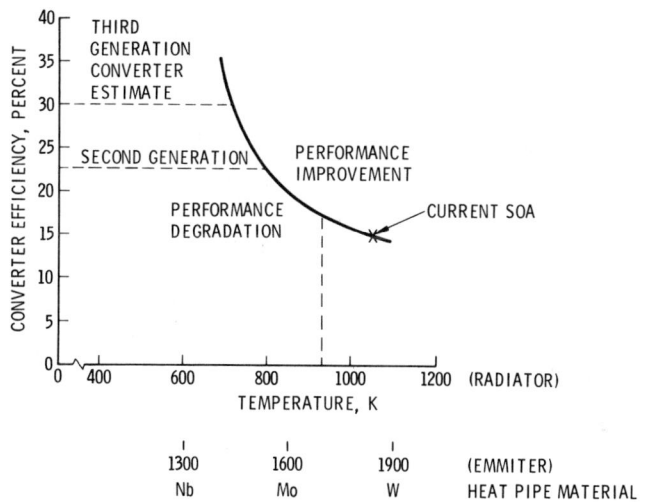

Fig. 4. Thermionic Converter Efficiency/System Mass Tradeoff as a Function of Operating Temperature

YEAR	1	2	3	4	5	6	7	8	9	10	11	12	13	14
$M	2	5	8	15	20	30	35	40	55	60	70	100	90	20

⌈MISSION DEFINITION
 NUCLEAR SYSTEM DEFINITION
⌊THERMIONIC CONVERTER DEVELOPMENT

⌈MISSION DEVELOPMENT
 POWER SUBSYSTEM DEVELOPMENT
⌊THRUST SUBSYSTEM DEVELOPMENT

⌈ENGINEERING MODEL FAB/TEST

⌈PROTOTYPE FAB/TEST

⌈FLIGHT PROGRAMS

⌈FLIGHT

Fig. 5. NEP Estimated Development and Cost Schedule (Exclusive of Thrust Subsystem)

AAS 75-165

TUG AND PAYLOAD-TO-ORBITER INTERFACE REQUIREMENTS

Edward H. Bock*

This paper presents results of a study performed under Contract NAS 8-31012 to identify, evaluate, and develop Tug plus payload-to-orbiter accommodation requirements. Shuttle interfaces required for Tug accommodations are primarily involved with supporting and servicing the Tug during launch countdown, inflight, and postlanding; deploying and retrieving the Tug on orbit; and maintaining control over the Tug when it is in or near the Orbiter.

Interface functional requirements for the complete mission including abort were defined. Each interface area was investigated to determine the best physical and operational method of accomplishing the required functions with an overriding goal of establishing simple and flexible interface requirements suitable for Tug, Tug payloads, and other cargo.

INTRODUCTION

The Space Transportation System flight vehicle, the Space Shuttle, consists of the major segments shown in Figure 1. Included as part of this transportation system is a propulsion stage called the Space Tug (Figure 2), which is carried into low-earth orbit by the Space Shuttle in the Orbiter cargo bay. The Tug extends Shuttle capability by placing payloads into higher orbits, such as geosynchronous and interplanetary trajectories, so that more payload users may be accommodated.

The Space Tug/Shuttle Interface Compatibility Study,* performed under Contract NAS 8-31012 for NASA Marshall Space Flight Center, was structured to compile, screen, evaluate, and recommend suitable Orbiter interface provisions for Space Tug integration. Figure 3 identifies typical Orbiter interfaces associated with Tug accommodation. The Shuttle Orbiter, as currently configured, includes some general payload accommodations applicable for Space Tug, but a detailed investigation of specific interface requirements had not previously been undertaken. Tug interface requirements needed immediate definition

---

*Design Specialist, General Dynamics Convair Division, San Diego, California.

*"Space Tug/Shuttle Interface Compatibility Study," Report No. CASD-NAS 78-017, June 1975, General Dynamics Convair Division.

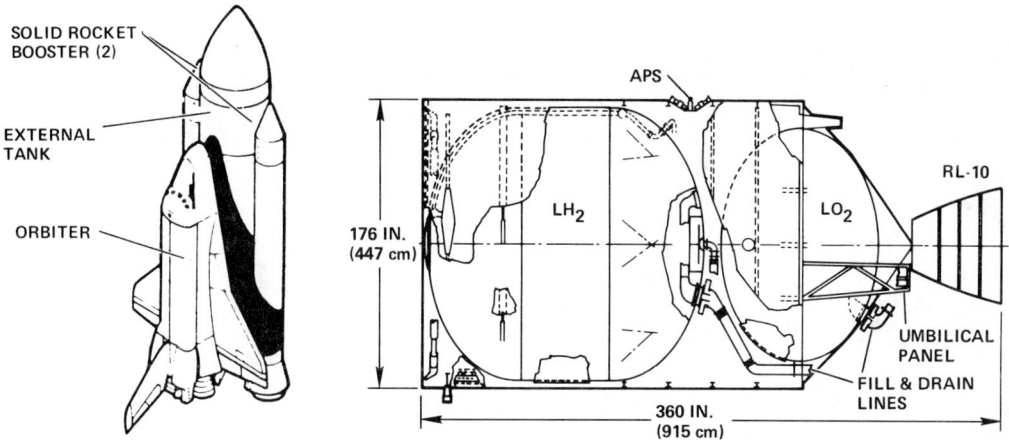

Figure 1. Space Shuttle Configuration.

Figure 2. Space Tug Configuration.

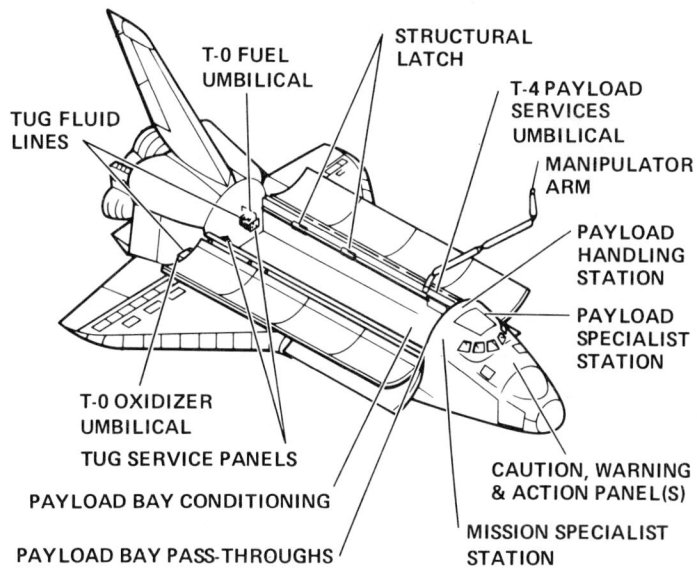

Figure 3. Tug-Related Orbiter Interface Provisions.

and consideration in conjunction with other payload interface requirements for early incorporation into the Shuttle Orbiter.

SUMMARY

Alternative interface concepts for satisfying Tug and payload functional requirements were investigated. Subsystem interfaces were grouped into six categories by technical discipline, as shown in Figure 4.

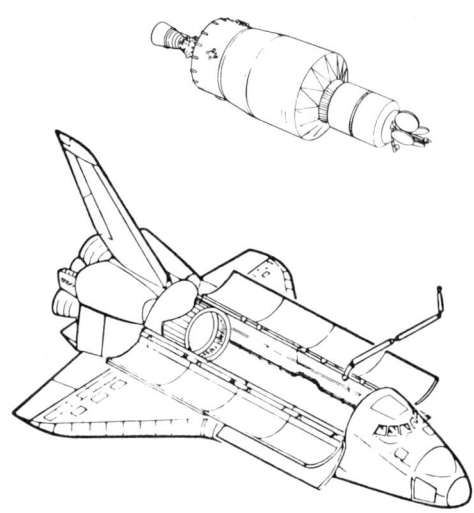

SUBSYSTEM	ACCOMMODATION
STRUCTURAL	SUPPORT & HANDLING
MECHANICAL	DEPLOYMENT/ RETRIEVAL
FLUID	SERVICES & ABORT DUMP
ENVIRONMENTAL	CONDITIONING & PURGES
AVIONIC	MONITOR & CONTROL
SAFETY	CAUTION & WARNING

Figure 4. Interface Subsystem Categories.

Major specific interface conclusions generated by study technical analysis are:

Structural Interface

The Tug and its peripheral equipment should incorporate a six-point structural support arrangement. The use of the six-point redundant support concept eliminates Tug/payload deflection and dynamic response problems associated with determinate support schemes.

Mechanical Interface

The Tug deployment adapter in conjunction with the Orbiter Remote Manipulation System (RMS) provides excellent Tug deployment and retrieval capability.

Fluid Interface

Tug service lines are sized for simultaneous propellant dump during Orbiter abort. LH_2/LO_2 dump is safe and compatible with all abort modes. The Orbiter must provide propellant settling thrust (RCS/OMS thrusters or axial dump of Tug propellants) for low-g (on-orbit) abort modes. Implementation of an Orbiter remote GH_2 vent capability is still required.

Environmental Interface

The Orbiter-supplied cargo bay prelaunch conditioning system is adequate for Tug and its payloads.

Avionics Interface

The Tug should take maximum advantage of Orbiter-supplied standard payload avionics equipment. Use of Orbiter-supplied avionics support equipment offers reduced integration costs and operational benefits.

Interface Safety

Detailed Tug/Orbiter interface safety analysis specified caution and warning philosophy, developed implementation approaches, identified specific Tug caution and warning areas, and defined the crew procedures and equipment to be used in the event of a caution/warning occurrence.

The recommended system concept for supporting and deploying Tug from Orbiter employs a cylindrical load-carrying structure called a deployment adapter. The deployment adapter contains all Tug-peculiar mechanisms required for transfer of Orbiter/ground services and support of deployment, retrieval, and abort operations. Because the deployment adapter is a cylindrical structure to provide efficient axial load distribution, a rotational deployment feature is incorporated to allow Tug removal during deployment without infringing on the Orbiter cargo bay volume available for Tug payloads. By using the deployment adapter concept, Tug umbilical and deployment mechanisms can be attached and checked out before Tug installation into the Orbiter. The entire Tug, adapter, and umbilical support are installed as an autonomous unit into the Orbiter. In addition to the cylindrical deployment adapter structure, peripheral equipment includes monitor and control panels and software, mechanisms, umbilical panels, and fluid/electrical umbilical kits. Figure 5 shows this equipment grouped into three categories.
Based on these study conclusions, recommendations for Tug configuration and Orbiter payload accommodations, changes were identified for NASA consideration.

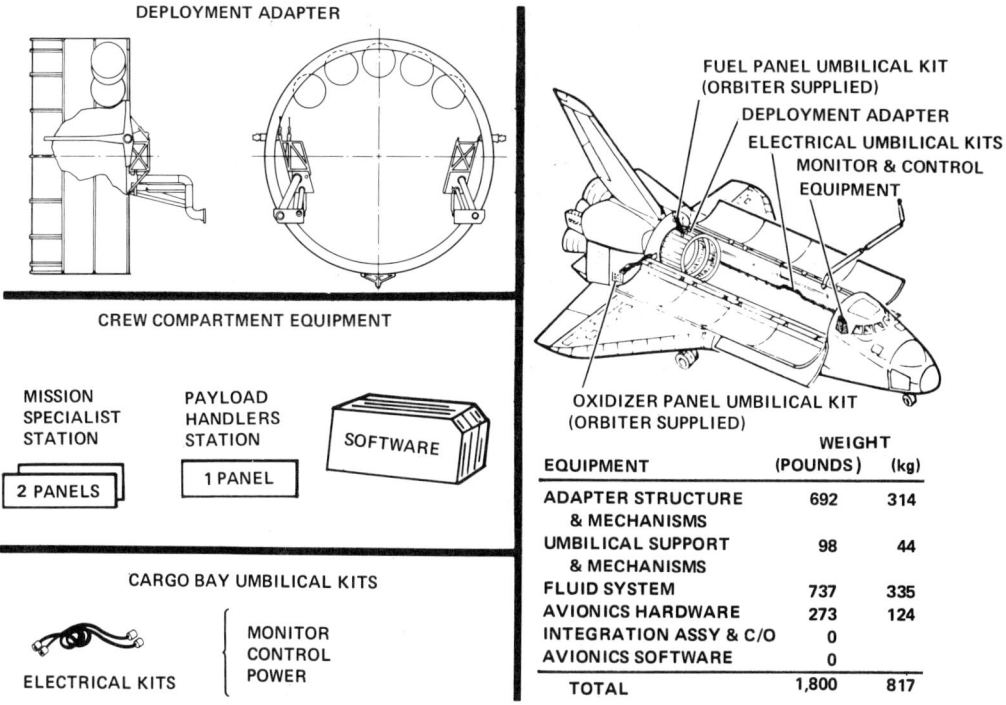

Figure 5. Tug/Orbiter Peripheral Equipment.

AAS 75-166

TUG PAYLOAD INTERFACES

Fritz Runge[*]

The design of any new space transportation carrier, such as the Space Tug, must be optimized to provide the following three unique services to the payload community:

1. Improved capability to fulfill current payload requirements

2. Revolutionary capability to provide new payload services

3. Flexibility to accommodate forecasted changes in payload needs.

The incorporation of such capabilities in a new carrier design requires a detailed joint-situation study by carrier and payload developers. A number of Tug/payload interface studies have been conducted in recent years by McDonnell Douglas and General Electric. This paper presents some conclusions reached during the most recent one, the IUS/Tug Payload Requirements Compatibility Study performed for NASA Marshall Space Flight Center under Contract NAS8-31013.

The range of subjects covered in the study is illustrated in the following charts. Subjects addressed cover (1) the key roles of Tug in its support of payloads, (2) payload requirements for Tug support, (3) Shuttle/Tug/payload interface categories, and (4) satisfaction of payload requirements based on trade studies of approach options.

---

*Program Manager — Payload Integration, Advance Space and Launch Systems Directorate, McDonnell Douglas Astronautics Company.

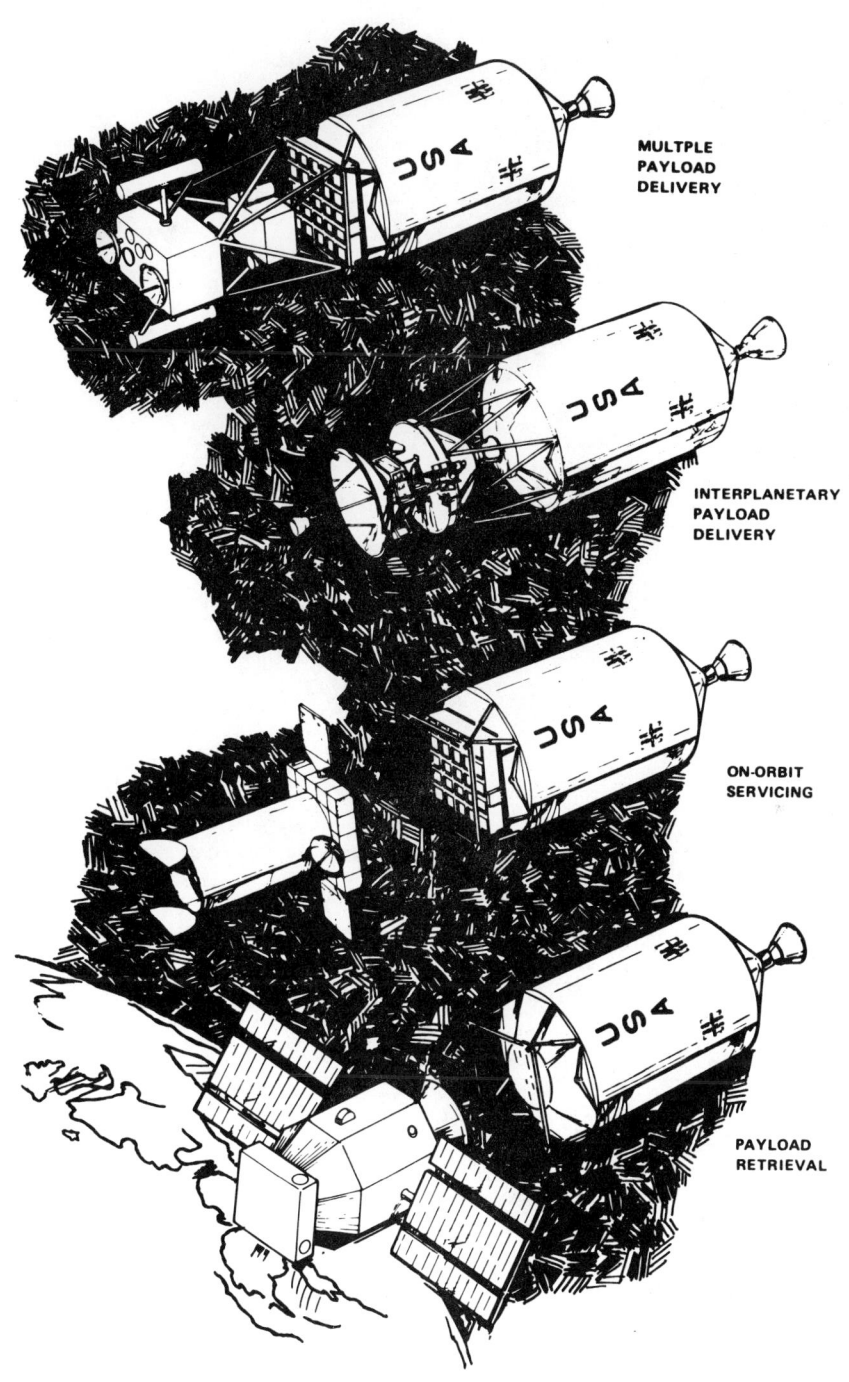

Key Roles of Tug

Table 1

SYNOPSIS OF TUG PAYLOAD REQUIREMENTS

Operational Requirements

The payloads considered in this study require support in the areas of installation and protection, which includes deployment, retrieval, maneuvering, activation, inspection, and servicing. The driving requirement for Tug performance is the deployment of a total of 7,770 pounds (3,524 kg) in geosynchronous equatorial orbit to meet a triple payload requirement (all same address) which the Tug can meet. In fact, the Tug's capability was found to be sufficient to accomplish all missions in the mission model, utilizing kickstages and expendable operations in some cases.

The geosynchronous placement accuracy required by payloads is:

	Position (km)	Velocity (m/sec)
Radial	46	12.9
Tangential	46	12.9
Normal	62	17.2

The separation tipoff rate must be less than 0.1 degree in any axis. The stability capability must be 36 arc sec/sec with a pointing accuracy of ± 1,800 arc sec. The Tug must be capable of spinning the payload at separation up to a rate of 100 rpm and despinning a satellite for retrieval.

For thermal protection of the payload, the Shuttle and Tug must be capable of rotating up to 6 rpm to vary the location of sun impingement on the payload or, during operations such as separation from the Shuttle, orientation such that the sun does not impinge on the payload.

Prior to Tug/payload return to the Shuttle and following payload retrieval by the Tug, the payload must be in a safe condition (not be hazardous to the Shuttle or to Tug/Shuttle operations). The Tug must be capable of commanding the payload to perform safing operations.

Ground handling requirements include provisions for handling payloads in the vertical orientation. This requirement stems from current operating modes and related designs.

Structural/Mechanical Interface

The structural/mechanical interface must accommodate payloads weighing up to 18,526 pounds (8,403 kg) and a product of weight and center of gravity up to 2.837×10^6 in-lb (3.2×10^5 Nm). Payload sizes vary in length from 5 to 25 feet (1.52 to 7.6 m) and in diameter from 4.3 to 14.6 feet (1.3 to 4.5 m). The interface diameter varies from 3 to 14 feet (0.9 to 4.3 m).

Avionics Interfaces

Payloads require electrical power at 28 ± 4 volts DC at a level of 1,150 watts average and 3,400 watts peak (for 2 minutes). The total energy requirement for a deployment mission is about 13 kWh and for a round trip mission about 18 kWh. There shall be a Tug/payload ground system, exclusive of structure, across the interface. Provisions for charging payload batteries, including monitoring of charge, are required during the prelaunch period.

Communications requirements include the following

Type of Data	Rate
Real time digital – scientific	43 kbps
Real time digital – engineering	10 kbps
Real time analog	95 kHz bandwidth

Provisions for monitoring the spacecraft status during all mission phases shall be made. Specific measurements (caution and warning) on hazardous equipment shall be made along with provisions for hardwire transfer of such data. Verification measurements shall be monitored for all commands.

Provisions are required for storage of data at rates up to 43 kbps of digital data and 0.22 MHz of analog data by the Shuttle.

Provisions shall be made for the issue and transfer of commands to the payload during all phases of the mission.

Checkout of payloads either by automatic or manual means shall be provided. This may take the form of data transfer to the ground or by on-board systems (on Tug or Shuttle).

Fluids Interface

Provisions for the venting of hazardous fluids are required. Hazardous propellants include N_2H_4, N_2O_4, and MMH.

Environment

The atmosphere temperature during ground operations (prelaunch and launch) shall be controlled in a range of 58 to 70°F (288 to 294 K). Thermal cooling of RTG power systems to remove 48,000 Btu/hr (14,058 W) shall be provided to multiple spacecraft (DOD payload requirements).

Prior to launch, the atmosphere shall consist of clean air or nitrogen at a pressure not exceeding ambient by more than 1%. Cleanliness shall be maintained to 5000-class clean room conditions per Federal Standard 209A.

The maximum accelerations to be imposed upon payloads shall be limited to 3.0 g except where kickstages are required. For kickstage missions, the maximum acceleration shall be limited to 0.1 g for a period of up to 8 hours.

Certain payloads require the acoustic noise level to be no greater than 135 db.

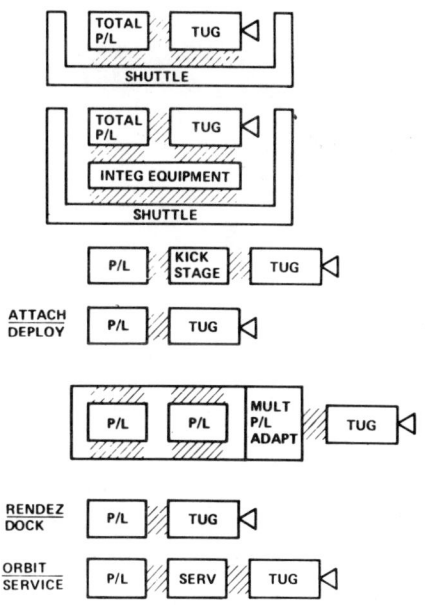

Shuttle/Tug Payload Interface Categories

CATEGORICAL LIST OF INTERFACE FUNCTIONS
- ATTACHMENT (PRIMARY/AUXILIARY)
- DEPLOYMENT
- POWER
- TELEMETRY
- PAYLOAD DATA
- CAUTION/WARNING
- THERMAL CONTROL
- CONTAMINATION PREVENTION
- RENDEZVOUS/DOCKING
- RETRIEVAL
- SERVICING
- CHECKOUT
- NAVIGATION UPDATE
- ORIENTATION
- VENTING
- DUMPING
- SPIN/DESPIN
- COMMAND/CONTROL

EXAMPLE

NUMBER	REQUIREMENT	OPTION	TUG	SHUTTLE	PAYLOAD
5.1.2	THERMAL COOLING OF PAYLOAD RTG POWER SYSTEMS SHALL BE SUPPLIED TO REMOVE UP TO 48,000 BTU/HR (14,058W)	1	NONE	SHUTTLE SHALL PROVIDE 48,000 BTU/HR OF COOLING FROM LAUNCH THROUGH DEPLOYMENT	INTERFACE WITH SHUTTLE COOLING SYSTEM
		✓ 2	TUG PROVIDE THERMAL COOLING UNIT ACCESSORY	INTERFACE WITH COOLING KIT	INTERFACE WITH THERMAL COOLING UNIT
		3	TUG PROVIDE 48,000-BTU/HR COOLING FOR PAYLOADS DURING CONTAINMENT IN SHUTTLE	NONE	INTERFACE WITH TUG COOLING SYSTEM
		4	TUG ROUTE COOLING THROUGH TUG FROM SHUTTLE	SHUTTLE PROVIDE 48,000 BTU/HR COOLING	INTERFACE WITH SHUTTLE COOLING SYSTEM THROUGH TUG
		5	TUG PROVIDE COOLING UNIT AND ROUTE LINES THROUGH TUG	INTERFACE WITH COOLING KIT	INTERFACE WITH THERMAL COOLING UNIT

✓ SELECTED APPROACH

Tug/Payload Interface Requirement Allocation

AAS 75-167

TUG RENDEZVOUS AND DOCKING WITH A SPACECRAFT — A REMOTE, MANNED APPROACH*

Michael J. Hurley[†]

INTRODUCTION

Payload rendezvous and docking represents the major capability difference between an interim and the full-capability Tug. Independent investigations (many in the open literature) have also demonstrated the economic advantages of on-orbit servicing of existing spacecraft whose fluid expendables and/or selected mission equipments cannot exploit the basic spacecraft's life potential. With the indicated order-of-magnitude potential savings to be incurred, Convair undertook an investigation into the possible short-term advantages to be realized through rendezvous and docking with current and future spacecraft.

SENSOR TRADES

The initial screening of candidate sensors for rendezvous and docking was based on a number of factors. Systems requiring actively cooperative targets were eliminated a priori; such systems would be inconsistent with servicing or retrieval of a failed spacecraft. Sensors operating at wavelengths greater than in the microwave region were rejected because of the impractically large apertures necessary to meet spatial resolution requirements. The range of wavelengths of passive sensors was determined by consideration of the spectrum of solar reflectance and target thermal emission. Active sensors were selected on the basis of the availability of reliable high output power.

Four generic types of sensors passed the initial screening: radars (radio detection and ranging systems) in the 2 to 30 cm wavelength region, ladars (laser detection and ranging system) in the 0.8 to 11 μm region, passive LWIR (long wavelength infrared) sensors utilizing target thermal emission from 6 to 16 μm, and passive sensors utilizing reflected solar radiation in the visible region (0.4 to 0.8 μm).

---

*Detailed applications of this research (and related studies) can be found in Report CASD-NAS75-012, "Space Tug Avionics Definition Study Final Report," Contract NAS8-31010, prepared by General Dynamics Convair Division, April 1975.

[†]Design Specialist, Systems Technology and Development, General Dynamics Convair Division, San Diego, California

Several candidates in each class of sensor were evaluated. Based on performance, reliability, weight, stage of development, system-level redundancy, and costs, low light level television (LLLTV) was selected for further study of remote, manned rendezvous and docking.

The novel LLLTV sensor system envisioned for remote, manned rendezvous and docking is more like a still camera than a broadcast TV. A single "still" frame is "exposed" by an electronic shutter (grid gate) and slow-scanned (destructively read) as it is being sent to the ground. This substantially reduces transmission bandwidth and is well within the requirements established by remote, manned operations.

OPERATIONAL SCENARIO

An actual scenario of the terminal phase of remote, manned rendezvous and docking might be as follows: at an assigned time, the supervisor situates himself at a TV monitor or large video projected display to perform the initial checks. Following a variable delay (typical of transmission delay between the ground and space networks), a picture appears. The supervisor periodically (or automatically) requests a new image and scans the last score or so of frames by means of a video disc recorder, attempting visual sighting by means of standard time-compression techniques or with the aid of known star sightings. Following visual sighting, the supervisor initiates the tracking mode, wherein he locates a range reticle on the target. This provides line-of-sight (LOS) angles for each display and LOS angular rates (which can be derived by difference techniques). These sightings support the injection burn to place Tug in close proximity to the spacecraft.

Following the insertion burn, visual sighting is once again established and the LOS information is provided for guidance corrections. Eventually, the target's known cross section permits the supervisor to make a crude range (hence range rate) measurement by adjusting a target silhouette or a range reticle ring to the cross section. Once target details can be discerned, a standard orientation can also be commanded by rotating the silhouette or range reticle index to line up with a desired target feature. These estimates get progressively better and allow closure to the near proximity of the spacecraft (e.g., to 100 ft, 30 m) for visual inspection.

Inspection entails a slow orbiting maneuver about the spacecraft, with the upper stage longitudinal axis essentially aligned with the upper stage-to-spacecraft vector. During inspection, the spacecraft's docking adapter is located, the spacecraft commanded to latency, and the orbiting rate adjusted to align with the docking adapter. Controlled closure then achieves docking.

The fundamental docking strategy for the remote, manned subsystem is to place the remote operator in a supervisor's role rather than a controller's role. This means that he can operate at a reduced task load, delegating much of the operation to spaceborne and/or ground computers. In essence, Tug provides task continuity and the basic docking operation; the supervisor operates as a feedback sensor (via positioning the reticle), removing accumulated biases, and does overall operation evaluation/decision-making.

The supervisor operates on each frame (frames are received approximately on 16-second centers) as illustrated in Fig. 1. The (silhouette or) reticle is envisioned as being ground-computer driven based on information known to the upper stage at exposure time. If the supervisor detects a discrepancy, he takes control of the reticle by pressing the mode SELECT switch, putting it in LOCAL control. He then positions (joystick), sizes (rotary potentiometer), and orients (large rotary control) the reticle to the spacecraft docking port (presuming this had shown a discrepancy). With the CROSS SECTION switch on PORT, a return to REMOTE control activates the ground computer, which interprets the measurements as PORT measurements and computes upper-stage pitch and yaw (from reticle location), relative roll (from orientation), and range (from size).

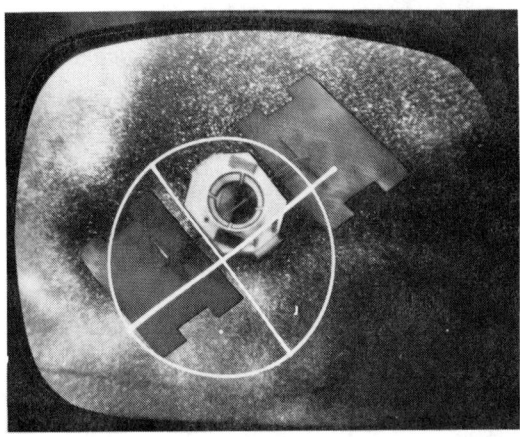

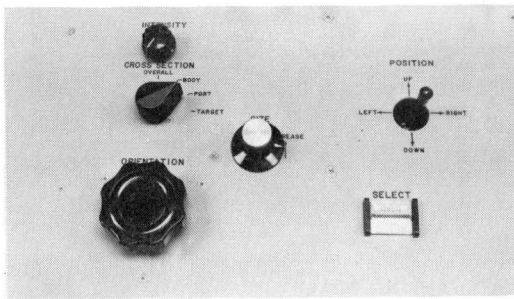

Figure 1. Remote, manned docking reticle positioning procedure.

If the discrepancy noted is with the target T (shown within the port), the crosshair need only be positioned on the T (joystick) to enable a computation of spacecraft – relative pitch (about the horizontal) and yaw (about the vertical). To indicate TARGET measurements, the design concept requires the supervisor to hold the spring loaded CROSS SECTION switch in this position while returning to REMOTE control.

REMOTE, MANNED RENDEZVOUS AND DOCKING FEASIBILITY DEMONSTRATIONS

The docking control law employed was a simple least-squares fit of the measurement data when transformed to Tug coordinates. The current simulation uses up to the last 10 measurements (selectable by the test conductor) to compute the current position errors (intercept P_E) and velocity errors (slope V_E) in all three axes, together with establishing the LOS and rotations about the LOS. If too many stages (measurement vectors) are employed in the digital filter, the system becomes too cumbersome and unresponsive to change; too few and it becomes highly susceptible to noise. Repeated success has been obtained with four to eight filter stages except at very large range, where too little target detail is available (discernible) from the simulator on which to range. (The present simulator employs a 30 degree (0.52 radian) field-of-view fixed lens rather than a zoom or turret lens to simulate the envisioned operational system. Hence, it lacks the ranging performance necessary to simulate terminal rendezvous.)

Repeated simulations of remote, manned docking were conducted on the simulator during simulation development. Only representative results have been obtained to date but these clearly establish the feasibility of remote, manned docking utilizing a low light level television (LLLTV) camera (Table 1). Initial displacements simulated ranged from 190 to 1500 ft (58 to 460 m); some approach orientations required orbiting the spacecraft although angles up to 30 degrees (0.5 radian) before alignment with the docking port was achieved. The time to contact (see Table) reflected the initial displacement, the requirement to orbit, and whether any stationkeeping was employed.

Table 1
REPRESENTATIVE SIMULATION RESULTS

Parameter	Units	Rqmt	Run Number 1	2	3	4	5
Initial Displacement	ft		190	190	1,000	190	1,500
Time To Contact	min		11.2	11.5	32.0	16.0	34.3
Position Error (Radial)	ft	1.0*	0.43	0.08	0.16	0.46	0.47
Velocity Error (Radial)	fps	0.3*	0.154	0.002	0.002	0.001	0.001
Angular Error (Radial)	deg	5.0*	2.03	2.45	1.35	3.96	7.75
Angular Rate Error	deg/sec	0.5*	0.863	0.001	0.006	0.003	0.004
Roll Index Error	deg		Included in Angular Error				
Number of Filter States			8	5	8	5	6
Required Orbital Arc	deg		0	0	30	0	20
APS Propellant Consumption	fps	10**	Oversimplified (Bang-Bang) Control System Invalidates Data				

*Obtained from Article 3.2.1.1.1.4.2 of MSFC 68M00039-1, Baseline Space Tug System Requirements & Guidelines, 7/15/74
**Obtained from Figures 2.1-7 & -10 of MSFC 68 00039-2, Baseline Space Tug Configuration Definition, 7/15/74

Two problems have been uncovered with this simulation. The first concerns the angular error (at docking). The TARGET measurement capability, although envisioned for an operational system, had not yet been added to the simulation due to a limited availability of computer space and since it was found that this capability could be emulated via the console's ORBIT control. However, this expedient caused additional delays, higher APS propellant consumption and occasional failures to "dock." The TARGET measurement capability is currently being added now that a larger simulation computer has become available. This will enable direct measurements on the docking "T" as soon as gross alignment on the docking port has been established.

The second problem relates to the maximum practical ranging with LLLTV. Since the simulator employs a camera with a fixed, 30-degree (0.5 radian) FOV, it was found that 1,500 ft (500 m) was a practical limit on ranging owing both to FOV restrictions and simulator errors. This result is in agreement with the ranging performance derived for LLLTV.

CONCLUSION

The use of LLLTV in a remote, manned configuration for rendezvous and docking with a solar illuminated spacecraft has been demonstrated (via calculations and simulation) as being feasible. Remote, manned docking was conducted in an environment identical to that envisioned for the operational system except for the application of stress on the console supervisor. Even though the observation measurement on the target "T" (located within the docking port) was not available to the console supervisor during simulation studies, approach, station keeping, orbiting (of the target), inspection, docking port location, controlled closure, and docking were all routinely accomplished from several hundred feet (circa 100 m) from the spacecraft. Although simulator restrictions (fixed field-of-view) precluded simulating postinsertion velocity capture and range lock, docking from ranges up to 1,500 ft (500 m) was successfully demonstrated.

It was concluded that remote, manned rendezvous and docking with an orbiting spacecraft is well within the current state of the art and represents a viable, low-cost alternative to performance-competitive autonomous techniques. Analysis has demonstrated that an order of magnitude cost savings is attendant with this approach, including operational support for the first score of rendezvous and docking missions.

ACKNOWLEDGMENTS

I am indebted to Mark F. Dorian*, who conducted the principal sensor trades, thus demonstrating the inherent capabilities of LLLTV, and to John E. Leib[†], who developed the computer simulation that was instrumental in demonstrating the viability of remote, manned (terminal) rendezvous and docking.

---

*Senior Staff Scientist, Optics Technology, General Dynamics Convair Division, San Diego, California
†Manager, System Design and Analysis, General Dynamics Convair Division, San Diego, California

AAS 75-168

SPACE TUG THERMAL CONTROL

T. L. Ward*

Introduction of a full capability Space Tug into the Shuttle mission spectrum in the 1980s will significantly broaden Shuttle's capability. To realize that capability fully, it will be essential that the Tug be designed to perform its mission within a broad range of thermal environments with currently planned mission durations up to seven days. The primary objective of this study was to develop forward and intertank compartment thermal designs and fuel cell heat rejection system that satisfy Tug requirements for low inclination geosynchronous deploy and retrieve missions. The key to this design was a system that was reusable and minimized ground refurbishment requirements. Figure 1 presents baseline Tug configuration used in the study.

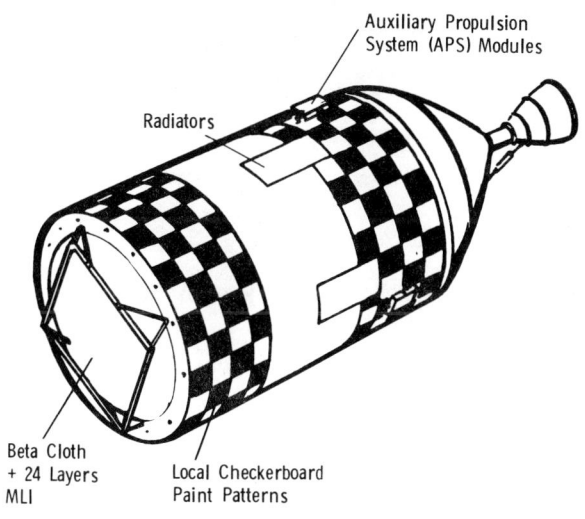

Figure 1 External Profile

---

* Staff Engineer, Martin Marietta Corporation, Denver, Colorado

Passive concepts were demonstrated analytically for the forward and intertank compartments. Each compartment used an external paint pattern tailored to the worst-case external thermal environments. The forward compartment, which contains the majority of the avionics equipment, was designed with circumferential fixed conductance heat pipes to reduce wide variation of skin temperatures resulting from constant attitudes and a multilayer insulation blanket included with the forward shield (beta cloth). The most significant thermal control problems were encountered in the forward compartment. The recommended solution is to mount the components on thermal conditioning panels with louvers mounted on the opposite side of the panel. The panel significantly reduces the need of heater power while the louvers reduce effects of low skin temperatures resulting in a near room temperature mounting surface for all components. The fuel cell requires an active heat rejection system such as a pumped fluid radiator or a heat pipe radiator system. Continued development of variable conductance heat pipes is key to their application to fuel cell problems.

Worst-case external heating environments were determined and used in the study. All mission phases were incorporated, the most significant one being heating of the Tug in the Orbiter after reentry and landing. Cargo bay purging was required to maintain both operating and nonoperating equipment temperature limits.

Component Cataloguing

Before initiating the thermal design study, potential components for application to Space Tug were identified and catalogued. The purpose in doing this was twofold: (1) to identify existing components that would satisfy mission requirements, and (2) to obtain thermal characteristics and constraints of the components to enable thermal design decisions to be made relative to their adaptability to Tug.

The approach chosen to identify, handle, and document these data was to develop a generalized data bank containing thermal and general information for each component catalogued. A FORTRAN IV program containing four major subroutines was written to compile the catalogues using the data bank data as input data. The two catalogues contain equipment thermal requirements, and equipment physical characteristics and constraints, respectively.

Parametric Studies

Several parametric studies were conducted using small thermal models to evaluate the overall design, identify worst-case thermal environments, and isolate major problem areas. The first parametric study was directed toward identifying the forward and intertank compartment overall thermal balance and selecting the external coating properties. The major variables included external heating, radiation to space, compartment internal heat load, and external surface properties. All of the heat dissipated within each compartment was uniformly distributed on the skin nodes. The tank insulation and/or forward beta cloth shield completed the compartment enclosures. The radiation average internal sink temperature derived from this study was plotted as a function of external paint properties for various anticipated compartmental heat loads.

The influence of the honeycomb structure on compartment temperatures was reviewed. The change in compartment sink temperature as a function of conductance through the honeycomb was evaluated, because conduction is the primary mode of heat transfer. Significant ΔTs result when the honeycomb conductance is less than 3.5 watts/meter^2 °K (2 Btu/hr-ft^2-°F). Not accounting for the bondline influence on the effective conductivity through the honeycomb results in an optimistic value of 3.4 watts/meter^2 °K. This value results from an aluminum core honeycomb skirt 1.5 cm thick. The use of a nonmetallic core would significantly reduce conductance and would result in higher ΔTs. Hence, the choice of the honeycomb structure for Space Tug will have a significant influence on the thermal design and could impact the basic passive concept chosen.

Detail Studies

Design studies were continued to include components in the modeling activity to enable prediction of a time history for each component from liftoff to landing. The timelines associated with the operation of various subsystems were the key to this activity. The data management subsystem operated from liftoff to after landing while the remaining avionics equipment operated from 3.7 to 98.6 hr except for the inertial updating and rendezvous equipment, which was turned on and off at appropriate times during the mission. Landing occurred at 100.8 hr.

Results indicated that, without exception, peak temperatures were achieved after landing occurred, based on internal cargo bay liner temperatures. It was also noted that a cargo bay purge initiated approximately ½ hr after landing was essential to avoid exceeding most of the nonoperating temperature limits.

The cold case design condition assumed that no external heating was applied to the vehicle. In this condition, the two compartments demonstrated similar trends of approaching steady-state conditions after 10 hr. Most of the equipment was observed to cool below the minimum design case temperatures and most of the equipment that operated continuously could be warmed up sufficiently by reducing the exterior coating emissivity. A coating change, however, would not handle those components that are used only a few times during the mission. For example, the laser radar, its electronics, and the TV were some of the critical items that required heater power in excess of 400 watts to maintain allowable temperature ranges.

At this point, a different component mounting and layout was considered desirable. The application of the thermal conditioning panel developed by MSFC was considered ideal here. The panel is a flat plate that can be designed to satisfy structural and dynamic requirements, while providing an isothermal mounting plate for the equipment. Heat pipes are an integral part of the panel design, and provide the means of reducing panel temperature gradients. This provides an ideal means of sharing heat between components. In the cold case environments, the application of louvers on the skin side of the panel enables the panel to operate at a temperature level that satisfies all of the mounted equipment temperature limits.

The radiators were sized to reject the maximum heat load during maximum external heating. The four radiators were sized at 8.05 meter^2 (22 ft^2) or 2.01 meter^2 (5.5 ft^2) per panel. The maximum heat load of 825 watts accounted for the fuel cell and radiator pumps of which 744 watts were derived from the fuel cell under a maximum electrical load of 1500 watts.

Silver-coated Teflon tape was selected as the radiator coating because of its desirable optical properties, stability of properties, and ease of maintenance and application. The cold case was also explored to ensure against fluid freezing at minimum heat load conditions. No external heating was applied for the cold case with 362 watts of heat to be rejected. This relates to a minimum fuel cell load of 600 watts electrical resulting in 281 watts of heat to be rejected less pump power. The cold case resulted in minimum fluid temperatures of 228°K (-50°F), which is well above the freezing temperature of Freon E-1.

CONCLUSIONS

A preliminary thermal design for the full capability Space Tug was evaluated and presented to NASA. The design minimizes the need of heaters for thermal control while integrating the structural mounting of components with the thermal design through the use of the NASA-developed thermal conditioning panels. The passive design provides a degree of latitude in the thermal integration of Tug that has been lacking before the thermal conditioning panel development, and offers many advantages to future spacecraft. The fuel cell heat rejection system was designed around a pumped fluid system based upon the current state-of-the-art. Continued development of variable conductance heat pipes is expected to allow the pumped system to be replaced with a lighter heat pipe system in the future.

A standardized cataloguing technique for identification of components was created. The resultant catalogues were very helpful throughout the study, and provide a basis for future application in other spacecraft.

ACKNOWLEDGEMENTS

The author wishes to express his appreciation to J. M. Connolly and S. H. Eichembaum of Martin Marietta Corporation, Denver Division, for their contributions to the study.

This paper is based upon work performed under the National Aeronautics and Space Administration, Marshall Space Flight Center, Contract NAS8-29670. The author wishes to express his appreciation to Mr. J. D. Loose for his guidance and assistance while serving as the contracting officer representative of this contract.

AAS 75-169

NDT FOR SPACE TUG THIN GAGE MATERIALS+

Ward D. Rummel*

Nondestructive Testing (NDT) or Nondestructive Evaluation (NDE) requirements for advanced programs constitute challenges in application to thin materials and configurations and in assuring reliable detection of all detrimental anomolies. Fracture mechanics design criteria assumes the existence of flaws in fabricated structures and predicts the functional performance of a structure on a maximum flaw size which can be reliably detected and eliminated by production testing. Several programs have been conducted to determine NDE reliabilities for crack detection. NDE reliability programs and their application to thin gage materials are discussed.

INTRODUCTION

Recent implementation for fracture control design and assurance has imposed requirements for assessment of nondestructive materials evaluation sensitivity and reliability. These requirements have been imposed in the National Aeronautics and Space Administration in the form of contractual requirements. Although nondestructive materials evaluation methods (ie x-radiography, liquid penetrant, magnetic particle, ultrasonic and eddy current) have been used routinely in industry for approximately thirty five years, little quantitative data on flaw detection sensitivity and reliability are available. Lack of such data can be appreciated if one considers the nature of development of nondestructive materials evaluation technology.

+ Paper No. 75-169 prepared for the twenty first annual meeting of the American Astronautical Society, August, 1975.

* Staff Engineer with Martin Marietta Aerospace, Denver Division

Nondestructive evaluation (NDE) techniques have been traditionaly used to detect flaws which had in the past caused failure of materials or hardware <u>after</u> a failure has occurred. As familiarity with the techniques was gained, the same techniques were applied routinely to similar materials or hardware to <u>prevent failure</u>. The challenge to the NDE engineer was "how small a flaw can you find?" Technology development was thus directed toward <u>finding smaller flaws</u>. Some measures were taken to assure detection reproducibility and reliability, but in general, high reliability was assumed, and exceptions were attributed to human factors.

Fracture mechanics technology has changed design concepts from the assumed flaw free material base to flaw tolerance in materials which contain sub-critical flaws. Critical flaw size has become a familiar design parameter. With acceptance of flaw tolerance in design, the new challenge to the NDE engineer is "how large a flaw can you miss?" To answer this question, some measure of the reliability of flaw detection methods must be established. Work to establish detection reliability has been performed by Rockwell International on the United States Air Force B-1 program and by Lockheed, General Dynamics/Convair and Martin Marietta for the National Aeronautics and Space Administrations, Lyndon B. Johnson Space Center. Data from the Martin Marietta studies are the most comprehensive set available and are the basis for design and acceptance criteria for Space Shuttle. The results of the Martin Marietta program are reported in detail in NASA CR-2369, February 1974.[1]

FATIGUE CRACK DETECTION RELIABILITY

In summary, this program consisted of fabricating 118 aluminum alloy panels containing 328 fatigue cracks and evaluating these panels in the "as machined", after etching and after proof test conditions by x-radiographic, liquid penetrant, ultrasonic and eddy current techniques. Each nondestructive test (NDT) evaluation was repeated by three different operators to randomize inspection data. This procedure resulted in just under 1000 observations for inspection sequence.

---

(1) Ward D. Rummel, Richard A. Rathke, Paul A. Rathke, Paul H. Todd, Jr. and Sandor A. Frecska, <u>The Detection of Fatigue Cracks by Nondestructive Testing Methods</u>. NASA CR-2369, February, 1974.

A statistical plot of crack detection reliability versus crack size was made for each data set to reduce this data to concise form. A 95% reliability at 95% confidence level was selected for data presentation. Sample size to these criteria is 60 observations with no failures to detect. To plot this data, observations were arranged in decenting order by crack dimension. A data set was obtained by counting down 60 observations from the longest crack dimension, calculating the arithmetic detection reliability for the set and plotting this value at the largest crack dimension observed in the set. This procedure purposely basis the data conservatively in the longer crack dimension. This process was repeated consecutively for cracks of decreasing size. The resultant data plots are statistically valid to the 95% reliability/95% confidence level criteria for large cracks where detection is expected. For small cracks, which may be missed by inspection, the 60 point sample is not statistically valid. To maintain plotting continuity, the 60 point sample size was machined for all plots. Figures 1, 2, 3 and 4 are examples of such plots.

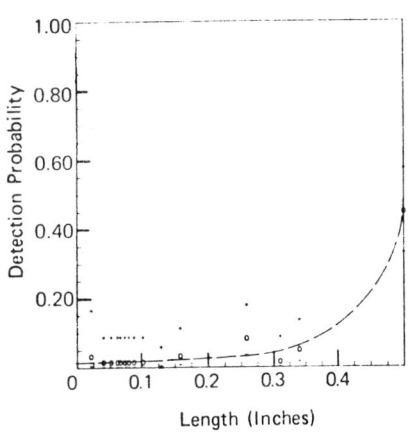
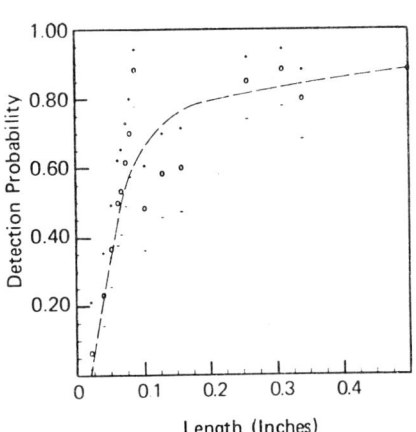

Fig. 1 - Crack Detection Probability of the X-radiographic Inspection Method at 95% Probability and 95% Confidence.

Fig. 2 - Crack Detection Probability of the Penetrant Inspection Method at 95% Probability and 95% Confidence.

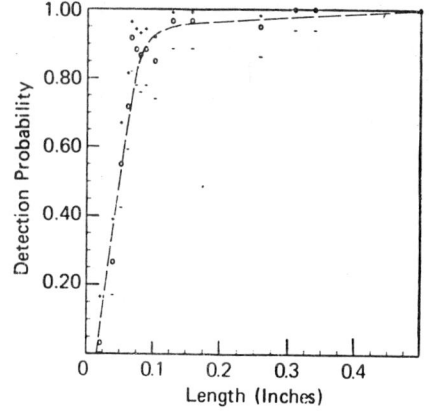 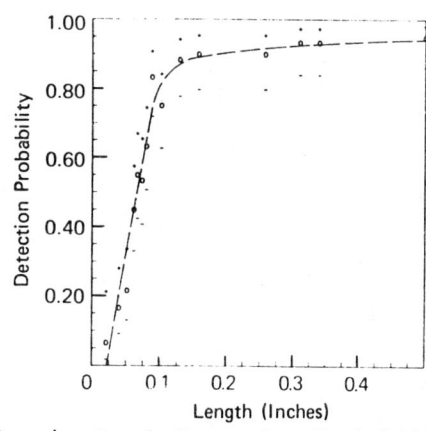

Fig. 3 -Crack Detection Probability of the Ultrasonic Inspection Method at 95% Probability and 95% Confidence

Fig. 4 -Crack Detection Probability of the Eddy Current Inspection Method at 95% Probability and 95% Confidence

The area of primary interest on these plots is the inflexion point, which constitutes the threshold detection sensitivity level for inspection method. At this point, the data is statistically valid and may be used as a basis for selecting inspection methods and for establishing acceptance criteria. Data analyzed by this method was used as a basis for selecting inspection methods and for establishing acceptance criteria. Data was used as a basis for establishing acceptance criteria for the Space Shuttle Orbiter.

SPACE SHUTTLE ORBITER REQUIREMENTS

For Space Shuttle Orbiter, an initial surface flaw size 0.150 inch long by 0.075 inch deep is assumed using standard NDE methods.[2] If special NDE methods are used, an initial surface flaw size of .050 inch long by 0.025 inch deep may be assumed. Capability to detect the smaller flaws must be demonstrated to qualify and use special NDE methods.

APPLICATION TO THIN MATERIALS

Martin Marietta Aerospace, Denver division, has successfully produced titanium alloy pressure vessels with wall thicknesses down to 0.020

---

(2) "Space Shuttle Orbiter Fracture Control Plan," Rockwell International Space Division, SD73-SH-0082A, September 74.

inches thick. Our experience in this production over the past ten years has resulted in respect and concern for special handling and in the development and application of well disciplined NDE procedures for hardware acceptance. Techniques used for our NDE reliability assessment for detection of tightly closed fatigue cracks in 0.060 inch aluminum specimens are similar to those used in our thin titanium tank acceptance. We would then expect a similar or a predictable change in NDE sensitivity for thinner specimens. We would, for example, predict that:

1. X-radiographic (x-ray) detection sensitivity would be decreased using the same techniques.

2. Penetrant detection sensitivity would be unchanged.

3. Ultrasonic detection sensitivity would be unchanged or improved using the same techniques, and

4. Eddy current detection sensitivity would be increased.

Our experimental work on thin sections shows that:

1. X-radiographic detection sensitivity may be improved by decreasing the kilovoltage and by choosing techniques which improve contrast as well as resolution.

2. Penetrant detection on thin materials may be improved due to decreased tightness of cracks of a given size.

3. Ultrasonic detection sensitivity may be improved by use of alternate energy propogation modes. Tolerances for application of such techniques are tighter to maintain the improved sensitivities.

4. Eddy current detection sensitivity is increased by use of higher frequencies to maintain full material penetration.

NDE detection sensitivity changes cited for transition to thin materials are qualitative and hence do not provide the boundaries necessary for design analysis. In the short term, experimental assessment of detection reliability on actual samples is necessary. A long term technology challenge is for the development of rigorous analytical methods to predict detection sensitivity/reliability with changing material and test conditions.

CONCLUSIONS

In conclusion, reliable NDE methods for flaw detection are required for current and future high technology hardware programs. NDE detection reliability data is being developed which shows that meaningful and reliable detection is possible. Such data also graphically illustrates that no single, blanket technique is applicable for all situations.

Reliable flaw detection in thin materials is possible. Attention must be given to the specific application and to the NDE method variables to attain reliable detection. Foreseeable hardware technology challenges can be met by judicious application of the correct method.

AAS 75-142

TUG AVIONICS SYSTEM OVERVIEW

Maurice T. Raaberg[*]

and

James I. Newcomb[†]

INTRODUCTION

The avionics system for the full-capability Space Tug to be developed by NASA for initial operations in late 1983 will be driven by the requirements listed in Table 1. These requirements have a dramatic effect on the avionics needed for the Space Tug. Performance requirements to deploy 8,000 lb of payload into or retrieve a 3,500-lb payload from geosynchronous orbit are supported by minimizing the avionics system weight. Safety and reliability requirements establish dual redundancy as the minimum level for all subsystems. Autonomy, and payload retrieval and servicing, are supported by new avionics sensors, techniques, and software. Long mission durations (typically 6½ days) require a compatible power system.

Table 1
SPACE TUG DEMANDS ON AVIONICS

Driving Requirements		Avionics Impacts
Performance	Deploy 8,000 Lb Retrieve 3,500 Lb	Low System Weight
Mission Duration	6½ Days	Electrical Power Capacity
Payload Retrieval & Servicing	Rendezvous & Docking	R&D Sensors, Control Techniques
Autonomy	Minimum Gnd Support	Navigation Update Checkout, Redundancy Management
Safety & Reliability	0.97 Mission Success	Subsystem Redundancy — Dual (Min)
IOC 83	First Flight (1978 Develop Start)	Implementation Using 1978 Technology

---

[*]*General Dynamics Convair Division, San Diego, California*
[†]*NASA George C. Marshall Space Flight Center, Huntsville, Alabama*

One of the most important factors is the 1983 schedule for the first operational flight. The 1978 Phase C/D go-ahead will allow the Tug program to take advantage of technology advances in the implementation of these avionics requirements to attain adequate power system capacity, necessary redundancy, new functions capability, and minimum total system weight.

AVIONICS SYSTEM DESCRIPTION

The Space Tug avionics system is shown in Figure 1. Its configuration features six major subsystems integrated into an advanced avionics system through a digital data bus technique under the control of a modular central computer. The dual data bus is depicted by the broad dark and light arrows connecting the remotely located digital interface units (DIU) with the computer through a computer interface unit (CIU). Those are the major components of the data management subsystem, which interfaces with and controls all of the functional elements on the Space Tug. The other five avionics subsystems are (from left to right) the communications subsystem, highlighted by the three electronically steerable phased arrays; the rendezvous and docking subsystem, with the scanning laser radar (LADAR) and TV; the guidance, navigation, and control subsystem incorporating a

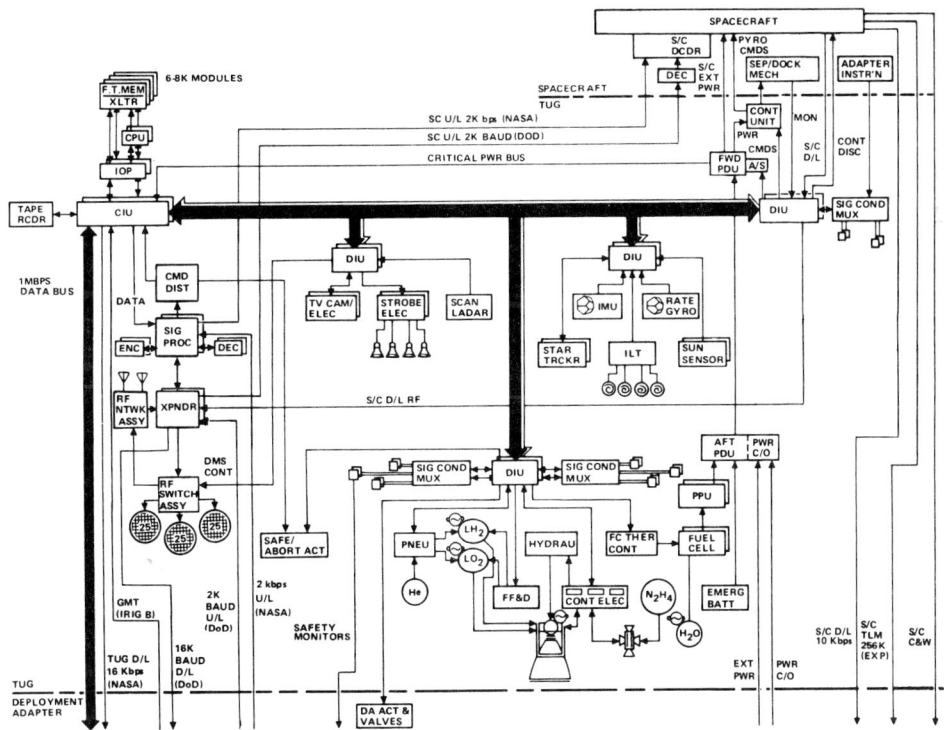

Figure 1. Tug avionics system baseline.

dodecahedron laser gyro inertial measurement unit (IMU); one of three signal conditioners and sensors of the instrumentation subsystem; and, below, the electrical power system using dual fuel cells and power processing units (PPU) and two power distribution units (PDU), one aft and one forward. The aft DIU interfaces with non-avionics systems providing valve controls for venting, fluid fill and drain, and main engine ignition and shutdown; and on-off control for helium pressurization and propellant mixing in the main tanks.

Functions associated with the Tug-to-Shuttle interface are shown at the bottom of the diagram. For example, the safety monitors are hardwire connections directly from the instrumentation sensors to the Orbiter's caution and warning system. The same safety data (from redundant instrumentation sensors) is redundantly supplied to the Orbiter and/or ground system via the telemetry downlink out of the CIU, after the data has been processed through the appropriate signal conditioner, DIU, and the data bus.

The Tug-to-payload interface is shown in the upper right corner of the figure. A forward DIU accommodates the primary control to and data input from the payload. Power is supplied to the payload from the Tug whether it be from the Tug's fuel cells or from some external power source.

The avionics system incorporates the latest technology concepts and components. It is a highly autonomous system capable of updating the Tug's position, velocity, and attitude knowledge by means of onboard sensors (interferometric landmark tracker (ILT), startracker and sun sensor). The system supports the vehicle reliability requirement of 0.97 for missions up to 6½ days with dual redundant subsystems; rendezvous and docking with a payload is provided by a hybrid system – LADAR providing the range data and television the attitude relationships. Dual lightweight fuel cells supply over 2 kilowatts of continuous power for support of Tug and payload subsystems.

The Tug avionics hardware is installed on the vehicle in both forward and intertank locations. The forward equipment area accommodates the data management, GN&C, rendezvous and docking, and communication subsystems. The four-sector installation arrangement provides for subsystem dedication and a common mechanical reference as required within the GN&C and rendezvous and docking subsystems. The intertank area accommodates the electrical power subsystem.

SUBSYSTEMS SUMMARY

Data Management Subsystem (DMS)

The DMS provides the central processing and command functions on board the Tug. All Tug subsystems interface with and are controlled by the DMS. The computer is a modular unit adapted from the SUMC development program. It utilizes a 48K fault-tolerant memory with internal redundancy and an error detecting and correcting translator.

The CIU is a dual redundant data preprocessing and control device. It performs buffering and formatting of systems data and directs bus traffic to and from the DIUs.

The DIUs are dual redundant at each of the four data bus remote terminals. Control of user sequencing and limited preprocessing of user data is accomplished in the DIU.

The recorder is a NASA standard 320M bit unit interfacing strictly with the CIU. It stores status/maintenance data for playback when needed to augment the telemetry coverage.

GUIDANCE, NAVIGATION, AND CONTROL SUBSYSTEM (GN&C)

The GN&C hardware is composed of sensors that provide input to the GN&C processing software in the DMS. The single output device in the GN&C subsystem is the engine control electronics, which electrically interfaces the main engine steering and APS discrete control.

The dodecahedron IMU achieves the equivalent of triple redundancy with only six laser gyros and six pulse rebalanced accelerometers. The IMU provides the prime vehicle state determining measurements.

The startracker and sun sensor work together to update the vehicle attitude information. Accurate vehicle position and velocity measurements required to update the guidance set are provided by the functionally redundant ILT. This, along with the attitude and time, provides a complete vehicle state update.

RENDEZVOUS AND DOCKING SUBSYSTEM (R&D)

The R&D subsystem is used to implement spacecraft retrieval and servicing.

The single LADAR provides the prime range and range rate measurements for the R&D acquisition and docking maneuvers. Functional dual redundancy is provided by the TV system. The prime function of the dual redundant LLLTV and its electronics is spacecraft inspection and attitude relationships for docking. A ground-based operator views the TV image and provides feedback data for control of the Tug all the way to "dock and latch."

COMMUNICATIONS SUBSYSTEM (COMM)

The COMM subsystem interfaces directly with the CIU. When Tug is deployed, it provides the uplink and downlink communication with the Shuttle and the ground. The COMM system functions as the main hardware interface to the Orbiter.

The antennas consist of three 25-element electronically steerable phase array antennas and associated directivity phase controllers. The phase controllers are driven directly from the DMS. Dual switched antennas provide local omnidirectional uplink/downlink coverage.

The transponder is a dual redundant cross-strapped unit with network mode select. It provides for carrier generation and modulation, ranging turnaround, command data detection, demodulation, and signal preprocessing. Any of the three phased arrays can be driven from either of the redundant transponders. This dual redundant unit is mode selectable to provide decoding of either NASA or DOD command formats. An internal data shunt provides access to the spacecraft engineering uplink. All uplink data is available at both the Tug signal processor and spacecraft decoder and is acknowledged/decoded when the proper vehicle is addressed. To meet secure communications requirements on DOD missions, data encryption/decryption devices are provided.

ELECTRICAL POWER SUBSYSTEM (EPS)

Vehicle prime 28 vdc power is supplied by the EPS. Prime power is generated by dual lightweight, thermally integrated fuel cells that operate from propellant grade reactants taken from the main propellant tanks. Power sequencing, distribution, and safing mechanisms are included in the EPS.

The dual redundant fuel cells use reactants from the main propellant tanks. The power generating subsystem waste heat is utilized to maintain both APS (N_2H_4) and hydraulic fluid temperature by integral recirculators and heat exchangers. Additional heat is rejected by way of four panel radiators mounted on the Tug intertank shell.

The battery functions as a contingency safety backup. This third backup function is sized around the worst case maximum loiter time upon Tug return to the Orbiter vicinity.

Prime power distribution is effected locally in both the forward and intertank areas. The intertank controller/distributor contains the Tug power changeover switch.

SUMMARY

The avionics system supports the advanced capability planned for the Space Tug. Projected technology advancements were instrumental in attaining an integrated system definition satisfying the requirements of performance, mission duration, payload retrieval and servicing, autonomy, and the safety and reliability required to operate in the manned Shuttle environment.

Subsystem requirements, selection rationale, and baseline configurations were developed and integrated into an optimized avionics system for the Tug.

This system, including its installation and interface characteristics, provides a valid basis for future phases planned within the Space Tug program.

AAS 75-143

LIGHTWEIGHT FUEL CELL
POWERPLANT FOR TUG

Lawrence M. Handley[*]

INTRODUCTION

There is a growing interest within the Government and among potential prime contractors in the type of prime electrical power supply to be developed for Shuttle-related spacecraft like the Space Tug and Interim Upper Stage. The fuel cell is an obvious candidate; for missions requiring more than 10 to 20 kilowatt-hours, the superior energy density of a fuel cell system results in important weight savings which are reflected in greater payload capability. Among the various fuel cells at United Technologies Corporation a lightweight powerplant based on advanced cell technology developed for the NASA Lewis Research Center is a viable option. Such a powerplant is the subject of this paper.

REQUIREMENTS

Typical requirements for a fuel cell powerplant in a Tug vehicle can be summarized as follows. Output voltage must remain within a 24.0 to 32.5 volt range over the normal power range of 0.5 to 2.0 kilowatts, but can dip to as low as 18 volts to supply 3.5-kilowatt peaks. Cryogenic propulsion grade reactants are supplied at low pressure. Contamination with helium pressurizing gas is possible. Oxygen pressure is always higher than hydrogen pressure. Total required operating time is 2500 hours. The powerplant accumulates operating time in missions averaging about 100 hours each. In case of an emergency, the powerplant must be able to supply the peak power of 3.5 kilowatts for up to eight hours.

---

[*] Manager, Government Fuel Cell Programs for Power Systems Division of United Technologies Corporation, South Windsor, Connecticut 06074

These requirements have implications for the powerplant design. The short missions force the designer to minimize fixed weight at the expense of reactant consumption. With careful design, the power section (cell stack) can be made to weigh less than the ancillary components needed for management of reactants, water, and heat.

CELL TECHNOLOGY

The basis for the lightweight power section is the alkaline cell with passive water removal. The cell consumes low pressure hydrogen and oxygen in separate reactions at the anode and cathode. Excess water vapor produced at the anode diffuses across the hydrogen cavity to the water transport plate, a porous matrix filled with aqueous potassium hydroxide. The driving force for water migration is the higher KOH concentration in the water transport plate relative to the cell itself. The water transport plate is kept "dry" by re-evaporation of water into a 4-psia cavity. The low pressure is maintained by venting to space or by an external condenser.

Some idea of the progress which has been made in alkaline cell technology as a result of Government interest and support can be obtained from Figure 1, which compares the specific weight of the cell stack for a

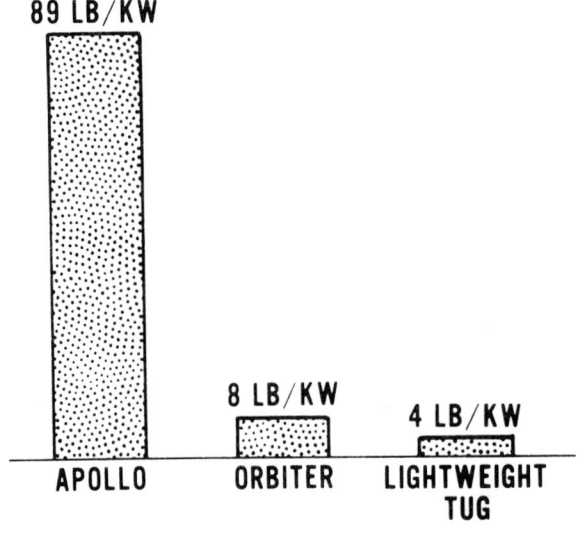

Figure 1   Cell Stack Specific Weight at Maximum Power Rating

lightweight Tug powerplant with the stacks of the Apollo and Orbiter powerplants. At its peak power of 3.5 kilowatts, the Tug stack has a specific weight less than one twentieth of the Apollo stack and about one half of the Orbiter stack at their maximum power ratings.

Figure 2 shows the components of the lightweight cell. Work on most components was initiated in IR&D programs, but substantial development of them all was accomplished in a four-year advanced fuel cell program

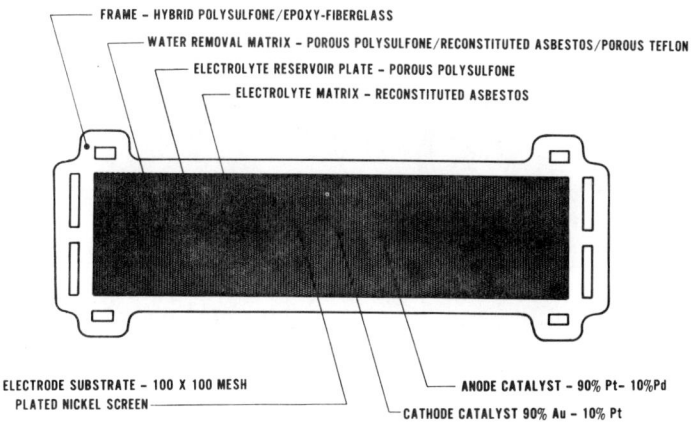

Figure 2   Lightweight Cell Components

sponsored by the NASA Lewis Research Center. Additional experience with some of the components was obtained in programs sponsored by the Air Force Aero Propulsion Laboratory and the NASA Johnson Space Center. The matrix and electrode materials are used in the PC17C Shuttle powerplant (under development) and in the PC15B powerplant to be delivered to the Navy for the Deep Submergence Rescue Vehicle. Most of the endurance testing of lightweight cells has been carried out under the LeRC-sponsored program. Thirty-eight individual 12 X 1.375 inch cells of various configurations and materials have been run for a total of 85,000 hours at the Tug powerplant operating pressure. Twelve of these cells accumulated more than 2500 hours each; five of these ran for more than 5000 hours; and one ran beyond 10,000 hours.

ANCILLARY COMPONENTS

The power section requires ancillary components for reactant conditioning, distribution and venting, water removal, coolant circulation, and electrical control. Cryogenic hydrogen and oxygen reactants are warmed in preheaters. The oxygen pressure is dropped to the level of the hydrogen before the gases enter the cell stack. Accumulated inerts are vented overboard periodically through redundant purge valves. Water vapor produced in the cells is vented overboard through a thermally biased pressure regulator. A condenser-separator can be included if there is a need to collect the water. A centrifugal pump circulates FC-40 coolant through the cell stack and the preheaters. A vehicle freon loop cools the FC-40 in an interface heat exchanger. Automatic electrical controls are used to start up and shut down the powerplant and actuate the purge valves. Many similar ancillary components have been developed for other fuel cell applications such as the Apollo Service Module and Space Shuttle.

POWERPLANT CHARACTERISTICS

Based on this previous experience with cell stack and ancillary components, a weight estimate of 48.0 pounds has been made for the Tug powerplant. Figure 3 compares the anticipated output voltage during the

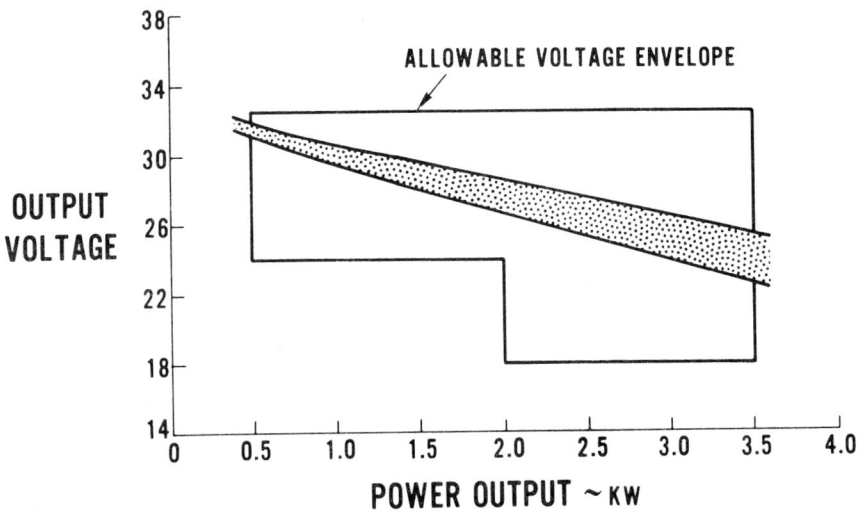

Figure 3   Output Voltage During 2500-Hour Operating Period

2500-hour operating period with the allowable voltage envelope. The powerplant meets the requirements with room to spare.

CONCLUSION

A fuel cell powerplant has been defined which meets typical requirements for a Tug or IUS vehicle and is substantially lighter than other powerplant options. The cell technology has been established by NASA and Air Force programs. Ancillary components similar to those needed in this unit have been developed for other existing powerplants. A logical next step is to put the cells and components together into a powerplant for development and demonstration. Successful completion of such a program will provide the confidence to move ahead to flight qualification.

AAS 75-144

SPACE TUG LASER GYRO IMU

Robert F. Morrison, Sperry Gyroscope
Bobby F. Walls, NASA, MSFC

Forthcoming space missions are going to require low cost equipment that can perform satisfactorily with very high reliability. Each of these missions employs some form of inertial measurement equipment that in the past has generally been special purpose. That is, it was tailored for each specific application, thereby resulting in high installation and maintenance costs. Recently the emphasis has been directed toward developing low cost building blocks from which the desired inertial functions can be derived.

This paper describes a National Aeronautics and Space Administration (NASA) program aimed at satisfying future space mission cost and reliability objectives. Specifically, an inertial measurement unit (IMU) employing strapdown laser gyros and accelerometers, is described. There is a two fold uniqueness to his approach; the utilization of six gyros and six accelerometers configured in a dodecahedron array and the gyros are laser gyros.

Studies have shown that IMU reliability can be maximized by configuring six gyros and six accelerometers such that their sensitive axes are normal to the faces of a dodecahedron. In this configuration all axes are uniformally separated by approximately 64 degrees. In an inertial system implementation groups of three sensor outputs are symmetrically placed about each of three reference axes so that reference axis angular rate and linear acceleration can be derived by combining data from three sensors. In this way a failed sensor can be automatically determined and eliminated from all subsequent inertial computations. Two failures of this type can be automatically accommodated, which is the key to increased reliability. A third failure can also be tolerated provided the failed axis can be determined.

The other unique area, the utilization of the laser gyro, has been under development for several years by Sperry Gyroscope for NASA Marshall Space Flight Center (MSFC). Substantial progress has been made that demonstrates the capability of the laser gyro to provide the required performance and reliability. NASA MSFC has accumulated several years of test data, including helicopter flight test of a laser gyro navigation system and reliability data, that provides the background for the laser gyro IMU described herein.

This paper presents a detailed description of a dodecahedron configured IMU that incorporates six Sperry Model ASLG-15 laser gyros and six Kearfoot Model 2401 accelerometers in a unique, integrated array. This uniqueness stems from a design approach that interleaves all six laser gyros on a common structure that also contains mounting provisions for the six accelerometers. The result is an extremely compact sensor assembly that has the high reliability and low cost potential of the laser gyro.

INTRODUCTION

The low cost and high reliability potential of the laser gyro was the impetus for this program.

Six years ago the National Aeromautics and Space Administration (NASA) initiated a laser gyor program directed towards the development of low cost, high reliability inertial systems for varied space applications. The NASA Marshall Space Flight Center was assigned the responsibility for this program and under contract to the Sperry Gyroscope Company, successfully developed a flight worthy laser gyro inertial measurement unit (IMU). A key program milestone was recently reached when a three axis laser gyro IMU was flown in a NASA helicopter as part of an aircraft navigation system. This success has prompted the current phase of the program, the development of a redundant strapdown laser gyro IMU where six laser gyro rate sensors and six linear accelerometers are configured for maximum reliability. The specific objective of this phase is to demonstrate, via flight test, the system's operation, performance and redundancy management capability.

The flight test program recently completed demonstrated that a strapdown laser gyro IMU can perform in a relatively rugged flight environment. Fig. 1 is a photograph of the inertial measurement unit. It consists of three single axis laser gyros, Sperry Model ASLG-15, and three linear accelerometers, Kearfott Model 2401 mounted such that their sensitive axes are orthogonal. The structure on which these sensors are mounted is thermally controlled, via thermoelectric coolers, to provide a 90°F base plate temperature. The sensor housing is attached to the vehicle via a three point vibration isolation system designed to attenuate vibrations above 100 Hz. The IMU was integrated into an aircraft navigation system that included a NASA developed digital computer. Reference 1 describes this system in detail.

Extensive laboratory and mobile ground tests were performed by the Marshall Space Flight Center (MSFC) that indicated navigation accuracies of about 3 NM/hr. Fig. 2 is a photograph of the NASA navigation van installation.

The system was transported to NASA Langley for installation in an SH-3 helicopter. The helicopter was flown over a radar instrumented course and its navigation accuracies measured for several different flight profiles. Table 1 is a summary of the results of six flights. Five of the six flights were better than 5 NM/hr. The sixth flight was significantly worse and a power supply malfunction was suspected.

Table 1

Flight	Position Error Rate (NM/Hr.) CEP	Nav. Time (Hrs.)	Flight Profile	Comments
1	5.1	2.2	Checkout	Flown over checkpoints
2	3.8	1.4	North-South	
3	4.8	1.3	East-West	Gusty winds
4		1.5	NE-SW	No ground track data
5	3.3	0.9	Circles/Spins	
6	12.0	2.0	Racetrack	Gusty winds, low temperature & low battery voltage

This phase of the flight test program was completed in April 1975. A follow-on test program is planned for later this year to investigate alternative strapdown system algorithms.

Based on the success of the flight test program, NASA has baselined a strapdown laser gyro IMU for the Space Tug Navigation System. Cost was the primary consideration in the selection of the laser gyro for the tug application. Historically, in order to maximize inertial system reliability two and three systems were paralled for redundancy. This approach, obviously expensive was also wasteful of size, weight and power and did not achieve maximum

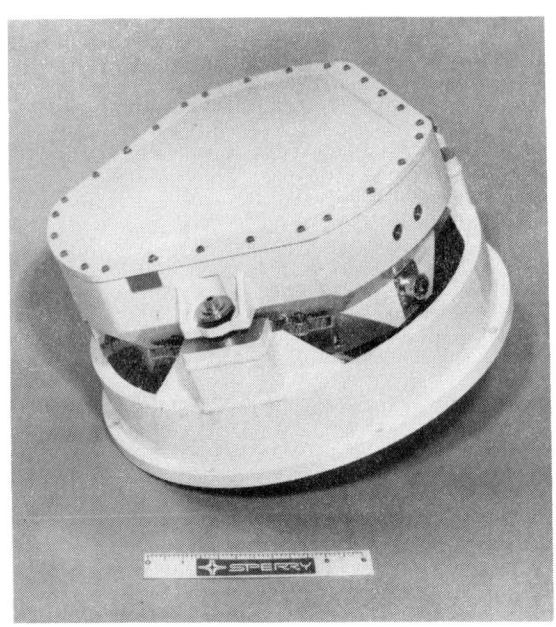

Fig. 1 Sperry Model 8300
Strapdown Laser Gyro IMU

Fig. 2 NASA Navigation Van
Installation

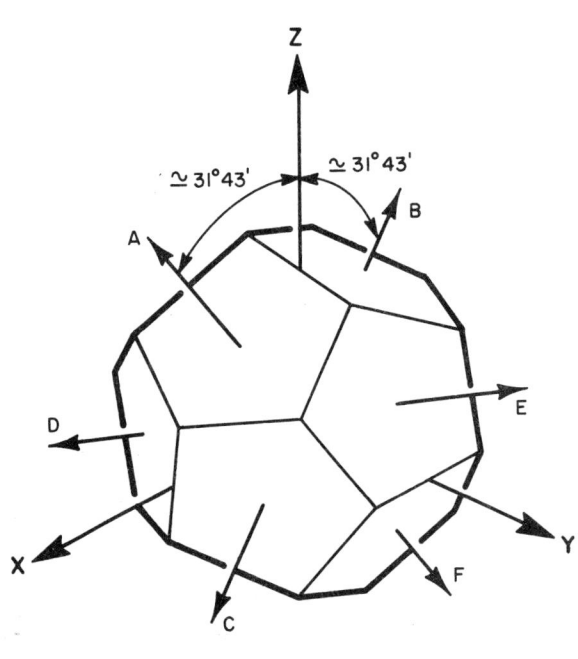

X,Y,Z — VEHICLE BODY AXES
A,B,C,D,E,F — SENSOR SENSITIVE AXES

Fig. 3 Redundancy Approach
Dodecahedron

reliability. Studies conducted by the Massachusetts Institute of Technology (MIT) relative to optimizing reliability showed that the most effective configuration utilizes six sensors configured in a dodecahedron array (Reference 2). This configuration, utilizing strapdown sensors, orients the six gyros (and six accelerometers) such that their sensitive axes are all symmetrically displaced by approximately 64 degrees (normal to the faces of a dodecahedron) as shown in Fig. 3. Vehicle body axes (X, Y, Z) are selected as shown so that body rates are derived from the resolution of three rate sensors. Thru software monitoring of the sensor outputs two sensor failures can be detected and automatically isolated so that a fail operational-fail operational-fail safe system results. A third failure can be tolerated and still retain full operational capability, however, the redundant management program will not automatically fault isolate.

Fig. 4 is a reliability comparison between alternative IMU configurations; non-redundant (one three axis strapdown laser gyro), dual redundant (two three axes strapdown laser gyro), triply redundant (three three-axes strapdown gyros) and the dodecahedron. As shown for one third fewer sensors the dodecahedron configuration can operate approximately four times longer for the same probability of success.

In order to complete the dodecahedron navigation system an appropriately programmed digital computer is required. Accordingly, NASA MSFC has developed, in parallel with the laser gyro, a small ultra-reliable modular computer (SUMC) for future space applications. A prototype model of this computer, designed by MSFC, was used in the three axis laser gyro navigation flight test program. The final version, principally faster, is being built for use in the dodecahedron flight test program. The strapdown algorithms and navigation software program used on the three axis flight test program will be used again. However, a redundancy management program will be added. The applicable redundancy management program was developed by MIT under contract to NASA. Fig. 5 is a block diagram showing the software functions programmed into the dodecahedron SUMC computer.

IMU incremental angle (laser gyro outputs) and incremental velocities (accelerometer outputs) are inputs to the computer program. These data, in sensor axes, undergo initial resolving into body axis coordinates. The body referenced gyro data is used to derive vehicle attitude relative to an earth reference coordinate system. The resulting attitude matrix (direction cosines) is then

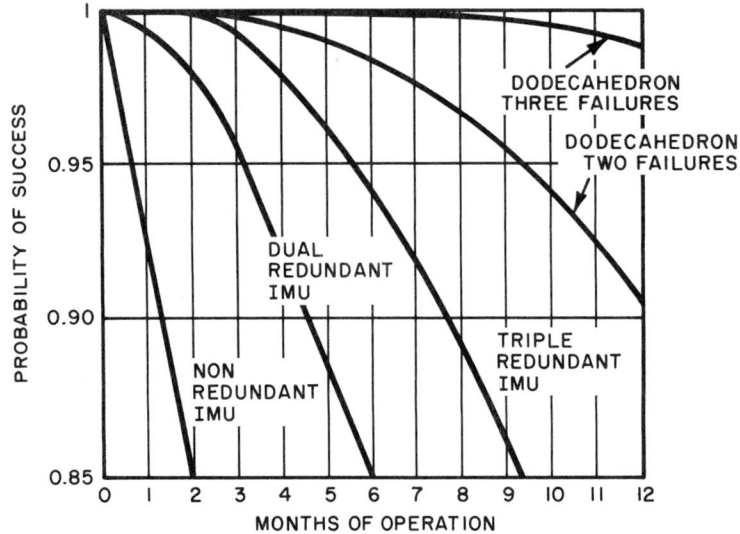

Fig. 4 Reliability Advantage Dodecahedron

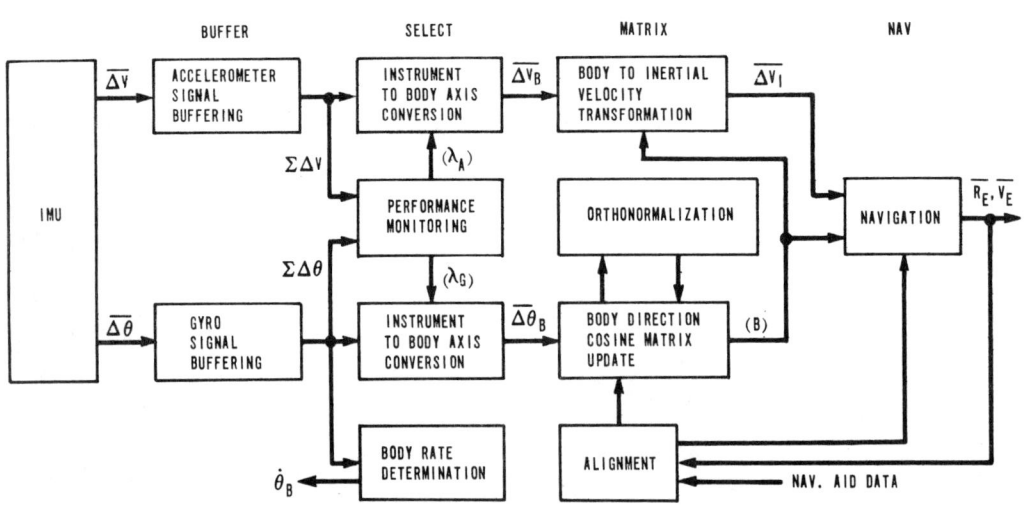

Fig. 5 Inertial System Block Diagram

used to resolve the body referenced acceleration into the earth coordinate system for navigation computations.

The redundancy management portion of the program evolves essentially around the performance monitoring function. This function compares the outputs of the sensors on a reasonableness basis (e.g., indicated body rates from the rate sensors are compared and the odd ball, if any, identified) so that data from a failed sensor can be eliminated from the computation. In addition, the IMU is designed with built in test equipment (BITE), as described below, to increase the failure detection probability. Further, the multi-sensor data enhances system performance by virtue of the statistics relative to several sources of the same information.

The next milestone in the NASA program is the flight demonstration of the dodecahedron navigation system. NASA MSFC is coordinating the program and will integrate a redundant strapdown laser gyro IMU and a SUMC computer into a flight demonstration system. Sperry Gyroscope is the contractor that is designing and building the IMU. IBM is building and programming the NASA designed SUMC computer. The discussion that follows presents a detailed description of the IMU.

REDUNDANT STRAPDOWN LASER GYRO IMU

General Description

The redundant strapdown laser gyro IMU is a six gyro, six accelerometer configuration designed to provide incremental body angular rates and linear accelerations in digital form for processing by the SUMC computer into earth referenced navigation data. Minimizing cost, an overall program objective, is addressed thru the utilization of the laser gyro in a strapdown configuration. Maximizing reliability, a specific design objective, is achieved by the six sensor dodecahedron configuration and careful electronics design.

Fig. 6 shows the three major assemblies that comprise the redundant strapdown laser gyro IMU; sensor assembly, electronics assembly and power supply. The sensor assembly contains the twelve inertial sensors enclosed in a thermally controlled, evacuated and magnetically shielded case. The angular rate sensors are Sperry Model ASLG 15 type laser gyros configured in a unique arrangement where the optical paths of the six gyros are interleaved to minimize size and weight and are oriented in the prescribed dodecahedron orientation. These gyros have the characteristics tabulated in Table 2.

Table 2
LASER GYRO - SPERRY MODEL ASLG-15

Random drift (1σ)	0.01°/hr
Day-to-day drift repeatability	0.1°/hr
Scale factor (nominal)	3.3 arc-sec/pulse
Scale factor repeatability	0.005%
Scale factor linearity	0.01%
"G" sensitivity	NIL
Temperature sensitivity	< 0.01°/hr/°F
Magnetic sensitivity	0.03°/hr/earth-field
Threshold	0.01°/hr
Life	>30,000 hrs.
Angular rate range	0.01°/hr to 100°/sec
Angular acceleration range	$\pm 1000°/sec^2$
Linear acceleration	20g

The accelerometers used in the sensor, Kearfott Model 2401, are mounted on the gyro optical cavity such that their sensitive axes are nominally aligned with the gyro axes. The cavity structure is made of aluminum and is thermally controlled to minimize gyro and accelerometer temperature sensitivity. Table 3 shows the characteristics of the accelerometer.

Table 3
ACCELEROMETERS - KEARFOTT MODEL 2401

Zero "G" bias	307.2 kHz
Scale factor (nominal)	17,066.67 pulses/sec/G
Scale factor stability	60 ppm
Bias repeatability	$\pm$0.0004 G
Bias stability	0.000015 G
Temperature sensitivity	0.00001 G/°F
Non-linearity (second order)	0.00001 G/G^2
Pivot cross coupling	0.00001 G/G^2
Acceleration Range	$\pm$18G
Accelerometer threshold	0.018 ft/sec

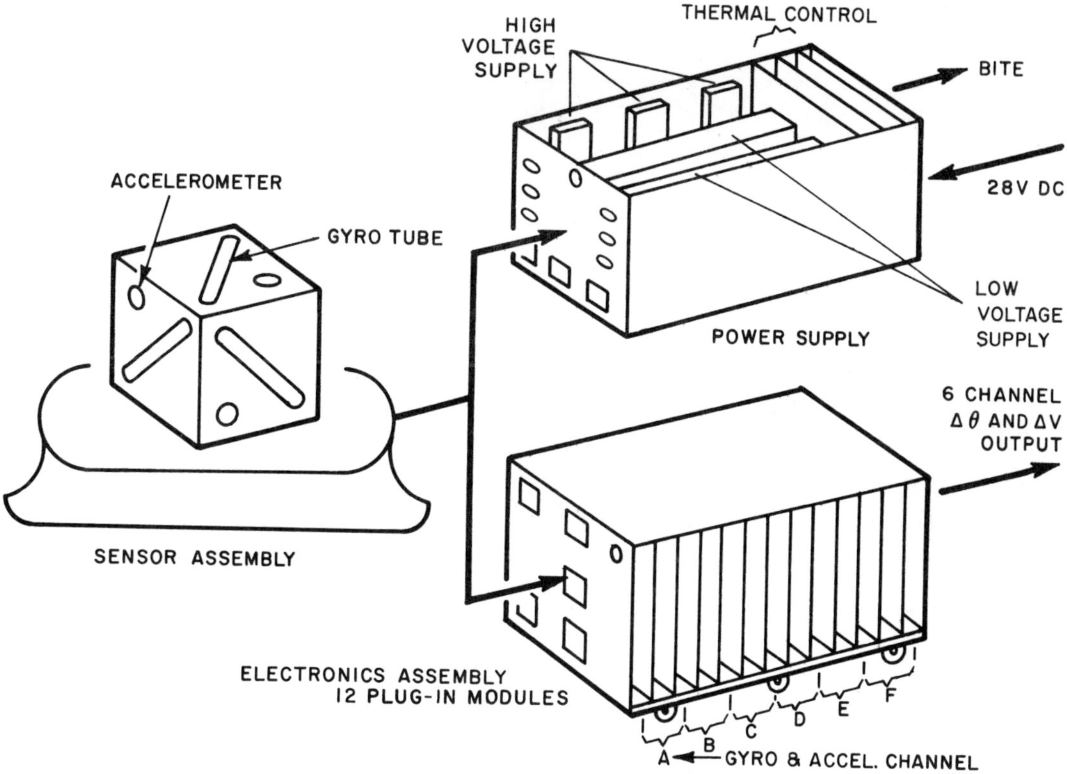

Fig. 6 Redundant IMU Implementation

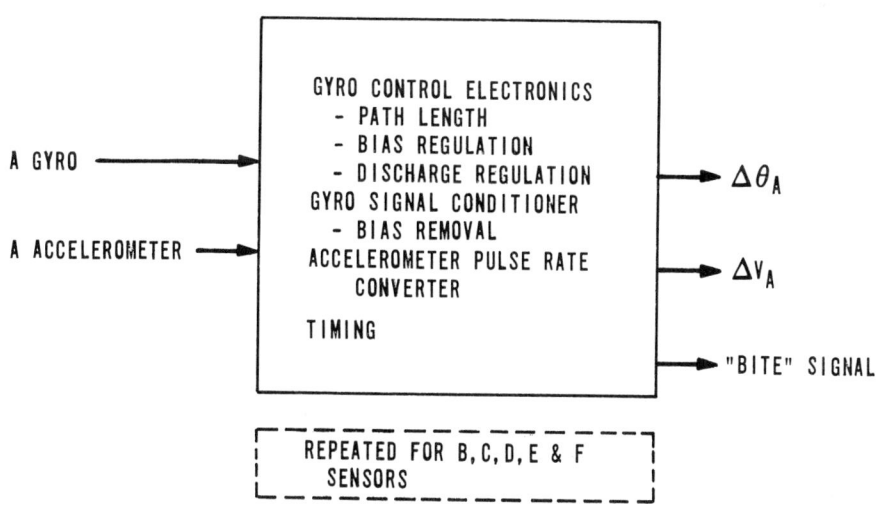

Fig. 7 Channelized Signal Flow

The electronics assembly contains the analog and digital circuitry for the control and signal conditioning of the gyros and accelerometers. These electronics consist of plug-in modules designed for independent control of each sensor axis. In addition, each module contains continuous monitoring of key sensor signals (built-in-test) that provides an output discrete and a visual (light) indication in the event of a failure.

The third assembly, the power supply, contains the power conditioning required for sensor and electronics operation and the active thermal control circuitry employed in controlling the sensor operating temperature. Functionally, the power supply is divided into three areas; low voltage supply, high voltage supply and thermal control circuitry. The low voltage supply and the thermal control circuitry are dual redundant, whereas there are six high voltage supplies, one for each gyro. This assembly also contains continuous monitoring of key voltages and signals for failure indication. Both the power supply and the electronics assembly contain external visual indicators that identify which assembly has failed so that the appropriate module can be identified and replaced. Table 4 summarizes the physical characteristics of the IMU.

Table 4

IMU PHYSICAL CHARACTERISTICS

Component	Size (Inches)	Weight (Lbs)
Sensor Assembly	10 Dia. x 10	30
Six Model ASLG-15 Laser Gyros		
Six Model 2401 Accelerometers		
Dodecahedron Configuration		
Electronics Assembly	7 x 9-1/2 x 13	15
Six Gyro modules		
Six Accelerometer Modules		
Power Supply	6 x 7 x 13	15
Two Low Voltage Supplies		
Six High Voltage Supplies		
Two sensor cavity thermal control circuits		
Two sensor case thermal control circuits		

Redundancy Implementation

The major consideration in the design of the IMU was achieving a fail operational-fail operational-fail safe configuration. This requirement dictates that the system can have two failures without compromising system performance. Historically, this has been achieved by employing triplicate redundancy with its inherent physical and cost penalties. In the redundant laser gyro IMU the two failure tolerance has not been completely adhered to. The higher failure rate elements, the sensors, have been channelized such that three failures can be tolerated without compromising system performance. The lower failure rate elements, power supplies and thermal controls, are only dual redundant or single failure tolerant.

The channelized sensor configuration is achieved by independent excitation and control of each sensor in the sensor assembly. The control electronics, in the electronics assembly, are modularized by axis as shown in Fig. 7. These control modules accommodate a specific gyro and accelerometer pair (one sensitive axis) such that all inputs and outputs are independently received and transmitted. In addition, a separate failure indication discrete is sent to the digital computer for continuous monitoring of the operation of each channelized module. These modules are repeated six times. One for each of the six sensor pairs.

Special consideration was given to the dual redundant power supply. The supply was designed to tolerate either an open or short circuit on both input and output as well as an internal failure without interfering with IMU operation. The short and open circuit protection was achieved by appropriate utilization of a combination of diodes, voltage regulators and fuses. Fig. 8 is a schematic diagram of the IMU power distribution. The primary 28 volt DC lines (dual redundant) are cross connected to the two IMU low voltage power supplies and to the six high voltage power supplies (these supplies are DC to DC converters). The power input fuse array effectively isolates each primary supply from a short circuit failure in any of the eight power supplies. The diode array is used to cross couple both primary inputs and still be tolerant to a primary power short circuit failure.

The IMU power supply outputs are cross coupled with diodes similar to the primary supplies. That is, a low voltage output short circuit or open circuit failure will not interrupt power as each redundant output is isolated via the diodes. Each of the two low voltage power supplies have the capability of independently powering the entire IMU, so that an internal failure of one supply will not interfere with IMU operation.

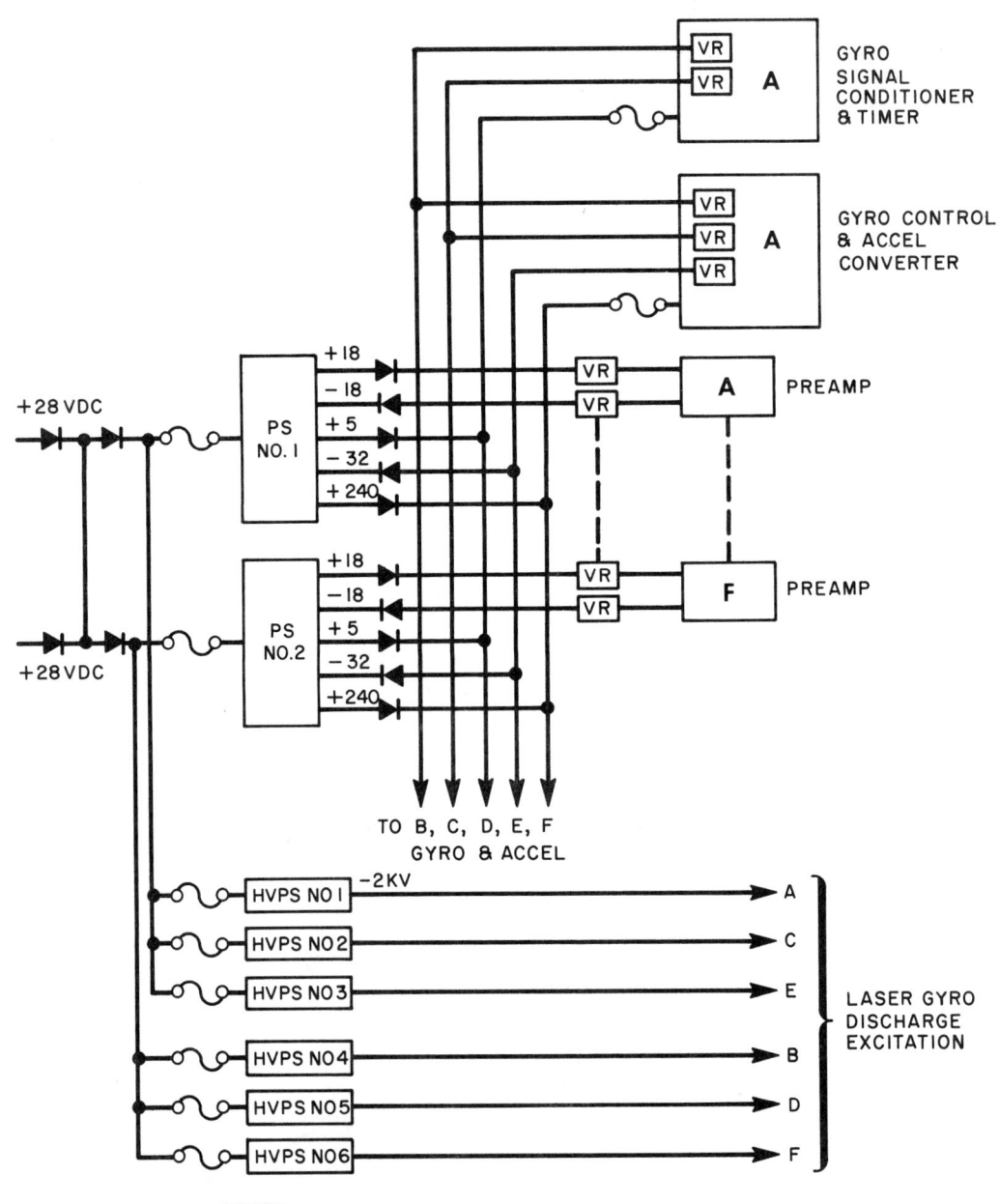

Fig. 8 Power Distribution

All of the secondary low voltage excitations are protected from load short circuit failures by either a voltage regulator (VR) or a fuse. Voltage regulators, effective open circuits for a load failure, were used for all circuits where voltage ratings permitted. In the higher voltage circuits fuses were employed to isolate load short circuit failures.

This power supply configuration effectively assures continuous IMU operation for any single short or open circuit failure.

Built-In-Test (BITE) and Fault Isolation

To assist in redundancy management and to facilitate maintenance the IMU provides continuous monitoring of key circuit parameters coupled with a visual indication of a failed module. The built-in-test (BITE) provisions have been implemented on a replaceable module basis. The operation of the channalized gyro and accelerometer control and signal processing circuits are monitored by comparing specific parameters to a reference. If a parameter fails to satisfy the established criteria a failure signal is generated (digital discrete) that is sent to the digital computer where it is used to effectively disable the failed sensor channel. Each module has circuits that "ANDS" the parameters being monitored so that any parameter failure will provide a failure indication. The failure indication, in addition to being sent to the digital computer, lights a failure indicator on the failed module to facilitate repair. Fig. 9 shows this IMU BITE and failure indication implementation.

To provide additional assistance in establishing IMU operational status, both the electronics and power supply assemblies contain externally visible failure indicators. The module failure indicators are not visible when each assembly is installed with covers in place. Therefore, all of the individual module failure output circuits are "ANDED" within each assembly and used to excite an assembly failure indicator which indicates an internal failure.

Sensor Configuration

The sensor configuration developed for the redundant strapdown laser gyro IMU is a very unique arrangement that is worthy of some additional discussion. The laser gyro is the key element. Fig. 10 is a functional schematic diagram of the laser gyro.

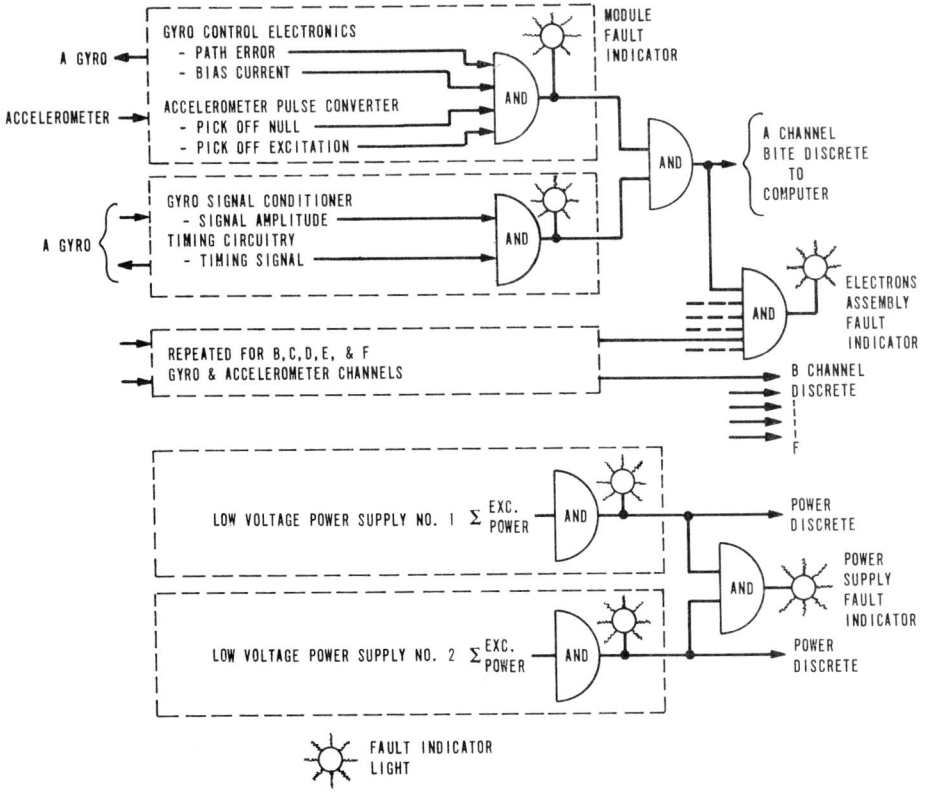

Fig. 9 Built-In-Test (BITE) and Fault Isolation

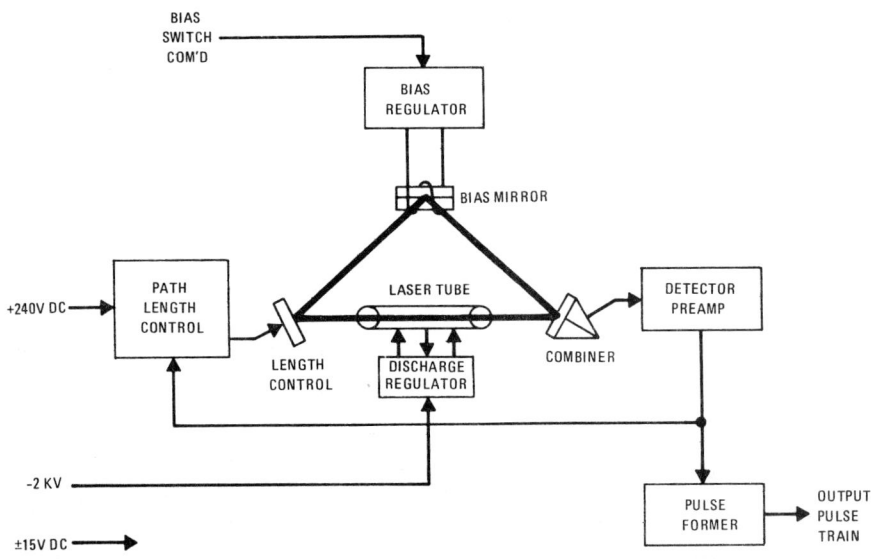

Fig. 10 Ring Laser Gyro Functional Schematic Diagram

The ring laser gyro is a strapdown inertial angular rate sensor. The outputs of this sensor are processed using digital microcircuits to provide a pulse train proportional to body axis angular rate. This type of gyro consists of an optical cavity around which counter-rotating laser light beams travel. It includes a gas discharge tube containing a low-pressure mixture of helium and neon, which serves as the laser energy source, and a combination of light refracting mirrors used to form a triangular light path, as shown in the figure.

The laser tube, located in one leg of the triangle, emits nearly monochromatic light from both ends. The beams are confined to a closed triangular path defined by reflectors located at the three corners of the triangle. Each of three corner elements, the bias cell, the length control, and the combiner, are different and have dual functions. The bias cell, in addition to "reflecting" the light beam around a corner, contains an electromagnetically controlled optical bias device that effectively alters the path length for the oppositely propagating beams to produce a frequency difference between them. This introduces an optical bias into the ring output, ensuring the existence of at least some minimum output and thereby avoiding the rate threshold caused by frequency lock-in at low rates.

The length control element, in addition to directing the light beam in the appropriate direction, provides control of the optical path length. This corner reflector is mounted on a piezoelectric actuator that adjust the mirror in response to detected changes in cavity length. The actuator provides the path length control that stabilizes the laser gyro scale factor.

The combining mirror directs the beams to close the optical path. It also allows a small portion of each beam to be transmitted out of the cavity to a beam splitter (a partially transmitting, partially reflecting optical element). Each of the two beam splitter outputs contains both clock wise and counterclockwide frequency components. One of these output beams is focused on a photodetector whose square-law response heterodynes the cw and ccw beams, producing a beat frequency that is proportional to the cw and ccw frequency difference. The photodetector output is an electrical signal varying sinusoidally at the beat frequency. The magnitude of the beat frequency is a function of the applied optical bias and the input angular rotation rate. This output signal is processed digitally to remove the applied bias component, leaving an output that is an exact measure of the ring angular rotation input.

One of the key characteristics of the laser gyro is that its operation depends on the ratio of its area to perimeter, which, in general, makes the laser gyro larger than comparable conventional sensors. In order to minimize the size, as well as the weight and power impact, a configuration was designed that provided interleaving the optical paths of all six laser gyros. Fig. 11, a photograph of a model of the redundant IMU, shows the gyro and accelerometer arrangement. The basic structure is a cube of aluminum that has triangular optical paths meachined into it such that when the basic laser gyro optical elements (3 mirros and gas discharge tube) are affixed they form six independent gyros with each gyro sensitive axis displaced in accordance with the dodecahedron array. Fig. 12 shows this sensitive axis orientation.

Provision is made to mount the six accelerometers on the aluminum block so that their sensitive axes are nominally aligned with the gyro axes. The exact position of each sensor axis is established via a complete sensor assembly dynamic alignment procedure. A body axes reference is provided via a permanently affixed optical cube on the outside of the sensor assembly. The positions of the six gyro sensitive axes, relative to this optical cube reference, are established via three single axis rate tests. An alignment compensation matrix is determined that precisely resolves each sensor output into appropriate body reference axes via a computer software program. Simularly, the accelerometer axes are established via a three position static test.

CONCLUSION

For the last several years NASA has pursued the development of low cost, high reliability inertial systems that would satisfy a broad spectrum of future space missions. Two specific developments, sponsored by the NASA Marshall Space Flight Center, have recently culminated in a successful flight demonstration of a strapdown inertial navigation system that contained two new component developments; a small ultra-reliable modular computer (SUMC) and a laser gyro inertial measurement unit (IMU). The SUMC is a digital computer that employs state-of-the-art large scale integrated circuits configured in a functional modular breakdown. The laser gyro IMU is a three axis strapdown sensor that included three linear accelerometers to provide incremental angular rates and linear acceleration outputs.

Fig. 11 Redundant Laser Gyro IMU

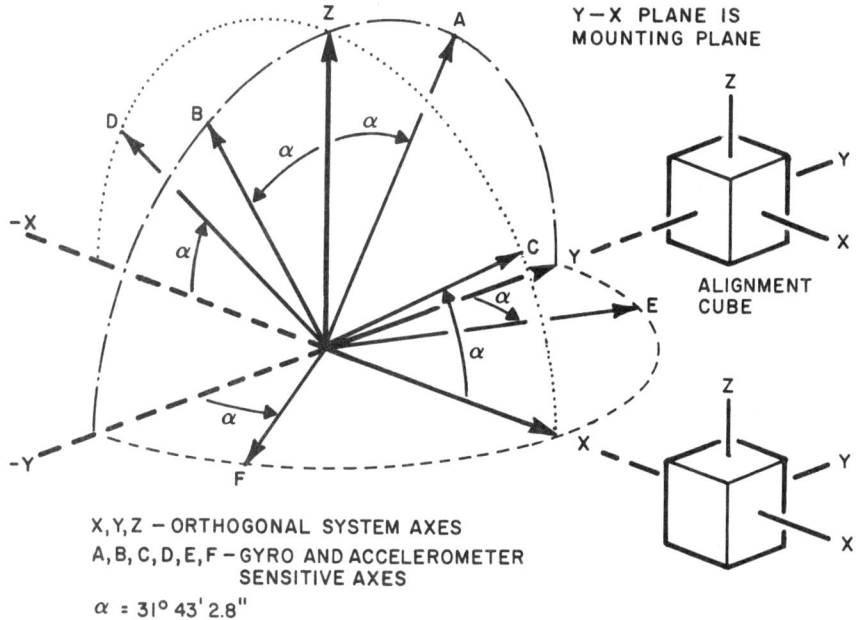

X,Y,Z — ORTHOGONAL SYSTEM AXES
A,B,C,D,E,F — GYRO AND ACCELEROMETER SENSITIVE AXES
$\alpha = 31° 43' 2.8''$

Fig. 12 Gyro and Accelerometer Sensitive Axis Orientation

Existing space vehicle inertial systems have employed complete system redundancy to provide the necessary high reliability. Analyses have shown that employing six sensors arranged on the faces of a dodecahedron results in maximizing reliability, such that a minimum amount of hardware (lowest cost) produces a system capable of operation after three failures, (a fail operational - fail operational - fail safe system).

A natural outgrowth of both the successful operation of the low cost, high reliability laser gyro and the dodecahedron redundancy analyses is the development of a redundant (dodecahedron) strapdown laser gyro inertial measurement unit. Accordingly, a program has been initiated that will culminate in the flight demonstration of such a system in 1976. A successful demonstration will be a major milestone towards the achievement of low cost, high reliability inertial systems for the space missions of the 80's.

REFERENCES

1. Paper, by R. Morrison, H. Garret, B. Walls, "A Strapdown Laser Gyro Navigator" Presented at National Aerospace and Electronics Conference, May 13 - 15, 1974

2. Paper, by G. Gilmore, "A Non-Orthogonal Gyro Configuration" Jan. 1967

AAS 75-145

INTERFEROMETRIC LANDMARK TRACKER APPLIED TO PRECISE SPACE TUG AND PAYLOAD NAVIGATION SYSTEM *

D.H. Aldrich and
W.F. Hubbarth[+]

INTRODUCTION

Self-contained spacecraft navigation systems that do not rely on a dedicated ground support system for earth reference data have the potential for great accuracy and rapid convergence from large initial errors. Such autonomous navigation systems acquire earth reference data continuously and process it in real time to establish spacecraft position, velocities, and attitude; as a result mission operations are independent of ground tracking locations. Advances in digital processing hardware and software technology and earth reference sensors have made such autonomous space navigation systems practicable. Consideration of these factors, with respect to current and future space missions, resulted in the initiation, in 1971, of an Autonomous Navigation Technology Program by the USAF Space and Missile Systems Organization.*

Figure 1 shows the system concept being studies by IBM for the USAF Space and Missile Systems Organization under the Autonomous Navigation Technology Program. The key element in this system concept is the Interferometric Landmark

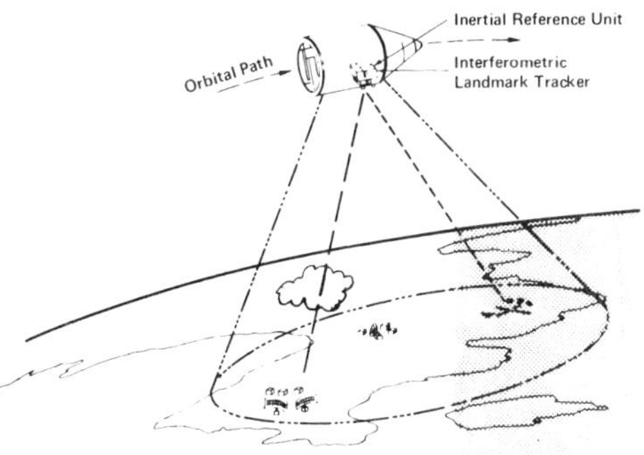

Figure 1. E/F-Band Emitters

---

*USAF/SAMSO Contract F04701-71-C-0339
 USAF/SAMSO Contract F04701-73-C-0221
 USAF/SAMSO Contract F04701-75-C-0046

+ IBM Corp., Federal Systems Division, Owego, NY 13827

Tracker (ILT), which provides accurate angular tracking of radar landmarks whose position is known (±100 ft) in latitude, longitude and altitude. This strapped-down sensor, composed of two orthogonal phase interferometers operating in the E/F frequency band, provides a 120° field-of-view (FOV) and permits continuous tracking of one or more radars without mechanical or electronic scanning. This performance is independent of cloud cover, season or time of day. The ILT angle data is processed by a Kalman filter to obtain vehicle navigation and attitude. Sequential tracking of landmarks provides sufficient measurement data in the spacecraft to accurately determine these nine vehicle states.

ILT DESCRIPTION

The ILT system block diagram (Figure 2) indicates the four-channel configuration chosen. The four 2.1" diameter antennas can be mounted directly on the vehicle structure since the two interferometers need not be in the same plane nor exactly orthogonal if baseline length (typically 40 inches) and the angle between interferometers is known to 1 part in 10^4. During typical ILT operation, the computer selects an available landmark and, by means of the signal processor, tunes the local oscillator to the landmark frequency. The receiver then down-converts all intercepted signals to the receiver IF; the IF signal phase differences from the four receiver channels are then converted to sine and cosine of pitch and roll phase and sent to the signal processor for conversion to digital words. The signal qualification and frequency control circuits: (1) provide coarse and fine measures of signal frequency, (2) compare wide and narrow band signal amtplitudes,

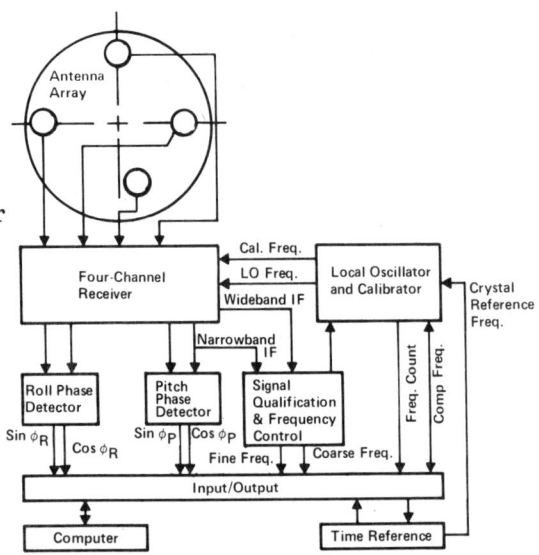

Figure 2. ILT Block Diagram

and (3) examine coarse frequency for signal qualification (spurious signal rejection). The input/output samples, holds, converts (A/D), and buffers the phase and frequency data to permit operation with a serial data bus, which also services the inertial reference unit. The average ILT output data rate is less than 200 bits/second.

The system functional configuration and physical characteristics are shown in Figure 3. The ILT provides the landmark tracking and angular sensing for attitude determination. The Harpoon Midcourse Guidance Unit (MGU) provides the short term inertial reference, computation support and provides interfaces with other vehicle subsystems. The MGU is a moderate accuracy inertial unit with short term gyro performance in the order of 0.25°/hr drift. This configuration also provides the basis for system guidance. Software estimates are 12,000 words (32 bit) and 15,000 operations/second.

ILT/SYSTEM PERFORMANCE FOR A TYPICAL SPACE TUG MISSION

The precision of an ILT autonomous navigation system for a typical space tug mission is shown in Figure 4, for the sequence of events on a transfer to synchronous orbit, placement of a spacecraft after a circularization burn, a return transfer and circularization at low earth orbits (LEO) for rendezvous with the shuttle.

SUMMARY AND CONSIDERATIONS

The ILT system provides a means of Autonomous space navigation applicable in the Shuttle era to many spacecraft, including the Space Tug. Studies and mission simulations have shown that this system, when combined with modest inertial elements, provides high precision navigation. It offers a means of achieving a low cost navigation, guidance and attitude determination capability since the relaxing of inertial component requirements and elimination of a separate attitude sensor can yield significant cost savings and improved reliability. The currently available ILT technology provides a valuable capability for designers of spacecraft in the Shuttle ERA.

ACKNOWLEDGEMENTS

The authors wish to thank Dr. Mohammed Islam and Mr. E.E. Kroman for the analysis of navigation system performance. The critique by Mr. N.F. Toda and his helpful suggestions during the analysis contributed substantially to this report.

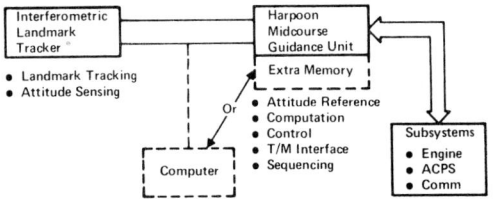

ESTIMATED SYSTEM CHARACTERISTICS (SIMPLEX)

System Component	Weight (lb)	Volume (in³)	Average Power (watts)	Peak Power (watts)	MTBF (hrs)
Interferometric Landmark Tracker (ILT) (Includes Time Reference Unit)	20.0 (No Mounting Hardware)	1080 (12 x 10 x 9)	8.0	93	>16,000
Harpoon Midcourse Guidance Unit	25.4	680 (12 Dia x 6)	96 (76 ARA)	127 (76)	> 3,000

Figure 3. System Functional Elements

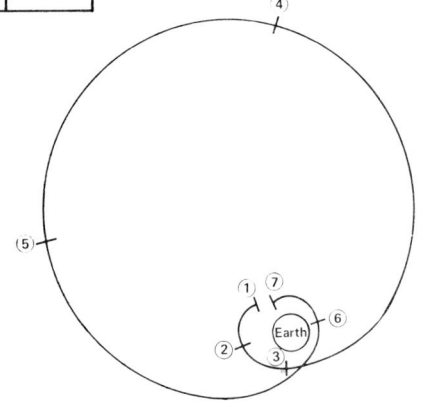

Event (Start)	Navigation Errors (RSS 1 Sigma)		
	Position (ft)	Velocity (ft/sec)	Attitude Arc Seconds
① Release	17,168	7.36	323
② Transfer to Syn Orbit	38	0.04	16
③ Midcourse Correction	98	0.13	8
④ Syn Orbit Circularization	1980	0.19	12
⑤ Transfer to LEO	3870	0.28	13
⑥ LEO Circularization	414	0.76	22
⑦ LEO Rendezvous	51	0.05	45

Figure 4. ILT Tug Mission Performance

AAS 75-146

LSI COMPUTER FOR 1980's SPACE MISSION

W. A. Clapp, RCA[*]
J. E. Saultz, RCA[+]
J. B. White, NASA/MSFC[≠]

A low power, high reliability computer -- the SUMC-III-C -- is being designed and fabricated by RCA under contract to NASA/MSFC (Contract Number NAS-8-29072). The computer is one member of the NASA/MSFC Space Ultrareliable Modular Computer (SUMC) family (Fig. 1). NASA/MSFC originated the concept of a modular, high reliability, computer family in the late 1960's to meet the needs of a wide range of missions through modularity in word size, throughput, and reliability. Several family members have been built, both by NASA/MSFC and by outside contractors. The SUMC-III-C is directed at such missions as Large Space Telescope and Space Tug. It is being fabricated using custom-designed CMOS/SOS LSI arrays, with critical computer paths packaged on thick-film hybrid substrates. This technology provides 300 to 600 logic gates per array with a typical stage delay of 3 to 5 ns. The SUMC-III-C takes full advantage of RCA developments supported by NASA over the past ten years in the major technological areas of MOS, computer-aided design, hybrids and memories.

Several unique, and forward-looking features have been incorporated in the computer system. Probably most important is that it is fully S/360 compatible at the user level. Programs developed for S/360 can be executed on the

---

[*] Manager, Applied Computer Systems Laboratory, RCA Advanced Technology Laboratories, Camden, N. J.

[+] Manager, Market Development, RCA Advanced Systems and Technology, Camden, N J.

[≠] Chief, Computer Development Branch, Data Systems Laboratory, NASA Marshall Space Flight Center, Huntsville, Alabama.

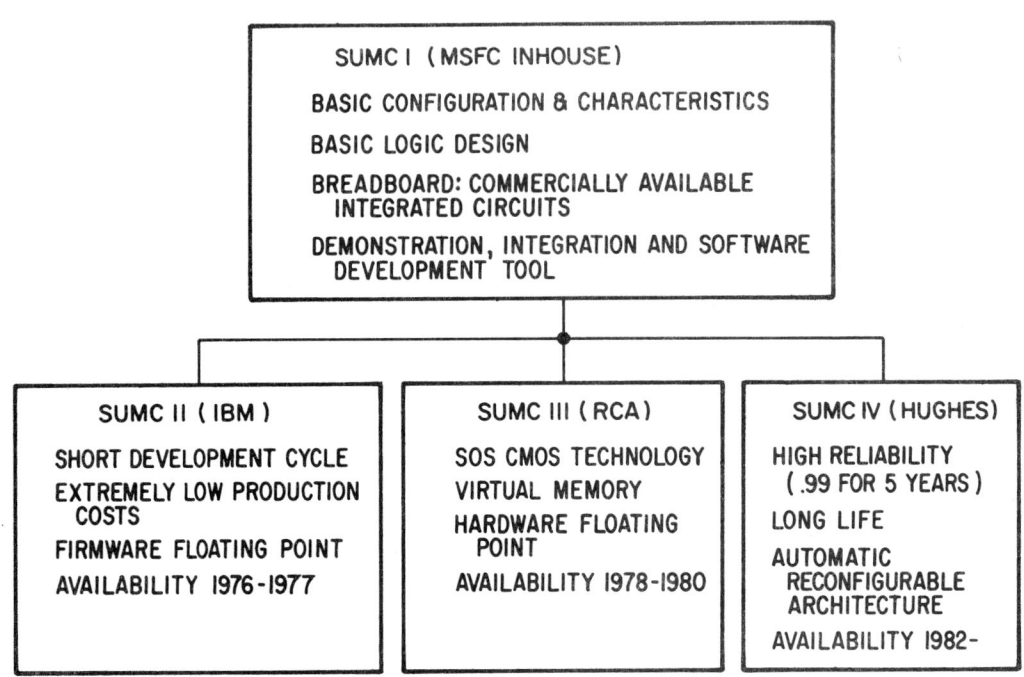

Fig. 1 Present SUMC Family

SUMC-III-C; and programs written for the SUMC-III-C can be debugged on the S/360 computers. A feature included for the first time in an aerospace computer is that of virtual memory. The virtual memory is implemented in the operating system, but enforced by hardware to minimize time delays. In addition to the actual virtual memory operation, two additional checks are enforced by hardware: (1) memory access type, and (2) degrees of privilegedness. Also included is a completely separate I/O processor which provides several channels of direct memory access capability.

The central processor organization is based upon a multiplexed 32-bit data path including hardware floating point implemented with 8-bit building blocks. The CPU and HSIU have been partitioned into 14 custom LSI building blocks (Table 1). The computer is controlled via a two-level microprogram. The data path organization and the building block concept have been proven in the four separate SUMC computer implementations.

Table 1

CUSTOM LSI CHIP TYPES FOR SUMC-III-C

No.	Chip Type	Chip Quantity in Typical System Usage	No. of Devices	No. of Pins
1	CAM/RAM	2	2962	70
2	Floating Point MUX	4	570	48
3	Sequencer/Iteration Counter	10	1060	64
4	8-Bit Register/Parity Generator	16	1158	64
5	8-Bit Adder	15	1604	65
6	9-Bit 4 × 2 MUX	27	536	64
7	Adder Control	4	580	48
8	Scratchpad Address/Interrupt	4	1574	64
9	Address Translation System Control	4	In Design	
10	I/O Register	5	2150	53
11	Bus Buffer	2	1580	64
12	Sequencer Control Unit Decode	2	In Design	
13	Random Logic	2	In Design	
14	Channel Control	1	In Design	
	TOTAL	98		

The flight unit will dissipate less than 55 W (including power supply inefficiencies) for the CPU, IOP, and 16K words of main memory. Typical I/O rate of 1×10^6 bytes/second per channel is provided with typical instruction times of 1.6 μs memory-to-register-add and 9.0 μs memory-to-register-multiply. Package size will be approximately 15.7 × 13.2 × 6.2 inches.

A new operating system -- the Modular Operating System for SUMC (MOSS) -- has been developed in concert with the hardware design. MOSS has been designed for real-time operation. It supports virtual memory, fault and error recovery, and concurrent control of multiple real-time and batch applications.

MOSS supports HAL/SM, Fortran IV, GOAL, and the S/360 assembler user interfaces.

Both hardware and software will be debugged by first quarter 1977 in a ground environment to ensure satisfactory operation. All of this capability will then be available for the space missions of the 1980's.

SPACELAB AND AUTOMATED PAYLOAD PROGRAMS

SPACELAB AND AUTOMATED PAYLOAD PROGRAMS

Program Chairman　　　　　　　M. E. Turner, Martin Marietta Corporation

Spacelab Operations
　Chairman　　　　　　　　　　H. Fletcher Kurtz, Jr., NASA/MSFC
　Co-Chairman　　　　　　　　　Richard Sylvester, Martin Marietta Corporation

Spacelab Payload Planning
　Chairman　　　　　　　　　　William Armstrong, NASA Headquarters
　Co-Chairman　　　　　　　　　William P. Pratt, Martin Marietta Corporation

Spacelab Integration
　Chairman　　　　　　　　　　Joel Levy, General Electric Company
　Co-Chairman　　　　　　　　　Robert L. Rosenthal, Martin Marietta Corporation

Spacelab Science Payloads
　Chairman　　　　　　　　　　Wilfred Scull, NASA/GSFC
　Co-Chairman　　　　　　　　　John Roach, Ball Brothers Research Corporation

Spacelab Life Sciences and Applications
　Chairman　　　　　　　　　　Richard Johnston, NASA/JSC
　Co-Chairman　　　　　　　　　Maurice Larue, Martin Marietta Corporation

Spacelab Technology Payloads
　Chairman　　　　　　　　　　W. Ray Hook, NASA/LRC
　Co-Chairman　　　　　　　　　Richard Herring, Ball Brothers Research Corp.

Communications Payloads
　Chairman　　　　　　　　　　Eugene Ehrlich, NASA Headquarters
　Co-Chairman　　　　　　　　　Peter McManamon, U.S. Dept. of Commerce

Multidiscipline Payloads
　Chairman　　　　　　　　　　Kenneth A. Stone, General Dynamics Corporation
　Co-Chairman　　　　　　　　　J. Kent O'Kelly, Martin Marietta Corporation

Automated Payloads
　Chairman　　　　　　　　　　Donald Hicks, Ball Brothers Research Corporation
　Co-Chairman　　　　　　　　　John Smialek, Martin Marietta Corporation

Automated Planetary Spacecraft
　Chairman　　　　　　　　　　Louis D. Friedman, JPL, Calif. Inst. of Tech.
　Co-Chairman　　　　　　　　　William T. Scofield, Martin Marietta Corporation

AAS 75-236

SPACELAB TASK ALLOCATION - A PRELIMINARY
DETERMINATION OF ONBOARD VERSUS
GROUND OPERATIONAL PRIORITIES

W. J. Harris[*]
R. E. Holmen[+]

The diversity of operational activities inherent in Spacelab's role as a general-purpose space-based laboratory requires a reassessment of flight task allocations for onboard and ground activities. The type of ground support utilized in previous manned spacecraft programs is economically impractical for Spacelab because a far greater number of flights will be required. Effective ground support will be a key element in achieving the versatility needed to support Spacelab. Before effective ground support methods can be established and analyzed, however, operational tasks must be allocated.

Using a matrix identification technique, the McDonnell Douglas Astronautics Company determined operational tasks and established onboard versus ground priorities for each of several criteria - for example, safety, efficiency, complexity, and required response time. The relationships of the various criteria were then considered in order to isolate individual task priority ratings for onboard performance.

Consideration of onboard crew time availability for selected operationally representative missions then established the preliminary task allocations for both Spacelab and ground activities. Permutations of both crew availability and criteria relationships were examined to determine their impact on task allocations.

Although task allocation activities are primarily limited at present to consideration of experiment- and mission-dependent subsystem tasks, the techniques used and solutions derived are applicable to all Spacelab flight operations tasks.

---

[*] Houston Flight Operations, McDonnell Douglas Astronautics Company, Houston, Texas

[+] Advanced Space and Launch Systems, McDonnell Douglas Astronautics Company, Huntington Beach, California

INTRODUCTION

The diversity of operational activities inherent in Spacelab's role as a general-purpose space-based laboratory requires a systematic basis for allocation of specific mission tasks to the onboard crew or the ground support team. A technique was developed to allow an objective evaluation of the relative desirabilities of task performance location. This technique was composed of a three-phased approach. First, a list of potential operational tasks anticipated for Spacelab missions was developed. Next, a set of criteria was established that reflected all significant reasons for allocating task performance location. Finally, a technique was developed for combining those criteria to allow tasks to be arranged in order of their priority.

A detailed description of the various spacelab systems was then prepared and operational tasks defined to cover the following mission phases:

      I.   Boost phase, preactivation, and descent
     II.   Spacelab activation
    III.   Subsystem management
     IV.   Experiment activation
      V.   Experiment operation
     VI.   Experiment deactivation
    VII.   Mission management and contingency planning
   VIII.   Emergency action.

DEVELOPMENT OF TASK ALLOCATION CRITERIA

To provide a systematic basis for allocation of the identified operational tasks, it became necessary to develop criteria allowing evaluation of the relative desirability of performing a task on the ground or on board. The following task allocation criteria were defined:

Criteria	Consideration
Crew safety	Crew safety cannot be compromised.
Work load	Minimize routine work, maximize time available for experimentation and observation.
Cost	Minimize implementation cost.
Efficiency	Take simplest approach.
Weight/space	Minimize Spacelab weight.
Accessibility	Minimize awkward operations.
Complexity	Minimize onboard complexity.
Response time	Enhance completion of time-critical tasks.
Capability	Perform task where support capability is best.
Constrained	Allocation driven by external considerations.

TASK EVALUATION USING TASK ALLOCATION CRITERIA

Using the task allocation criteria and the Spacelab operations task list, each task was evaluated by independently considering each criterion. This evaluation assigned a quantitative value for allocation of the task to a specific location when viewed in the light of a single criterion. Due to their diverse nature, the criteria for task allocation obviously were not of equal weight. For this reason, a strict numerical or statistical combining of the task evaluations for each criteria would not be valid. Consequently, various degrees of weighting for each of the criteria were prepared by MSFC and modified through analysis.

A mission which was considered representative of a Spacelab mission (involving a lab and a pallet) was chosen and the weightings applied against the tasks for that mission using onboard crew time availability data supplied by NASA MSFC.[1] Using the various criteria and varying available crew time by ± 10% to establish sensitivity, it was determined that the tasks which should be performed on board are generally directly associated with running the experiments themselves. The effect of variations in available crew time are to vary the results with respect to whether data evaluation, consumable allocation planning, and generation of flight program updates should be done on board or on the ground.

---

[1] Operations Development Division letter EL-14 (35-74) dated Oct. 30, 1974.

AAS 75-237

NASA/ESA CV-990 AIRBORNE SIMULATION OF SPACELAB*

D. Mulholland,** C. Neel,** J. DeWaard,[†] R. Lovelett,[††] L. Weaver,[§]
and R. Parker[§§]

Beginning in the 1980 time period an advanced space transportation system involving a reusable Space Shuttle vehicle will be employed to carry a laboratory (Spacelab) into earth orbit to conduct experiments from the space environment.

Similarities between the method of experiment accommodation and operations planned for Spacelab and the successful methods used by the NASA–Ames Airborne Science Office to conduct experimentation aboard aircraft led to an interest in applying the airborne techniques to Spacelab. The resulting program, called ASSESS (Airborne Science/Spacelab Experiments System Simulation), has involved a deep study of the airborne techniques applicable to Spacelab and several simple simulations, which led up to the complex NASA/ESA Joint Mission using the CV-990 airborne laboratory.

Although it is impossible to fully simulate Spacelab operations using aircraft, many of the ingredients of an airborne science operation are Spacelab-like and can serve to develop appropriate Spacelab operational procedures. Foremost is the ability to conduct authentic science, which automatically induces all of the hard personal interest and drive to achieve basic goals. These factors in turn provide authenticity for application of the results to Spacelab. The mission did not address physiological or psychological factors.

The overall objective of the Joint ASSESS Mission was to evaluate a simplified management and implementation concept for conducting Spacelab-like experiment operations. Contributory objectives were evaluation of experiment design, impact of operational requirements and

---

*Prepared for technical papers that will later be published in the Proceedings of the American Astronautical Society
**Ames Research Center, NASA, Moffett Field, CA 94035
[†] ESA Headquarters, Paris
[††] NASA Headquarters, Washington, D.C. 20546
[§] Marshall Space Flight Center, NASA, Alabama 35812
[§§] Johnson Space Flight Center, NASA, Houston, Texas 77058

procedures on Spacelab design, evaluation of payload operations and integration of experiments and equipment, and analysis of factors affecting selection and training of payload specialists.

Mission guidelines included selection of three U.S. and three European experiments to conduct authentic science on the CV-990 airborne laboratory, use of Ames ASO practices as a point of departure, maximum participation by principal investigators (PIs), selection of two European and two U.S. experiment operators (EOs) to be confined onboard with a Mission Manager for a five-day Spacelab simulation period, limited test equipment and tools to be carried along, and limited communication.

Guidelines were established by representatives of several NASA installations and ESA. The mission was implemented and managed totally by the Mission Manager and a small support staff, and utilization of several NASA Ames standard support organizations. Very limited documentation was needed. The PIs were fully responsible for their own experiments throughout but with appropriate imposition of safety and interface requirements. The EOs had a wide variation in background experience, and included a scientist-astronaut from NASA and an inexperienced young graduate student from England. All had strong backgrounds in physics and science. The PIs were responsible for EO training. Experiments were constructed at the PI's home laboratory and integrated on the CV-990 at Ames. One readiness review was held about two months before flight.

The joint mission provided valuable scientific and engineering data that will not only enhance scientific knowledge in the fields of upper atmospheric studies and IR astronomy, but will also provide sound guidelines for the design and operation of future Spacelab experiments. Science data will be published appropriately by the PIs. The EO approach for proxy experiment operation was successful, although many problems were encountered with the experiments, as might be expected. Baseline information on selection and training requirements was obtained.

The mission illustrated that a low-cost program for experiment preparation, testing, and integration can operate successfully under proper management approach. The Mission Manager concept worked very well. It was evident that an aircraft can serve as an excellent platform for optimizing the methodology, design, and operations aspects of experiments conceived for Spacelab. As expected, however, the added complexity of Spacelab-type operations does require more rigid and formal arrangements than those normally associated with airborne payloads. Specific evaluations were made in several areas such as subsystems, experiment support equipment, use of a centralized data system versus minicomputers, EMI, communications, experiment design, and PI/EO relationships, all of which can benefit Spacelab planning.

AAS 75-238

SPACELAB PAYLOAD ACCOMMODATION

Donald M Waltz[+]

Spacelab, when operated with the NASA Space Shuttle System for earth-orbital flights starting in 1980, will be a highly versatile, general purpose laboratory to perform space-based experiments in several disciplines.

This paper reviews the ability of Spacelab to accommodate a variety of scientific, applications, and technology payloads.

SPACELAB DEFINITION

Spacelab consists of two basic elements - a pressurized, man-tended laboratory type module and an unpressurized pallet - which can be used separately or in combination. The three typical flight configurations of Spacelab are: Module only, Module + pallet and Pallet only.

Figure 1 shows the pressurized module plus on pallet sector combination and also indicates summary details of a pallet only configuration.

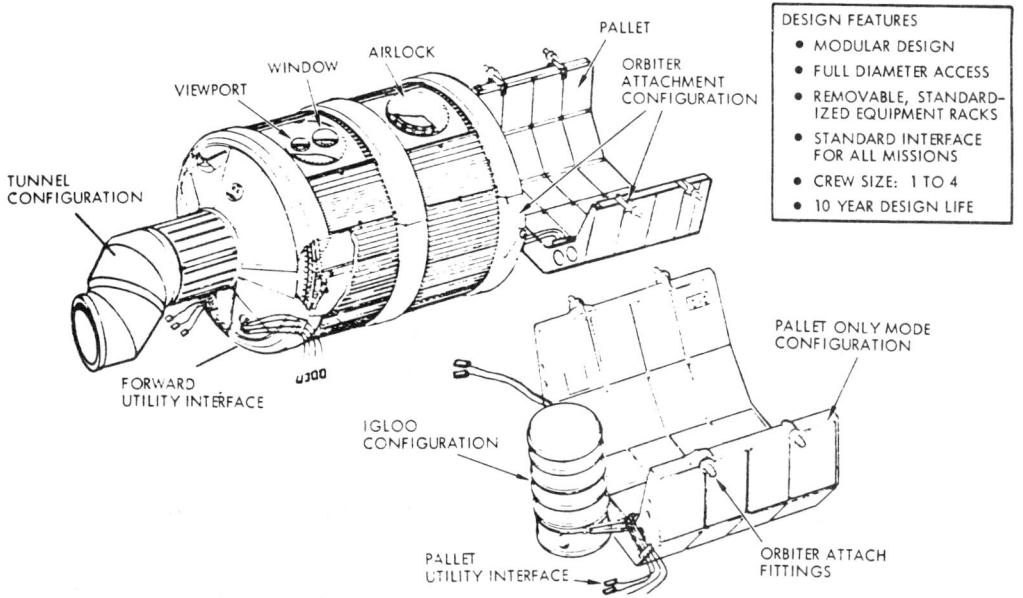

Figure 1. Spacelab External Features

[+] Space Vehicles Division, TRW Systems Group, Redondo Beach, Calif.

The module provides a controlled pressurized environment for the users and their equipment, and supplies basic services such as power, thermal control, and data management, together with certain basic support equipment such as standard racks, scientific airlocks, etc. which may be used as required. The pallet is an unpressurized platform to which instruments such as telescopes, sensors and antennas may be mounted which require direct exposure to space. The pallet provides some basic services, such as power, data distribution and thermal control.

The Spacelab payload designer can select from 13 different module/pallet arrangements for payload equipment layout.

Current NASA planning shows a Space Shuttle model consisting of 572 flights of which 226 will involve some combination of Spacelab elements.

SPACELAB CAPABILITY

Figure 2 summarizes the Spacelab user support capability. The values for the capability parameters vary depending on the Spacelab configuration.

SPACELAB USER SUPPORT CAPABILITY

- CREW SIZE
- EXPERIMENT WEIGHT
- TOTAL PRESSURIZED VOLUME
- AVERAGE POWER (EXPERIMENTS)
- ENERGY
- POINTING ACCURACY
- DATA TRANSMISSION (DOWNLINK)
                    (UPLINK)
- DATA RECORDING
- ENVIRONMENT CONTROL

- 1-4 PAYLOAD SPECIALIST
- 5500 Kg TO 9100 Kg
- 5 TO 22 M^3
- 4 TO 5.2 KW
- 420 TO 590 KWH
- 1 ARC SEC
- 50 MBPS DIGITAL; 5 MHz ANALOG/VIDEO
- 2 KBPS COMMANDS
- 30 MBPS DIGITAL; 2.5 MHz ANALOG; 6.0 MHz VIDEO
- 4 TO 6.6 KW HEAT REJECTION

OTHER SUPPORT AVAILABLE

- VIEW PORTS
- EXTRA VEHICULAR ACTIVITY
- CONTROLS/DISPLAYS

- AIRLOCKS
- BOOMS
- FILM STORAGE
- THERMAL CONTROL

- MANIPULATORS
- COMPUTER

Figure 2. Spacelab Capability

ACCOMMODATION ANALYSIS

Primary inputs to Spacelab accommodation analysis are Spacelab configuration defintion, payload descriptions, and assessments of payload requirements. When these are formatted against the 10 experiment Sortie mission disciplines the result is the tabulation of payload data shown in Figure 3.

DISCIPLINE	NUMBER OF PAYLOAD EXPERIMENT CLASSES	SPACELAB ACCOMMODATION MODE			CARRY ON
		MODULE	MODULE AND PALLET	PALLET	
ASTRONOMY	33			X	X
HIGH ENERGY ASTROPHYSICS	16			X	
SOLAR PHYSICS	14			X	
ATMOSPHERIC AND SPACE PHYSICS	6		X	X	
EARTH OBSERVATIONS	16	X	X	X	
EARTH AND OCEAN PHYSICS	10		X	X	
SPACE PROCESSING APPLICATIONS	16	X	X	X	X
LIFE SCIENCES	5	X	X		X
SPACE TECHNOLOGY	24	X	X	X	X
COMMUNICATIONS/NAVIGATION	13	X	X	X	

Figure 3. Spacelab Payloads Accommodation Summary

AMPS PAYLOAD ACCOMMODATION

Employing the Shuttle Orbiter/Spacelab as the carrier/laboratory system, the objectives of the Atmospheric, Magnetospheric and Plasma in Space (AMPS) program is to:

- Study the earth's atmosphere and magnetosphere
- Answer questions about man's total natural environment
- Conduct basic plasma physics experiments in space.

The AMPS program lists an impressive set of experiments. Planned at about 4 flights per year program from 1980 through 1990, some 80 instruments associated with about 50 different experiments will be required to conduct research in the four general categories of: atmospheric sciences, wave interaction, particle interaction, and plasma interaction and flow.

A full complement of AMPS payload equipment would use the Spacelab-Short-Module plus a 9 meter pallet configuration. This Spacelab combination when outfitted to take full advantage of Spacelab resources will be similar in concept and operation to a large laboratory in space. During the planned 7 day Sortie missions, scientists will use a host of perturbing sources and diagnostic sensors to make both in situ and remote atmospheric, magnetospheric and plasma physics measurements.

AMPS laboratory will grow in response to the needs of the using scientists. The variety of instruments and perturbing sources will increase on later flights; individual instruments will become more versatile. For example, the first AMPS payload includes a simple electron accelerator for use in particle-injection experiments. On later flights, beam-shaping and modulating techniques will improve its versatility, and ion and plasma accelerators will be added.

AMPS payload may fly on Shuttle verification flights 3 and 5 and on Spacelab missions assigned to Shuttle flights 8 and 19.

SPACE PROCESSING APPLICATIONS PAYLOAD ACCOMMODATION

The Objectives of a NASA program for Space Processing Application (SPA) Spacelab payloads, which relate to materials science and manufacturing in space, are:

- Develop the technology needed to make space manufacturing possible

- Use space laboratories to solve important problems in materials science and technology

- Develop working relationships between the government, industry and the international scientific and industrial communities to implement the program

- When appropriate, produce products in space for use on the ground.

There are four general interest areas of SPA experiment conduct. They are: (1) biological preparations, separation, and purification, (2) solidification of metals, alloys, composites, and electronic materials, (3) ceramics and glasses, and (4) other space environment dependent processes that affect the physical or chemical properties of materials.

Space Processing anticipates a start-up phase which begins modestly and grows into a routine R&D program phase. More immediate is the present identification of early missions which are outlined in the 1979-1981 period. Presented on Figure 4 are the objectives of SPA payloads on Shuttle flights 4/6 and 8/14.

SPA MISSION 4/6 OBJECTIVES

- CRYSTAL GROWTH AND METALLURIGICAL PROCESSES THAT EXCEED SKYLAB & ASTP CAPABILITY
- TEST OF NEW APPARATUS E.G. CONTAINERLESS POSITIONING FURNACE WITH EB HEATING
- PURIFICATION OF HIGH MELTING POINT MATERIALS IN CONTACTLESS POSITIONING FURNACE FACILITY
- CHECTOUT OF APPS* POWER AND THERMAL SUBSYSTEMS
- APPS/STS THERMAL CONTROL INTERFERENCE TESTS
- PAYLOAD/STS INTERFACE VERIFICATION TESTS

* Auxiliary Payload Power System

SPA MISSION 8/14 OBJECTIVES

- EVALUATE METHODS AND EQUIPMENT FOR THE SEPARATION AND PRESERVATION OF BIOLOGICAL SAMPLES IN LOW GRAVITY
- MOBILITY MEASUREMENT OF PREPARATIVE QUANTITIES
- SOLID LIQUID INTERACTIONS IN CRYSTAL GROWTH
- VERIFY AND EXTEND RESULTS OF THE SKYLAB & ASTP MISSIONS

Figure 4. Objectives of SPA Early Shuttle Mission Payloads

CONCLUSIONS

Spacelab offers the potential for conducting meaningful and effective space research and development activities for a variety of payloads.

Three types of issues result from accommodation analysis for the disciplines which plan to have payloads on Spacelab flights:

1) Issues related to basic feasibility within the Spacelab physical capability.

2) Considerations pertaining to the Shuttle/Spacelab induced environment

3) Performance margins between Spacelab subsystem capabilities and payload requirements.

Both AMPS Space Processing Applications payloads can be accommodated very well by the Spacelab and are both strong payload candidates for early year Shuttle/Spacelab missions.

AAS 75-239

CONCEPT VERIFICATION TEST: EVALUATION OF SPACELAB/PAYLOAD OPERATION CONCEPTS

R. O. McBrayer and H. H. Watters[*]

Past concepts of integrating experiments, experiment operation on-orbit, and ground support of orbital experiment operations must be changed significantly for the Shuttle era of low cost, rapid turn-around payloads and greater user involvement. The Concept Verification Test (CVT) Project began experimenting in April 1973 with an early representation of Spacelab, using a proposed Spacelab operational concept as an initial approach. Four simulations to date in the CVT/General Purpose Laboratory have involved approximately 50 scientific users from NASA Centers, other government departments, and universities. These simulations have provided the operational environment to evaluate: concepts for collecting and implementing experiment integration and operations requirements; techniques for allowing the user to integrate experiments before delivery to the test facility; on-board duties for Payload operation; User/Payload Specialist interface; and proposed Spacelab operations concepts. In general, these simulations have provided a forum for simultaneous operations evaluation by Spacelab/User personnel in a setting most reflective of Spacelab and Experiment development. As the CVT Project brings additional, higher fidelity simulators to an operational status, Spacelab/Payload Operation concepts are being further analytically defined and iterated by evaluations in existing CVT simulators. Future CVT Simulations will verify the actual Spacelab/Payload operation concept in a timely phase of the Spacelab Program.

---

[*] Man/Systems Integration Branch, NASA, Marshall Space Flight Center

INTRODUCTION

Space Shuttle, currently being developed, is expected to represent the "coming of age" of U.S. manned space operations. Central to the operational flexibility of this reusable craft is its 4.6m (15 ft) x 18m (60 ft) cargo bay. This bay can accommodate various payloads, configured to match the requirements of given flights. Primary among currently planned cargos for the Shuttle is the European-developed Spacelab, a multi-purpose, modular, manned or unmanned experiment carrier.

Spacelab itself can assume several configurations, depending upon the way in which the several basic modules are assembled. One version might be a small 4.3m (14 ft) diameter x 2m (6.6 ft) manned pressurized section combined with a larger space-exposed "pallet" upon which certain remotely operated vacuum compatible instruments might be mounted. Another configuration could be a somewhat larger manned pressurized section, in which a number of experiments are mounted and conducted.

Thus, Spacelab might assume a wide range of configurations. For example, on one flight, a Spacelab might accommodate a Life Sciences payload; on another, a Material Sciences payload. On still other flights, experiments from several disciplines may be intermingled.

Moreover, Spacelab is to accommodate, with minimal restriction, the needs of scientific users; an ever greater share of responsibility for experiment integration and operation is to rest with what has been termed the "user community." It is envisioned that qualified users will be afforded the opportunity to conduct their own research on orbit, if they so elect. More often, it is expected that users will have a major role in the selection and training of flight experiment operators or "payload specialists" who will be their orbiting science representatives.

It is these payload specialists who will bear primary responsibility for the scientific success of their assigned missions.

Current Shuttle/Spacelab planning calls for a three-man Orbiter crew and a payload crew consisting of a variable number (1-3) of researchers (payload specialists). One member of the flight crew, the "mission specialist," is also oriented toward payload operations. His payload duties include contingency maintenance/repair of payload support systems, EVA as required, and assistance with experiment operations as necessary.

Spacelab, then presents a range of questions which are, for NASA, new: How can NASA best accommodate the users, allowing them maximal freedom to structure and conduct their experimentation? What guidelines govern the combination of disparate disciplines on a given flight? How effectively can a few flight researchers conduct the protocols of a larger number of ground-based investigators? By what mechanism are payload specialists selected? These and a variety of related questions began a search for new integration and operation concepts for Spacelab. One highly relevent model which had addressed operations in a way analogous to that envisioned for Spacelab was that developed by the Ames Research Center Airborne Science Office. This procedure had been used for several years in conducting research flights to various parts of the world. As other papers in this session point out, the ASO program continues and is increasingly oriented toward evaluation of Spacelab concepts. The model provided by ASO was also adopted by the CVT project as a starting point for the development of specific Spacelab operations concepts.

The Concept Verification Test (CVT) project provides elements which approximate the overall provisions of Spacelab. Although the internal arrangement of the simulators differs from the current Spacelab configuration, CVT provides potential researchers with representative experiment resources such as 28 VDC and 115/220 VAC power, fluids, vacuum, thermal conditioning, data systems, and instrumentation. Like Spacelab, CVT seeks to provide research users with a supportive environment in which to integrate and conduct their experiments. Also like Spacelab, CVT places primary responsibility for experiment operations in the hands of the researchers. The CVT personnel are primarily <u>facilitators</u> of user requirements and <u>observers</u> of the flow of test operations.

The CVT Project converted existing simulators to physical representations of Spacelab and initiated development of a total mission simulation capability. This "building block" approach includes an early physical representation of Spacelab and a separate breadboard of a data system functionally similar to Spacelab. The experience gained with these two simulators (and evolving definition of Spacelab) is being combined into functionally high fidelity Spacelab and Shuttle Interface Simulators. Thus, operations concepts and simulator fidelity (functional and physical) mature concurrently.

This paper presents a description of the CVT Project, a summary of results from four CVT simulations completed to date, and planning for operations evaluation in future CVT simulations.

CVT DESCRIPTION

The CVT Project was created to support, through simulations, the pre-design and development phases of payload carriers (excluding Shuttle Orbiter except for payload interface), payloads, and to support potential users of the carrier in the development of the experiments. CVT coordinates the hardware, software, and operational concepts and requirements of different payload carrier developers, payload developers, and users. Detailed mission simulation plans are formulated, simulators are provided, operations are conducted, and the simulation results are distributed.

The CVT Project affords an opportunity, early in the design and development phase, to accomplish the following:

1. Exercise operational concepts for payload carrier, payload, and user integration in pre-mission ground operations, mission operations, and post-flight mission operations.

2. Identify interface problems between the payload carrier, payload, and user. The level to which interface problems are identified is directly related to the fidelity of the simulations.

3. In selected cases, identify payload carrier subsystem integration problems.

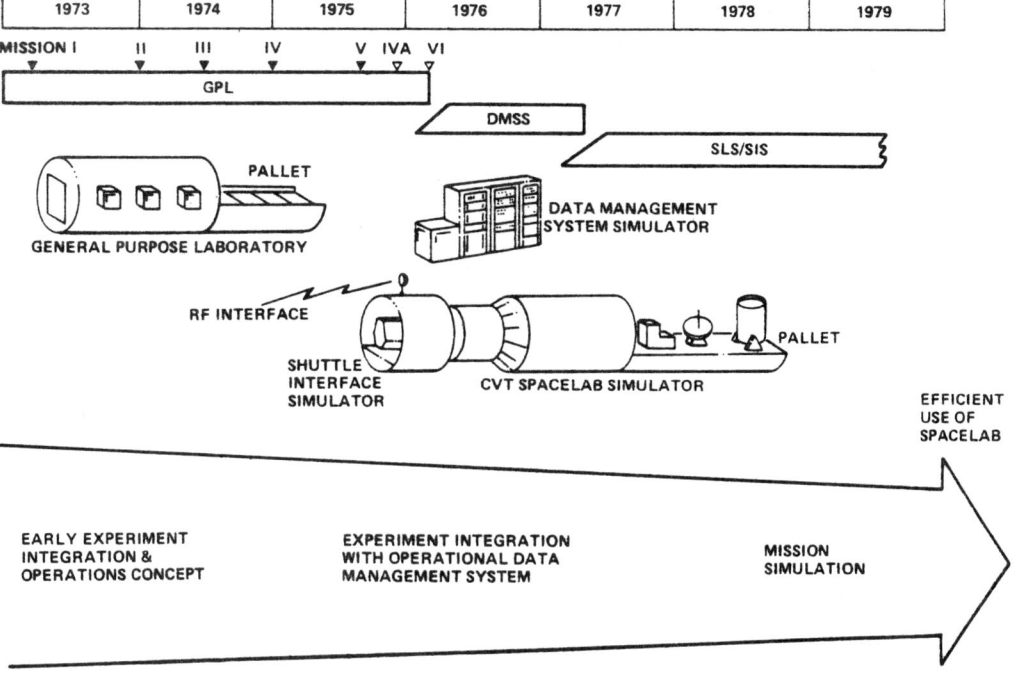

Fig. 1   CVT Support to Spacelab/Spacelab Users

To accomplish these objectives, a series of simulators has been planned, funded, and is in varying stages of completion. Fig. 1 summarizes the scheduling of these CVT Simulators in support of Spacelab users. A brief description of these simulators is provided in the following:

1. General Purpose Lab (GPL) - The GPL is a 4.3m (14 ft) diameter, 7.3m (24 ft) long cylinder with a 4.6m (15 ft) long pallet attached to one end of the cylinder (Fig. 2). The MK I GPL used in Tests I through III had an off-set floor which divided each deck, longitudinally into seated and standing workspace. The MK II GPL began service in Test IV and is more representative of Spacelab with a floor 1.12m (44 in.) below the cylinder centerline. The GPL summary description (Table 1) indicates identical capability for the MK I and MK II except for the floor location. Stowage, systems monitoring panel, audio and video communications, and ambient air conditioning are provided in the GPL.

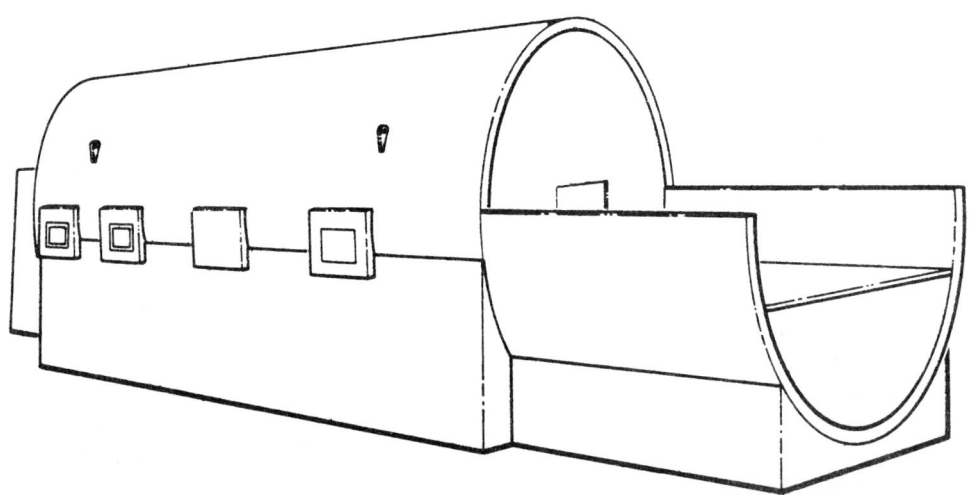

Fig. 2 CVT General Purpose Laboratory

Table 1
CVT/GPL LABORATORY SUMMARY DESCRIPTION

SIZE:
   INTERNAL VOLUME, 4.1 m dia x 7.3m LONG
   PALLET, OR EXTERNAL MOUNTING PLATFORM: 3.66m wide x 4.88 m long

INTERNAL
CONFIGURATION: MK I   MK II
         TWO DECKS, UPPER DECK OFFSET   FLOOR 44" BELOW CYLINDER
             FOR STANDUP HEADROOM   CENTERLINE

THERMAL CONTROL:
   EXTERNAL AIR SUPPLIED AND VENTED; COOLED BY OUTSIDE AND INSIDE
   AIR CONDITIONERS. TEMPERATURE REGULATED BY THERMOSTATICALLY
   CONTROLLED DUCT HEATERS.

VACUUM PROVISIONS:
   MSFC FACILITY VACUUM SYSTEM (3 TORR)
   SUPPLEMENTED BY SMALL VACUUM PUMP RATED AT 0.1 TORR

FLUIDS:
   GASEOUS NITROGEN @ 100 PSIA (68.9 NEWTONS/SQUARE CM)
   GASEOUS "MISSILE GRADE" AIR @ 100 PSIA (68.9 NEWTONS/SQUARE CM)
   HOT AND COLD FACILITY POTABLE WATER.

POWER:
   28 VDC
   110, 60Hz, SINGLE PHASE
   220, 60Hz, SINGLE PHASE
   120/208, 400Hz, THREE PHASE

MEASUREMENTS: FACILITY MEASUREMENTS SUCH AS POWER USEAGE, TEMPERATURE,
              HUMIDITY, PLUS 100 UNASSIGNED CHANNELS AVAILABLE FOR EXPERIMENT USE.

With the MK II GPL, a new concept for integrating experiments was initiated. This concept utilized a "skid," four of which were accommodated in the GPL per mission. There are two kinds of skids which can be used in various combinations depending upon user requirements. The experiment hardware skids can provide any and all power, data, and fluids available in the GPL. These skids each contain two independent power supply panels. If the user's experiment requires more power than is available on the skid, the power supply of an unused skid can be used. Experiment hardware skid instrumentation is also used to supplement pallet experiment equipment requiring any power other than the pallet's standard 110 Vac, 60 Hz, 20 A circuit. Workbench skids are also available to the user. These skids can be modified to accommodate special requirements such as sinks; however, they contain no instrumentation. This can be provided by using the instrumentation from an unused experiment hardware skid. These skids then present a standard interface to the experiment, the User's Laboratory, and the GPL.

An Experiment Operations Center and Test Control Room support the 8-12 hour/day, 5-day missions.

2. Mission Development Simulator (MDS) - The MDS is a computer driven manned telescope simulator consisting of a simulated crew cabin with telescope eyepiece, optical tower, test operations console, and hybrid computer support equipment. A rhomboid arm, driven by the hybrid computer system, passes over photographic transparencies of the earth, taken on U-2 aircraft missions and Skylab missions. The operator in the simulator crew cabin (Fig. 3) can control the rhomboid arm to acquire and identify sites during a simulated orbital pass.

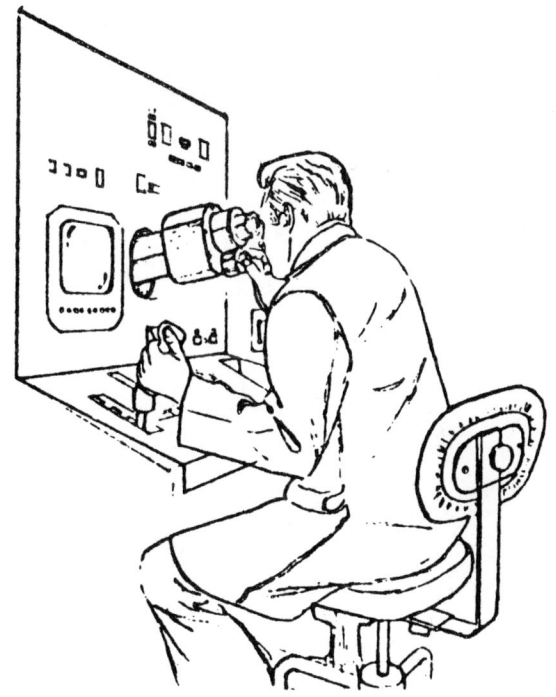

Fig. 3   CVT Mission Development Simulator

3. Data Management System Simulator (DMSS) - The DMSS is a breadboard of the data bus system planned for the CVT Spacelab Simulator and Shuttle Interface Simulator, and is functionally similar to the Spacelab data management system. The DMSS consists of a Space Ultrareliable Modular Computer (SUMC), data acquisition and distribution equipment, and controls and displays (Fig. 4). This simulator will be used to verify the CVT Spacelab Simulator data management system, and it will then be made available to support the development of experiment software for potential Spacelab Users.

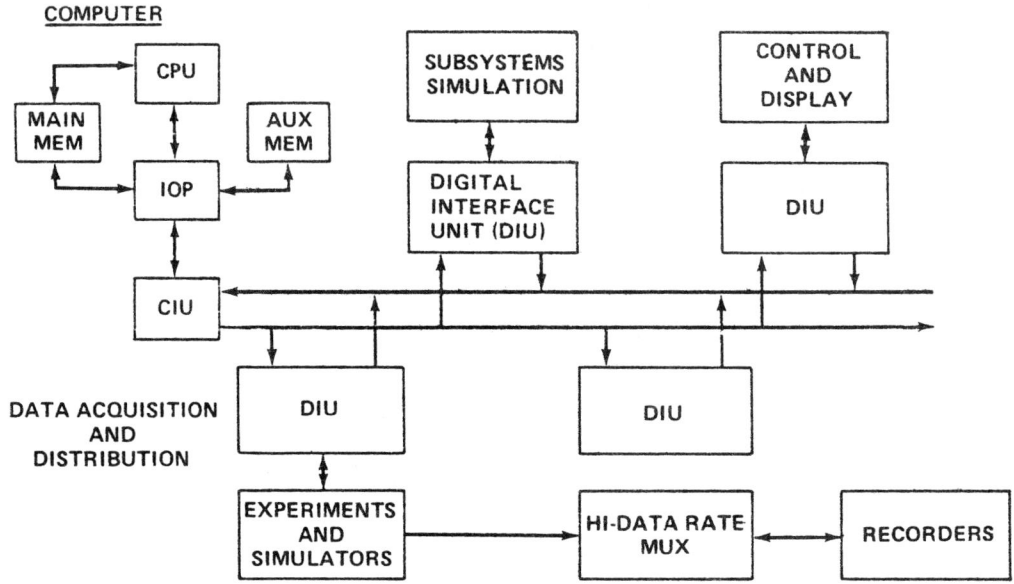

Fig. 4   CVT Data Management Systems Simulator (DMSS)

4. Spacelab Simulator (SLS) - The SLS is a 4.3m (14 ft) diameter, 9.2m (30 ft) long cylinder with a removable floor 1.12m (44 in.) below the cylinder centerline (Fig. 5). The floor is designed to accept flight-type Spacelab racks or CVT skids. Three 3.3m (10 ft) Spacelab type pallet sections are also provided. The STS is more functionally and physically representative of the Spacelab than the GPL. A data bus data management system, multifunction displays and controls, SUMC, electrical power, and experiment support equipment are functionally representative of Spacelab. A Payload Operations Control Center and Test Control Room will support SLS missions.

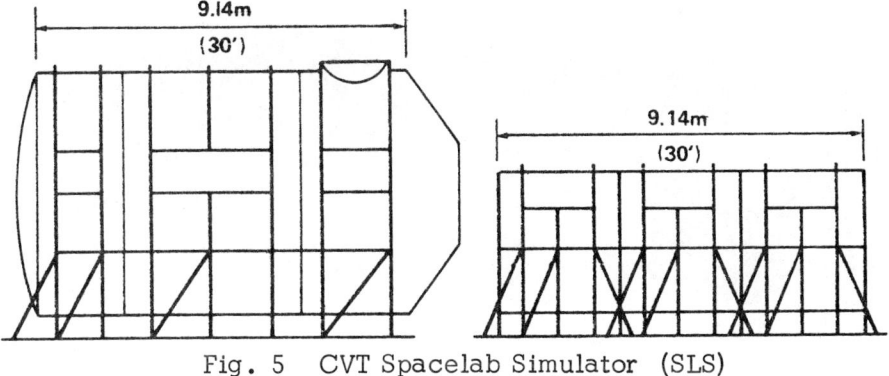

Fig. 5 CVT Spacelab Simulator (SLS)

5. Shuttle Interface Simulator (SIS) - The SIS is a representation of the Shuttle Orbiter Aft Flight Deck containing functional representations of the Mission Specialist station and selected interfaces between the Orbiter and Payload (Fig. 6). The SIS permits operation of the SLS, SLS with pallet, or pallet only configurations.

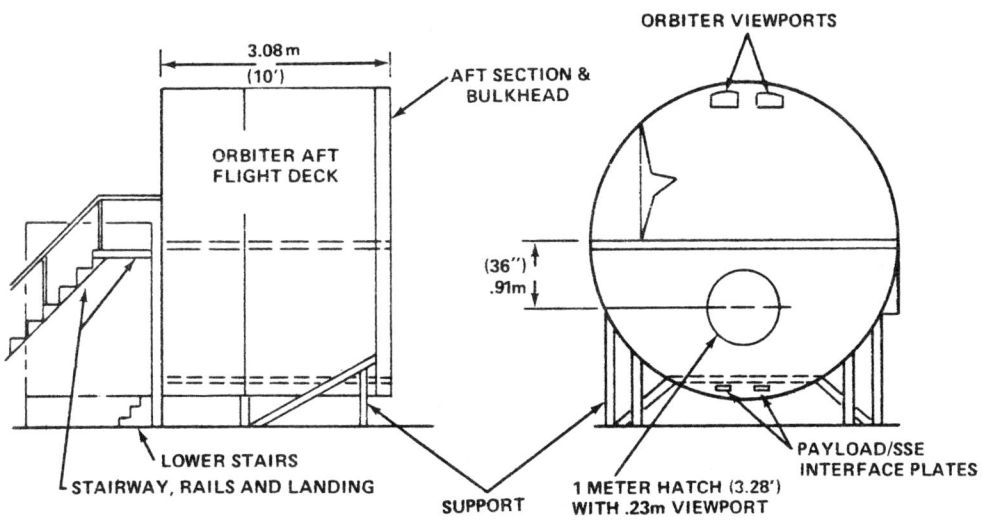

Fig. 6 CVT Shuttle Interface Simulator (SIS)

The experience gained from each simulation (i.e. GPL Mission I) is integrated into the succeeding simulation (i.e. GPL Mission II) and from simulator (i.e. GPL & DMSS) to simulator (i.e. SLS). This approach gives a logical progression of experience and simulator fidelity from the pre-Skylab era through Spacelab operations in the 1980's.

GENERAL PURPOSE LABORATORY SIMULATION OPERATIONS

The simulators just described and user experiments provide the test bed for evaluating Spacelab/Payload operations concepts. The involvement of the users allows simultaneous concept evaluation by the users and operations personnel.

As previously noted, initial CVT integration and operations in the GPL were closely modeled after the Ames Airborne Science Office mission concept (Fig. 7).

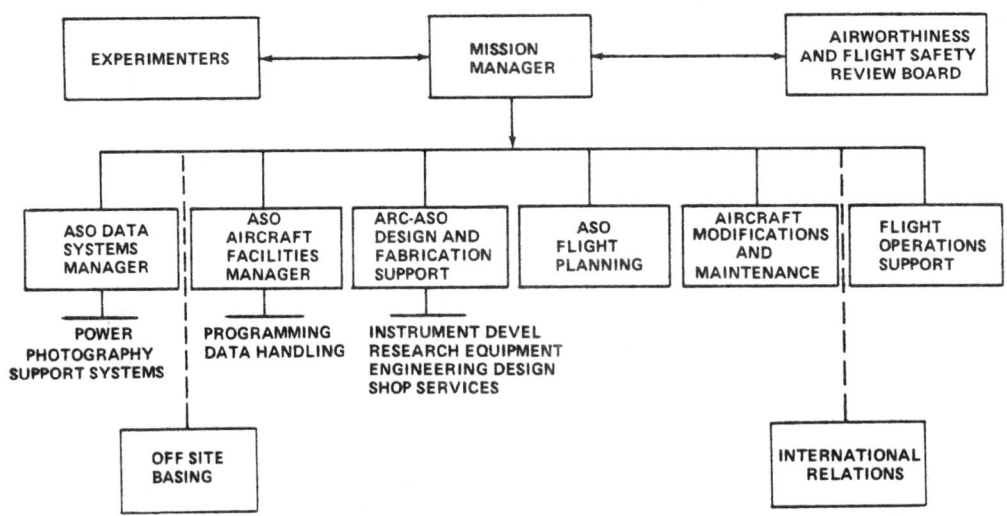

Fig. 7 Ames Airborne Science Office Mission Manager Concept-Mission Preparation Elements

A small team of CVT personnel (Fig. 8) integrated the requirements and hardware before the test, and operations of the experimenters during the test.

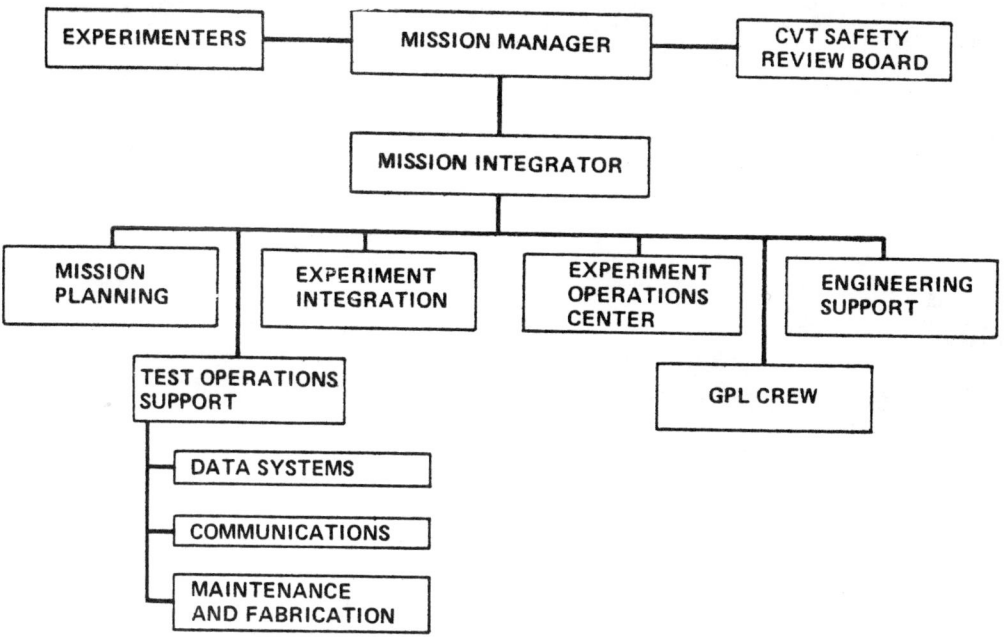

Fig. 8 Concept Verification Test Mission Manager Concept Mission Preparation Elements

Test plans, test procedures (general), and schedules were the primary documents. The single CVT departure from the ASO experiment integration procedure has been the use of off-line integration capability in the form of integration fixtures. In early tests (tests I and III) a full scale mockup of a section of the GPL and GPL interfaces for the experiments was fabricated and sent to the users. In test IV a skid was used as an integration fixture. The skid has a defined interface with the users' laboratories, and it has all of the interfaces contained in the GPL that are available to the experiment so that complete integration of the experiments could be accomplished in the users' laboratories. The sequence of activities for the use of a skid in GPL simulations is depicted in Fig. 9.

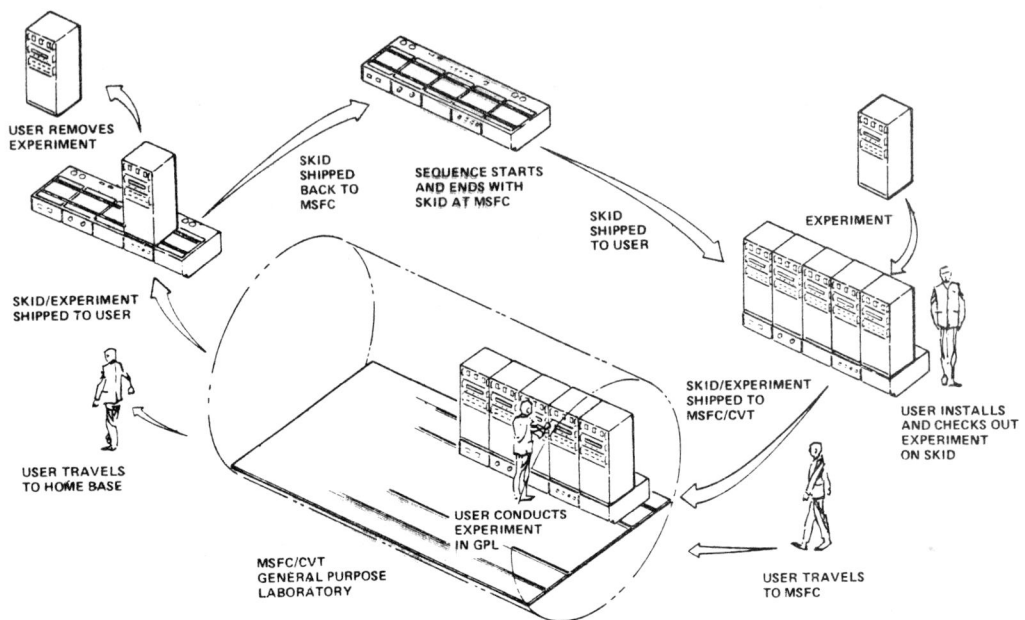

Fig. 9. Skid/Experiments/GPL Activity Sequence

To date, four, one-week (8-12 hrs per day) tests, or mission simulations, have been conducted. The first, conducted in April 1973, was in some respects a "shakedown" test of the GPL MK I, but it also directly addressed questions concerning mixed discipline payloads. Experiments fell within four groups: bioresearch, materials sciences, atmospheric cloud physics and ionospheric disturbances. A crew of four scientists, assisted by a fifth person, a systems engineer, performed all research.

The second test in January 1974, was, in many ways, a replication of the first. It, too, was a mixed discipline simulation. Material sciences, cloud physics and ionospheric disturbance experiments were again on board as were new experiments investigating superfluid helium and high energy astronomy. The high energy astronomers sought to determine degree of training which would enable relatively naive experiment operators to collect and interpret research findings.

The third test, in July 1974, was dedicated to life sciences experiments. Jointly structured and conducted by Ames Research Center (ARC) and Marshall Space Flight Center (MSFC) personnel, this ambitious simulation used a wide range of specimens and protocols. Three on-board researchers conducted ten experiments on behalf of thirty researchers from across the United States.

The management approach for CVT/GPL testing began to change with Mission III. A more significant programmatic role was recognized for the Mission Manager. This modification was initiated to more closely simulate the Spacelab Mission Manager concept, which requires broad coordination between user groups, other Centers, and NASA Headquarters.

The fourth test, in December 1974, was dedicated to materials science experiments. Ten experiments were performed by three on-board researchers, and a set of common experiment support equipment for the space processing discipline was evaluated. This was a "shakedown" test for the GPL MK II and the use of skids as an integration tool. This payload resulted from participation by these same users in CVT/GPL Tests I and II. In addition, a Mission Integrator was introduced. The Mission Integrator coordinated day-to-day integration of experiments and the GPL and brought the mission together at the detail level. The Mission Manager retained overall cognizance and responsibility for the mission. A top level mission timeline was initiated for Mission IV to determine compatibility of experiment operations and resource availability; power requirements projected for this mission were unusually high.

To support real time operations, an Experiment Operations Center was established to monitor and support payload operations through simulated uplink and downlink audio and video communications. The users, if they are not on board as Payload Specialist, monitored operations and interacted with the Payload Specialist directly from the Experiment Operations Center. A Test Control Room monitored

GPL systems, supported resolution of GPL system problems, and served as communications traffic control. The on-board Mission Specialist provided status information on GPL systems and experiment operations to the Test Control Room and Experiment Operations Center. The on-board Payload Specialist conducted experiment operations and communicated with the Experiment Operations Center only when experiment operations dictated the need.

Concurrent with these changes in approach to CVT/GPL testing, documentation has increased. A Mission Definition Document is prepared in the early stages of planning to outline the overall mission objectives, identify available experiments, establish a schedule, and identify any unusual resource requirements. This document is used by the CVT Project Manager to formally establish the new mission and identify the Mission Manager and Mission Integrator. The Mission Integrator and support team prepare a Mission Requirements Document that contains detailed mission objectives, experiment descriptions, a matrix of GPL capabilities and experiment requirements, general timelines, and mission schedules for approval by the Mission Manager.

Even though the hardware designs of many of the experiments were not prototypical of eventual "zero'g" design, experiment operations were in every sense real, not simulated switch functions and cardboard panels. Crystals were grown, animals were scrutinized, clouds were formed. The on-board researchers were in no sense "test subjects"; they were full participants in payload operations. This user participation has been most valuable in developing meaningful adaptions of Spacelab/payload operations concepts.

Detailed objectives and results for each of the four GPL tests are summarized in Table 2.

Table 2
SUMMARY TO GPL TESTING-TEST 1 THROUGH TEST IV

TEST	DESCRIPTION	OBJECTIVES	RESULTS
I	APRIL 1973 FIVE DAY (8 HRS/DAY) DURATION MULTI-DISCIPLINE PAYLOAD EXPERIMENTS BIORESEARCH BREADBOARD ATMOSPHERIC CLOUD PHYSICS IONOSPHERIC DISTURBANCES CRYSTAL GROWTH	1. IDENTIFY EXPERIMENTS/SPACELAB DESIGN CONSIDERATIONS 2. IDENTIFY NON-SCIENTIFIC CREW DUTIES 3. EXAMINE CV990 INTEGRATION AND OPERATION CONCEPTS 4. INVESTIGATE INTEGRATION FIXTURE CONCEPT 5. INVESTIGATE PAYLOAD SPECIALISTS CONCEPT 6. ACHIEVE SIMULATOR OPERATIONAL STATUS	1. A. VIBRATIONS (PERSONNEL MOVEMENT) WERE TRANSMITTED TO THE CRYSTAL GROWTH MELT. B. SIMULTANEOUS UTILIZATION OF N_2 AND VACUUM SYSTEMS BY DIFFERENT EXPERIMENTS REQUIRE MORE PRE-TEST ANALYSIS C. METABOLIC STUDIES REQUIRE BETTER ISOLATION OF THE SPECIMEN. D. A FUME HOOD WAS USED BY THE BIORESEARCH BREADBOARD AND CRYSTAL GROWTH EXPERIMENTS. 2. THE CREW CHIEF PERFORMED THE FOLLOWING NON-SCIENTIFIC DUTIES: COORDINATED WITH THE TEST CONDUCTOR TO REPAIR GPL SYSTEMS. COORDINATED WITH THE TEST CONDUCTOR ON EXPERIMENT ACTIVITY CHANGES REPOSITIONED TV CAMERAS FOR BETTER VIEWS OF EXPERIMENT OPERATIONS. 3. A. THE CV990 TIMELINE FOR EXPERIMENT APPROVAL BUILDUP AND DELIVERY WAS FOLLOWED FOR THE BIORESEARCH BREADBOARD. B. PRE-TEST BUILDUP OF THE CLOUD PHYSICS EXPERIMENT ON BOARD THE GPL WAS AN UNACCEPTABLE INTEGRATION TECHNIQUE. C. THE CONCEPT OF A SMALL TEAM OF PERSONNEL INTEGRATING EXPERIMENTS AND FOLLOWING THROUGH THE MISSION SAVES PAPERWORK AND PROVIDES A GOOD EXPERIMENTER INTERFACE. 4. A CVT/GPL INTEGRATION FIXTURE PROVIDED THE MECHANISM FOR DEFINING EXPERIMENT/GPL AND EXPT./EXPT. INTERFACES 5. ONE SCIENTIST WAS ABLE TO SUCCESSFULLY CONDUCT II BIORESEARCH PROTOCOLS FOR 13 PRINCIPAL INVESTIGATORS. 6. STANDARDIZED EXPERIMENT INTERFACES FOR FLUIDS AND ELECTRICAL POWER WERE ADDED TO THE GPL AS A RESULT OF TEST I.
II	JANUARY 1974 FIVE DAY (8HRS/DAY) DURATION MULTI-DISCIPLINE PAYLOAD EXPERIMENTS ATMOSPHERIC CLOUD PHYSICS IONOSPHERIC DISTURBANCES HIGH ENERGY ASTRONOMY METALLIC ALLOY PREPARATION SUPERFLUID HELIUM	1. IDENTIFY EXPERIMENT/SPACELAB DESIGN CONSIDERATIONS 2. IDENTIFY NON-SCIENTIFIC CREW DUTIES 3. EXAMINE CV990 INTEGRATION AND OPERATION CONCEPT 4. INVESTIGATE PAYLOAD SPECIALIST TRAINING CONCEPT 5. INVESTIGATE PRINCIPAL INVESTIGATOR/ PAYLOAD SPECIALIST OPERATIONAL INTERFACE	1. A. ADDITIONAL INSTRUMENTATION FOR USING CRYOGENICS WITHIN THE GPL WAS REQUIRED. B. AN ONBOARD GPL TOOL KIT WAS ADEQUATE FOR GPL AND EXPERIMENT MAINTENANCE 2. THE CREW CHIEF PERFORMED THE FOLLOWING NON-SCIENTIFIC DUTIES: COORDINATED WITH THE TEST CONDUCTOR TO REPAIR GPL SYSTEMS. COORDINATED WITH THE TEST CONDUCTOR ON EXPERIMENT ACTIVITY CHANGES REPOSITIONED TV CAMERAS FOR BETTER VIEWS OF EXPERIMENT OPERATIONS. TAKE DOCUMENTARY PHOTOGRAPHY 3. FLEXIBILITY IN TEST OPERATIONS AND EXPERIMENTS PARTICIPATION IN THE TESTING ALLOWED REAL TIME MODIFICATION OF CLOUD PHYSICS AND METALLIC ALLOY PREPARATION EXPERIMENTS PROCEDURES TO COLLECT MORE DATA THEN PRE-TEST EXPECTATIONS. 4. THE HIGH ENERGY ASTRONOMY EXPERIMENTERS TRAINED 9 INDIVIDUALS WITH VARYING ASTRONOMY BACKGROUNDS TO SUCCESSFULLY OPERATE THEIR EXPERIMENT ONBOARD THE GPL. 5. AUDIO AND VIDEO COMMUNICATION BETWEEN THE HIGH ENERGY ASTRONOMY EXPERIMENTERS IN THE TEST CONTROL ROOM AND PERSONNEL OPERATING THEIR EXPERIMENT ONBOARD THE GPL WAS ADEQUATE.

Table 2

SUMMARY TO GPL TESTING-TEST 1 THROUGH TEST IV
(CONTINUED)

TEST	DESCRIPTION	OBJECTIVES	RESULTS
III	JULY 1974 FIVE DAY (8 HRS/DAY) DURATION LIFE SCIENCE DEDICATED PAYLOAD 30 EXPERIMENTERS PROVIDED EXPERIMENTS OPERATED BY 3 PAYLOAD SPECIALIST ONBOARD THE GPL.	1. IDENTIFY EXPERIMENT/SPACELAB DESIGN CONSIDERATIONS 2. IDENTIFY NON–SCIENTIFIC CREW DUTIES 3. EXAMINE CV990 INTEGRATION AND OPERATION CONCEPT 4. INVESTIGATE PAYLOAD SPECIALIST CONCEPT 5. INVESTIGATE INTEGRATION FIXTURE CONCEPT 6. INVESTIGATE GROUND FACILITY CONCEPTS FOR SPECIMEN HOLDING 7. INVESTIGATE PRINCIPAL INVESTIGATOR/PAYLOAD SPECIALIST OPERATIONAL INTERFACE	1. A. THE USE OF RADIOISOTOPE TRACERS IN EXPERIMENTATION SIGNIFICANTLY COMPLICATES DESIGN AND OPERATIONS. (1) FUME HOOD FLOW RATES OF 5.67m 3/min (200 FT. 3/MIN) MUST BE FILTERED AND REUSED IN THE SPACELAB .43m 3/min (15 FT. 3/MIN) MUST BE FILTERED AND REUSED IN THE SPACELAB. B. SPECIMEN CAGE MODULE COOLING IS A SIGNIFICANT DESIGN PROBLEM. 2. THE CREW CHIEF PERFORMED THE FOLLOWING NON–SCIENTIFIC DUTIES: COORDINATED WITH THE TEST CONDUCTOR TO REPAIR GPL SYSTEMS. COORDINATED WITH THE TEST CONDUCTOR ON EXPERIMENT ACTIVITY CHANGES REPOSITIONED TV CAMERAS FOR BETTER VIEWS OF EXPERIMENT OPERATIONS TAKE DOCUMENTARY PHOTOGRAPHY 3. THE CVT/GPL INTEGRATION FIXTURE OMPLEMENTS THE CV990 CONCEPT BY SIGNIGICANTLY REDUCING EXPERIMENT INTEGRATION TIME. 4. THREE PAYLOAD SPECIALIST CONDUCTED 10 EXPERIMENTS FOR 30 EXPERIMENTERS 5. A CVT/GPL INTEGRATION FIXTURE PROVIDED THE MECHANISM FOR DEFINING EXPERIMENT/GPL INTERFACES AND EXPT./EXPT. 6. THE BIOCLEAN ROOM WAS ADEQUATE FOR PRE AND POST TEST SPECIMEN HOLDING ADDITIONAL FACILITIES FOR SEPARATING SPECIES AND DIFFERENT EXPERIMENT SPECIMENS WILL BE REQUIRED FOR SPACELAB 7. PAYLOAD SPECIALIST AND EXPERIMENTERS RECOMMENDED DEDICATED EXPERIMENTER CONSOLES ON THE GROUND FOR CONSULTATION.
IV	DECEMBER 1974 FIVE DAY (8-12 HRS/DAY) DURATION MATERIAL SCIENCE DEDICATED PAYLOAD THE THREE PAYLOAD SPECIALIST ON BOARD THE GPL PROVIDED ALL TEN EXPERIMENTS IN THE PAYLOAD	1. IDENTIFY EXPERIMENT/SPACELAB DESIGN CONSIDERATIONS 2. EVALUATE SPACE PROCESSING COMMON OPERATING EQUIPMENT 3. ASSESS MISSION MANAGER TEAM FUNCTIONS AND PROCEDURES 4. IDENTIFY NON–SCIENTIFIC CREW DUTIES.	1. A. SKIDS WERE USED AS INTEGRATION FIXTURES BUT NOT AS SHIPPING FIXTURES B. HIGH TEMPERATURE FURNACES WERE SUCCESSFULLY WATER COOLED C. PORTABLE LIGHTING WAS REQUIRED FOR REMOTE AREAS OF THE GPL. D. A COMBINATION DESK AND WORK BENCH WAS NEEDED. 2. ITEMS SUCH AS ACID CONTAINERS, AND THE FUME HOOD NEED ADDITIONAL DESIGN FOR ZERO G. 3. A. GPL DATA SYSTEM/EXPERIMENT INTERFACES REQUIRE ADDITIONAL DOCUMENTATION. B. FREQUENT CHANGES WERE MADE TO MISSION TIMELINES WITHOUT IMPACTS. C. PHOTOGRAPHIC DATA NEEDS TO DEVELOP ONBOARD FOR POSSIBLE EXPERIMENT REDIRECTION. D. TV WAS USED TO UPLINK CHANGES TO THE MISSION TIMELINE. E. TRAINING OF PAYLOAD AND MISSION SPECIALIST ON GPL SYSTEMS WAS INADEQUATE. 4. THE MISSION SPECIALIST PERFORMED THE FOLLOWING NON–SCIENTIFIC DUTIES: PROVIDED GPL AND EXPERIMENT STATUS TO THE "GROUND" MAINTAINED OPERATIONS LOG REQUESTED DATA FROM THE "GROUND" FOR THE PAYLOAD SPECIALIST DOCUMENTED THE MISSION PHOTOGRAPHICALLY.

RESULTS

Trends can be observed in these test results (summarized in Table 2) that affect experiment and Spacelab planning for payload design and operation. The more significant trends are outlined below:

1. In attempting to prepare the simulator for the arrival of experiment hardware, engineers did not always fully grasp experimenters' needs. Experimenters, for their part, were not always articulate in expressing their needs in terms engineers could readily implement. Problems with the accommodation of biological specimens -- before, during and after test operations -- although generally successful, were more challenging than many of the MSFC engineers had anticipated. For all parties this experience suggests the need for a fairly standardized interface language or means of communication to span the science-engineering interface (e.g., standard check lists oriented toward particular experiment disciplines).

2. Some type of "integration fixture" is necessary that allows the users to integrate experiment to experiment, and with Spacelab interfaces before delivery of the group of experiments for integration with the remaining payload. Integration fixtures used in CVT have substituted for considerable interface documentation and have allowed problems in experiment-to-experiment integration to be resolved by the users in their laboratory. This has resulted in considerable savings of time, documentation, and frustration when the complete payload is integrated.

3. Despite occasional communication gaps, a wide variety of experimenters were facilitated with few real operational problems and with little documentation (i.e. highly cost effective). At the very conceptual outset of each test, a small team was assigned to follow all

phases through test completion. This experience confirms the "Mission Manager" concept, wherein a dedicated team devotes its energies to one particular mission and remains with this mission through completion.

4. Experimenters outside the GPL and Payload Specialist (experimenters inside the GPL) stressed the need to communicate directly. In the third test particularly, on-board researchers frequently communicated with support personnel and experimenters who were positioned in the GPL control room. Usually, these consultations related to necessary changes to preplanned protocol. A separate "science" communications loop for Spacelab is suggested.

5. Detailed research timelines were not prescribed nor enforced during testing by CVT personnel. Several researchers noted that they had during the test significantly modified their pre-test plans -- and had, in their opinions, improved their results. Real time changes such as these become increasingly difficult with more rigid scheduling. This suggests that Spacelab flight plans must retain large blocks of unstructured time, designated simply: "experiment operations."

6. Although maximal schedule flexibility for researchers is important, there is, nonetheless, a need for experimenters to communicate major changes to other involved test personnel. On several occasions, test support people were planning around test procedures which had, in fact, been changed. Also, lack of detailed procedures resulted in several important duties going unfulfilled. There exists some minimal structure which must be accepted.

7. Several on-board researchers commented that had they been more familiar with the details of other concurrent research, there would have been less mutual interference and a chance for increased mutual assistance. For Spacelab, this suggests that all flight researchers

should be familiar with neighboring experiments. Cross training would become even more important if one flight experimenter should become temporarily disabled.

8. CVT multi-purpose support facilities, although carefully sized and preconfigured, frequently failed to support specific user needs. Special simulator modifications are made for each test and experimenters added conditioning equipment to their hardware. A record of initial systems capabilities and all subsequent modifications has been kept; it appears that Users will be called upon to assume considerable responsibility for modulating Spacelab-supplied resources to meet their specific needs.

9. Just as Skylab crewmen reported difficulty keeping abreast of stowage locations, GPL participants noted problems with stowage. Stowage, although hopefully straightforward on Spacelab, must, nevertheless, be well controlled. In the GPL tests, it was possible to send a technician aboard with the forgotten petri dish. It will not be this easy on Spacelab.

10. Finally, there were several expected problems which failed to materialize:

    o There were no major problems resulting from a small number of discipline-oriented researchers representing a much larger community of principal investigators.

    o There was little, if any, experimenter criticism of the general simulator environment, except as specific provisions (or lack of provisions) impinged on the job to be done. There were no comments about wall colors, decor, etc. This strongly agrees with Skylab findings.

o There was none of the "friction" between simulation and support personnel which is frequently reported in other simulations. Perhaps this is because personnel inside and outside the simulator shared the same overall goals.

o Although, as previously noted, researchers would have desired greater familiarity with neighboring disciplines, there were no major problems associated with the conduct of mixed discipline payloads.

To be sure, many of the above observations may change as testing progresses -- particularly as tests become more ambitious, become longer, and as round-the-clock operations are added.

PLANNED SIMULATIONS

Future GPL missions will explore the following Spacelab/payload operations concepts:

o An on-board computer terminal that can be used to alter mission timelines during the mission.

o Examine the ground operation flow of Spacelab racks in experiment integration.

o Examine a concept for Mission Specialist and Payload Specialist selection.

o Examine a concept for Mission Specialist and Payload Specialist mission dependent training.

o Examine Spacelab stowage concepts.

o Continue adaption of the Mission Manager Concept for Spacelab.

The activation of the Data Management System Simulator (DMSS) in early 1976 and the Spacelab Simulator (SLS) and Shuttle Interface Simulator (SIS) in early 1977 provide new and higher fidelity simulation capabilities. The DMSS provides the capability to concentrate on data system interactions with experiments.

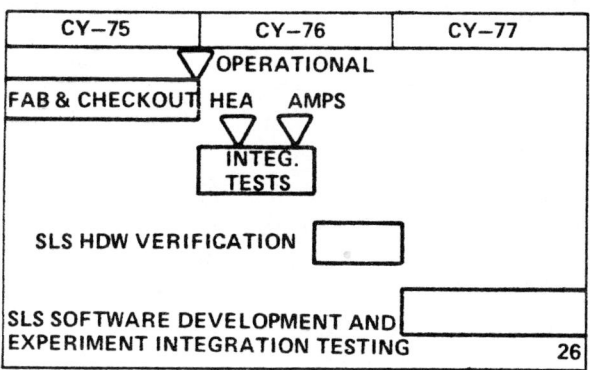

Fig. 10   CVT Data Management System Simulator Planned Activity

Figure 10 indicates the schedule of activities planned for the DMSS. A High Energy Astronomy Experiment (HEAE) will utilize the DMSS as a central computer system to operate the experiment. In previous GPL missions, the experiment used a PDP-11 mini-computer for operation. The results of this DMSS simulation will provide information to compare operation of the experiment with a central versus mini-computer. The AMPS program envisions a simulation to identify divisions between computer control versus direct payload specialist control of experiment operations. In addition, concepts for experiment application software integration with the DMSS operating software will be evaluated.

Figure 11 indicates the schedule of activities planned for the Spacelab and Shuttle Interface Simulators. The activation of these simulators provides the first opportunity to conduct mission simulations within a physical management similar to Spacelab and an integrated data bus system that simulates Spacelab. In addition, the existing GPL Experiment

Operations Center will be modified to simulate a Spacelab Payload Operations Center. SLS/SIS Missions I and II are planned to simulate the experiment complements for the first two Spacelab missions.

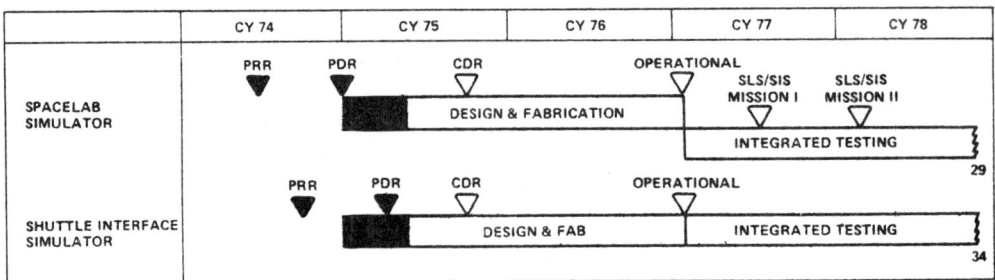

Fig. 11 CVT Spacelab and Shuttle Interface Simulator Planned Activity

The overall objectives and time phasing of CVT simulations to support the development of Spacelab/payload operations is depicted on Table 3.

Table 3
PLANNED CVT SUPPORT TO SPACELAB/PAYLOAD OPERATIONS

OPERATIONS SIMULATIONS OBJECTIVES	PHASE I (1975 – 1977)	PHASE II (1977 – 1979)	PHASE III (1979 – 1982)
● EXPERIMENT SELECTION CONCEPTS	EVALUATE	REFINE PROCEDURES	
● PAYLOAD SPECIALIST SELECTION CONCEPTS	TEST AND SELECT	REFINE PROCEDURES	
● P/S TRAINING OPTIONS	TEST AND SELECT	REFINE PROCEDURES	
● ASPECTS OF EXPERIMENT INTEGRATION	EVALUATE APPROACHES	REFINE TECHNIQUE	
● DOCUMENTATION REQUIREMENTS	DEVELOP SYSTEM	REFINE SYSTEM	REEVALUATE BASIC CONCEPTS – EXPLORE OPTIONS EMPHASIS TO BE DETERMINED
● POC OPERATIONS			
• ON-BOARD VS GROUND FUNCTION OPTIMIZATION	DEVELOP DESIGN CONCEPT	DEVELOP PROCEDURES	
• REAL TIME MISSION PLANNING	EVALUATE OPTIONS	DEVELOP PROCEDURES	
• MCC/POC INTERFACE		DEVELOP PROCEDURES	
• COMMUNICATIONS MANAGEMENT		DEVELOP PROCEDURES	
• DATA HANDLING & DISPLAY		EVALUATE DETAIL REQUIREMENTS	
• INTEGRATION OF FLIGHT PLANS		EVALUATE	
● CREW OPERATIONS			
• ORBITER CREW/MISSION SPECIALIST/ PAYLOAD SPECIALIST INTERFACES		REFINE INTERFACES	
• PROCEDURE REQUIREMENTS	EVALUATE MISSION FLOW	DEVELOP PROCEDURES	
• SKILL REQUIREMENTS	EVALUATE REQUIREMENTS	REFINE REQUIREMENTS	
• MAN/MACHINE OPERATIONS ALLOCATION		REFINE ALLOCATIONS	
• ON-BOARD DATA PROCESSING		REFINE RECHNIQUES	

Spacelab operating concepts (mission management, integrated flight planning, orbiter crew/mission specialist/payload specialist responsibilities) and design (location of controls and displays, degree of remote control and operation, etc.), will be established in the 1975-1976 time frame to support the Spacelab ORR and the CDR scheduled for mid-1976 and early 1977, respectively. In the 1977-79 time frame, detailed operating procedures will be developed and verified. Procedure development will blend with crew training in the later part of this period. Priority of preparations for the first flights and lead times will limit significant conceptual changes. Following the first flights in 1980, operating philosophy and techniques for later flights will be reassessed, based on early flight experience and on definitive requirements for operational period payloads.

AAS 75-295

ANALYSIS OF EXTENDED-DURATION SORTIE MISSIONS
Robert C. Ring* and Wilton C. Lide†

INTRODUCTION

Many Spacelab payloads could significantly benefit from an extension of the duration of sortie missions beyond the nominal seven days. However, longer sortie missions require the addition of Shuttle or Spacelab mission extension kits, which results in a significant reduction in allowable scientific payload weight. An analysis of extended-duration sortie missions was performed as a special emphasis task of the Spacelab Payloads Accommodation Study, a contracted effort by General Dynamics Convair Division for NASA's Marshall Space Flight Center.

ANALYSIS OF EXTENDED-DURATION SORTIE MISSIONS

Payload Benefits

Some payload benefits derive from consideration of an experiment time constant (e.g., the 15-day solar cycle for viewing the total sun). Other benefits are (simply translated) — more time, more data. Payload benefits from extended-duration sortie missions were identified in three categories: viewing, space processes, and life sciences.

Net Scientific Payload Weight Available

To determine the scientific payload weight available for extended-duration missions, the weight impact of extending the Orbiter and Spacelab mission must be assessed, since the payload is charged for all mission-extension provisions.

The Orbiter average power for an extended-duration mission is 12.5 kW, powered down for single-string avionics operation. The Orbiter can supply 7 kW continuous power to its payload — in this case the Spacelab plus scientific payload. At this power level, 15 energy kits (10 located in the cargo bay) are required to support a 30-day mission. The landed weight of these kits totals 6,121 kg — chargeable to the payload.

---

*Manager, Spacelab Payloads Accommodation Study, Space Shuttle Payload Integration (610-04), General Dynamics Convair Division, San Diego, California

†Contracting Officer Representative, Spacelab Payloads Accommodation Study, Spacelab Program Office (NA-21), Marshall Space Flight Center, Huntsville, Alabama

The total weight of crew equipment and consumables chargeable to the payload for a 30-day mission is: 758 kg with a crew of four (one payload specialist) and 1,623 kg with a crew of seven (four payload specialists). It is estimated that up to 3.5 m^3 of volume will be required in the Orbiter to stow this equipment.

A preliminary analysis of Reaction Control System (RCS) propellant requirements suggests that additional tankage may be required for 30-day missions. Additional study of extended-duration mission payload operations is needed. No weight for additional RCS tankage was included in the analysis of payload weight available.

The scientific payload weight available as a function of mission length was determined for three configurations of Spacelab: long module, small module plus 9-m pallet, and 12-m pallet. Available payload weight was computed by subtracting from the Orbiter payload landing weight constraint (14,515 kg) the total of the following items: basic Spacelab equipment, Spacelab mission-dependent equipment, transfer tunnel, and payload-chargeable Orbiter equipment. The results of this analysis are shown in Figures 1 through 3, where available Spacelab scientific payload weight is plotted as a function of mission length for each of the three Spacelab configurations. All curves shown account for an Orbiter average power of 12.5 kW. The top curve for each Spacelab configuration represents the payload minimum crew, minimum power condition: a crew of four and only enough power to operate the Spacelab subsystems — no scientific payload power. The middle curve shows the effect of including enough payload power so that the total of payload and Spacelab power is 7 kW — the maximum available from the Orbiter. The bottom curve in each figure represents a payload maximum crew, maximum power condition: a crew of seven and 7 kW total payload plus Spacelab power. Payload weight available for other crew and power requirements can be interpolated from the curves in Figures 1 through 3.

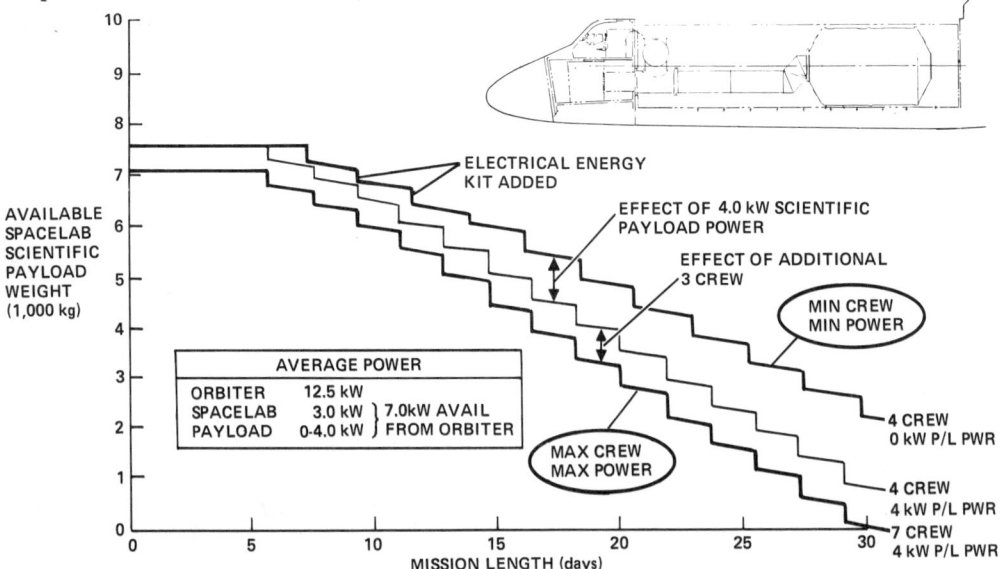

Figure 1. Landed payload weight available for extended-duration Spacelab mission (long module configuration).

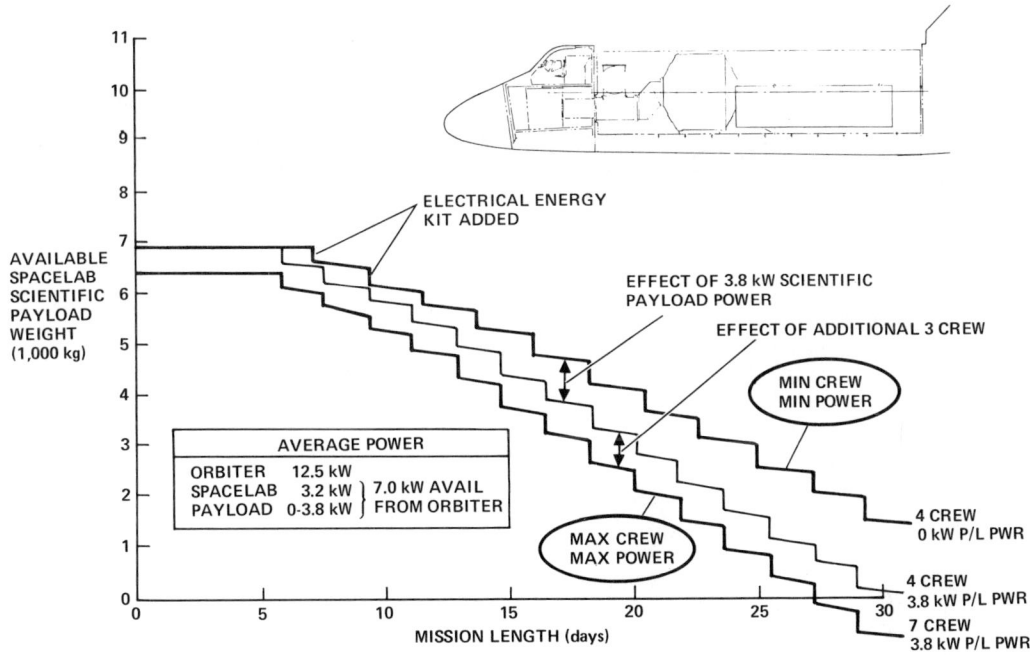

Figure 2. Landed payload weight available for extended-duration Spacelab mission (short module plus 9-m pallet configuration).

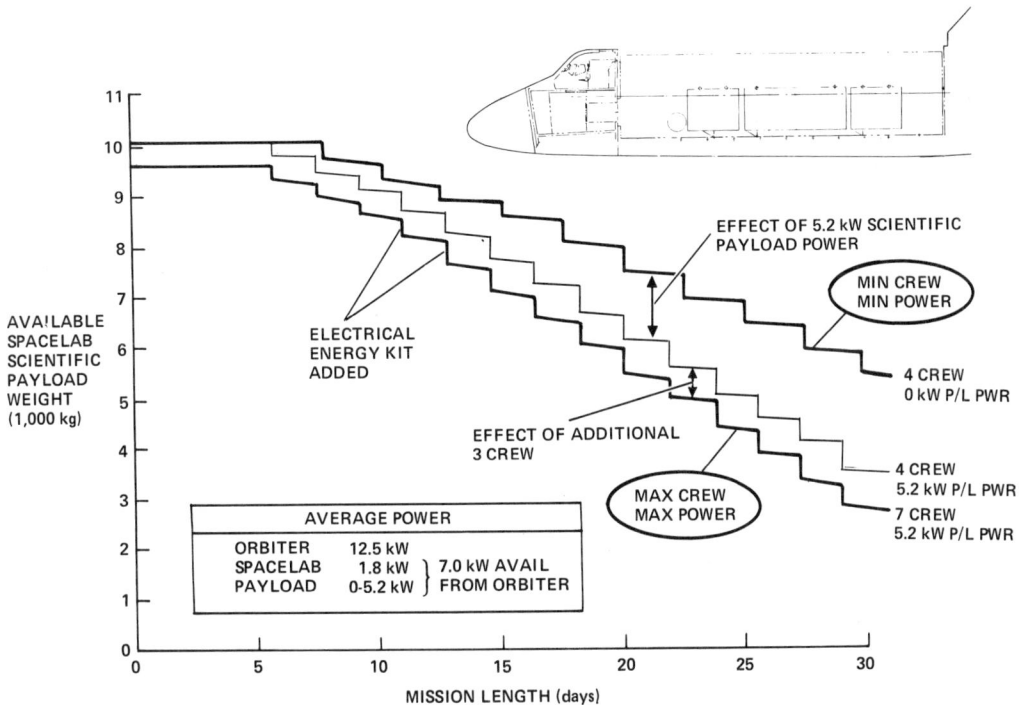

Figure 3. Landed payload weight available for extended-duration Spacelab mission (12-m pallet configuration).

Impact on Spacelab Design and Operations

A reliability analysis of Spacelab subsystems was made to determine if any Spacelab design changes are needed to incorporate additional redundancy to maintain a high probability of mission success for extended-duration missions. Study results show that Spacelab design/reliability is not very sensitive to extended missions of up to 30 days duration if the safety-critical items identified from the mission criticality analysis are made redundant for the baseline seven-day mission. The estimated weight impact to make these items redundant is 73 kg. One mission-critical item is a candidate for built-in redundancy to maintain a high probability of success on extended-duration missions: Spacelab CDMS — Experiment I/O unit and data bus. Estimated weight impact for this item is 15 kg.

Impact on Payloads

Thirty-day missions at reduced payload weight are still useful to payloads by off-loading equipment to concentrate on longer term, time-variant research. Missions of about 15 days duration would be scientifically beneficial to payloads — and offer significantly increased payload weight over a 30-day mission.

Alternatives to Baseline Systems and Operations

There is a big payoff in increased payload weight if the mission average power can be reduced. Overall, however, there appears to be little potential for a significant reduction in mission average power.

The baseline weight of electrical energy kits to be mounted in the Orbiter cargo bay includes weight for mounting structure and plumbing equal to 20% of the loaded weight. This amounts to 142 kg per kit; each kit is mounted independently. Several mounting arrangements for energy kits on integrated structures were assessed for preliminary feasibility and potential weight savings. A truss structure mounted in the rear of the cargo bay appears to provide the best volumetric efficiency and the minimum impact on payload center of gravity. The potential weight savings for this mounting arrangement — over the baseline — is 760 kg for a 30-day mission at 19.5 kW average power.

Alternative approaches to the baseline fuel cell reactant tanks were considered. Throwaway tanks and larger tanks — both supercritical storage and low-pressure storage — were assessed. Overall, throwaway tanks appear to provide the greatest weight saving with minimal development cost and schedule impact.

Sources of power, other than the baseline Orbiter fuel cells, were assessed for applicability to an extended-duration sortie mission. A power system based on the use of deployable solar arrays was selected for detailed analysis. This system retains the Orbiter fuel cells and baseline reactants (1,836 kW-hr) to provide about 2.3 kW average power over a 30-day mission. The solar array (based on Hughes Aircraft Corporation Flexible Rolled-Up Solar Array work) provides the balance of power needed during sunlight conditions plus enough power to charge a set of NiCd batteries that provide power during orbital dark periods (34 minutes per orbit). For a 30-day mission, the solar array system has a weight advantage for all mission powers. For a mission average power of 19.5 kW, the solar array system has a weight advantage after 13 days.

CONCLUSION

Significant payload benefits are achievable on an extended-duration sortie mission. Spacelab scientific payload weight, however, is limited for a 30-day mission due to the addition of payload-chargeable mission-extension kits. Electrical energy (cryo tank kits) is the primary cause of reduced payload weight.

Spacelab subsystem design/reliability is not very sensitive to sortie mission duration if safety-critical items identified from the mission criticality assessment are made redundant for the baseline seven-day mission.

Thirty-day missions at reduced payload weight are still useful to payloads by off-loading equipment to concentrate on longer term, time-variant research. Fifteen-day missions would be very beneficial to payloads and offer greatly increased payload weight over 30-day missions.

Although there appears to be little potential to reduce sortie mission average power, several other alternatives to baseline systems and operations offer weight advantages. An integrated mounting structure for energy kits in the Orbiter cargo bay and throwaway fuel cell reactant tanks are possibilities to achieve increased payload weight. A preliminary assessment of an alternative power/energy system using deployable solar arrays shows a significant weight gain for payloads.

AAS 75-244

SPACE TRANSPORTATION SYSTEM PAYLOADS DATA AND ANALYSIS[†]

J.D. Peterson[*] and H.G. Craft, Jr.[**]

ABSTRACT

Space Shuttle payload planning activities are a continuing effort at Marshall Space Flight Center, under the direction of the Payload Studies Office. The purpose of this activity is to develop and maintain a current Space Transportation System (STS) payload data bank for NASA and non-NASA (but excluding DoD) payloads through the 1980's, with emphasis (depth and detail) on those payloads occurring during the early years of Shuttle operation. The study includes both Spacelab payloads and automated payloads. Payload requirements imposed on the launch site facilities and on the communications/data networks are also included.

Payload descriptions and requirements data have been developed for all STS payloads in the NASA payload model. The basic descriptions were developed by the Shuttle Payload Planning Working Groups for each scientific, applications, and technology discipline. The Space Transportation System Payloads Data and Analysis (SPDA) activity provides the engineering support to describe and document the payload requirements and to assess their compatibility with elements of the STS.

As the data scope expands and payload descriptions evolve through conceptual and definition studies to the hardware phase, the need for automated storage and retrieval of the payload data has become evident. The SPDA data bank is now being automated for a multi-site, on-line information system, with remote data terminals at NASA payload centers.

---

[*]*Design Specialist, General Dynamics Convair Division, San Diego, California*
[**]*Payload Studies Office, NASA Marshall Space Flight Center, Huntsville, Alabama*
[†]*SPDA work performed by General Dynamics has been carried out under Contract NAS 8-29462*

BACKGROUND

In the early 1970's, the advent of the Space Shuttle system established the need to define a total spectrum of payloads, including not only the manned applications and technology activities described in the Blue Book but also future automated spacecraft systems. The latter class includes spacecraft delivered to low earth orbit by the Space Shuttle and those launched to geosynchronous orbit or escape trajectories by a combination of the Shuttle and an additional propulsive stage (IUS or Tug). To satisfy this need, MSFC was assigned the responsibility for assembling an integrated manned/automated Shuttle System payload document for NASA and non-NASA payloads (but excluding DoD) through 1990 with emphasis (level of detail) placed upon those payloads occurring during the early years of Shuttle operation. This document would provide a single, authoritative data base containing payload descriptions and support requirements for all candidate STS payloads included in the NASA mission model. It would contain descriptive data that would provide an adequate data source for future STS and Spacelab design efforts and a basis for Shuttle/payload interface tradeoffs to determine the most cost-effective means of meeting payload requirements. Innovative payload design concepts made possible by the Shuttle were to be incorporated, such as simplicity, low cost, repair, and reuse. Documentation formats were to be developed that would provide timely data in a form best suited to the needs of the users, in a manner that could be easily updated.

To accomplish these objectives, an approach was established that combines the efforts of: (1) a NASA management team whose function is to provide guidance for each of the 12 disciplinary areas, and; (2) a contractor team to collect data, analyze it for accuracy and completeness, and document the data in a clear and concise format.

To develop a credible future payloads program that is supported by the scientific and technical communities, an *ad hoc* organization for Shuttle payload planning was established. The organization consists of a Policy Group, a Steering Group, and Working Groups in each of the science, applications, and technology disciplines.

These working groups had their beginning in the Space Shuttle Sortie Workshop held at Goddard Space Flight Center in 1972. Discipline working groups were established that covered space sciences, applications, technologies, and life sciences. In the fall of 1972, membership of the working groups was broadened to include non-NASA users and to consider all modes of Shuttle use.

Later, several of the groups were combined where they had overlapping objectives. The steering group was established to coordinate activities both among the working groups and with the Shuttle System planners. The working groups continue to meet periodically on an as-needed basis.

SPDA ACTIVITIES

SPDA activities have been under way continuously since January 1973. To date, three iterations of payload documentation and supporting analyses have been completed, as shown in Figure 1. The figure also illustrates the temporal relationship of SPDA to the activities of the prime data users, the Shuttle, and Spacelab programs.

The procedure followed in each of these iterations is illustrated in Figure 2. For each payload in the mission model, the payload objectives, descriptions, and support requirements are established and coordinated with the appropriate Headquarters Program Office. Payload descriptions are prepared using data from on-going or completed payload definition studies to the maximum practical extent. Where source data is incomplete it is augmented by additional analysis and conceptual design. Payload requirements are screened for compatibility with established STS performance and payload accommodation capabilities, and those that exceed specified limits or constraints are identified as "tall-pole" requirements. Each is then reviewed to ascertain if it is, in fact, a firm requirement or if an alternative approach that will not degrade payload performance or usefulness could be followed that would be compatible with STS capabilities.

Typical alternative solutions include physical modification to payload size and/or weight, changes in operational sequences to reduce power or data rates, or the addition of support equipment kits such as additional tape recorders, peaking batteries, special instrument-pointing mounts, or additional electrical energy (fuel cell reactants). It should be noted that the primary purpose of "tall-pole" verification is not to solve payload accommodation problems but to validate the payload requirements data.

Preliminary copies of the summary (Level A) and detailed engineering description (Level B) payload data sheets are then published and distributed to the appropriate working groups. Following a review cycle, working group comments and requested changes are incorporated in the data sheets and updated Level A and B data documents are published. The Level A documents also contain summary data charts of key parameters that are useful for identifying tall-pole requirements and for analyzing the sensitivity of payload accommodation compatibility to changes in STS capability; e.g., power, data rates, pointing accuracy, stability, etc.

Key parameters from the Level A and B documents are input to the STS/Payload Planning Data Bank at MSFC. Currently, all engineering descriptive data from the Level A data sheets and all non-graphic data from the Level B data sheets are stored.

Special-emphasis studies to support NASA planning and STS/Spacelab design efforts are conducted throughout the activity.

The basic elements of payload documentation are the Level A and Level B data sheets. Level A sheets contain summary-level data that briefly presents the payload purpose, major instruments and measurement objectives, orbital data, equipment weight and volume,

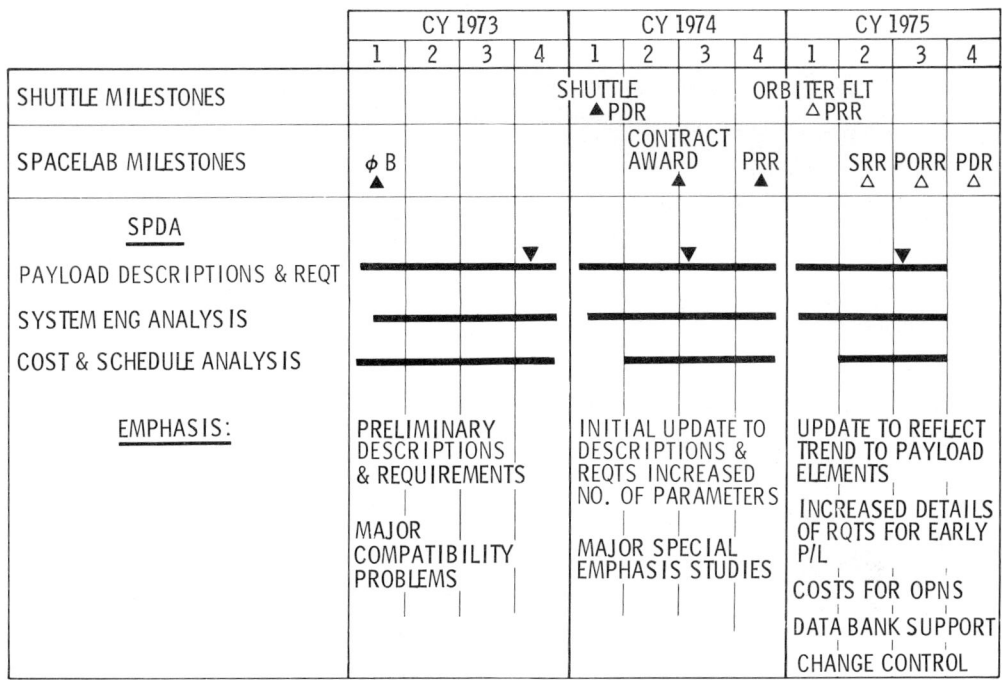

Figure 1. SPDA evolution.

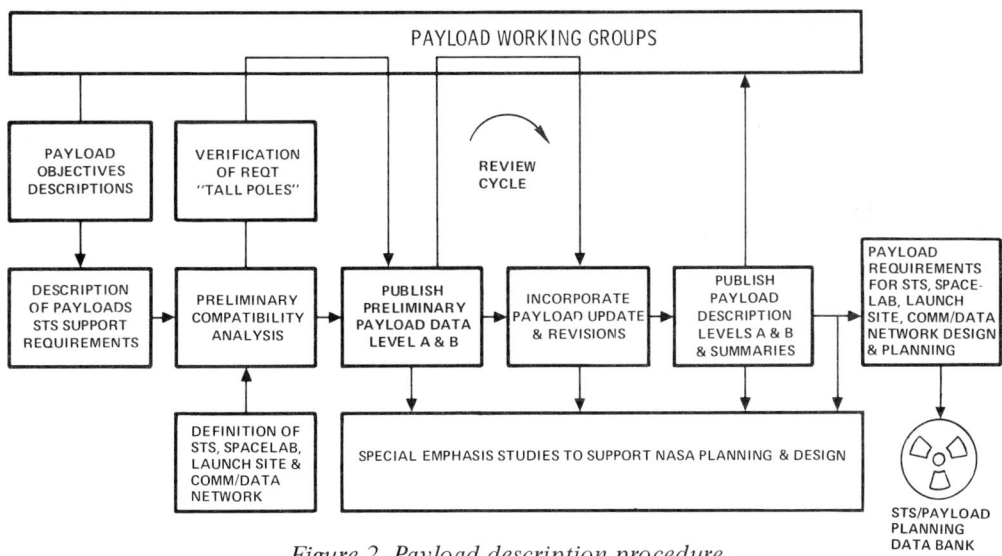

Figure 2. Payload description procedure.

resource and environmental requirements, and special support requirements. It is intended to provide summary-level payload data for preliminary mission planning purposes; Level A data sheets are provided for all candidate STS payloads endorsed by the payload working groups.

More detailed engineering data on payload descriptions and requirements is contained in the Level B data sheets. A set of 25 sheets comprises the current format for sortie payloads.

Automated payloads are described using a set of 21 sheets. These sheets provide very detailed descriptions of payload equipment physical characteristics, input power requirements, environmental limits, crew requirements, operational cycles, data handling, pointing and stabilization requirements, and the physical, functional, and environmental interfaces of the payload/STS and/or Spacelab. Data is also included for unique ground support requirements and a safety analysis of potentially hazardous equipment items.

The increasing volume of summary and detailed payload data has made computerized data bank operations mandatory. Beginning in 1974, the summary (Level A) data bank was computerized so that special sorts and listings could be quickly provided for any desired parameters or groups of parameters. These were developed primarily to support special analyses and special-emphasis tasks, but had the additional benefit of helping to improve overall data quality and consistency. In 1975, the data software was expanded to provide histograms and cumulative distribution plots, and the capability to screen payload requirements for "tall-poles." In addition, a Level A data bank updating method has been implemented wherein the output of the Convair CYBER 70 computer is formatted for direct input to the MSFC Univac 1108 computer via a data exchange tape, thus avoiding the additional time, expense, and inevitable errors of another keypunch step.

The SPDA Level B data published in July 1974 has also been loaded into the MSFC computer. Key payload characteristics and requirements needed for mission planning can be called up on remote data terminals located at MSFC, NASA Headquarters, and several other centers. Hardcopy equipment is also available at each terminal. Future expansion of Level B data storage and retrieval capabilities is planned that will enable all descriptive data, including graphics, to be accommodated.

FUTURE PLANS

Planning is presently under way for the next iteration of payload descriptions and the procedures to be followed. Problems encountered in the past are being analyzed and methods will be implemented wherever possible to eliminate them. In future SPDA activities, payload specialists designated by the working groups will be responsible for payload documentation using the updated data sheet formats. Contractor efforts will assist payload specialists with documentation, will maintain the summary payload data bank and systems capabilities data bank, will perform screening for compatibility analyses to identify "tall-pole," and will edit data for input to the MSFC payload data bank. As before, special-emphasis studies will be conducted throughout the efforts.

AAS 75-245

PROGRAMMATIC ASPECTS OF GERMAN
SHUTTLE/SPACELAB UTILIZATION

Gottfried Greger[*]

German interest in Spacelab utilization is focused besides space science on technological research and applications. The tasks set for the various scientific and technological disciplines are integrated under a coordinated application-oriented program. Measures are taken to make the best technical and economical use of the unique opportunities offered by Spacelab as a versatile research, testing, and production facility. New approaches for the participation of a wide range of user groups in program-planning have been identified. Trade-off investigations and definition studies for experiments, payloads, modular material science labs, and first monodisciplinary and multidisciplinary missions have been initiated, resulting in a German mission model. Pilot experiments are being developed and operated.

INTRODUCTION

When the European governments decided two years ago to develop the manned space laboratory, the way was opened for European participation in a new phase of space flight technology with re-usable, manned systems and - at the same time - for attaching greater importance to the practical utilization of space technology. This intensified orientation of the European space program towards a closer cooperation with the United States of America on the further development of space transportation and orbital systems was connected with a stronger awareness of the essential aims for space research and application. This change

---

[*] Dr. G. Greger is Head of the Orbital Systems Section in the Federal Ministry for Research and Technology (BMFT), 53 Bonn, Stresemannstr. 2, Federal Republic of Germany

is, in the Federal Republic of Germany, strongly influencing aims and methods with regard to planning and preparation for the utilization of future re-usable orbital systems.

We are convinced that the pre-condition for planning the German utilization of Spacelab is that the future of German and - maybe - even European space activities can only be guaranteed if we succeed in integrating this technology into the overall technological framework of our industrialized economy. The starting point for preparation of the utilization phase of future orbital systems is the expectation that the above activities can be used on a broad scale for priority tasks in research and industry, and that the share of industry's participation in these activities can and must be increased. In the following I shall explain these expectations - which are, indeed, very high - and the measures resulting from them. I shall refer to studies and planning under way in the Federal Republic of Germany. The efforts being made within the European space organization, ESA, in which the German partner plays a active part, will be reported on elsewhere.

PRIORITIES FOR UTILIZATION

As far as the expectations I already mentioned are concerned, we have already, despite the relatively short period of preparation, identified several concrete priorities for utilization due to numerous requests presented to us by German research centers and industry. Besides an increase in efficiency and range of utilization possibilities for the previously existing tasks of space research and applications, the re-usable orbital systems enable for the first time to undertake technological basic research and applications in a variety of other areas which can be developed in time, making use of all the special and unique conditions in space, such as zero gravity. It is expected that decisive technological innovations will be achieved in different sectors of research and technology assisting our key industries to solve tasks critical to our long-term economic efficiency. Therefore, the measures promoted in this sector by the Federal Ministry for Research and Technology have been concentrated on:

1. projects of extraterrestrial research and life science in order to provide continuity to the German space program with the declared intention to put as many of these projects as possible on an international level

2. selected activities in the field of earth resources exploration and meteorology by means of orbital systems

3. and in particular tasks of technological basic research and its application, with the aim of essentially improving efficiency, productivity, and economical operation.

TECHNOLOGICAL RESEARCH AND APPLICATION

The diversity of possible questions, which can already be seen now, calls for restriction to, and concentration on, several priorities above all in the field of technological basic research and its application. I should like to take this opportunity and briefly describe the fields which have been included in our planning.

1. By means of studies undertaken in Spacelab, we will be able to eliminate effects on technical processes which are due to gravitation and which cannot be controlled on earth. In addition, further advantageous experimental conditions such as high vacuum and containerless study methods are available. These opportunities, which cannot be realized on earth, are to be used for the exploration of physical and chemical fundamentals and processes:

    a. for the improvement and development of materials, especially in the following categories:

        metals
        compound materials
        glasses
        ceramics and
        semi-conductors

    b. for the study and separation of new preparations in bio-medicine and pharmacy, as well as

        for solving specific problems related to process engineering.

    The results gained in space will then be used mostly for improving and increasing the efficiency of terrestrial processes. In special cases, those materials which cannot be produced on earth on account of the gravitational field could be produced in space.

2. As a manned research laboratory, Spacelab will furthermore serve as an experimental platform for studying and testing

a. technological developments for earth-bound application under extreme operational conditions, to the extent that these tasks cannot be fulfilled in normal testing facilities, or can only be carried out on earth at prohibitive expense;

b. processes, technologies, and systems supporting developments for the utilization of solar energy by means of space flight systems;

c. procedures and systems to improve communication and navigation;

d. elements of operational satellite systems, in particular for earth resources exploration and meteorology.

Furthermore, Spacelab will be suited for activities such as

e. the direct investigation of important meteorological phenomena (i.e., a survey on the aerosol pollution of the upper atmosphere) as well as for the

f. exploration of terrestrial sources of stray radiation in frequency ranges of communication engineering.

USER PROFILE

First preparatory studies were begun in 1970. A survey was carried out in mid-1974 asking approximately 300 German organizations for their ideas on the utilization of Spacelab. The experiment and payload proposals in the different sectors of utilization were then examined by discipline-oriented advisory and planning groups and further studied. The "user profile" was very interesting: among those 370 proposals for utilization which have presently been selected for further consideration, the universities have a share of 23%, non-university research 37%, and industry 40%. More than two-thirds of the proposals refer to technological applications, materials research, and process engineering being the most important ones. These proposals for utilization are the basis for our further program planning.

GROUND-BASED INVESTIGATIONS, ANALYSES, AND STUDIES

By means of systematic scientific and technical ground-based investigations and studies, users in the different fields of application are given the opportunity of defining the technical and economic potential for the utilization of the special conditions existing in space for their specific research and development tasks, as well as to develop the technology required for implementation. Without doubt, the focal task is that of integrating the opportunities offered by space flight technology into the different fields of earth-bound application.

We have tried to solve this complex integration problem in joint planning groups with experts from research and industry, since here scientific fundamentals, which are still unknown to a large extent, and concrete applications directly correlate. The time between research and innovation is to be kept as short as possible. In addition to the 21 studies started in 1974, 36 more have been planned in 1975 in the field of material science and space processing which will lead to proposals for improved, or new experiments. Just to quote some examples from the promotion program: studies are planned in general process engineering, for chemical and pharmaceutical processes and for chemical and medical fundamentals. Furthermore, studies are planned in technological fields, as well as for the improvement of material combinations and electronic materials up to the testing of levitation measuring techniques and devices. These studies are especially oriented towards commercial application at a later stage, since industry should then contribute to the financing of the flight programs.

I should like to quote only one example: In the field of mechanical engineering, the largest German enterprise MAN has identified three development priorities which - according to the firm's opinion - will justify a more detailed study with regard to commercial exploitation later on:

> commercially applicable casting methods for high-quality components for machine construction (i.e., gas turbines);

> methods for the manufacturing of wear-proof moulded parts which are accurate to dimension (CVD-method and hot-working).

The most important and urgent proposals for utilization gradually emerge in the course of these various discipline-oriented studies. The necessary data are already collected within a data bank and reprocessed for further planning. An information center for the users is in the process of being established.

PLANNING

The planning and activities based on the above inquiry are being carried out at different levels. The aim is to show the long-term aspects and requirements of Spacelab utilization from the users' point of view with regard to the technical and financial feasibility, and to initiate short-term program activities for carrying out the first missions.

1. Feasibility studies have been initiated this year for selected experiments, in order to prepare the decision on the final selection of the experiments for preparatory flight programs. As in the scientific and technical preparation phase, responsibility will have to remain largely with the user himself. The user has to decide which pilot experiments must be carried out in pilot programs, or first Spacelab flights, in order to achieve the expected progress in his research or application program or to optimally use the operational opportunities in space laboratories for his specific purposes later on.

2. A reference mission model study carried out by industry in 1974 defined typical payload combinations which comply with the constraints set by the Spacelab system. A first mission model has been developed by iterative methods to assist further planning for Spacelab utilization, taking account of system-oriented aspects. The constraints are still lacking, such as further program development within the different fields of utilization (space technology is only an auxiliary means and not the actual aim of the program), the financial possibilities for the provision of these services and program development at the European level. Despite this, I believe that this attempt will be useful and even necessary to establish the overall framework for the program and to show the interdependence of the individual activities necessary to carry out such programs. The basis of these considerations was mainly the wish to meet the German needs forecast with minimum expense. This mission model envisages that from 1980 onwards an annual mission of monodisciplinary laboratory for materials science and space processing. In addition, multi-disciplinary technological missions and partial scientific payloads for participation in other flights will be accepted at greater intervals. I shall talk about the organization of these flights later, though they are, of course, not yet established programs.

Program planning and system engineering support

Parallel studies form the necessary basis for the planning of the above missions. In the interest of the user, who is above all concerned with uncomplicated and economic access to the transport system, these supporting studies will provide a basis for optimal cost-effectiveness as far as the development, operation, and evaluation of experiments and, of course, as far as the investments for infrastructure are concerned. I will quote some examples for those studies already under way:

a. an examination of the flow, model, and test philosophy, as well as of reliability requirements, in order to bring down the development and test expenditure to a minimum;

b. examination and formulation of a data management system for mono- and multi-disciplinary payloads, which is adapted to user requirements;

c. stock-taking and analysis of the requirements for the necessary test and integration facilities with regard to experiments and payloads for carrying out the above mission model;

d. development of selection criteria and of a training program for payload specialists; and

e. technological preparatory developments for supporting experiments and payloads, as well as standardization and modularization of auxiliary technical equipment.

FIRST MISSIONS

The 1975 promotion program set out to define the first concrete missions, which incorporate the results of both experiment definitions and system-oriented studies. We are concentrating studywork on the first disciplinary missions of a materials laboratory, accompanied by a detailed study on its modular structure, and multi-disciplinary technological missions. Compatibility between the payloads and the system is to be assured by analytical integration parallel with experiment and payload definition.

This should result in establishing compatible payloads for the first few mission projects.

In my remarks, I did not make special mention of the experiments proposed by the German side for the first Spacelab mission which will be carried out jointly by ESA and NASA. They are part of the preparatory activities which I could only describe very briefly here. Whether such experiments will be financed and developed jointly or whether they will be made available by interested partners will depend on further discussions within ESA.

PILOT PROGRAMS

The survey would be incomplete without mentioning the pilot programs. With regard to selected pilot experiments, the definition phase has already been completed and the development and testing phase has been started.

1. Apollo Soyuz Test-Program

    We are very grateful that NASA offered us the opportunity of participating. We successfully tested two important pilot experiments:

    a. The electrophoresis experiment (EPE), which has been carried out by Professor Kurt Hannig, renders it possible to eliminate disturbances caused by gravitation which - in the existing applications in technology, chemistry and, above all, medicine - have been detrimental to the effectiveness of this separation technique.

    b. The Biostack III experiment, which was developed by Professor Horst Bücker, concentrates on the effects of heavy nuclei of cosmic radiation, above all on the creation of developmental disturbances, mutations and delayed radiation effects; this experiment is not only important for assessing radiation hazards in space, but also for the development of methods required in the field of radiation biology, and in particular for that of radiotherapy.

    c. Professor von Baumgarten investigates the behaviour of small fish when their equilibrium is upset. This experiment is of importance for understanding not only the space sickness of astronauts, but also medical studies of the functions of the human labyrinthine system.

2. Sounding Rocket Program

    In this next step field, NASA accepted two German experiments concentrating on materials research:

    a. Professor Ahlborn and Professor Löhberg examine the miscibility gap of an indium-aluminum alloy and

    b. Professor Heye the manufacture of a lead-barium oxide-silver alloy, a special super-conductor being user for drawing thin wires in superconducting magnets.

    In these cases, the zero gravity which can be achieved for some minutes is sufficient to prove the existence of the expected advantageous effects. We will intensity and continue this program and hope that close cooperation with NASA can be established, since I should prefer a further extension of these possibilities to a separate rocket program of our own.

3. ASSESS-Program

Besides the investigation of scientific effects on pilot experiments, we are especially interested in gaining practical experience in operations. The approaches made possible by the ASSESS program in this area are of particular interest to us because airborne procedures could be transferred to the future Spacelab operations. We believe that these procedures could provide a viable approach to developing cost-effective operational methods. I hope that we will be able to continue this program in one form or another.

ORGANIZATION OF THE UTILIZATION PROGRAM

In conclusion, I should like to refer again to questions of organization. The preparation program promoted by the Federal Ministry for Research and Technology is -- as far as scientific and technical studies are concerned -- being carried out by the users at universities, research institutions and industrial firms. System-oriented studies for program planning are carried out by aerospace firms, mainly by MBB. The Deutsche Forschungs- und Versuchsanstalt für Luft- und Raumfahrt (DFVLR) plays a particular role: on the one hand, it participates as a research institute and provides experiments as a user; on the other hand, it places contracts with industry on behalf of the BMFT.

International cooperation is an essential component of German program planning. In founding ESA, European governments established an institution for the coordination of the different national activities and for carrying out projects pursuing joint objectives. In this framework, the Federal Republic Germany is one of the strongest advocates of Spacelab and its utilization. The establishment of ESA is an important step forward, but it does not release the member states from their obligation to make every effort to utilize the considerable investments made in the field of spaceflight for their specific national economic objectives. I hope that I have shown clearly that in this connection promotion programs are touched upon such as those on materials research, process engineering, medicine, and biology. These important and mainly national programs have to take account of other aspects of research policy than those involved

in spaceflight. For these programs, spaceflight will be used as an
auxiliary service and must even compete with terrestrial methods. These
are new perspectives which we have to consider. We will, therefore, seize
all the opportunities offered by Spacelab utilization and by cooperation
with our European partners, either within ESA or on a bilateral level,
and with the United States. In this context, we attach particular
importance to US-German cooperation. The success achieved during the
past ten years of joint effort can be viewed as an convincing indication
of the potential inherent in such a bilateral cooperation.

PROGRAMATIC ASPECTS OF GERMAN SHUTTLE/SPACELAB UTILIZATION
— PROGRAMME LOGIC —

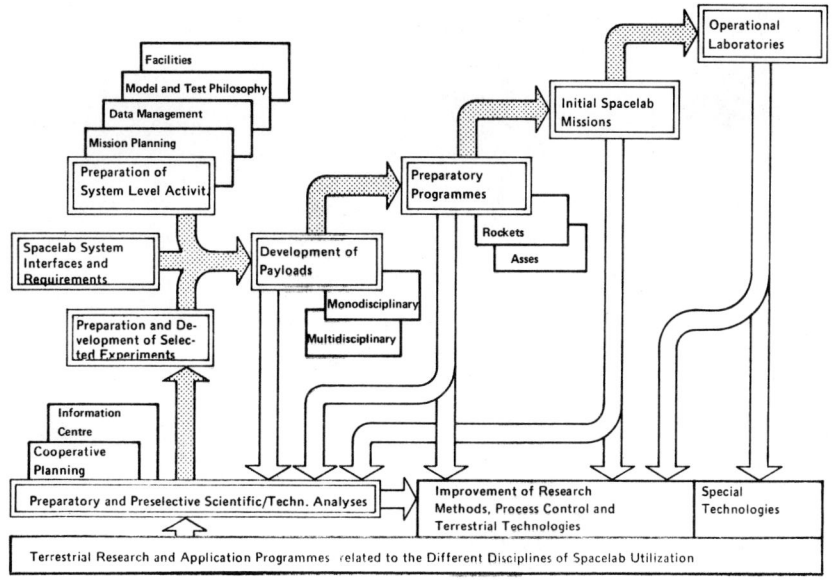

PROGRAMMATIC ASPECTS OF GERMAN SHUTTLE / SPACELAB UTILIZATION

TECHNOLOGICAL BASIC RESEARCH AND APPLICATIONS

---

1. ELIMINATE UNCONTROLLED EFFECTS ON TECHNICAL PROCESSES DUE TO GRAVITATION

    - IMPROVEMENT OF MATERIALS:
        . METALLURGY
        . COMPOUND MATERIALS
        . GLASS
        . CERAMICS
        . SEMICONDUCTOR MATERIALS

    - IMPROVEMENT OF SEPARATION METHODS
    - IMPROVEMENT OF PROCESS ENGINEERING

2. USE AS EXPERIMENTAL PLATFORM FOR STUDYING AND TESTING OF PROCESSES, TECHNIQUES AND SYSTEMS FOR

    - EARTH-BOUND APPLICATIONS
    - SOLAR ENERGY CONVERSION
    - COMMUNICATION AND NAVIGATION
    - OPERATIONAL SATELLITE SYSTEMS

PROGRAMMATIC ASPECTS OF GERMAN SHUTTLE / SPACELAB UTILIZATION
"USER PROFILE" IN F.R. GERMANY

BASIS: 370 REQUESTS FOR UTILIZATION

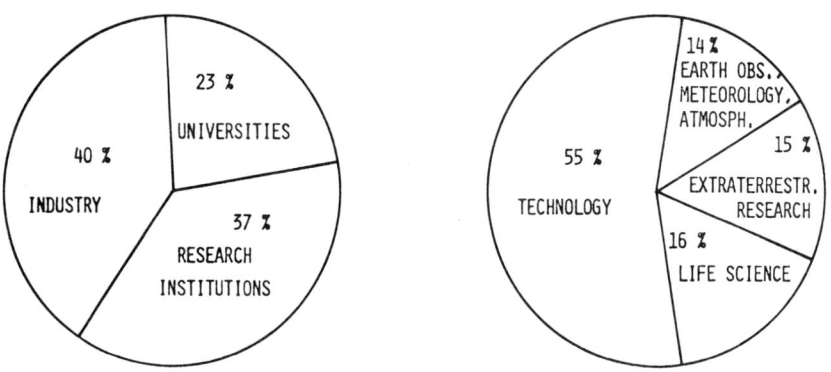

PROGRAMMATIC ASPECTS OF GERMAN SHUTTLE / SPACELAB UTILIZATION
PROGRAM PRIORITIES

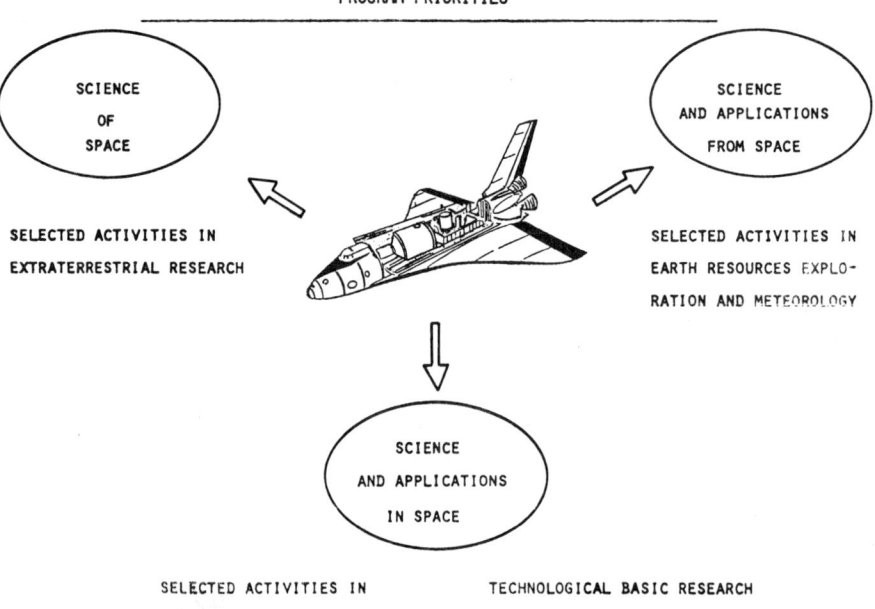

PROGRAMMATIC ASPECTS OF GERMAN SHUTTLE / SPACELAB UTILIZATION

EXAMPLE FOR INDUSTRIAL PROPOSAL FOR MECHANICAL ENGINEERING
WITH COMMERCIAL BACKGROUND

BASIS: STUDY OF MAN

- CASTING METHODS FOR HIGH-QUALITY COMPONENTS FOR MACHINE CONSTRUCTION (E.G. GAS TURBINES)
- MANUFACTURING OF SPECIAL COMPOSITE MATERIALS
- METHODS FOR MANUFACTURING OF WEARPROOF MOLDED ARTICLES WHICH ARE ACCURATE TO DIMENSION

PROGRAMMATIC ASPECTS OF GERMAN SHUTTLE / SPACELAB UTILIZATION

EXAMPLE FOR INDUSTRIAL PROPOSAL FOR MECHANICAL ENGINEERING
WITH COMMERCIAL BACKGROUND

BASIS: STUDY OF MAN

- CASTING METHODS FOR HIGH-QUALITY COMPONENTS FOR MACHINE CONSTRUCTION (E.G. GAS TURBINES)
- MANUFACTURING OF SPECIAL COMPOSITE MATERIALS
- METHODS FOR MANUFACTURING OF WEARPROOF MOLDED ARTICLES WHICH ARE ACCURATE TO DIMENSION

AAS 75-246

FUTURE PAYLOAD TECHNOLOGY REQUIREMENTS
Howard M. Ikerd* and Larry R. Alton**

The overall objective of the Future Payload Technology Study was the identification and description of technology items that must be advanced beyond the current state of the art in order for early shuttle-era NASA payloads to meet their currently defined objectives. The purpose has been to provide data that will effectively assist the NASA payload technology planning and budgeting effort. The payloads selected for this study were those included in the 1973 Payload Model, which NASA scheduled for delivery by the Space Transportation System in 1980s. Emphasis was on those payloads scheduled for flight in the early to mid 1980s.

The purview of the study team's activity consisted of the definition of technology advances needed for an overall mission model standpoint as well as those for individual payloads. The technology advances relate to the mission scientific equipment, spacecraft subsystems that functionally support this equipment, and other payload-related equipment, software, and environment necessary to meet broad program objectives.

In the interest of obtaining commonality of requirements, it appeared most useful

---

* Engineering Staff Specialist, General Dynamics Convair Division
San Diego, California 92138

** Research Scientist NASA Ames Research Center
Moffett Field, California 94035

to structure the study according to technology categories rather than in terms of individual payloads. The study was carried out within the classification of the following categories:

Collectors	Environmental Protection
Sensors	Cryogenic Control
Generators	GN&C
Systems	Propulsion
Special Devices	Attitude Control/Measurement
Inertial/Electromechanical	TT&C/Data
Life Sciences	Electrical Power
Contamination	Instrument Electronics
Structural and Spacecraft Mechanical	Software
Environmental Control	

Some applications and desirable characteristics of equipment, particularly sensors, are in the classified literature as are some current state-of-the-art data. However, this study was restricted to the open literature and unclassified knowledge.

The team planned its activity to ascertain the best available and most credible information that will effectively assist the NASA technology effort in closing the gap between the current state of the art and the required state of the art for each item within the technology categories. For each of these items it was attempted to determine:

Advancement required based on payload objective.

Current state of the art as it relates to the advancement required.

Description of technology relating to the critical parameters.

Degree of benefit to the payloads.

Acceptable technology maturity, advancement, or confidence demonstration.

Potential problems, options, and alternatives.

Technology requirement schedule to support need date.

Expected advancement in the state of the art by the need date, if NASA expends no special effort beyond currently planned level on that specific technology item.

To provide uniformity of terminology and consistency of results within the study and to obtain definitions of what constitutes a satisfactory technology advance, the level of maturity of the state of the art was indicated by ten levels. The levels are listed in Table 1. Ascending numerical values or levels were assigned to provide a common reference and to facilitate identification and use of the levels. In the application of this scale it should be recognized that the difficulty in going from one step to another will depend on the specific item as well as which step is being made; however, the scale is useful in highlighting the overall technology gaps. These levels of the state of the art were used to assess three areas that are keys to the defined technology requirement. These are:

1. Current State of the Art: To what level has the technology which more nearly fits the requirement been carried to date.
2. Unperturbed Advancement: To what level is the technology expected to be by need date if NASA expends no special effort in this area beyond current plans.
3. Required Advancement: To what level must the technology be carried to make it acceptable for its intended use or commitment to a program.

Table 1

LEVEL OF STATE-OF-ART DEFINITION

LEVEL	LEVEL DEFINITION	GENERAL AREA
1	Basic Phenomena Observed & Reported	Theoretical & Laboratory
2	Theory Formulated to Describe Phenomena	
3	Theory Tested by Physical Experiment or Mathematical Model	
4	Pertinent Function or Characteristic Demonstrated; e.g., Material & Component	
5	Component or Breadboard Tested in Relevant Environment in Laboratory	
6	Model Tested in Aircraft Environment	Prototype & Operational Models
7	Model Tested in Space Environment	
8	New Capability Derived from an Operational Mode (A lesser Model Operating in Space)	
9	Reliability Upgrading of an Operational Model	
10	Lifetime Extension of an Operational Model	

A summary of the findings of this study in terms of the levels of technology just described is presented graphically in Figure 1. The state-of-the-art level versus cumulative percent of technology items is shown in Figure 1A. The lower curve gives the current level of the state of the art, while the middle curve indicates the additional expected normal advance by currently planned effort. The larger advance, which must be provided by NASA, is indicated by the separation between the center curve and the upper curve, or levels to which the technology is required to be advanced if the payloads are to perform their expected missions. The upper curve shows that only about 22 percent of the technology items can be satisfied in the laboratory, level 5, and be ready for application or commitment to a program, whereas the remainder require some type of demonstration in an aircraft environment (level 6 or higher).

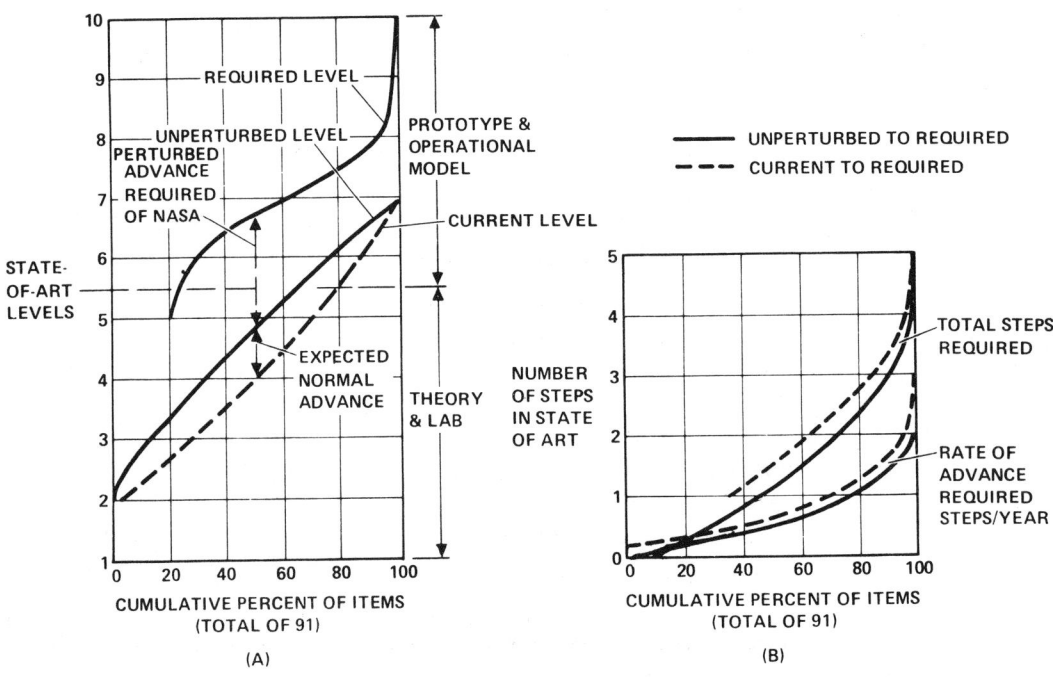

Fig. 1 Summary Results

The second set of curves (Figure 1B) presents the number of steps of advancement between current and required versus cumulative percent, as well as that between unperturbed and required advancement. There is a small but significant difference between the two. Here the differences were taken before calculation of percentage. The upper two curves show the overall magnitude of the advance required. For the current to required, 34 percent of the items require only one step, whereas the upper 3 percent require five steps, and the average number of steps for all the items is 2.2. For the unperturbed to required, the average number of steps is 1.75.

The lower curve provides insight into how fast the technology must be advanced. It was derived by ratioing the number of steps to the number of years beginning in 1975 and counting up to the year in which the technology will be needed to support the payload development for flight in the early 1980s. A rate of one or more levels per year is considered critical and occurs for about 20 percent of the items.

The degree of difficulty in advancing the state of the art will depend not only on the number of levels to be advanced but where in the chain of advancement one is operating; and probably more importantly, it will depend on the specific item itself. In any event, since the unperturbed advance falls short of the required advance in 84 out of the 91 items, NASA should provide the major effort for the technology required of the payloads, otherwise the project schedule may be unduly delayed, cost increased, or the planned research may fail.

The highlights of the study are listed below:
1. Current requirements and schedule require major technology impetus by NASA.
2. Many of these technologies must be available in three to four years to support payload schedule.
3. About half of the technology advancements strongly suggest a test in space.

AAS 75-247

COMPUTER-AIDED SCHEDULING OF SPACELAB
GROUND OPERATIONS

J. K. Willoughby[*]

INTRODUCTION

The complexities of Spacelab ground operations will require the use of improved computer-aided scheduling of activities. However, computer software to perform operational support scheduling and resource allocation has in the past been costly to develop and maintain. For example, some of the scheduling software used to support Skylab operations took much longer to develop than anticipated and never did perform all of the functions that it was originally intended to perform. Since the operational complexities associated with Spacelab operations are greater than in previous programs, the task of developing the required support software has to be viewed with concern about excessive costs. This paper discusses briefly three techniques for reducing the development costs of scheduling and resource allocation software typical of that required for Spacelab operations.

Software Library

Cost savings can be realized by eliminating unnecessary reprogramming and checkout. In the past two years, analyses have been performed to identify the logical functions that are common in scheduling programs and to develop a method for standard usage of those functions. Ordering, checking constraints, performing interval algebra, finding earliest availability of resources, and updating resource assignments are some of the logical functions found in typical programs to solve operations

---

[*] Staff Engineer. Dr. J.K. Willoughby is affiliated with Martin Marietta Corporation Denver Division, Denver, Colorado

problems. An analysis of functions that should belong in a scheduling library led to the functions described in Ref. 1.

Preliminary evaluations of the library's utility on typical problems indicate that substantial savings in program design, coding, and debugging time can be expected by utilizing the standardized logic. Table 1 shows the percentage of library code to executive code used in writing four scheduling programs for typical NASA applications. The results, though preliminary, show promise of a substantial cost reduction in large-scale program development.

Table 1

USE OF STANDARD LIBRARY ROUTINES
TO REDUCE SCHEDULING SOFTWARE COSTS

SCHEDULING PROGRAM NAME	LINES OF CODE IN MAIN PROGRAM	NO. OF CALLS TO LIBRARY ROUTINES	LINES OF CODE IN ROUTINES CALLED*	RATIO
Heuristic Scheduler	47	9	1051	22:1
Tracking Facilities Resource Allocation	24	3	370	15:1
Crew Assignment Model	26	1	287	11:1
One Year Flight Support Scheduler	387	14	1731	4.5:1
			WEIGHTED AVG	7:1

* Statements in a routine that is called several times are only counted once

Man-Computer Solution Strategy

A particular subset of the library routines can be used in a powerful man-computer solution strategy to generate activity timelines and resource utilization profiles. This subset is referred to here as the Project Scheduling System. Project scheduling is an extension of PERT which includes not only predecessor/successor relationships between jobs but also resource requirements and utilization constraints. The resources are considered to be pools of identical entities that are substitutable one for another to satisfy requirements for specified quantities. A man-computer solution strategy that is applicable to Spacelab operation can be built around the project scheduling capabilities. By proper use of the man as an algorithm iterator, complex problems such as those found in Spacelab operations can be solved even though the algorithms themselves are functionally inadequate for the problems. Fig. 1 shows a man-computer strategy that uses project scheduling to generate start times for all jobs so that predecessor/successor constraints and resource utilization constraints are satisfied. The outputs of the project scheduling routines can be iterated if more complex temporal relations must be satisfied. After a satisfactory result is obtained, an explicit allocation of resources is accomplished. If no allocation is possible, the human is involved with modifying the project scheduling result until explicit allocation can be accomplished successfully. Examples of the convergence of this technique have been presented by O'Doherty[2]

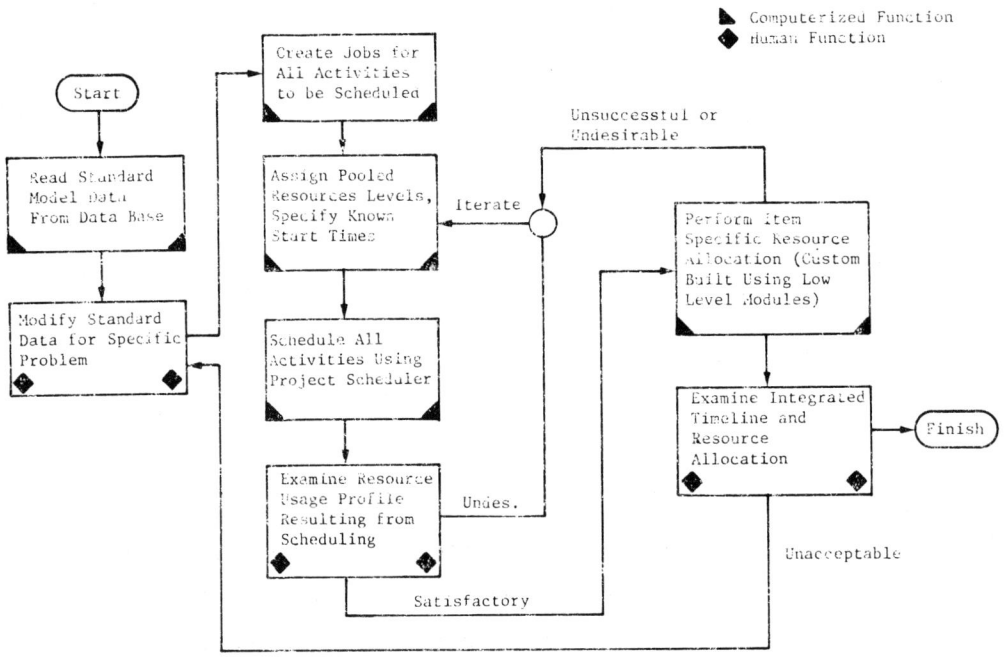

Fig. 1 A Man-Computer Solution Strategy Based On Project Scheduling Capabilities

Programming Language

The costs and development times for Spacelab operations software can also be reduced by using a special-purpose programming language designed to be compatible with the information structure and logical constructs found in typical operation problems. A language called PLANS (Programming Language for Allocation and Network Scheduling) has been designed to allow functional designers of scheduling and resource allocation algorithms to do their own programming and modification, rather than merely providing inputs to another level of detail (e.g., FORTRAN) program design and coding[3]. In addition, PLANS provides for extreme data independence thus allowing many problem variations to be accommodated without modification of the source code.

Preliminary data comparing PLANS with other languages for coding typical scheduling and resource allocation problems shows a favorable comparison. Since the number of coding lines to cost dollars is relatively independent of language level, a reduction in coding volume required would contribute substantially to the reduction of software development costs associated with Spacelab operations. Table 2 shows results from a simple compara-

Table 2

SUMMARY OF RESULTS OF A BENCHMARK COMPARISON

	PLANS			FORTRAN		
	Programmer 1	Programmer 2	Average	Programmer 3	Programmer 4	Average
Coding Time (min)	29	97	63	154	205	180
No. of Executable Statements	22	32	27	105	108	107
No. of Nonexecutable Statements	10	11	11	33	41	37
No. of Logical Errors Found and Corrected	0	3	2	5	4	5

tive study in which PLANS and FORTRAN code written to perform logically similar functions. These comparisons indicate that the potential of PLANS to reduce program development and modification costs is encouraging.

CONCLUSION

The development of a standard program library, the exploitation of appropriate man-computer solution strategies, and the use of an appropriate programming language for scheduling applications should contribute to the development of less expensive and more useful software to support Spacelab operations.

ACKNOWLEDGEMENTS

The author has reported the work of many people, his own, but also that of several colleagues. In particular, the ideas on project scheduling are those of Robert O'Doherty and the PLANS language design is due to Dr. H. Rudy Ramsey.

REFERENCES

1. J. B. Armstrong, D. R. Cochran, R. J. O'Doherty, R. E. Paulson, H. R. Ramsey, C. C. Reynolds, and J. K. Willoughby, "Detailed Design Specifications for the Language and Module Library," Scheduling Language and Algorithm Development Study Phase II Interim Report. Martin Marietta MCR-75-169, April 1975.

2. R. J. O'Doherty, "The Use of Project Scheduling for More General Problems: An Application to the Space Transportation System," ORSA/TIMS National Meeting, Chicago, Illinois, April 30, 1975.

3. H. Rudy Ramsey, "PLANS: Human Factors in the Design of a Computer Programming Language," Human Factors Society, 18th Annual Meeting, Oct. 15, 1974.

AAS 75-240

THE INTEGRATION OF COMMERCIAL PAYLOADS INTO SPACELAB

H. L. Bloom[*] and K. R. Taylor[†]

Historically, space program payload integration has necessitated resolution of such classical problems as payload accommodation size and weight constraints vs. payload physical characteristics, analysis of adherence of safety standards, etc. To such problems, the advent of <u>commercially</u>-oriented payloads adds new problems, and adds new dimensions to historical problems.

Review of the factors involved in space commercial enterprises identified in the study documented in Ref. 1, quickly established that the root of these new and expanded problems lies in economics; in particular, in three categories:

1. Industrial Security (competition, patents, anti-trust, etc.)

2. Cost and Charges (financing, plants, space launch & transport, liability, etc.)

3. Plans, schedules, decisions and priorities (space launch and transport, development, etc.)

INDUSTRIAL SECUTIRY

How do questions related to economic problems of industrial security influence integration of payloads into Spacelab?

Primarily, it is a matter of balancing the minimal amount of proprietary information that the User is willing to disclose about his materials, phenomenology,

---

[*] Advanced NASA Programs, Space Division, General Electric Company
[†] Applications and Technology, Marshall Space Flight Center, NASA

process, equipment, designs, etc. against the payload details that the integrator feels he must know in order to establish intra-payload compatibilities, payload/Spacelab accommodation utilization, interface control, safety, etc.

What can be done to reassure potential Users that their investments (dollars, time, expertise) will be protected in spite of the requirements imposed by integration of their payloads with other payloads, and into Spacelab?

In the long run, answers to this question will evolve from experience. For now, we list in Fig. 1 some possible approaches to reducing such concerns.

USER CONCERNS	POTENTIAL APPROACHES TO REDUCE CONCERNS	IMPACT ON INTEGRATION	STATUS/REMARKS
PROTECTION OF PROPRIETARY METHODS DURING PRE-LAUNCH INTEGRATION ACTIVITIES	• LIBERALIZATION/CLARIFICATION OF NASA PATENT AND RIGHTS-IN-DATA REGULATIONS	• MAJOR DRIVER OF NUMBER AND TYPES OF COMMERCIAL PAYLOADS TO BE INTEGRATED • INCREASES NUMBER AND TYPES OF REQUIRED CAPABILITIES	• REGULATIONS PRESENTLY UNDER REVIEW • NASA HAS CEDED RIGHTS IN SEVERAL CASES • FINAL DECISION NEEDED EARLY
	• RELAXATION/MINIMIZATION OF DATA ACQUISITION AND DATA FLOW FOR INTEGRATION	• REDUCES PAPERWORK • INCREASES OPERATIONAL RISK PROVISIONS	• WHAT IS "MINIMUM" DATA, "MINIMUM" FLOW? • PRESENT TRANSPORTATION SYSTEMS HANDLE SENSITIVE CARGO
	• APPLY MILITARY/INDUSTRIAL-TYPE SECURITY SYSTEM	• INCREASES COSTS, TASK DURATION • INCREASES TASK DIFFICULTY	• DIFFICULT TO ENFORCE • RELATIVELY EFFECTIVE FOR NEEDED SHORT LEAD-TIME
ON-ORBIT SECURITY	• AUTOMATED, "SEALED" EQUIPMENT	• INCREASES RISK PROVISIONS • MINIMIZES PROVISIONS FOR ON-ORBIT MAINTENANCE, REPAIR	• AUTOMATION IS NORMAL INDUSTRY MODE
	• MINIMAL DATA TRANSMISSION	• MINIMIZES COMMUNICATIONS PROFILING	• REDUCES REAL TIME EXPERIMENT REDIRECTION • NORMAL FOR COMMERCIAL PRODUCTION
	• SECURE COMMUNICATIONS	• INCREASES COST, COMPLEXITY	• AVAILABLE TECHNIQUES

Fig. 1. Potential Approaches For Reducing Typical User Concerns Over Commercial Payload Security

COSTS AND CHARGES

For commercial Users, Ref. 2 points out that the expense of Space Operations will likely be critical to the profitability of their ventures. On that basis, potential Users will require realistic estimates of the costs of space operations, and the charges NASA expects to impose.

What has this to do with Payload Integration?

First, a clear, equitable cost allocation method, coupled with a reasonable charge policy will encourage a large, broad spectrum of Users, thus requiring deep, broad integration capabilities.

More important, the suggested cost method provides incentives and dis-incentives for encouraging and/or discouraging use of Shuttle resources, so that a significant portion of the payload optimization effort, normally a part of payload integration, is carried out by the User.

Other User concerns regarding costs and charges include the need for protection of investment via insurance of payload, flight delays or schedule changes that may call for indemnification of costs, potential liabilities incurred through accidents involving his equipment. In all of these situations, the payload integration function will include primary responsibility for establishing, a priori, the boundaries (physical, functional, schedule, etc.) that delineate responsibilities among payload entities, between payload and Spacelab, and between the User and the NASA.

Fig. 2 lists some typical approaches in which cost and charges data may be developed, and how those data may affect the integration effort.

Plans, Schedules, Decisions, Priorities

Timing becomes critical to the commercial User, both because of competitive pressures, and because invested "sunk funds" generate interest costs.

Technical and administrative decisions for commercial products are highly interactive, both in the alternatives for decision and in timing. Changes or delays in decisions are, therefore, sources of concern to Commercial Users, since such

USER PLANNING AREAS	POTENTIAL APPROACHES TO PROVIDE PLANNING DATA	IMPACT ON INTEGRATION	STATUS/REMARKS
COSTS OF SHUTTLE UTILIZATION	• ALLOCATION OF COSTS BASED ON RESOURCES USED • INCENTIVE/DIS-INCENTIVE WEIGHTING • RE-ALLOCATIONS BASED ON EXPERIENCE	• OPTIMIZATION BY USER SIMPLIFIES INTEGRATION • EARLY DECISIONS WILL LEAD TO EARLY, NUMEROUS PAYLOADS TO SCHEDULE AND ACCOMMODATE.	• SUGGESTED COST ALLOCATION MODEL DERIVED FROM PRESENT TRANSPORTATION INDUSTRY METHODS (INCENTIVES, RESOURCES USED, AVERAGING, ETC.) • USERS CONCERNS ALREADY IN EVIDENCE
	• AVERAGING OF SHUTTLE OPERATIONAL COSTS ON (AT LEAST) ANNUAL BASIS • AVERAGING ALL MISSIONS (HI, LO ALTITUDES, INCLINATIONS)	• INCREASES "ACCOUNTING" PAPERWORK • INCREASES COMPATIBILITY ANALYSIS EFFORT • MINIMIZES UNDER-UTILIZATION OF HIGH COST (HI ALTITUDE, HI INCLINATION) MISSIONS	
CHARGES TO USER	• EARLY DEFINITION OF OFFICIAL POLICY (SEE FIGURE 11) LONG RUN COST RECOVERY EQUITABLE TO LARGE AND SMALL USER, ETC.)	• EARLY DECISIONS AND "ACCOUNTING" IMPACTS, AS ABOVE	• WHAT IS "EQUITABLE"? IN MANY PRESENT SERVICE BUSINESS, SMALL USERS SUBSIDIZE LARGE USERS!
	• "BROKERING" OF MISSIONS/ACCOMMODATIONS TO SHARE CHARGES	• "BROKER" MUST HAVE SOME INTEGRATION CAPABILITY • INCREASES COMPLEXITY OF USER/NASA INTERFACES AND RESPONSIBILITIES IN LEVEL III, II INTEGRATION	• "BROKER" IS IN VULNERABLE SPOT • NASA IS PRESENTLY "BROKER", BUT COMMERCIAL ASPECTS MAY NECESSITATE CHANGE
PROPERTY INSURANCE, INDEMNIFICATION FOR DELAYS, ACCIDENT LIABILITY, ETC.	• EXTENSION OF PRESENT COMMERCIALLY UTILIZED COVERAGES, AS MODIFIED BY NASA HISTORICAL PERFORMANCE	• ESTABLISHES, ALLOCATES BOUNDARIES OF RESPONSIBILITIES BETWEEN PAYLOADS, PAYLOAD/SPACELAB	• COMMERCIAL COVERAGE MAY BE COSTLY. GOVERNMENT GUARANTEES HAVE PRECEDENTS

Fig. 2. Potential Approaches for Providing Typical User Planning Data on Shuttle Utilization Economics

changes or delays exert impacts throughout subsequent technical and administrative aspects.

Inclusion of <u>Commercial</u> experiments in Spacelab requires reassessment of prioritization methods, and, will, no doubt, evolve new priority criteria. Commercial Users will seek a clearly defined, equitable priority system, which allows them to plan commitments for equipment, facilities and other resources.

The solutions to much of these problems lies in adequately prepared, intelligently communicated, and competently carried out, plans and schedules.

NASA has demonstrated these capabilities in programs involving the Aerospace community. Many of the same techniques utilized in those programs will apply in commercially-oriented programs, with some modifications. Bearing the brunt of resolving the problems of plans, schedules, many decisions, and priorities is a function of payload integration. Fig. 3 lists some typical solutions of these problems.

EARLY USER PROGRAM INTEGRATION AREAS	POTENTIAL APPROACHES FOR EARLY USER PROGRAM INTEGRATION	IMPACT ON INTEGRATION FUNCTION	STATUS/REMARKS
PLANS, SCHEDULES	• MUTUAL EDUCATION JOINT STUDIES PERSONNEL EXCHANGE INTENSIVE PUBLIC RELATIONS PROGRAMS USER HANDBOOK/BUSINESS PRACTICES HANDBOOK	• MAJOR EFFORT, NEW CAPABILITIES NEEDED (PR, HANDBOOK EDITING, ETC.)	• TYPICAL STUDIES ON-GOING (BUS, TERSSE) • SHUTTLE, SPACELAB ACCOMMODATIONS DOCUMENTS GOOD BASELINE – NEED USER ORIENTATION.
	• MAKE USERS PARTY TO PLANNING, SCHEDULING	• INCREASES NASA/USER INTERFACES, WORK FORCE • REQUIRES CREDIBLE STS (AND ELEMENT) PLANS AND SCHEDULES • REQUIRES UNIFORM PAYLOAD (AND ELEMENT) PLANS AND SCHEDULES	• TYPICAL STUDIES (BUS, TERSSE) • SCIENTIFIC COMMUNITY WELL ON-BOARD
DECISIONS	• AS ABOVE	• "TRACKING" AND ASSESSMENT OF EFFECTS OF CHANGES INCREASES COMPLEXITY AND AMOUNT OF WORK	• INDUSTRY HAS ANALOGOUS PROGRAMS • COMPUTERIZATION COULD HELP • CREDIBILITY OF, PARTICIPATION IN, COMMERCIAL SPACE PROGRAMS HINGES ON THIS EFFORT • NEAR-REAL TIME 2-WAY COMMUNICATIONS VITAL
PRIORITIES	• DEVELOP METHODOLOGY EARLY • PUBLISH, EXERCISE • ENGAGE IN ACTIVE REVIEWS, CRITIQUES • UPDATE/REVISE, REPUBLISH	• ENCOURAGE/PERFORM DEVELOPMENT • EXERCISE, REVIEW REQUIRES TIME AND WORK FORCE • FIRST INTEGRATOR/USER INTERACTION	• MANY SCHEMES EXIST (SPACE STATION, ETC.) • ACTIVE REVIEW VITAL-SEEK DEVIL'S ADVOCATES, IF NECESSARY • PUBLICIZING VITAL

Fig. 3 Potential Approaches for Integrating Commercial Payload Program Plans, Schedules and Decisions, and Establishing Payload Priorities

References

1. Study for Identification of Beneficial Uses of Space (Phase III), Contract NAS 8-28179, Final Report (to be published)

2. Scarff, D.D., Bloom, H.L., A Businessman Views Commercial Ventures in Space, AIAA Paper 73-78, AIAA 9th Annual Meeting, Jan 8-10, 1973

AAS 75-241

LOW COST INTEGRATION TECHNIQUES
FOR SPACELAB PAYLOADS*

C. A. Braunwarth[†] and T. C. Aepli[‡]

CONDENSATION

Existing operational programs are used as models for standing Spacelab integration approaches. Emphasis during analytical integration is on optimized communications between integrator and experimenter. Physical integration benefits from interface hardware designed to facilitate and support Spacelab integration and operation. A low cost program approach seeks to optimize resource investment per flight to lessen initial investment requirements.

INTRODUCTION

Integration is a complex process that begins long before flight hardware delivery and continues through hardware assembly and checkout. One of the greatest challenges of the Spacelab/Shuttle era is to develop an integration concept that performs this process with a minimum of documentation, manpower, and repetitive testing.

INTEGRATION APPROACHES

Spacelab activities will take place in an era of operational manned spaceflight that will be quite different from the research and development situation of the

---

* Prepared for the Twenty First Annual Meeting of the Americal Astronautical Society, "Space Shuttle Missions of the 80's," Denver, Colorado, August 1975

† Systems Engineer, Advanced NASA Programs, General Electric Space Division, Valley Forge, Pa. (Room M-7223 Phone 215-962-1298)

‡ Systems Engineer, Advanced NASA Programs, General Electric Space Division, Valley Forge, Pa. (Room M-7223 Phone 215-962-3870)

past. Operations and procedures that produced such an outstanding record of success in Apollo, Skylab, and ASTP must be modified and streamlined to handle the greatly increased flight rates planned for Spacelab and Shuttle.

Integration Models

The operational complexity of Spacelab appears to fall between that of an automated spacecraft such as Nimbus and the CV-990 research aircraft. Nimbus and CV-990 lessons learned are quite similar and they can be used to structure an approach to Spacelab integration. At the heart of the concept is reliance on teamwork and individual responsibility, which involves accepting the risks of trusting people to do their jobs, rather than resorting to extremes of supporting and protesting documentation. What makes it work is focussed authority, a single point where decisions are made in the time frame when decisions are needed.

Analytical Integration

Analytical integration begins with grouping experiments into candidate payloads and ends with receipt of flight hardware at the integration site. It includes everything the integrator does to "set the stage" for physical integration. The key element is effective communication between experimenter and integrator, so that the experimenter knows exactly what is being provided and the integrator knows exactly what he must accommodate. The primary vehicle for this is the Accommodation Handbook. This document should not only levy requirements but should also offer solutions. The handbook should guide and structure the epxerimenter's responses so that he can supply the right information to the integrator at the time it is needed with minimum disruption of his experiment development activities. Experimenter response documentation takes the form of stage release requirements data at set times during the experiment development cycle, followed by a safety conformation document at flight hardware delivery.

Physical Integration

Physical integration begins with delivery of experiments to the Spacelab integration site and involves test and checkout activities of four basic types: experiment tests, performed prior to assembly into the Spacelab experiment train; integration tests, performed during assembly; systems tests, performed to check out the assembled experiment train; and confidence tests, performed before and after shipment to the launch site. Most experiment level testing is optional, to be determined by experimenter or integrator according to the requirements and conditions of each experiment. The implication is, however, that facilities for experiment level testing, including environmental tests, should be available at the integration site. Integration tests and system tests make up the major share of the test and checkout effort, and although they are constrained to a rather standard sequence at the top level the key to their effective implementation is flexibility - adapt the level and extent of testing to the needs of the specific payload group, and utilize analysis in lieu of testing whenever feasible and cost-effective. Simulated orbit operations provides a means to combine test, training, planning, and validation and is a central part of the test and checkout approach. Confidence tests are not complicated and are considered necessary for generating baseline data for go/no go prelaunch checks.

INTERFACE DESIGN

The integrator's responsibilities in interface design fall into three areas: (1) configuration of Spacelab subsystems to properly support the experiment payload, (2) design or acquisition of interface support equipment when Spacelab capabilities require augmentation, and (3) management of experiment interfaces to ensure compatibility with Spacelab and support equipment capabilities. The first two areas are obviously the province of the integrator, while the third is the subject of some controversy. The objectives of low cost integration are not served if the integrator involves himself in all facets of experiment design, but a certain involvement in interface design, particularly by recommending proven design solutions in the experimenter, helps insure compatibility both with the Spacelab system and between the experiments themselves.

Modular Hardware

Spacelab experiment racks and pallet sections are modular hardware that can be assembled into a great variety of flight configurations. Few if any experiments, however, will require an entire pallet section or even a full experiment rack. Smaller interface modules, of a size that can be shipped to the experimenter, offer the potential of simplifying experiment interfaces and reducing both experiment and integration costs. The standard 19 inch experiment racks can be fitted with rack modules or drawers of various heights that contain standard electrical connectors and standard interface circuits as appropriate. Modular shelf sections can be used in similar fashion on pallet, and in addition can provide standard thermal control interfaces and dynamic isolation. Other modular hardware can be used to adapt experiments to the Spacelab airlock or to provide auxiliary mounting space.

Standard Interface Circuits

The Nimbus program has found that experiment integration is greatly simplified by identifying standard interface circuits to be used by the experimenter. In this way random noise, spurious signals, unwanted feedback, and signal misinterpretations can be kept to a minimum. Instead of simply identifying standard circuits, Spacelab may provide the actual circuit on a card which the experimenter builds into his hardware.

Support Equipment

Certain specialized functions not provided for by standard Spacelab subsystems can be best accomplished using special support equipment. For example, many experiments require deployment out of the Shuttle payload bay for purposes of viewing or to sample the local environment. For these it makes sense to provide a choice of tried and proven deployment devices rather than have each experimenter provide his own. The same can be said for gimbal systems and electronic actuators.

CONCLUSION

The low cost integration techniques presented in this paper depend upon optimized information exchange between integrator and experimenter, and an extensive stable of interface hardware designed to simplify Spacelab integration and operation. Optimized information exchange can be modeled on existing operational programs using minimum documentation and manpower requirements as direct measures of minimum cost. Interface hardware requires a substantial initial investment which can be justified only by significant potential savings over the life of the program, so it is necessary to optimize the resource investment of each flight to attain a truly low cost program.

AAS 75-243

SPACELAB PAYLOAD AND PROGRAM PLANNING
IN GERMANY

Horst Schreiber [+]

The preparation of the Spacelab Utilization Program in the Federal Republic of Germany is based on 370 experiment and utilization proposals which are now available from the German user community covering all disciplines in research and technology. The paper covers substance and status of the German payload and program planning done so far. The outlook may give starting points for potential cooperation with other users and agencies in the field of Spacelab utilization.

INTRODUCTION

The plans of the Federal Republic of Germany for the utilization of the reusable space laboratory for scientific, technological, economical and social purposes are self-evident and consistent with the financial contribution made by Germany to the Spacelab development. Since about one year the preparational work for the utilization program is carried out intensively so that we are now in the position to have a program approach and to present our preliminary payload planning.

[+] Manager, Spacelab Utilization Office, DFVLR-BPT, 5000 Koeln 90, FR Germany

PROGRAM TOPICS

Within the scope of the German Space Program which is represented by the Bundesminister fuer Forschung und Technologie [+] (BMFT), Spacelab utilization will be sponsored in the following areas and disciplines:

- Space Science
    - Astronomy & Astro Physics
    - Solar Physics
    - Atmospheric Physics
    - Interplanetary Space Physics
    - Life Science

- Near-Earth Space Applications
    - Earth Observation
    - Meteorology

- Technical & Economic Applications
    - Basic Research in Physics & Chemistry
    - Material Science and Technology
    - Space Processing
    - Space Technology
    - Solar Energy
    - Communications & Navigation

In addition to these disciplines the development and provision of orbital systems is also included, i.e. the contribution to the Spacelab development (via ESA) and the planning and preparation of the utilization program structure.

[+] Federal Ministry for Research and Technology

ROLE OF DFVLR

For program preparation and execution the BMFT makes use of the Deutsche Forschungs- und Versuchsanstalt fuer Luft- und Raumfahrt (DFVLR). This establishment has highly skilled personnel and various facilities in 5 field centers for the purpose of applied research, technical and scientific services, and project management. Accordingly DFVLR has the following share in the German Spacelab utilization program:

- Participation as User

  embodied by the pertinent research institutes

- Program Implementation

  represented by the Project Executive Division (formerly GfW)

- Program Support

  achieved in the fields of

  Payload expert selection & training

  Integration, test & simulation

  Payload mission operations

  Data processing & evaluation.

DFVLR is also offering its capability and resources to the European Space Agency (ESA) in context with the execution of European projects using Spacelab. A corresponding agreement is being negotiated between ESA and DFVLR which shall become effective for the first Spacelab mission.

EXPERIMENT AND UTILIZATION PROPOSALS

The call for ideas initiated by BMFT last year was well responded by German universities, institutes and industry. Up to now 370 experiment and utilization proposals are in hand of DFVLR / Office of Spacelab Utilization. They split up as follows:

- Earth Observation & Meteorology   40 Proposals
- Space Technology & Energy   58
- Communications & Navigation   31
- Material Science & Processing   115
- Extraterrestrial Physics   65
- Life Science   61
                                   Total 370

All proposals are registered in Level-A-data sheets similar to NASA practice. For general information a status report is issued and periodically updated which shows the list of proposals (theme, objective, proposing institute and name of investigator) grouped into disciplines.

The experiment proposals from the German user community reflect certain interests and indicate special preferences which are of importance for the payload planning. The following explanations may characterize this for some disciplines.

Among the proposals concerning **Earth Observation** the interest in sensor technology is well to the fore. In this case Spacelab is preferably considered as an experimental platform for the development and pre-operational test of multispectral scanners and cameras as well as microwave sensors. Although the use of operational satellite systems for collecting earth observation data is still considered preferable to Spacelab, some German proposals describe the evaluation of earth observation data gathered by Spacelab missions. In particular for

- Studies of coastal phenomena
- Studies of characteristics of waters in arid zones and the prospection of subsurface water reservoirs in the tropics
- Exploration of natural resources in the tropics and geodynamic studies.

The experiment proposals in **Atmospheric Physics** plan especially the use of Laser/Lidar systems and Infrared Radiometers but also Photometers and Spectrometers. Primary objectives are

- Determination of vertical/horizontal density profiles
- Aerosol size distribution and local and global transport mechanism in the troposphere and stratosphere
- Determination of trace gas concentration
- Studies of cloud physics
- Measurement of vertical temperature and humidity profiles.

In the area of <u>Communications/Navigation</u> the proposals can be grouped according to four points of main effort, namely

- Development and test of navigation techniques such as one-way navigation
- Development and test of wide-band communication systems, e.g. with Laser and pseudo noise modulation
- Investigation of electromagnetic wave propagation, e.g. at very low frequencies or SHF up to 70 GHz
- Test of transmission techniques.

At most of these proposals the anticipation of a possible commercial utilization is obvious.

The discipline of <u>Material Science and Processing</u> concerns the largest user family and consequently represents the sponsoring emphasis within the German utilization program. The individual themes can be grouped as follows:

- Glasses and ceramics
    e.g. production of new glasses and ceramics
- Metallurgy and compounds
    e.g. eutectics, compound materials, influence of convection on solidification, separation process of immiscible alloys, high and ultrapure metals
- Chemistry
    e.g. continuous flow electrophoresis, phys./chem. processing
- Fluid and gas dynamics
    e.g. diffusion, convection, fluid dynamics
- Crystals
    e.g. crystal growth from the melt, solutions and gases

For these applications the utilization of the zero-g-condition in space has primarily the objective to support ground research and development work with findings which can only be achieved in the space laboratory.

The proposals from the area of Extraterrestrial Physics do form an important group but not in terms of figures; surely because the traditional user of orbital systems does not consider the Spacelab as an exclusive observation platform. The Spacelab experiments proposed by German scientists can be outlined as follows:

- Investigation of solar radiation and flares using X-ray and UV spectrography and Gamma spectroscopy
- Investigation of the physics of the upper atmosphere and space by active and passive diagnostical methods
- Investigation of cosmic radiation and particles by X-ray and Gamma spectroscopy and detector systems
- Astronomical investigations using infrared and UV spectroscopy.

The pronounced international collaboration in these disciplines will presumably result in a coordination of experiment proposals and also in a cost saving division of effort.

The experiment proposals concerning Life Science can be divided into two groups. Based on the well-known Biostack concept the development of this type of experiment will serve for advanced investigations of animal and plant objects and bacterial spores. Furthermore studies of zero-g-effects on biological objects (Lifestack) and biological basic research (Biostack Monitor) are subjects of proposals. The second group of proposals is aiming at medical research, in particular experiments investigating vestibular reactions and sensations of man and animal. Germany has achieved significant results in this field through aeromedical research in the past, apart from the participation in the scientific evaluation of Apollo and Skylab results of NASA. The planned active experiments shall be carried out mainly in connection with the DFVLR engagement in selection and training of payload experts.

PAYLOAD PLANNING AND DEVELOPMENT

First of all the registered proposals must be individually refined in order to become appropriate candidates for the purpose of payload planning. Depending on the condition of each proposal, preliminary idea or final concept, an iterative process towards the experiment definition must be promoted:

- Preliminary investigations and definition studies
- Concept and feasibility analyses
- Experiment specification and mission requirement definitions.

These tasks should primarily be carried out by the proposers themselves in order to avoid unnecessary transfer of ideas and - not least - to obtain motiviation of the users. As far as system engineering and interface aspects are concerned, pertinent support by DFVLR and the aerospace industry will be provided to them. According to the working schedule of the German program, mid-1976 the definition and predevelopment work for the first selected experiments will be finished, so that the subsequent phase of hardware/software development and instrumentation procurement can be started. Up to this date it does not matter whether or not the defined experiments have already been appointed to a certain SPACELAB mission, since mission flexibility of this new facility shall be a feature by definition.

Recently a German payload proposal has been submitted to ESA for the first NASA/ESA SPACELAB mission in 1980. This proposal contained almost 40 experiments which have been selected according to the special objectives jointly established by NASA and ESA for that mission, and for which a preliminary accommodation assessment was made by DFVLR. It is to be understood as a maximum offer for a German payload contribution which the BMFT is ready to sponsor.

PROGRAM PLANNING AND IMPLEMENTATION

Apart from individual experiments which could be included in ESA or NASA payload planning, the great user interest and the amount of experiment proposals require national effort in the areas of payload integration and mission implementation. From mid-1974 to mid-1976 several studies and analyses are being carried out by German industry and institutes, defining all program and system aspects of the German Spacelab Utilization program, in particular

- Overall program planning
- Model missions
- Organization models
- Payload integration concepts
- Mission projects preparation
- Test and integration facilities
- Data management system
- Payload expert selection
- Model and test philosophy.

Whereas the system aspects of this program are specifically covered by a parallel paper of this meeting, the preparatory work done for payload and mission planning shall be explained here. The most important study result in this respect is the MISSION MODEL, which presents an approach for the basic BMFT program together with a working hypothesis for the ESA program. This model takes into account the German utilization ideas and financial possibilities of the government as well as increasing industrial user contribution, the last is considered vital to the success of the program. It reflects clearly the program topics of the sponsoring ministry. It shows in particular the emphasis in economic and technological application such as material science and space processing, which obviously can lead to dedicated missions carried out autonomously (with NASA mission support), or even on a bilateral or multilateral basis with NASA and other agencies. In the areas of extraterrestrial science, life science, and earth observation, only partical payloads can be provided by German users, so that either the contribution to ESA-conducted special projects or the participation in appropriate NASA missions must be considered.

It is not intended to further develop the mission model for the purpose of long-term planning. The prime purpose, however, is to use this mission model as an orientation aid in the program development, and especially as a reference basis during the preparatory work for mission projects. A study contract has now been placed to the German aerospace firms for the preparation of such mission projects. The subjects are four different types of payloads as indicated in the mission model, with the following disciplines:

(1) Space processing / Technology / Antropotechniques
(2) Solar physics
(3) Earth observation / Meteorology / Communications / Navigation
(4) High Energy Astrophysics.

The study objectives are to obtain preliminary mission definitions based on feasibility analyses for payload accommodation and mission implementation. Moreover, the required program activities, resources, and investments will be identified and defined. This work will be finished by mid-1976. The results will, together with the other supporting study output, enable the program authorities to review the necessary effort on program, mission, and payload level, and to make decisions concerning further actions within the SPACELAB utilization program.

AAS 75-297

SPACELAB RESOURCES

Hans M. Kappler[x]

In June 1974 the Spacelab System Design and Development Phase under the overall direction of ESA/ESTEC as customer and VFW-Fokker/ERNO as industrial Prime Contractor and lead company of an European Consortium has been initiated.

Spacelab is designed in order to provide a facility for basic scientific research and technological development in space for many disciplines, it is a reusable, modular space-laboratory which will be accommodated in the cargo bay of NASA's Shuttle-Orbiter[4] for 7 - 30 days missions.

SPACELAB SYSTEM

The Spacelab consists of two basic elements - a pressurized module and an unpressurized pallet - which can be used separately or in combination[xx]. The module provides a controlled pressurized environment for the users and their equipment and supplies basic services such as power, thermal control and data management, together with certain basic support equipment such as standard racks, scientific airlocks etc., which may be used as required.

The pallet is an unpressurized platform to which instruments such as telescope and antennas may be mounted which require direct exposure to space. The pallet also provides some basic services such as power, data acquisition and thermal control.

---

[x] Dr. H.M. Kappler is the Spacelab Payload Advisor at the Spacelab Prime Contractor VFW-Fokker/ERNO, 28 Bremen, Huenefeldstr., West Germany

[xx] More detailed information about the Spacelab-System is given in the "Spacelab Payload Accommodation Handbook" Preliminary Issue May 1975 [5]. In the same sense the information given in this paper has to be considered preliminary.

Spacelab is carried to and from orbit by the Space Shuttle. It remains attached to the Orbiter throughout the flight. Major external design features of Spacelab in a typical "Module plus Pallet" configuration are shown in the Figure. This configuration consists of a two-segment module and one pallet segment. The module diameter is slightly over 4 meters and each cylindrical segment is approximately 2.7 meters long. The pallet segments are approximately 2.9 meters long and 4 meters wide. The lower portion of the Figure shows a pallet-only configuration, there Subsystem equipment is accommodated in a small pressurized

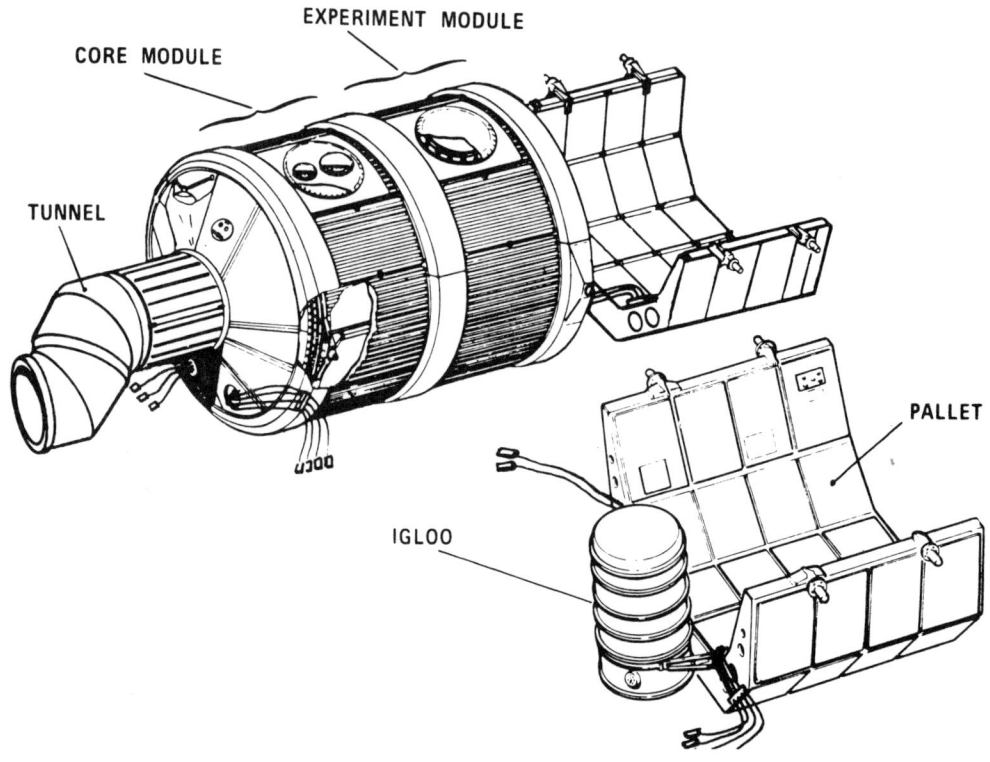

External Design Features of Spacelab

container, the so-called Igloo. The forward located module segment (core segment) contains subsystems equipment in the "Control Center Rack" and the subfloor area. Crew work space is provided at the "Work Bench Rack". The rest of the space in the core segment, about 60% of the rack volume, is available for experiments. The experiment segment is dedicated entirely to experiment installation and operations.

SUBSYSTEMS

The module structure provides various mechanical provisions to accommodate experiments and experiment related equipment by the following structural means.

1. The floor provides mounting provisions for standard experiment racks and/or experiment equipment, it is designed for a load carrying capability of 300 kg/m in the central portion and of 500 kg/m along the outer regions.
2. Standard 19" experiment racks are mission dependent Spacelab subsystem equipment and can be removed if required. Design load for the racks is 300 kg/m^3.

The pallet structure accommodates equipment for direct exposure to space, see Fig.

The total Spacelab load carrying capability (up to 9100 kg) and the volume available for experiment accommodation depend on the system configuration, the characteristics of the orbit and the mission duration.

The modular elements of Spacelab can be arranged in various flight configurations to suit the needs of specific mission/payload requirements. There exist 13 typical configurations. The electrical power and distribution subsystem receives its primary power from a dedicated power source in the Orbiter: 7 KW average and 12 KW peak for nominally 15 minutes every 3 hours at a nominal voltage of 28 V DC. However, the supply of peak power is related to the Orbiter heat rejection capabilities, which are attitude dependent. The conditioning and distribution of electrical power is strictly separated between subsystems and experiments. Power available to experiments (3.6 - 5.1 KW typically) depend on the configuration containing a specific set of mission independent subsystem equipment and on the amount of mission dependent equipment.

The Environmental Control Subsystem consists of the environmental control life support subsystems and the thermal control subsystem.

Rack mounted equipment in the module is cooled by avionics cooling loop (3 KW max.), the cabin air cooling loop can take up to 1 KW max. and an experiment dedicated heatexchanger is designed for 4 KW max. Equipment on the pallet is cooled by cold plates (1 KW max. per plate).

The command and data management subsystem provides a variety of services to Spacelab experiments including data acquisition, monitoring, formatting, processing, displaying, caution and warning, recording and transmission in addition to providing command and control capability for Spacelab experiments.

The common payload support equipment consists totally of mission dependent equipment. The aft end cone provides the capability for installing the 1.5 m long airlock, the aft viewport and a feed-through panel located at the lower part of the bulkhead. There is a flanged cutout of the top of each cylindrical segment of the module to allow installing of either the viewport and/or the optical window or the 1 m long airlock.

The Spacelab Computer Software comprises the software used for Spacelab and ground support equipment, integration, testing and operations. This includes subsystem testing, integration, checkout, on-board data handling for subsystems, on-board data handling support for experiments and checkout for the CDMS portion of the experiment interfaces.

SUMMARY

Summarizing the resources provided by Spacelab to the user - structure, its volume and weight carrying, capability, electrical power distribution, environmental control, command and data management and the common payload support equipment - it becomes obvious, that this reusable system is able to serve as a general purpose laboratory for application in various disciplines. Due to the availability of this system it will be possible, on the one hand to reduce the cost and the complexity of experimentation in space compared to the "classical" space experimentation of the recent years, on the other hand space experimentation will be available and attractive for a larger user community and new scientific and application disciplines.

BIBLIOGRAPHY

1. Memorandum of Understanding between ESRO and NASA, 14. August 1973.

2. NASA/ESRO Joint Programme Plan for Spacelab. ESRO-SL-74-2, NASA-MF-74-2.

3. Spacelab System Requirements, Issue 5, November 1974, ESTEC Ref. No. SLP/2100.

4. Space Shuttle System Payload Accommodations: Level II Program Definition and Requirements, Volume XIV, ISC 07700.

5. Spacelab Payload Accommodation Handbook, Preliminary issue May 1975, ESTEC Ref. No. SLP/2104.

AAS 75-260

ASTRONOMY SPACELAB PAYLOADS

Richard Ott[*]
Gary Wengrow[+]

Four typical astronomy payloads have been defined by NASA's Goddard Space Flight Center (GSFC) for early Shuttle flights. The payloads combine instruments from ongoing balloon and rocket programs with facility-class instruments designed expressly for the future Astronomy Spacelab Payloads (ASP) project. Three of these payloads, described in this paper, are dedicated to specific disciplines: high-energy astrophysics, solar physics, and stellar astronomy. The fourth payload consists of elements from each of these three areas.

The high-energy astrophysics (HEA) payload (Fig. 1) is composed of two gamma-ray instruments, two X-ray instruments, and a cosmic-ray electron spectrometer. These are mounted on Spacelab pallets, with the two X-ray instruments utilizing instrument pointing system (IPS) gimbal mounts to achieve the required pointing accuracy and stability. During a seven-day HEA mission, 93 X-ray and gamma-ray observations will be made of each of three representative targets: Cygnus XR-1, Crab Nebula, and the galactic center. Of the 150 hours of cosmic-ray detector operations, 83 percent will be free from earth occultation.

The solar physics payload (Fig. 2) incorporates four small instrument pointing system (SIPS) gimbal mounts that house 10 fine-pointing instruments, including a helioscope facility, a coronagraph, and various spectrographs and spectroheliographs. A single pallet segment contains two coarse-pointed instruments: a gamma-ray spectrometer and an X-ray burst detector. Since the coarse-pointed instruments are sensitive to trapped particles, the solar physics mission is constrained to orbit inclinations of less than 33 degrees and to

---

[*]Goddard Space Flight Center
National Aeronautics and Space Administration

[+]Space Division
Rockwell International

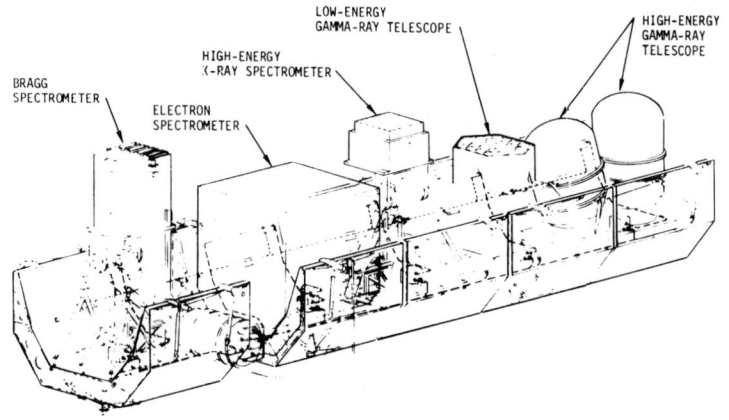

Fig. 1 HEA Payload

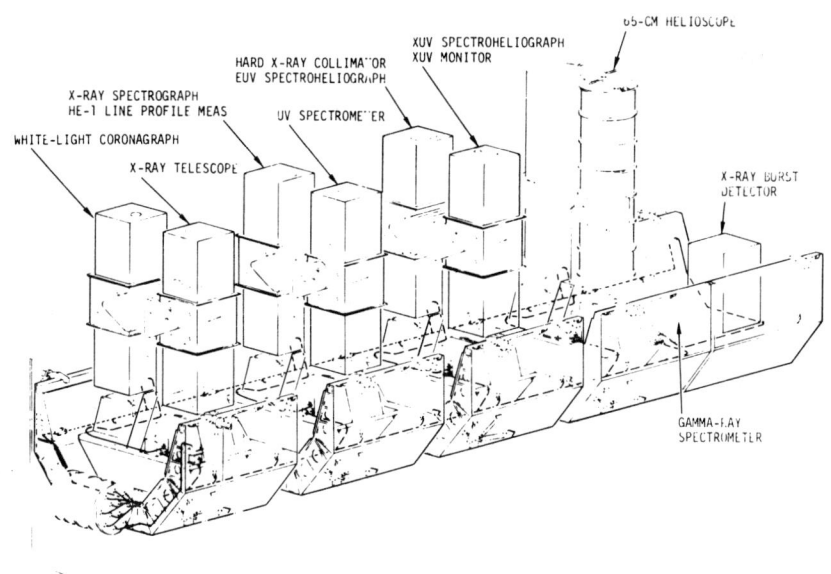

Fig. 2 Solar Physics Payload

altitudes of less than 370 kilometers. Seventy-percent solar visibility can be obtained with a midwinter launch, providing over 100 hours of observations during a seven-day mission.

A 1-meter facility-class telescope, mounted on a Spacelab IPS, is the principal instrument of the stellar astronomy payload (Fig. 3). Its focal-plane instrumentation consists of a field camera, an on-axis spectrograph, and a planetary camera. Real-time user interaction will be provided by a TV monitor of the telescope field. Other instruments on this payload are derived primarily from existing rocket payloads, including a pair of Schmidt cameras. A

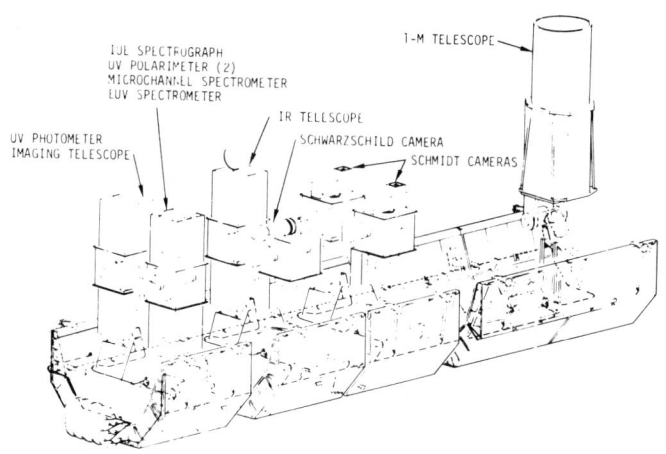

Fig. 3 Stellar Astronomy Payload

spectrograph of the type selected for the International Ultraviolet Explorer is also included. Of the 285 target observation opportunities on a seven-day mission, 95 will be made in total shadow. Each observation will be 20 to 30 minutes in duration. Orbit geometry and launch parameters have been selected to minimize exposure of instruments to the enhanced radiation background of the South Atlantic Anomaly (SAA) during orbital darkness.

Engineering analyses have been performed on these representative payloads to ensure that they are compatible with Shuttle and Spacelab weight, load, and center-of-gravity constraints. Design layouts have been prepared and interface system concepts developed for pointing and stabilization, command and data handling, and other subsystems.

PHYSICAL COMPATIBILITY

Each point design has been analyzed to ensure physical compatibility with Shuttle. Lift-off, normal entry, and abort weights are shown in Table 1. All lift-off weights are compatible

Table 1
PAYLOAD WEIGHT SUMMARY

Payload	Weight (kg)		
	Lift-off	Normal Entry	Abort
High-energy astrophysics	16,167	13,956	14,352
Solar physics	14,702	12,806	13,202
Stellar astronomy	14,570	11,678	12,074

with Shuttle's capability to reach the required orbit, and all entry weights are less than the 14,528-kilogram limit. It should be noted that each weight includes a reserve for growth. Longitudinal, lateral, and vertical centers of gravity are all within allowable limits.

Attach-fitting loads have been calculated for each payload-to-pallet and pallet-to-orbiter combination and compared with Shuttle limit loads. All such analyses have resulted in compatible loads, generally with large margins of safety. However, analysis of payload dynamic response is needed to ensure compatibility.

POINTING AND STABILIZATION

The baseline pointing and stabilization approach for all payloads uses the orbiter's vernier RCS to provide coarse pointing and stabilization. The orbiter can accept error signals from payload-provided target tracking systems to eliminate pointing bias introduced by misalignments between the payload bay and the orbiter navigation base, located in the crew compartment. Fine pointing will be achieved through SIPS or Spacelab IPS, depending on instrument size. With either system, an automated or crew-controlled search routine will be required to acquire targets with precision. Fine stabilization will be attained by closed-loop control, using a computer, target trackers, stable platforms, etc.

Alternatives to the baseline mode have also been examined. These include free drift (discussed below), use of control moment gyros (CMG's), and cold-gas RCS systems. The principal reason for considering these alternatives is the reduction of contamination from RCS effluents. An additional benefit, at least for free drift and CMG's, is improved stability of the pointing base.

Several free-drift modes have been examined. The most attractive are quasi-inertial (QI) modes, in which the orbiter attitude and rate are established at the beginning of an observation period in order to provide extended periods of near-inertial pointing.

The QI modes will experience large angular excursions relative to the local vertical, and hence will result in large gravity-gradient torques on the vehicle. Some insight into the behavior of the attitude motion under these circumstances may be obtained from the pitch-axis phase plane solutions presented in Fig. 4. Response to the gravity-gradient torques is cyclical for trajectories under the stability boundary, with oscillations occurring about the stable attitudes $\theta = 90$ degrees and 270 degrees. The motion is secular for phase plane trajectories above the stability boundary, and it is in this regime that the most promising QI modes exist. The initial conditions designated QI No. 3 in Fig. 4 are designed to provide a

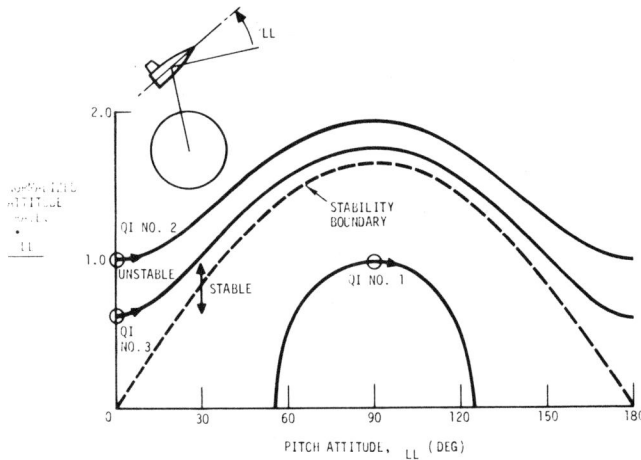

Fig. 4 Free-Drift Phase Plane Trajectories

time-averaged rate in local-level coordinates approximately equal to orbit rate. This mode will provide inertial attitude rates of less than 0.05 deg/s for periods up to 70 minutes.

COMMAND AND DATA HANDLING

Three mechanization approaches have been considered for the command and data-handling system. In the "laboratory" approach, all operating and monitoring functions would be performed with dedicated components. In the "universal" approach, all functions would be carried out with shared equipment, except those requiring dedicated components for reasons of safety. The "hybrid" approach would provide dedicated components for key experiment operating and monitoring functions and use shared components for all other functions. Based on detailed experiment requirements, preliminary system sizing parameters have been determined for each of the three approaches for each of the three payloads.

The laboratory approach, in most cases, would require more control and display panel area and payload bay wires than are available. However, the universal approach may underutilize available capability. The system-mechanization concept adopted for this analysis is therefore a hybrid. Experiment data, at low and moderate rates, will be carried on the Spacelab data bus and will interface with the orbiter communications system through an orbiter multiplexer-demultiplexer (MDM). Some dedicated control and display components use the Spacelab data bus, while others are hard-wired to experiment equipment. High-rate data (up to 50 mbps) must be recorded or transmitted to the ground by the orbiter Ku band system through the TDRS.

CONCLUSION

Feasible payload and mission concepts have been defined for astronomy Spacelab payloads. These provide the flexibility to accommodate a variety of astronomical instruments, including derivatives of sounding rocket and balloon payloads as well as dedicated facility-class telescopes. NASA is continuing the analysis of these payloads and missions in order to develop additional insight into their design and operations and to incorporate the results of a number of supporting studies.

ACKNOWLEDGMENT

The authors wish to thank H.L. Myers, D.W. Peebles, R.E. Oglevie, and D.E. Jorgenson of Rockwell International for their contributions to this paper.

AAS 75-261

ATMOSPHERIC X-RAY EMISSION EXPERIMENT FOR SHUTTLE*

R. A. Goldberg[†]
K. L. Hallam[†]
J. G. Emming[‡]

The Atmospheric X-ray Emission Experiment (AXEE) for Shuttle is designed to measure the spatial, temporal, and energy distribution of X-ray aurorae produced by precipitating electrons. It will provide vital data on solar-terrestrial relationships that may lead to defining the transfer mechanism that causes certain terrestrial weather events and climatological behavior. The instrument concept is based on a spatially sensitive multiwire proportional counter, combined with collimators to produce X-ray images of the aurorae. It will be mounted on an instrument pointing system to provide the required attitude control and is operated by the Spacelab payload specialist who has full control over its observing and data taking modes.

INTRODUCTION

In this work, we are proposing the Atmospheric X-ray Emission Experiment (AXEE) as our experiment for Shuttle/Spacelab, to investigate certain physical processes responsible for energy transfer from the thermosphere to the stratosphere, and thereby candidate for linking solar-terrestrial weather correlations. AXEE will measure the spatial, temporal, and energy distribution of auroral X-rays produced by precipitating electrons impacting on the upper atmosphere. Such electrons, induced by solar disturbances, produce bremsstrahlung X-rays which radiate isotropically. This permits topside study of radiation which simultaneously penetrates to stratospheric depths, where modification of critical

---

*Prepared for the Proceedings of Twenty-First Annual Meeting of the AAS, August 26-28, 1975, Denver, Colorado.
[†]Goddard Space Flight Center, Greenbelt, Maryland 20771
[‡]Ball Brothers Research Corporation, Boulder, Colorado 80302

parameters such as ozone, conductivity, and temperature is possible. Ideally, this instrument will be part of a Shuttle Solar-Weather Facility, wherein a more comprehensive study of the physical processes can be accomplished.

INSTRUMENT REQUIREMENTS

The measurement requirements based on the scientific objectives are listed in Table 1. The instrument is sensitive in the energy range from 5-150 keV and will provide images of the X-ray aurora between 5 and 30 keV. The sensitivity of 0.1 photons cm^{-2} s^{-1} is achieved with an effective collection area of 500 cm^2, which takes into account both background effects, and a source caused by a precipitating electron flux of 10^6 electrons cm^{-2} s^{-1} over an area of 200 km diameter at a 2000 km range.

Table 1

MEASUREMENT REQUIREMENTS-SUMMARY

Parameter	Range Limitations	Instrument Requirements
Energy Range	5-150 keV	Imaging Below 30 keV
Sensitivity	0.1 Photons/cm^{2}s (=10^6 Electrons/cm^{2}s)	Effective Collective Area = 500 cm^2
Energy Resolution	Bremsstrahlung Spectrum	FWHM = 20%
Spatial Resolution	200 km at 2000 km Range	2°
Field of View		
• Diverging Array	180° Scan	20° - 30°
• Parallel Array	2°	2°
Time Resolution	1-60 seconds	1-60 seconds
Dynamic Range	10^6 - 10^{11} electrons/cm^{2}s	6 decades of Sensitivity

INSTRUMENT DESCRIPTION

The AXEE detector system includes a sandwich of three individual proportional counters, the outer two of which are used as anticoincidence shield detectors. These are filled with an argon based gas.

The central detector is a multiwire proportional counter with spatial readout in two directions using delay lines, and is filled with a xenon based gas. Gas pressure in all counters is between 3 and 5 atmospheres. Pulse-height discrimination and analysis is used to provide 20% energy resolution of the source spectrum.

The collimator arrays are of two different types. A diverging hole array, combined with the spatially sensitive proportional counter, directly provides an X-ray image between 20° and 30° field of the source scene. Mechanical scanning will permit a 100° horizontal search in 6 to 9 image steps. When reversed in direction, the opposite side of the detector views through a parallel array collimator and yields more detailed information concerning the energy and intensity structure of a localized source. Here, imaging is obtained through mechanical raster scanning of AXEE. These two modes are combined in one instrument to provide substantial flexibility for coverage of a wide range of source configurations, sizes, and intensities.

An exploded view of the instrument is illustrated in Figure 1. Table 2 summarizes the most important instrument characteristics.

SHUTTLE INTERFACE

AXEE will be mounted on an independent pointing system to achieve the pointing and tracking requirements imposed by its two operational modes. This will also significantly decouple it from Shuttle motion and orientation, and permit greatly enhanced flexibility and independence during operation. We have examined various pointing systems under consideration for Spacelab use and concluded that the intermediate size Small Instrument Pointing System (SIPS) could best meet our requirements.

The SIPS will interface to a Spacelab pallet and can be operated in conjunction with the Spacelab computer and Shuttle inertial reference system. The SIPS enclosure will also provide a thermally controlled environment, adequate for AXEE operational requirements. When mounted to the SIPS, AXEE would appear as in Figure 2.

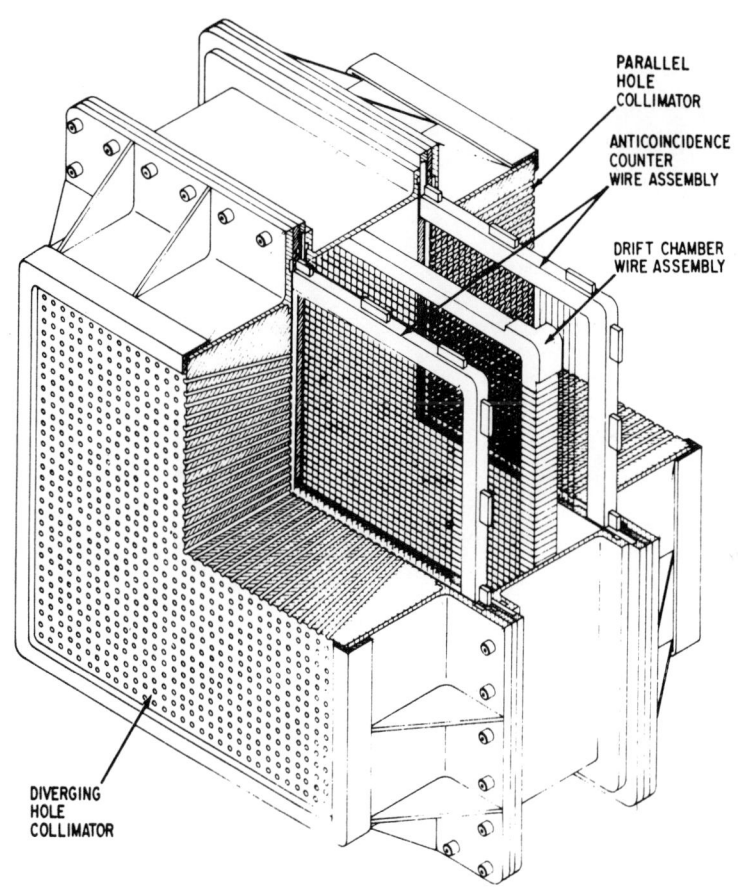

Fig. 1 AXEE Imaging Assembly

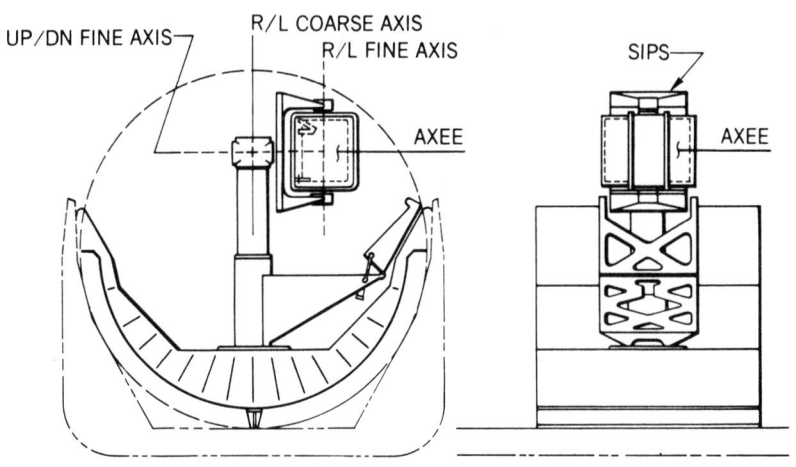

Fig. 2 Small Instrument Pointing System (SIPS) Mounted on Spacelab Pallet

Table 2

AXEE CHARACTERISTICS

Parameter	Dimension
Overall Dimensions	44.3 by 44.3 by 35.4 cm^3
Physical Detector Area	1000 cm^2
Effective Detector Area	~500 cm^2
Weight (excluding SIPS)	192 Kg
Power (excluding SIPS)	10.2 Watts
Energy Range	5 to 150 keV
Energy Resolution	~20 percent FWHM
Angular Resolution	1°-5°
Field of View:	
• Diverging Array	20-30°
• Parallel Array	2°
Pointing Arc	90° x 360°
Data Rate	20 kbit s^{-1}

The payload specialist will observe the X-ray sources on a two-dimensional video monitor at the payload specialist station. From these image displays he can determine the occurrence and location of specific sources. Adequate imaging will be relayed to ground to permit near real time consultations between the payload specialists and ground based scientists.

The system will be under full control of the payload specialist who will be able to override its automatic scanning features and directly control parameters relating to the observing modes.

FINAL COMMENTS

A visible and ultraviolet auroral camera will accompany AXEE on its missions. We also plan to group AXEE with other instruments designed to simultaneously study recent solar and atmospheric (terrestrial) parameters during the same mission. This grouping will form a Solar-Terrestrial Weather Observing Facility to be used for comprehensive investigations of physical phenomena linking the Earth's weather and climate to solar activity.

AAS 75-262

SPACELAB ULTRAVIOLET-OPTICAL TELESCOPE FACILITY
Murk Bottema*

When the initial plans for astronomy from space in the Shuttle era were laid, the Shuttle Astronomy Payload Planning Working Group recommended a telescope of 1-m aperture for the ultraviolet-visible wavelength region as a prime complement to the Large Space Telescope (LST)[1]. The scientific programs towards which this telescope can be applied were discussed in the National Academy of Sciences Woods Hole conference of 1973[2]. Although the light-gathering power and the angular resolution are smaller than that of the free-flying LST, a 1-m telescope operated during the Shuttle sortie missions (initially with a duration of 7 days, later possibly extended to 30 days) offers a much greater flexibility for varied observational programs, adaptation to changing astronomical interests, as well as a wider choice of image-recording techniques (including specifically photography and electrography), and an opportunity for early application of new instrument developments.

In 1974, the Goddard Space Flight Center (GSFC) initiated a series of studies to evaluate the feasibility of a 1-m Spacelab Ultraviolet-Optical Telescope (SUOT) to be operated from the Shuttle payload bay (so-called Spacelab pallet-only mode). These were conducted by Ball Brothers Research Corporation (BBRC)[3,4,5], in close cooperation with a Facility Definition Team (FDT) formed in December 1974. The FDT consists of representatives of diverse areas of research in astronomy and identified the scientific programs and instruments that would

---
*Ball Brothers Research Corporation, Boulder, Colorado.

be of prime interest for early Shuttle missions and of prime concern in the design of the facility[6]. These include:

1. high angular resolution imagery over wide fields;
2. far ultraviolet spectroscopy;
3. precisely calibrated spectrophotometry and spectropolarimetry over a wide wavelength range;
4. planetary studies, including high-resolution synoptic imagery.

The above activities led to a phase-A concept for SUOT, consisting of a 1-m aperture, f/15 modified Ritchey-Chrétien telescope, that can accommodate at least 2 major scientific instruments on each mission. One of these is placed at the conventional cassegrain focus, a location that is particularly suited for large spectrographs, polarimeters, far UV cameras (wavelengths down to 90 nm) and other instruments for which a minimum number of reflections is essential. By inserting a diagonal mirror, the telescope beam can be diverted to a radial focal plane, which is primarily intended for imaging with an intensified film camera, but can also be used for direct photography or electrography. Refractive correctors are available to produce a well-corrected, flat field of 0.5° diameter in the wavelength region 210 nm to 600 nm. The achievable angular resolution in this field is about 0.15 arc-sec (Rayleigh criterion), which is equivalent to diffraction-limited performance at 600 nm. However, the actual resolution in the recorded image depends to a large extent on the resolution capability of the detector. In terms of modulation transfer characteristics, the telescope resolves about 2.3 cycles per arc-sec at 50% modulation, which corresponds to about 32 cycles per mm in the focal plane. This roughly matches the resolution capability of intensifiers and photographic emulsions now available.

As presently envisioned, SUOT will be pointed by means of the Instrument Pointing Subsystem (IPS), under development for

Spacelab. Residual pointing errors are reduced by means of an internal image-motion compensation (IMC) system, that articulates the secondary mirror. Separate IMC guide-star sensors are available for each of the focal planes. This assures accurate reacquisition if the two instruments are used alternatingly in the same mission. The IMC sensors are located in an annular field, O.D. 0.8°, I.D. 0.7°, surrounding the data field for most direct coupling to the target image.

The facility includes, as permanent fixtures, an f/3.75 TV camera, intended primarily for inspection of the data field and as an aid in acquisition, as well as a 1.5 min of arc field, high-resolution camera. The main purpose of the latter is to conduct synoptic planetary observations on each mission, whenever possible.

The above concept allows an almost unlimited variety of instruments to be flown on SUOT. Because of the short duration of the missions and the redundancy in scientific equipment, the demands on reliability can be less stringent than in most other space programs, a situation very much like that in sounding-rocket work. In this manner, SUOT promises to become a most valuable and cost-effective general-purpose facility for space-astronomy in the Shuttle era.

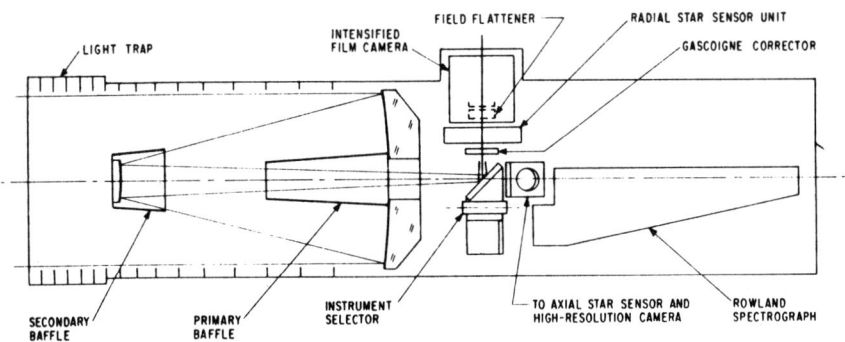

The phase-A concept is based on a Ritchey-Chrétien telescope with a 1-m aperture, f/2 primary mirror and a secondary mirror with magnification 7.5. The Ritchey-Chrétien produces images that are free from systematic imperfections, save for third-order astigmatism and field curvature. These limit the usable field to 0.27° (curved image plane) or to 0.13° (flat image plane), unless expanded by correctors. For the 0.5° flat field, these consist of a CaF_2 Gascoigne corrector[7], modified to cope with the high dispersion in the UV, and a fused silica field flattener.

The main driver in the selection of the **telescope f-number** was the field diameter that could be achieved with the above corrector system. Also considered were the technical feasibility and cost of the primary mirror, as well as an acceptable upper limit for the obscuration by the secondary baffle (37% of the free primary mirror diameter), necessary for full-field baffling.

REFERENCES

1. Final Report of the Payload Planning Working Groups, Vol. 1, "Astronomy," NASA, GSFC, May 1973.

2. "Scientific Uses of the Space Shuttle," National Academy of Sciences, 1974.

3. "Preliminary Feasibility and Definition Study of the Spacelab Ultraviolet-Optical Facility Telescope," BBRC Final Report, November 1974.

4. "Telescope Optical Design Concept Study," BBRC Final Report, July 1975.

5. "SUOT Facility Feasibility Study," BBRC Final Report, September 1975.

6. "Spacelab UV-Optical Telescope Facility," Interim Report of the Facility Definition Team, Astronomy Spacelab Payloads Study, GSFC, April 1975.

7. S.C.B. Gascoigne, Applied Optics $\underline{12}$, 1419 (1973).

AAS 75-263

ADAPTATION OF AN EXISTING COSMIC RAY IONIZATION SPECTROMETER EXPERIMENT TO SPACELAB

U.R. Alvarado*, J.F. Ormes+, and C.V. Stahle*

This paper examines the technique of adapting an existing experiment to a Shuttle sortie mission. A cosmic ray balloon experiment was studied to determine the feasibility of this cost-saving technique, which is applicable to a large spectrum of existing experiments, and to determine the programmatic impact and key problems. The main areas investigated include the determination of required modifications, steps in integration to the Spacelab/Shuttle, and the impact on orbital support and operations. One of the main problems in equipment adaptation is the acoustical loading during Shuttle boost; an environmental cover design presented herein shows a potential method to attain the required acoustic attenuation.

INTRODUCTION

The full economic benefits of the Space Shuttle can only be realized through a significant reduction in payload costs. A method of reducing costs in space is to use experiment-equipment currently in ground based laboratories, in investigations where an extension into the space environment is advantageous. The benefits can be very significant; for instance, the cost of sophisticated sensors for spacecraft use can range from $2.5M to $30M. The cost of refurbishing the experiment which is the subject of this paper is less than $0.5M.

---

* General Electric Company Space Division, Valley Forge, Pennsylvania
+ NASA-GSFC High Energy Astrophysics Division, Greenbelt, Maryland

The Cosmic Ray Ionization Spectrometer (CRIS), used as an example, consists of an assembly measuring 1.5 meters in diameter, two meters in height and weighing 2270 kilograms.

The purpose of the experiment is to measure the charge spectrum and the energy spectra between 1 and 1000 GeV/nucleon for heavy nuclei ($Z > 3$) up through iron. It is expected that testing in space will require considerably reduced corrections to determine the chemical composition of the cosmic ray source.

REQUIREMENTS ANALYSIS FOR ADAPTATION TO SPACE MISSION

The process for planning the steps in adapting the experiment to the spacelab mission must determine the following inter-related aspects:

1. Re-design and modifications to the equipment (including need for new interface equipment and component/assembly testing)

2. Steps in the integration of equipment to the Shuttle Orbiter/Spacelab.

3. Launch, orbital and post-flight operations.

Analysis of the effects of the Shuttle environment was emphasized since it had the largest impact on the requirements for adaptation of the experiment. Table 1 shows a set of typical environmental requirements.

TABLE 1 - TYPICAL ENVIRONMENTS

1. Maximum Combined Accelerations (Normal Conditions):

Linear Acceleration - g During Landing			Angular Acceleration $\frac{Rad}{Sec^2}$		
X	Y	Z	X-X	Y-Y	Z-Z
-0.2 ± 2.6	0 ± 1.4	$+2.0 \pm 0.2$	± 0.2	± 0.2	± 0.1

2. Temperature Limits of Components: $263° - 293°K$ ($303°$ Peak)
3. Maximum Heat Dissipation (Electronics): 400 Watts
4. Pressure Inside Cover: 15 psia $\pm$ 10% (N_2)
5. Acoustic Loading Inside Shuttle Cargo Bay: 145 dB Max.

MODIFICATIONS TO THE EXPERIMENT EQUIPMENT

The cost of refurbishment was estimated at approximately $433,000 spent over a period of 18 months. This included modifications, new parts, test equipment, and substitute components. The most costly item in the refurbishment was the replacements of the 38 photodetectors with ruggedized units, costing a total of $96,000. A key factor in maintaining low refurbishment costs is the incpororation of a cover that accommodates the acoustic control and thermal control requirements. This feature made it possible to retain all of the main spectrometer components in their original configuration and geometric relationships.

Acoustic Control Concept

The acoustic control concept is to replace the present aluminum cover with one featuring a double wall joined by strips of viscoelastic epoxy to provide high stiffness and damping. The viscoelastic epoxy provides shear continuity between two aluminum walls 1mm thick for high stiffness and also act as a contrained damping layer to provide high energy dissipation at the resonant frequencies of the cover. Experimental evaluation of this viscoelastic laminated construction using a cylindrical cover three feet in diameter showed that an order of magnitude reduction in acoustic pressures can be obtained with lightweight construction. A 20 dB reduction in the overall acoustic level was obtained with attenuations in the low frequency range of 20 to 30 dB. This low frequency attenuatuion is particularly significant in that this is the frequency range of maximum predicted acoustic levels. Using the measured noise reduction, the acoustic levels within the cover are well below the levels of the Delta booster (Fig. 1) and should preclude damage to the CRIS components.

Thermal Control Concept

The thermal control concept selected is to use fan circulated nitrogen with radiation cooling and heaters. This concept, depicted on Fig. 2 provides the lowest cost, requires modest additions to the CRIS unit and has few interfaces with the Spacelab. Nitrogen within the container is circulated by four fans through the NIM-CAMAC electronics and other heat sources. Then the nitrogen is convectively and radiantly cooled as it flows past the domed top of the cover and through a two inch wide annulus formed by the exterior wall of the cover and a large cylindrical baffle over the CRIS components. The exterior of the container is covered with silica cloth (α/ϵ of .21/80) which is adequate for the maximum Sun load.

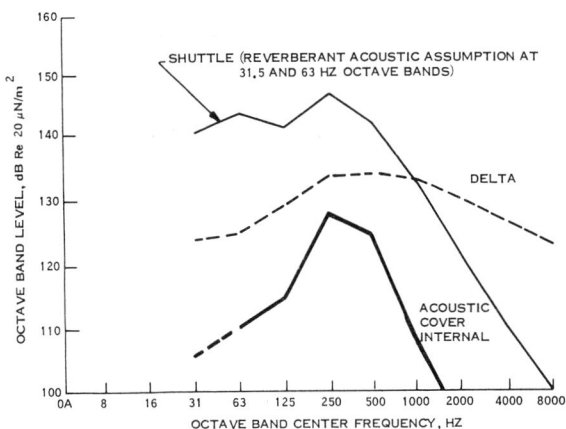

Fig. 1 Acoustic Cover Internal Noise Levels Compared with Shuttle and Delta

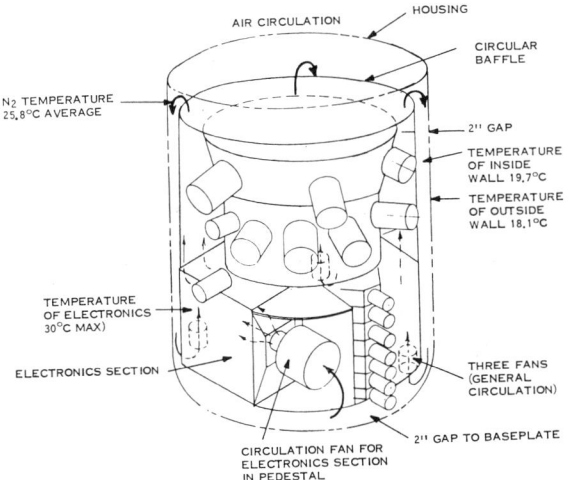

Fig. 2 Thermal Control

CONCLUSION

Safe and economical adaptation of existing experiments such as the Cosmic Ray Ionization Spectrometer are feasible through suitable design capable of coping with the Shuttle mission environment, particularly the random vibration resulting from acoustic loading during boost and reentry, and the thermal environment in orbit. The experiment can be made ready for flight within 24 months of the start of the adaptation program.

AAS 75-234

A LARGE COOLED INFRARED TELESCOPE FACILITY FOR SPACELAB

Stephen G. McCarthy, Hughes Aircraft Company
Lou S. Young, NASA/Ames Research Center
Fred C. Witteborn, NASA/Ames Research Center

Operation in earth orbit will benefit infrared telescopes because the severe limitations imposed by absorption and emission of the earth's atmosphere in the infrared are virtually eliminated. Another major advantage is that the telescope optics can be cooled to below 20°K to reduce photon noise from the telescope itself below the noise level of current and projected detectors. An optimized cooled IR telescope of 1m aperture diameter will offer sensitivities over the 1μm to 1 mm spectrum at least 1000 times greater than can be achieved from the ground in the atmospheric windows, and comparable sensitivities at wavelengths that are outside the windows. Hughes Aircraft Company, under contract NAS2-8494 to NASA/Ames Research Center, has examined the feasibility of large IR facilities for Spacelab. The results show that a cryogenically-cooled telescope of up to 1.6 m aperture diameter is feasible without new technology, although some development of detectors and scientific instruments to take advantage of the extreme sensitivity is indicated.

SUMMARY

The infrared emission from both stellar and non-stellar objects is believed to contain information that will contribute importantly toward understanding the basic processes of the universe. Some features, such as very cool gas clouds and protostellar objects, can only be observed in the infrared. Other areas of great astronomical interest, such as galactic nuclei, are typically obscured by dust clouds that are quite transparent in the infrared but which drastically attenuate other wavelengths. Even objects that can be seen at other wavelengths reveal some of their most important information at infrared wavelengths. Molecular spectra and the characteristic absorptions of some important solids are most prominent in the infrared.

In 1973, the National Academy of Sciences Space Science Board met at Woods Hole, Massachusetts to consider scientific uses of the Space Shuttle. One of the recommendations of the Space Science Board was that a detailed study be performed of a 1-meter-diameter-aperture class, cryogenically-cooled telescope. The NASA Ames Research Center, which manages several airborne infrared programs, undertook to determine what instrument facility design would be feasible on the Shuttle/Spacelab. An advisory group of infrared astronomers was formed to provide science guidance for the study. With their help, Ames defined a set of performance requirements and goals for a large, cooled telescope on Spacelab, to be known as the Shuttle/Spacelab Infrared Telescope Facility (SIRTF). Hughes Aircraft Company was selected to perform the engineering portion of the study.

Fig. 1 shows a concept of SIRTF as part of an integrated infrared payload. As an alternate, compatible payloads of other disciplines, or free flyers, could share the mission. Additional instruments with their cryogen supply are shown mounted on the Small Instrument Pointing System (SIPS). These instruments have not been defined, but are expected to involve parallel measurements of sources by SIRTF and the SIPS-mounted instruments. Another possibility is the use of second IR telescope at the front of the bay to provide a 15-meter-baseline interferometer.

The SIRTF telescope is mounted on the ESA-developed Instrument Pointing System (IPS) which in turn is mounted on a platform above the floor of the standard Spacelab pallet. This configuration allows SIRTF to utilize the full $60°$ half-angle cone capability of the IPS. Although the telescope itself is only 1.6 m in diameter, the cryogen supply tanks mounted on the telescope and the requirement to rotate the telescope $90°$ about its line of sight for polarization measurements makes the

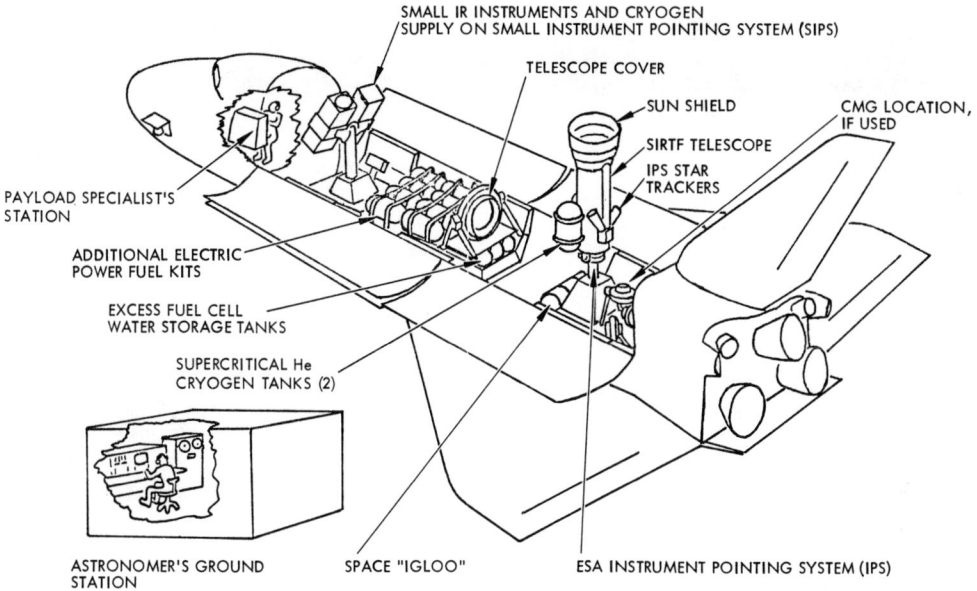

Fig. 1. Spacelab IR Astronomy Facility

effective telescope diameter almost the full 4.6 m (15 foot) diameter of the payload bay, and produces interference at the extreme angles. Use of the platform eliminates the need for raising the IPS after arrival in orbit. A special tie-down fixture to handle launch and re-entry loads will be required, in lieu of the standard IPS fixture.

The telescope is equipped with a retractable sun shield to reduce scattered infrared radiation from the sun and earth which would otherwise reduce SIRTF sensitivity by increasing the background; it also reduces the heat load which must be absorbed by the cryogenic cooling system, and thus reduces the quantity of cryogen which must be carried.

Two supercritical helium tanks, each of 1300 liters (45 cu. ft.) capacity, will provide sufficient cooling for missions up to 28 days; one tank will suffice for 14-day and shorter missions. The current favored configuration is to mount the cryogen tanks on the telescope; this eliminates the heat losses, pointing errors, and technical risks in transferring the cryogen from pallet-mounted tanks.

Preliminary design work on the Shuttle/Spacelab Infrared Telescope facility is largely complete. The study has shown that a 1-meter-class cryogenically cooled telescope is feasible on Spacelab, and that the performance and operational techniques required for infrared astronomy can be achieved within the Shuttle/Spacelab framework. Some additional study of a larger (1.6 m) telescope is planned, in response to interest expressed by astronomers; preliminary analyses indicate that accommodation of this size telescope on Shuttle/Spacelab is practical.

An important objective of the study was to determine whether development of large cooled IR facilities for space flight would require prior new technology development; it appears that all requisite technology is in hand for telescopes as large as 1.6 m in diameter. However, it will eventually be necessary to increase the sensitivity of detectors and scientific instruments at wavelengths upwards of 30 μm in order to take full advantage of the sensitivity that will be offered by Spacelab IR facilities.

Finally, the study has confirmed the very great advantages on cost, weight, performance assurance and operational flexibility that accrue from use of the Shuttle/Spacelab system; in addition, areas of Shuttle/Spacelab development that will require careful attention to maintain the required capability have been identified.

AAS 75-255

APPLICATION OF SPACE SHUTTLE TO FUNDAMENTAL AND APPLIED MICROBIOLOGICAL RESEARCH

Jerry V. Mayeux[*]

SUMMARY

An analysis has been made to identify the gravity sensitive mechanisms that may be present in the microbial growth system. As a result of that analysis it has been determined that natural convection (density gradient induced sedimentation and buoyancy) is indeed important in microbial systems. Absence of natural convection may provide an opportunity for new approaches for experimental screening of organisms which could lead to new developments in industrial microbiology or agriculture.

The analysis of gravity dependent cellular functions and gravity limiting environmental growth factor has led to the conclusion that in the absence of gravity induced convection:

1. Intracellular biochemical reactions would exhibit an abnormal periodicity due to the time delay for translocation of regulatory substrates or metabolites.

2. Multiphasic microbial growth systems which are stable, though not emulsified, can be prepared and maintained as a uniform mixture without the interference of natural convection.

---

[*] Ferma-Gro Corporation, Alta, Iowa 51002

3. Single cells suspended in a suitable liquid medium can be made to grow and reproduce to form discrete colonies, originating from a single parent cell or clumps of cells.

If intracellular biochemical reactions would exhibit an abnormal periodicity due to the time delay for translocation of regulatory substrates or metabolites, it would then be possible to approach many research problems in an entirely new manner. Some of the areas that can be presently identified are the use of single cell mamalian culture lines and protozoal models to study the aging process; use of similar cultures to study cell membrane function and deterioration (this may relate to aging); preparation of stable synthetic membrane models, research in genetic regulation and control mechanisms which could lead to better understanding of cellular biochemical interaction and metabolic disorders. This could possibly be used to provide new insight into metabolism of cancer cells. Kinetics of enzymes with multiple sites of activity could be better evaluated. Multiphasic microbial growth systems which are stable in the absence of convection would allow the scientist to expose cultures to various candidates substrates in order to identify the few cells in a population of billions which have the ability to transform the candidate material. These cells can then be transferred and enriched for subsequent return to the ground-based industrial laboratory for development. Materials which may be candidates would be organic compounds which are inefficient to produce; compounds which have a low efficiency of production because of the many isomers; pharmaceuticals that must be altered to increase their potency or biological activity, selection of cultures that transform recalcitrant molecules that present a waste disposal problem; or cultures that can produce extracellular enzymes that degrade solid waste materials. I'm sure that many of you can think of other experimental uses.

If single cells suspended in a suitable medium can be made to grow and reproduce to form discrete colonies, originating from a single parent cell or groups of cells, we have opened a new horizon for research with single celled organisms. Many of the applications of this technique would be related to selecting cultures for unique transformations with multiphasic systems. However, other cloning for use in basic genetic and metabolic studies would also be desireable.

One of the most fascinating applications for which this technique has potential relates to GENETIC ENGINEERING. This topic has been much abused recently, especially, as it relates to human embryos and potential pathogenic micro-organisms. However, a new field in genetic engineering is emerging. This is the use of single cell propogation of plants.

Scientists can now take the leaf of some plants and treat the leaf with enzymes to remove the cell walls and other structural material, thereby releasing single indivual leaf cells. These cells are then transferred to an appropriate culture medium in which they are allowed to propagate and form tissue called callus. The callus is then transferred to a medium containing various hormones which encourage cell differentiation. The callus then starts to form chlorophyll and eventually will produce leaves and roots, and is called a plantlet. The plantlet eventually develops into a flowering plant. All of this can be accomplished in a matter of a few months. It allows the laboratory scientist to screen and select disease resistant plants or plants with other unique characteristics. Millions of plants can thus be screened in less time compared to the conventional seed to plant to seed technique.

Plant scientists are now adapting microbiological techniques to single cell plant propagation and mutant selection. Techniques such as transformation (insertion of free DNA into a competent recipient cell) and conjugation (DNA transfer via actual cell to cell contact between the recipient and the donor cell). The latter technique has the greatest promise for genetic engineering in plants if it can be perfected. One of the limitations in this technique with plant cells is the ability to keep the two types of cells in contact sufficiently long for conjugation to occur. Since these manipulations are conducted in (suspecious parent) fluid nutrient media, convection has an opportunity to play the role of the villian by preventing conjugation.

If convection is one of the factors limiting conjugation, its elimination could make the technique possible. It might then be possible for plant scientists to accomplish one of their age old objectives, e.g., to transfer the nitrogen gas utilization ability of legumes such as soybeans to grain crops such as wheat, corn and rice.

AAS 75-256

LIFE SCIENCES MANNED PAYLOADS FOR SHUTTLE/SPACELAB[*]

Dennis B. Heppner, Ph.D.[**]
George L. Drake [**]
Chester B. May [+]

The Life Sciences Payload Definition and Integration studies[*] performed under the direction of NASA/MSFC by General Dynamics Convair during the 1970-1975 time period provides a foundation for future life sciences space research programs. These studies are an integral part of the current NASA planning activity for the Shuttle/Spacelab. Figure 1 is an overview of the life sciences payload studies and how they interface with current NASA payload planning and study activities.

Several studies during the post-Apollo era of the late 1960s and early 1970s gave scientific direction to the payload definition studies and provided the research program data base. Table 1 lists the more important of these studies. They covered the range of life sciences research interests and led to the sub-discipline classifications — biomedicine, biology, man-systems integration, and life support/protective systems — which generally describe the life sciences functional program elements. Several of these studies such as IMBLMS, Biotechnology and Blue Book provided exemplary life sciences experiments from which the methodology of the laboratory concept was developed.

The pre-Phase A studies of 1970 to 1974 covered three studies which were aimed at obtaining the broadest extent of scientific input and guidance possible. This broad base was used to define the engineering studies and preliminary conceptual designs which would serve as the background for the subsequent program phases. The first of the three studies was the definition phase wherein the details of the science requirements were developed. The second study considered the engineering and integration aspects of the science requirements in the form of preliminary conceptual designs of dedicated life sciences laboratories. The third

---
[*]NAS8-26468, -29150, -30288, -31368.
[**] General Dynamics Convair, San Diego, California
[+] NASA Marshall Space Flight Center, Huntsville, Alabama

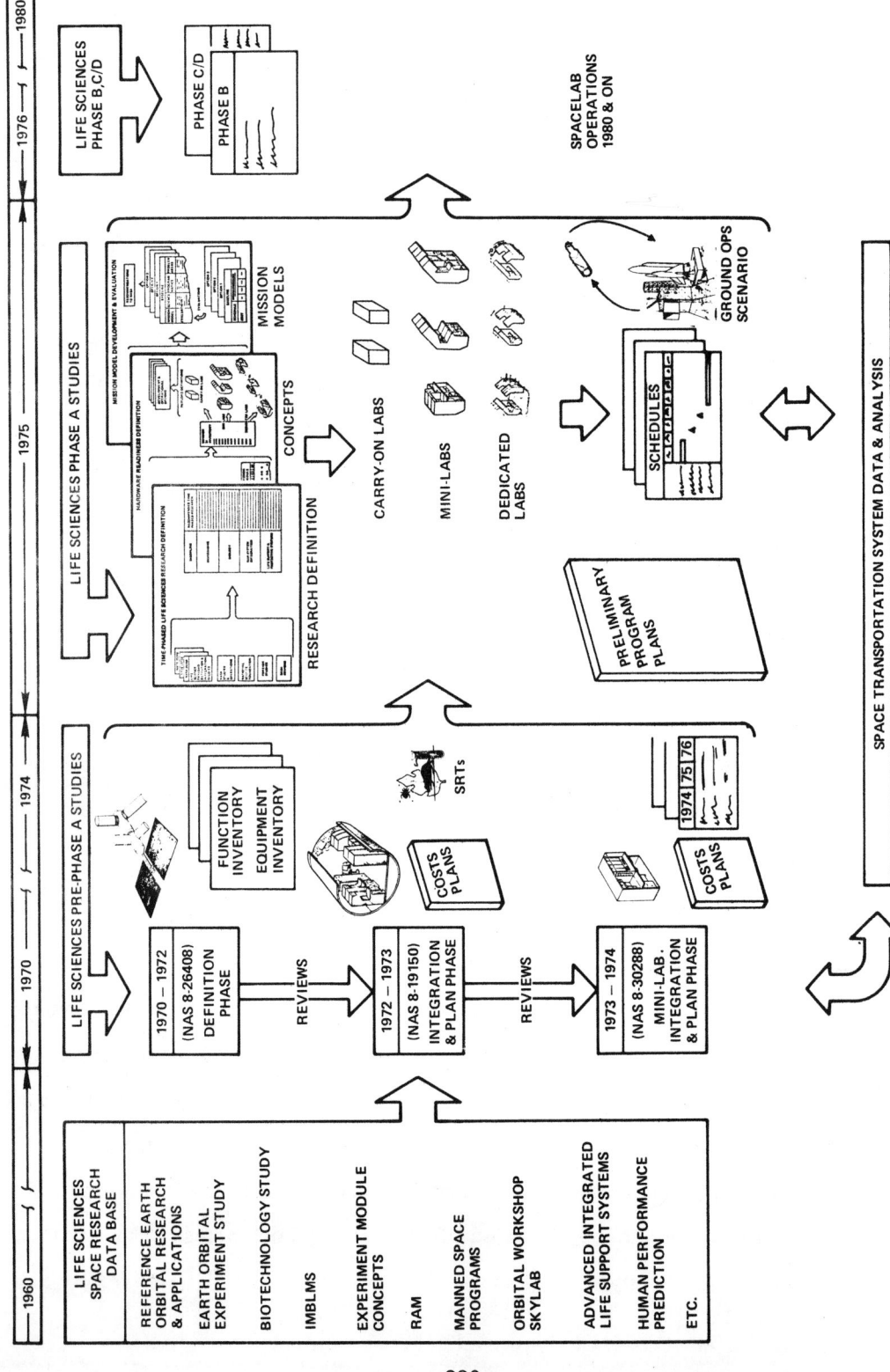

Figure 1. Life Sciences Payload Development Overview.

Table 1

PROGRAM DATA BASE

Reference Earth Orbital Research and Applications Investigations (Blue Book)	NASA/Convair
Earth Orbital Experiment Study	MDAC
Biotechnology Study	MDAC
IMBLMS	LMSC/GE
Experiment Module Concepts	Convair
Space Station/Base	MDAC/Martin & NAR/GE
Orbital Workshop - Skylab	Martin/MDAC
Space Shuttle	NAR
Human Performance Prediction	Bunker-Ramo
Advanced Integrated Life Support Systems	Ham Std
Teleoperator	Bell Aero
Research and Applications Module	Convair
Bioresearch Module	LTV/GE

study concentrated on preliminary conceptual designs of carry-on labs and mini-labs. The Space Transportation System Data & Analysis (SPDA) study provided the day-to-day interface with the Spacelab constraints that would be considered during the subsequent phases of the life sciences payload development program.

The current Phase A life sciences payload study will establish the desired research requirements. These, in turn, will be used to develop detail mission models for which program plans and costs are determined.

A first task in the study, initiated in February 1975, was to update the science requirements and recommend a time-phased research program. Inputs from the NASA Centers, recommendations of the National Academy of Sciences, as well as the Skylab and the Russian flight experience, were considered in ordering the research.

Payloads supporting the selected research areas are being defined. Generally, these payloads fall into the three classes shown in Figure 2 with their overall characteristics. The carry-on laboratories are true "suitcase" experiments —

CARRY-ON	MINI-LAB	DEDICATED LABORATORIES
• ORBITER CREW COMPARTMENT • LESS THAN 23 kg • MINIMAL INTERFACES – POWER • FLIGHTS OF OPPORTUNITY • 0 TO 7-DAY MISSION	• SHARED MISSION • GENERALLY LESS THAN 500 kg • ONE TO SEVERAL RACKS OF EQUIPMENT • SIGNIFICANT INTERFACES WITH SPACELAB – CDMS, POWER, THERMAL, ECS • SHARED P/L SPECIALIST • 7 TO 30-DAY MISSIONS	• UP TO 3,500 kg • FULLY DEDICATED SPACELAB MISSION • EXTENSIVE INTERFACES WITH SPACELAB – CDMS, POWER, THERMAL, ECS • UP TO 3 DISCIPLINE SPECIALISTS, 12-HR/DAY ON 7-DAY MISSIONS • 7 TO 30-DAY MISSIONS

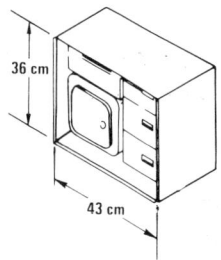 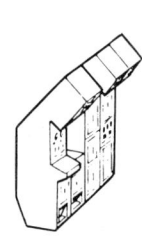 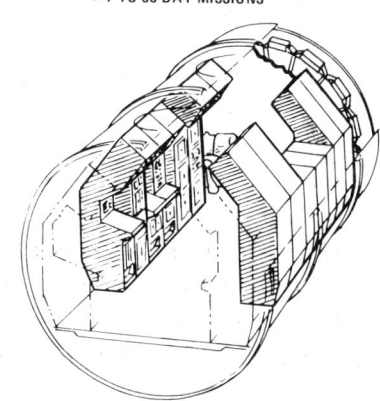

Figure 2. Life Sciences Payload Classes.

small, lightweight, with the minimum of interfaces with the supporting spacecraft. While basically intended to be flown early in the Shuttle program, they can be taken aboard any flight of opportunity. Mini-labs are flown on shared missions and can range in size from 23 kg to approximately 500 kg. They will have significant physical interfaces with the Spacelab and operationally will be integrated with the sharing disciplines. Besides providing for in-depth scientific research, mini-labs can be used as hardware testbeds, flight-testing equipment items that eventually will be used in the dedicated laboratories. Dedicated laboratories will use the entire pressurized Spacelab, generally, the long module. Defined payloads range from 1,000 to 3,500 kg. Interfaces with the Spacelab subsystems will be extensive. The experiment crew will consist of life sciences payload specialists with preliminary estimates showing three one-shift crewpersons needed for the all-up version. Dedicated payloads will cover all research areas of interest.

The primary theme of all the life sciences payload studies has been to emphasize the science requirement to perform meaningful research in space. The programs have evolved from the comprehensive capability of a space station-type laboratory to the small, individual research area carry-on laboratories. In all cases, however, the scientific requirements have maintained a prominent position in the determination of the conceptual laboratory designs.

The first dedicated laboratories weighed tens of thousands of kilograms; the dedicated laboratories under consideration today are in the order of a few thousand kilograms at most. Power, a valuable commodity in space, has seen an order of magnitude reduction from the original space station-type laboratory demands to the present. The early multi- and special module requirements have given way to the use of a single Spacelab module. The ultimate life sciences laboratories that emerge from these studies will meet the scientific requirements of the 1980s and be completely compatible with the real world operations of the Shuttle/Spacelab.

AAS 75-257

SHUTTLE BIORESEARCH LABORATORY BREADBOARD SIMULATIONS*

S. T. Taketa[†]

INTRODUCTION

To accommodate the Life Sciences payload requirements for Shuttle missions is a formidable task. The discipline requires living organisms for experimental subjects and the requirements are diverse and complex.[1,2,3] Laboratory breadboard simulations (Tests I and II) were conducted to test concepts and assess the problems associated with bioresearch support equipment, facilities, and operational integration for conducting manned research in earth orbital missions. This paper describes Test I and discusses the major observations made in Test II.[4] The tests emphasized candidate experiment protocols and requirements: Test I for biological research and Test II for crew members (simulated), subhuman primates, and radioisotope tracer studies on lower organisms. The simulations were conducted as part of the NASA Concept Verification Testing (CVT) Program in collaboration with the Marshall Space Flight Center (MSFC) and the Johnson Space Center (JSC) and the support of the NASA Office of Life Sciences and the Spacelab Program Office. The tests were conducted first at the Ames Research Center (ARC), then at MSFC. The emphasis at ARC (with input from JSC) was on science and payload integration and, at MSFC, on engineering to interface the payload with the simulator subsystems.[5,6] (The candidate experiments used in the simulations are not to be construed as having been

---

* Prepared for technical papers that will later be published in the Proceedings of the American Astronautical Society.

† Project Manager for the Ames Research Center Life Sciences simulations. Dr. Taketa is affiliated with the National Aeronautics and Space Administration, Ames Research Center, Moffett Field, CA 94035 (LF:236-5, (415) 965-6046).

selected for actual flight experiments. Henceforth, candidate experiments are referred to as experiments.)

CONCEPTS TESTED

The major conceptual breadboards tested were the bioresearch support equipment unit, the modular organism-housing unit, and the portable glove box-like unit attached to the organism-housing module for carrying out research functions. The methods and equipment necessary for sampling, preparing, and preserving biological materials, organism maintenance, etc., were emphasized. The role of a payload specialist (a trained scientist) to perform onboard research activities, the research support role of a mission specialist (an onboard engineer), etc., were assessed. Procecures and techniques applicable for use in the weightless environment of space were implicit considerations but not firm requirements.

TEST PROCEDURES, SIMULATOR, AND LABORATORY BREADBOARD

The test-related activities were performed by in-house personnel, except for the contractual support for the breadboarding, installation, and maintenance of research support hardware. The participating scientists provided detailed, time-lined experimental protocols, organisms, etc. The laboratory supplied the hardware breadboards and utilities. The test duration was 5 days; the daily testing period was 8 hr. The involved investigators carried out both practice and test runs at ARC; the author, playing the role of a payload specialist, conducted the experiments at MSFC with the support of K. A. Smith, mission specialist (simulated). The test at MSFC was conducted as part of a shared multidisciplinary payload with four experiments involving the physical sciences.[5] An agency aircraft was used for transporting organisms and equipment between ARC and MSFC.

The MSFC-supplied ARC simulator was a low-cost functional mock-up similar in dimension and configuration to the more sophisticated one used at MSFC — the lower deck of the Z-ceiling, double deck-configured General Purpose Laboratory (GPL) Simulator. (The upper deck was used by the physical scientists.)

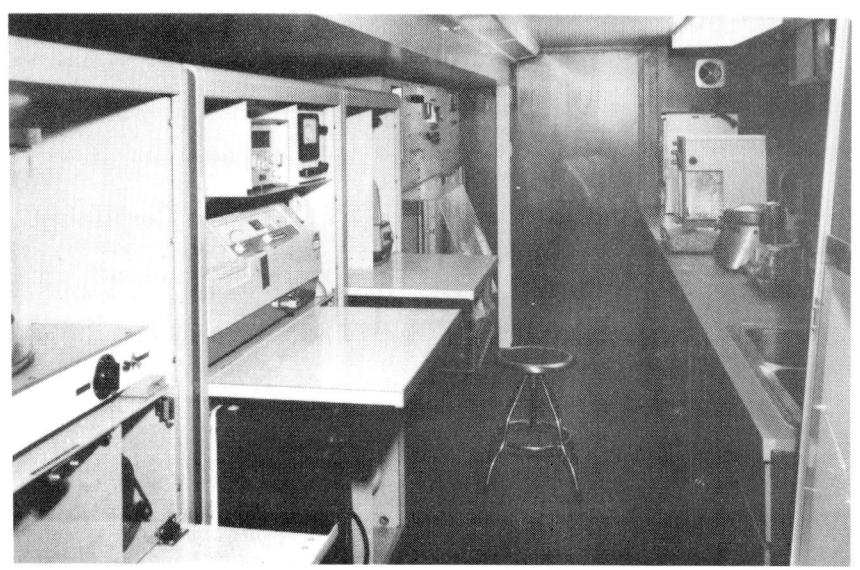

Fig. 1 Test I Laboratory Breadboard

The laboratory breadboard depicted in Fig. 1 shows the rack-mounted research equipment (near left), organism housing unit[3] (far left), and fume hood (near right) followed by a sink and workbench.

CANDIDATE EXPERIMENTS

The eleven candidate experiments used in Test I are identified in Table 1. Since the experiments were for demonstrations, only representative numbers of organisms and three organism-housing modules were used; the organisms were brought onboard as required, except for the rats, which were installed at the start of the test.

RESULTS AND DISCUSSION

The Test I simulation was highly successful with respect to test objectives. The problems encountered and the experience gained were valuable in preparing for and carrying out the more complicated and complex Test II simulation.[4,6]

A functional volume and configuration simulator of a payload-carrier such as the one used at ARC can be a useful, low-cost tool for integrating, testing, and iterating payloads. The Z-ceiling configured simulator

Table 1
TEST I CANDIDATE EXPERIMENTS

EXP. CODE	INVESTIGATOR	SHORT TITLE	ORGANISM	MAJOR FUNCTIONS DEMONSTRATED
R1a	R. GRINDELAND C. DOLKAS	BLOOD CHEMISTRY	RAT	BLOOD SAMPLING, PLASMA GLUCOSE ANALYSIS AND PRESERVATION FOR INSULIN ANALYSIS
R1b	S. T. TAKETA (R. C. SIMMONDS)*	HEMATOLOGY		RED AND WHITE CELL COUNTS, DIFFERENTIAL COUNT AND HEMATOCRIT
R2a	G. HARRISON	TISSUE ULTRASTRUCTURE	RAT	TISSUE BIOPSY, AND SURGICAL EXCISION AND PRESERVATION FOR MORPHOLOGICAL STUDY
R2b	S. T. TAKETA	LIVER LIPID CHEMISTRY		PREPARATION AND PRESERVATION FOR TOTAL LIVER FAT/FATTY ACID ANALYSIS
R3	S. T. TAKETA	RODENT HOUSING AND SURGERY	RAT	MAINTENANCE, SURGICAL PROCEDURES, AND WASTE MANAGEMENT
R4	J. OYAMA	METABOLISM	RAT	BODY TEMPERATURE AND HEART RATE (ECG) BY TELEMETRY, RESPIRATORY O_2 AND CO_2 AND DATA RECORDING ON TAPE
F1	K. SOUZA P. DEAL, et al.	AMPHIBIAN EMBRYOGENESIS AND DEVELOPMENT	FROG	MAINTENANCE, MATING, EGG COLLECTION, EMBRYO CULTIVATION AND PRESERVATION OF EGGS AND EMBRYOS
FF1	J. MIQUEL	INSECT BIOLOGY	FRUIT FLY	MAINTENANCE, MATING, EGG COLLECTION, ACTIVITY, AND PRESERVATION FOR MORPHOLOGICAL STUDY
PL1	J. SHAPIRA R. DEDOLPH (C. WALKINSHAW)*	PLANT BIOLOGY	LETTUCE	MAINTENANCE, MORPHOLOGY, AND PRESERVATION FOR BIOCHEMICAL ANALYSIS
M1	A. MANDEL (G. TAYLOR)*	MICROBIAL CONTAMINATION CONTROL	MICROORGANISM	MICROBIAL TECHNIQUES FOR CONTROLLING CONTAMINATION IN LAB
TC1	D. PHILPOTT	TISSUE CULTURE ANALYSIS	MOUSE FIBROBLAST	TISSUE CULTURE MAINTENANCE, GROWTH, SAMPLING, AND PRESERVATION FOR MORPHOLOGICAL STUDY

*JOHNSON SPACE CENTER COLLABORATORS

was used, since the now-defined Spacelab configuration was not known nor was a simulator with such a configuration available.

The scientists involved in the simulations were unanimous in supporting the concept of payload specialists. The involved payload specialists endorsed modular racks for mounting equipment and the research support role of mission specialists. The conceptual standardized organism-housing modules with built-in environmental control systems (ECS) and species-peculiar inserts and life support systems (LSS) appear worthy of further study and development for space research. The conceptual glove box was suitable for carrying out simple procedures, but not complex functions. The Test I workbench was unrealistically large for a Shuttle-sized laboratory; it occupied space needed for equipment and organisms. The number of experiments that can be accommodated in a dedicated Life Sciences Spacelab mission may be limited by equipment and/or organism requirements rather than workload. Neither animal waste odor (Tests I and II) nor radioisotope hazards (Test II) were experienced. The use of an agency aircraft greatly facilitated the transportation of test organisms and equipment.

From his experience in Test I, the author conceived a hooded workbench/ surgical table (HWB/ST) unit, which was breadboarded and evaluated in

Test II.[4] In addition to fulfilling the functions of three hardware units — glove box, workbench, and fume hood — the HWB/ST unit facilitates surgical procedures that may involve two or three operators.

CONCLUSION

The problems encountered and the experience gained from these simulation activities provide valuable background for subsequent testing of payload concepts in a Spacelab-configured simulator. The procedures and approach used could form the basis for developing preflight integration, testing, and logistics of flight payloads.

RECOMMENDATIONS

The recommendations are based on problems encountered and observations made in Tests I and II and assessment of requirements.

1. Develop flight experiments to provide sufficient lead-time to test and develop Shuttle payloads in an orderly manner.
2. For subsequent integration and testing of sciences payloads as were conducted at ARC: (a) Use a Spacelab-configured simulator; (b) Assign a full-time management team composed minimally of a project manager, an assistant, a project engineer, and a mission scientist; (c) Ensure active involvement of selected scientist in all phases of simulation activities; (d) Substantiate equipment specifications, dimensions, etc., during planning; (e) Require use of techniques applicable to weightlessness; and (f) Use number of organisms specified in experiment protocols.
3. Use well motivated and trained scientists as payload specialists.
4. Train mission specialists (engineers) in discipline-oriented science to provide research support.
5. Use modular racks for mounting research support equipment.
6. Emphasize onboard preservation of biological materials.
7. Design and develop standardized organism-housing modules with appropriate ECS, inserts, LSS, and other provisions to house commonly used organisms in all phases of flight experiment.
8. Provide means to control animal odor, debris, and potential radioisotope hazards in the development of Spacelab ECS and LSS.

9. Design and develop the HWB/ST unit for Spacelab use.
10. Improve the quality of the cordless microphone for Spacelab use.
11. Provide suitable organism-holding facilities at life sciences payload integration center and launch/recovery site.
12. Provide an aircraft at the user's disposal for long-distance transportation of experimental organisms, etc. for Shuttle missions.

ACKNOWLEDGMENT

The author gratefully acknowledges the valuable contributions made by the NASA Life Sciences Payload Integration Team, the participating scientists, and the ARC and MSFC engineers and support personnel. He owes special thanks to Robert Johnson and Gerald Hall and associates of the ARC Research Support Directorate for their dedicated and untiring support; to William Brooksbank, Jr., Jack Rowan, Robert McBrayer, Kenneth Smith and associates of MSFC for their wonderful cooperation in all phases of the simulation activities; and to Convair Aerospace Division, General Dynamics, for its contractual support and loan of the conceptual organism-housing unit.

REFERENCES

1. Reference Earth Orbital Research and Application Investigation (Blue Book), NHB 7150.1, vol. III, Life Sciences, NASA, Washington, D.C., Preliminary Edition, Jan. 15, 1971.
2. Final Report of the Space Shuttle Payload Planning Working Groups, NASA TM X-66842, vol. 4, 1973.
3. Life Sciences Payload Definition and Integration Study (Task C & D), vol. II: Payload Definition, Integration and Planning Studies, Report No. CASD-NAS 73-003, Convair Aerospace Div., General Dynamics, San Diego, California, Aug. 1973.
4. S. T. Taketa, R. C. Simmonds, and P. X. Callahan, "Life Sciences Laboratory Breadboard Simulations for Shuttle," Proc. Eighth Conference on Space Simulation, Silver Springs, Maryland, Nov. 3-5, 1975.
5. R. O. McBrayer and J. D. Steadman, "CVT/PCS Phase I Integrated Testing," NASA TM X-64770, 1973.
6. R. E. Shurney, E. Cantrell, G. Maybee, and S. Schnitt, "CVT/GPL Phase III Integrated Testing," NASA TM X-64909, 1975.

AAS 75-258

A LIFE SCIENCES SPACELAB MISSION SIMULATION

John A. Mason[*]
F. Story Musgrave[*]
Dennis R. Morrison[*]

The U.S. Space Shuttle Program will use a manned Spacelab as one of its payload carriers. This Spacelab is now under development by the European Space Agency (ESA). Previous experience in the preparation for manned space flights has shown that simulations are very useful in selecting approaches to the design of experiments and science equipment, and in defining various aspects of operational and human engineering.

During the first week of October 1974, a seven-day simulated life sciences mission was conducted at the Johnson Space Center, Houston, Texas, in a Spacelab simulator. Twelve experiments were selected to evaluate the equipment, crew training, operational procedures, and the interactions of the science team. Two crewmen took part in the simulation. The Mission Specialist was a scientist astronaut, and the Payload Specialist was a life scientist. Experiments included biomedical research, animal cardiovascular research, plant growth studies, microbiological studies, biochemical determination, and other similar studies. The crewmen remained in simulation seven full days for increased realism.

A major objective of the simulation was the evaluation of in-orbit Spacelab operations and those mission control support functions which will be required from the Payload Operations Center. Using the operational constraints of a typical Spacelab mission, the manned space flight operations team was able to evaluate planned inflight operational concepts for preliminary design of Spacelab systems, and the ground facilities required to support these flights.

---

[*] NASA - Lyndon B. Johnson Space Center, Houston, Texas

The payload operations included reduced manning concepts for payloads operations control, coordination of payload integration and testing, principal investigator interface requirements, air-to-ground communications, and inflight procedure changes and real-time flight planning. Significant insight was gained regarding new training approached which included having the crew train with the principal investigator in his laboratory and during integration testing. The test illustrated the increased capabilities generated by an air-to-ground science communication loop which would enable the principal investigator to talk directly with the crewman in the Spacelab during the execution of his experiment.

The Major Spacelab systems evaluated were ERNO concepts for experiment racks, Common Operational Research Equipment (CORE), commercial off-the-shelf equipment, experiment hardware interfaces with Spacelab, experiment data handling concepts, and Spacelab trash management. Important results highlighted the difficulties of performing inflight equipment repair within the ERNO rack configuration and the specific recommendations for various life sciences CORE equipment required for animal holding facilities, fluid handling, and onboard radioisotope use.

The simulation contributed greatly to the development of facility requirements needed for Life Sciences payload integration, biospecimen support, ground support of Spacelab operations, and mission simulations. Evaluations of payloads realted flight control functions and interfaces with Mission Control Center resulted in specific recommendations for the development of a science console and an integrated communications/data console in the Payload Operations Center.

This experience has provided a better perspective of typical payload preparation and inflight operations. The test demonstrated the value of fully integrated mission simulations, as part of the development efforts leading to the first Spacelab flights and as a basis for initial efforts to define a Life Sciences payload.

The majority of the simulation results are not confined to life sciences but relate to Spacelab inflight operations and required ground support in general regardless of the type of payload.

The major conclusion is that fully integrated mission simulations are extremely valuable as part of the development efforts leading to the first Spacelab flights. Furthermore, integrated simulations should be considered for training and operations testing as part of the staging and flight support of actual payloads in the Shuttle Era.

This experience has given a better perspective of payload operations and a basis for initial efforts to define a Life Sciences payload. The Payloads Test Program provides a method for continuing experiment definition and testing and should result in the selection of an optimum Life Sciences Payload for the first dedicated mission.

AAS 75-259

SPACE SHUTTLE TRACE GAS ANALYZER*

Wallace Dencker+

INTRODUCTION

This document describes the Trace Gas Analyzer (TGA) being developed for use on the Space Shuttle. The TGA will analyze the atmosphere to detect and measure any trace constituents that could present a hazard to the crew and scientists aboard the spacecraft. The TGA is an automated Gas Chromatograph-Mass Spectrometer (GCMS) system based on the technology and hardware of the GCMS developed for the Viking Mars Lander. The contract for the development of a breadboard TGA has been awarded to the Advanced Technology Operations of Beckman Instruments, Inc., Anaheim, California, with the Aerospace Division of the Perkin-Elmer Corporation, Pomona, California, as a major subcontractor. This same team designed and fabricated the Viking GCMS hardware.

BACKGROUND

In previous manned space flights, the materials and processes used in the fabrication of the spacecraft and its payload were rigorously controlled to minimize production of physiologically significant levels of toxic atmospheric contaminants through material outgassing. However, because the Space Shuttle Program is intended to provide a laboratory for experiments varying widely in purpose, scientific discipline, and experimental material required, it would be extremely cumbersome and costly to impose

---

*Being developed under Contract NAS9-14637, NASA-JSC.

+Mr. Dencker is affiliated with the Advanced Technology Operations (ATO) of Beckman Instruments, Inc., Anaheim, California. He was associated with the GCMS Program over a period of six years at the Jet Propulsion Laboratory, Pasadena; the Analog Technology Corporation, Pasadena; and ATO, where he was the GCMS Test Engineer. He is responsible for systems and test engineering on the TGA Program.

these same strict material controls. Moreover, it is a fundamental precept of the Space Shuttle Program that the Space Transportation System (STS) provide a means of conveying and conducting experiments in space at a cost comparable to unmanned systems. This approach dictates the use of off-the-shelf or minimally modified off-the-shelf payload equipment which, because it is constructed of uncontrolled materials, offers the potential for the production of toxic gases, especially when stressed by overloads, excessive heat, etc.

TGA SYSTEM

To accommodate the use of off-the-shelf equipment safely, NASA has recommended the use of pre-launch off-gas testing, an atmosphere regeneration system, and a trace gas analyzer. It is toward this latter requirement that the TGA development effort is directed.

The TGA incorporates unique concepts that adapt the Viking GCMS system for routine flight use. These include:

- Refillable carrier-gas tank.
- Two columns used alternately.
- Effluent divider to optimize MS performance and increase dynamic range.
- MS interface (separator) that removes carrier gas while allowing transmission of 100% of the sample.
- Separators that reclaim and recirculate the carrier gas.
- Miniaturized MS developed for extraterrestrial operation.
- Data system that compresses the data for transmittal.

The TGA system comprises four major subsystems (Fig. 1), all of which are independently replaceable for ease of routine maintenance.

Gas Chromatograph Subsystem

The GC Subsystem consists of a carrier gas supply, sample valve, two GC columns, and a GCMS interface composed of an effluent divider and two carrier gas separators. The carrier gas supply consists of a refillable tank that contains 12,000 scc of ultra-pure hydrogen pressurized to 1000 psi. Since this quantity of stored hydrogen would last less than two days of constant operation at the required flow of approximately 4 scc/min, a recirculating system is used to reclaim the carrier gas. This method extends the operating life to greater than twice the maximum mission length and avoids venting of the carrier gas into the spacecraft atmosphere.

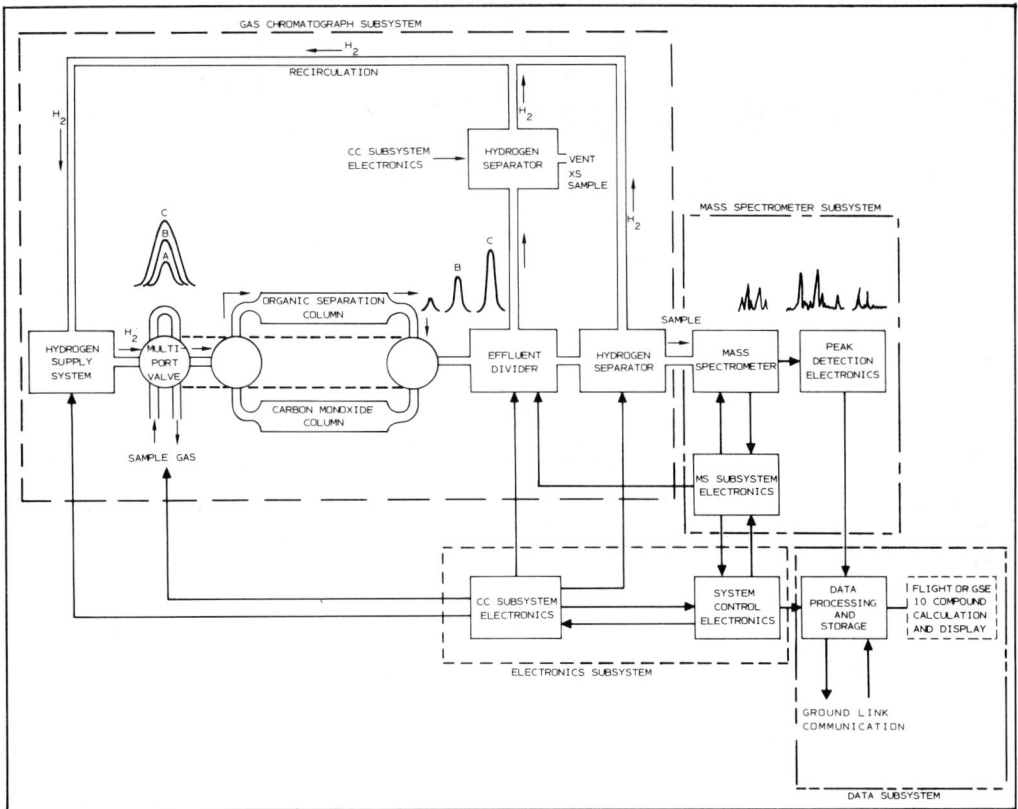

Fig. 1　TGA GCMS System Block Diagram

Because of the variety of materials that may be present in the atmosphere, two columns are used. One, an organic column, was developed under NASA contract (NAS9-14001) to separate complex mixtures of potentially dangerous components likely to be found in a spacecraft environment. The second column separates the lightweight materials to determine the concentration of carbon monoxide in the atmosphere.

The GCMS interface (effluent divider and separators) is required to allow maximum sample to the MS, yet protect the MS from pressure overloads and prevent the hydrogen carrier from being vented directly to the atmosphere. The effluent divider is constructed using miniature valves and restrictors that are calibrated in a dynamic flow system. The position of the valves (open or closed) determines the percentage of carrier flow, therefore sample, which is directed toward the MS. This splitting procedure increases the dynamic range of the instrument. Before the carrier gas and sample pass into the MS, they pass through an electrochemical silver-palladium separator where the hydrogen carrier is removed (99.999+% efficiency). The hydrogen is then returned to the equilibrium volume to be reused. The hydrogen and sample, which are directed to the vent line by the effluent divider, pass through a second separator where the hydrogen is removed and recycled. This system allows 100% of small samples to be transmitted into the MS for maximum sensitivity.

Electronics Subsystem

The Electronics Subsystem is composed of the temperature controller for heating the columns, valves, and connecting lines, and the sequencer for automatically controlling the operation of the instrument. This sequencer (a small computer) consists of incremental subroutines that can easily be varied to change the timing of the system operation. This subsystem contains provisions for monitoring component temperature, valve status, separator current, and other key operational parameters.

Mass Spectrometer Subsystem

The MS Subsystem consists of a miniature double-focusing MS and associated electronics. This is the same flight-qualified MS developed under the Viking Program. It contains a dual-filament ion source and has a mass range of 24 to 240 amu. The output of the MS is fed into a logarithmic electrometer that provides a dynamic range of seven decades. The logarithmic signal is fed into a filtering network and peak picking circuit and then to an analog-to-digital converter before going into the Data Subsystem. The peak picker circuit is used to compress the data to make efficient use of the data storage system (tape). The m/e ratio (mass number) at maximum MS peak intensity is also digitized and fed to the data system.

Data Subsystem

The Data Subsystem consists of a microprocessor and tape recorder for handling and storage of the mass spectral data. The breadboard system will output the stored data from the tape at a convenient time into a computer for reduction to determine GC peak intensity (sample concentration) and component identification.

AAS 75-248

AN ORBITING MOLECULAR SHIELD VACUUM FACILITY:

A MATERIALS LABORATORY IN SPACE

James W. Youngblood,[*] R. A. Outlaw,[*] Leonard T. Melfi, Jr.,[*]
and John R. McIlhaney[†]

INTRODUCTION

Some experiments in the fields of materials research and processing require simultaneous conditions of near zero gravity and very low ambient gas density. These conditions occur naturally in a quiescent spacecraft in relatively high Earth orbit. The low density may also be realized at lower altitudes within a properly designed molecular shield. Gas density of $\sim 10^4$ cm^{-3} should be achievable at an altitude of only 200 km. (This density occurs naturally at about 2500 km.)

This paper presents preliminary details of a proposed hemispherical molecular shield vacuum facility (MSVF) to be deployed from the Space Shuttle Orbiter. The facility would be designed for extension out of the payload bay on a retractable boom, or for placement in orbit as a free flyer. Since attitude control capability and expendable resources are available from the orbiter, this paper emphasizes the boom deployment mode; however, a free-flyer concept is also presented.

MOLECULAR SHIELD CONCEPT

At altitudes of 200 km or more, local atmospheric disturbances caused by the orbiter and the molecular shield are small; therefore, the atmosphere remains in near equilibrium and may be considered a Maxwellian gas drifting at orbital velocity (~ 8 km sec^{-1}) with respect to the shield. In such a gas, only a small fraction of the molecules have the proper

---

[*]Aerospace Technologists, NASA Langley Research Center, Hampton, Virginia 23665.

[†]Senior Design Engineer, LTV Aerospace Corporation, Hampton Technical Center, Hampton, Virginia 23665.

combination of spatial location, kinetic energy, and momentum components such that they can overtake a surface element whose normal is parallel to the atmosphere drift velocity. Therefore, only a tiny fraction of the drifting Maxwellian gas from the aft half space can enter the hemispherical molecular shield, implying that the atmosphere contribution to gas density within the shield is very low.

Five principal gas sources may contribute to the density within the shield: (1) free-stream ambient atmosphere, (2) outgassing from the shield inner surface, (3) gas released by experimental apparatus within the shield, (4) atmospheric gas scattered off the orbiter, and (5) gas released by the orbiter (outgassing, leaks, dumps, rocket effluents, etc.). It is likely that gas released from the orbiter will be the principal gas source. This potentially massive flux will require the shield to be deployed a substantial distance from the orbiter. The primary contributor to this flux is the effluent from the six vernier thrusters (~25 lb thrust each) which provide attitude control while on orbit. The vernier propellants are monomethyl-hydrazine (MMH) and nitrogen tetroxide (N_2O_4), which do not burn to completion and leave a reactive film of MMH and MMH-nitrate on exposed surfaces. Outgassing from film constituents can severely degrade the high purity apparatus within the shield and must be avoided. The induced gas density in the vicinity of the shield as a function of separation distance L has been estimated. For conceptual design purposes, at $L = 100$ meters the magnitude of the induced density is acceptably low.

PHYSICAL DESCRIPTION AND OPERATIONAL CHARACTERISTICS

The molecular shield is envisioned as a lightweight, thin foil, grid stiffened hemispherical shell. Hardware peculiar to various materials and processing experiments will be mounted inside the shield, while other apparatus (power supplies, experiment control, and monitoring devices, etc.) which do not undergo a preflight degassing treatment will be mounted externally on an adapter module. The UHV chamber, a grid stiffened enclosure providing ultimate pressures less than 10^{-9} torr, is used for preflight degassing of the shield, shield cover, and enclosed experimental apparatus. Once degassing is completed, the UHV chamber is installed in the payload bay.

Once on orbit, a payload specialist in the orbiter cabin will control and monitor extension of the boom and deployment of the shield. After checkout of the facility, the shield cover is jettisoned, and the experiment starts. Near the end of the mission, the deployment sequence is reversed and the molecular shield and experimental apparatus are stowed in the UHV chamber prior to deorbit. Upon return to Earth, experiment samples are available for analysis, and the shield and UHV chamber may be refurbished.

MSVF missions will be executed in circular orbits to facilitate orbit synchronous rotation and shield pointing. Orbit inclination will be chosen so as to maintain the orbit plane-sun line angle between 0° and 60° to eliminate periodic orbiter roll maneuvers for thermal control.

To maintain ultra-high vacuum within the shield, two conditions must be satisfied; namely (1) no point on the orbiter surface can be allowed line of sight access to the shield interior, and (2) shield axis must remain parallel (within $\pm 1°$) to the atmospheric drift velocity. To achieve this accuracy, it will be necessary to "fine tune" the shape and orientation of the configuration's inertia ellipsoid (i.e., to tailor carefully the boom deployment angle and the initial orbiter-shield configuration orientation) so that gravity gradient torques are minimized.

Attitude control for MSVF missions will be maintained by firing the RCS vernier motors. This control mode, with the inherent stabilization provided by the boom, will produce a relatively quiescent orbiter.

Boom Requirements

Preliminary analysis indicates that development of a 100-m boom is feasible for this application, provided that the quiescent state can be maintained for the duration of experiments. The rigid body motion time history of a representative configuration in a 170-km circular orbit has been simulated. In response to gravity gradient torques and RCS vernier firings at intervals of ~2000 sec, the pitch rate oscillated ~± 0.002 deg/sec about the orbit synchronous rate of ~0.068 deg/sec. Oscillation frequency was ~4 x 10^{-3} Hz which is acceptably low from the standpoint of boom resonance excitation. Angular error between shield axis and drift velocity did not exceed $\pm 0.2°$. Acceleration of the shield in response to vernier firings was ~3 x 10^{-3} g.

A FREE-FLYING MOLECULAR SHIELD FACILITY

A free-flyer version of the MSVF embodies design concepts already proven on existing satellites. It consists of the molecular shield assembly, the adapter module, a systems module, and the UHV chamber. The shield, adapter module, and the UHV chamber are identical to those employed in the boom-deployed version. The systems module houses power, attitude control, and telecommunications subsystems. The shield adapter and systems modules are joined by an extendable boom to minimize potential contamination of the shield environment by outgassing from the solar cell arrays.

Operational Characteristics

The satellite will be carried into orbit by the shuttle orbiter and deployed by the remote manipulator system (RMS). A three-axis reaction wheel attitude control system will maintain alinement of spacecraft axes in a local Earth-vertical/orbit velocity orientation for proper functioning of the molecular shield.

Once on orbit, the satellite, with the boom still retracted, is placed overboard by the RMS. After alining the boom with local vertical, the boom is fully extended and the solar arrays are deployed. After operational checks are completed, the satellite is separated from the orbiter. After the orbiter has moved away so that it poses no contamination threat to the satellite, the shield cover plate is jettisoned. Experiment operations will then be conducted automatically.

The free flyer will be deployed in a near-polar, Sun synchronous orbit to simplify the solar array drive. The arrays will be driven continuously about a single axis at the orbital rate so that they are always perpendicular to the Sun line. Power for the night portion of the orbit will be stored during the day by means of flywheel energy storage units, which will serve also as the reaction wheel attitude control devices. These integrated power and attitude control systems (IPACS) are under development at the Langley Research Center, and should be lighter in weight and require substantially less volume than separate attitude control and power (all battery) systems. The IPACS would be located in the systems module. Acceleration levels induced at the shield by attitude control torques would be substantially less than 10^{-3} g.

CONCLUDING REMARKS

The concept of an orbiting molecular shield vacuum facility is under intensive investigation at the Langley Research Center. This paper presents ideas concerning the implementation of such a facility. Although numerous unsolved problems in design, fabrication, and deployment of the shield have been recognized, the results of preliminary analyses indicate that development of the molecular shield concept is feasible.

AAS 75-249

ADVANCED EXTRAVEHICULAR MOBILITY UNIT TECHNOLOGY EXPERIMENTS

Gary Wengrow*

A study recently conducted for NASA's Ames Research Center by Rockwell International determined that significant reductions in Shuttle payload costs can be achieved through routine use of extravehicular activity (EVA). This paper summarizes the results of the study, describes potential EVA applications for the Advanced Technology Laboratory (ATL) all-pallet payload, and discusses briefly EVA system technology advances likely to lead to more EVA usage in support of Shuttle payloads.

In the Ames/Rockwell study, 13 representative Shuttle payloads were analyzed to identify functions that could be performed in an EVA mode rather than in the current baseline automated mode. The functions were subjected to detailed design and operations analyses, and costs were computed for both modes of operation. For each representative payload, an EVA-oriented mode resulted in net savings. On the basis of these results, EVA savings were calculated for total Shuttle payload program research, development, test, evaluation, and production through 1990. The civilian payload program savings amounted to $610 million. Spacelab payloads accounted for two thirds of the EVA savings: approximately $15 million could be saved, on the average, for each Spacelab payload program that adopts EVA as the baseline mode of operation for selected functions.

DESIGN AND OPERATIONS ANALYSES

Many mechanical functions planned for automated mechanisms can be performed in the EVA mode with significant reductions in development and fabrication costs. Several of these functions are common to almost all payloads analyzed. For example, many automated

---
*Space Division
Rockwell International

spacecraft use deployed solar panels. Automated solar panel deployment requires the following hardware elements (among others):

- Erection drive motors
- Solenoid-operated erection mechanism latches
- One or more solenoid-operated tie-down latches
- Command and monitor avionics
- Caution and warning avionics
- Orbiter controls and displays

In the EVA mode, these elements are replaced by the mechanism shown in Fig. 1. The panel is stowed in a reusable manually operated fixture during the boost phase, unstowed by the EVA astronaut, and slipped over a mounting pylon on the spacecraft. An electrical connector and alignment pin are built in, and the solar panel is latched with a pip pin. A cost comparison for the two modes reveals a savings of $240,000.

Integrated time lines were developed that compare the automated and EVA modes in delivering representative automated spacecraft and activating representative Spacelab payloads. Elapsed-time differences between automated and EVA operations are insignificant in most cases. However, compared to the automated mode, there is a significant delay before EVA operations commence. This delay results from the planned use of low-pressure extravehicular mobility units (EMU's), requiring the crew members to "prebreathe" oxygen for three and one-half hours prior to EVA to prevent the bends.

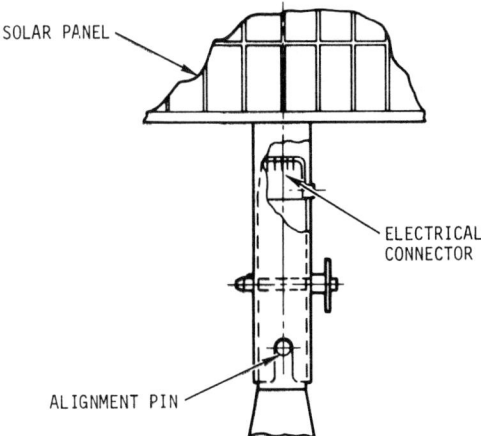

Fig. 1 EVA Solar Panel Mechanism

COST ANALYSES

The results of the representative payload cost analyses are summarized in Table 1. Baseline costs include development and fabrication of the first unit. Operations costs are not included. (It should be noted that these and all other cost data contained in this paper, while accurate in a relative sense, do not reflect the official position of either Rockwell International or NASA.)

As a percentage of baseline costs, EVA net savings are higher for Spacelab payloads than for automated spacecraft. Net savings represent the difference between costs eliminated through use of EVA and costs added for payload-peculiar EVA support equipment.

These results were projected to Space Shuttle payloads in the latest NASA traffic model and modified to reflect the flight schedule used by NASA manned space flight program offices for operations analyses. These projections are contained in Table 2. On the average, each automated spacecraft program that adopts EVA as its baseline operational mode would save about $4 million; each Spacelab payload program would save about $15 million.

Table 1
COST ANALYSIS RESULTS

Representative Payload	Baseline Cost ($M)	EVA Net Savings ($M)
Automated Spacecraft		
Earth Observation Satellite	230	8
Gravity and Relativity Satellite	46	2
Large Space Telescope	180	13
Minilageos	2.4	0.4
Magnetic Field Monitor	16.4	1.2
High-Altitude Explorer	31	0.6
Domsat	24	1.1
Geopause	44	1.2
Mariner Jupiter Orbiter	67	1.3
Spacelab Payloads		
Shuttle Infrared Telescope	73	7
Atmospheric, Magnetospheric, and Plasmas in Space	246	7
Advanced Technology Laboratory	150	14
Physics and Chemistry Facility	36	8

Table 2
COST ANALYSIS PROJECTIONS

Sponsoring Agency	Automated Spacecraft			Spacelab Payloads		
	Programs	Units	Savings ($M)	Programs	Units	Savings ($M)
Civilian	51	178	192	27	82	418
Department of Defense	44	154	166	–	–	–
Totals	95	332	358	27	82	418

EVA APPLICATIONS FOR ADVANCED TECHNOLOGY LABORATORY

Remote activation of this payload from the orbiter's aft flight deck requires two crew members to work approximately two hours. The EVA-oriented design, with manual deployment of extended devices, release of boost tie-down latches, etc., requires three crew members (two EVA, one in the cabin) for about 3.6 hours.

Fig. 2 illustrates the activation sequence to be followed by the EVA crew members–a sequence designed to minimize interference among payload instruments while ensuring clear translation paths for the crew. EVA work stations are numbered in sequence and correspond to the station numbers in the table included with Fig. 2.

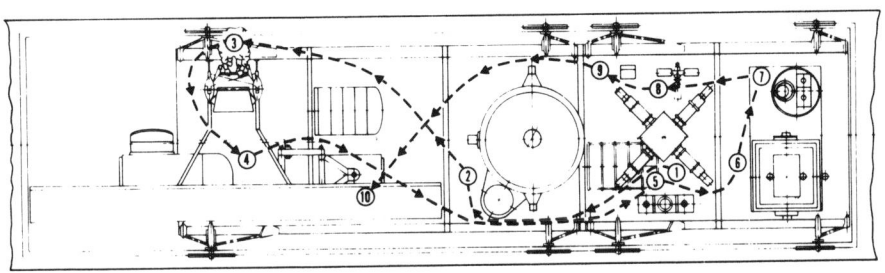

Fig. 2 ATL Activation Sequence

These relatively simple activities would result in net ATL program savings of over $14 million. Additional functions could be performed in the EVA mode that would increase these savings, even after allowing for design and fabrication of payload-related EVA equipment (handholds, protective covers, translation aids, etc.). It should be noted, however, that in order to realize these cost savings, EVA must be "designed into" Shuttle payloads: retrofit of EVA provisions would not be cost effective.

MAN RATING

To be "man rated," a space system must provide the following capabilities or features:

- Life support
- Vacuum and radiation protection
- Thermal and meteor protection
- Crew refuge
- Rescue provisions
- Safe work stations

Of these, only the last is a requirement imposed on Shuttle payloads; all of the others are provided by the orbiter and/or Spacelab. Furthermore, the impact of the safe-work-station requirement on Shuttle payloads can be substantially reduced by advanced technology EMU's, which incorporate elements providing improved mobility, visibility, and dexterity, compared to systems based on current technology.

CONTAMINATION

Current technology systems that control the thermal environment by sublimation of water introduce a contaminant into the orbiter cargo bay that must be considered in assessing the utility of EVA payload support. Technology is being developed that can effectively eliminate the EVA astronaut as a source of contamination. Systems under consideration can support up to one hour of EVA operation without use of sublimator-type thermal control systems. They provide an interface with an external heat sink, such as a coolant umbilical or a portable fusible system. Hamilton-Standard is studying fusible heat sinks for Ames, where in-house studies are also being conducted. Additionally, advanced suit closures (without zippers) and rotary joint technology can reduce suit leakage to one fourth or less of current rates.

REACTION TIME

A key performance improvement area is reaction time—the time between establishing an EVA requirement and initiating the effort. As discussed earlier, current technology systems operate at relatively low pressures (≤ 5 psia), requiring a crew member to prebreathe pure oxygen for approximately three hours before conducting EVA. Systems operating at higher pressures can eliminate the prebreathing requirement. Elements of high-pressure suits, such as gloves and rotary joints, have been developed and evaluated in the laboratory; and they should now be demonstrated in an operational environment.

Reaction time is strongly influenced by suit design as well as operating pressure. Present systems require nearly two hours for equipment preparation, suit donning, life support system installation, helmet and glove donning, system checkout, and airlock operations. To take advantage of higher pressure suits, the time required for these operations must be reduced. Possible time-saving measures include integrating the suit and life support system, simplifying the checkout procedure, reducing the number of closures required for suit donning (and replacing zippers with advanced closures), and reducing the number of connectors required during donning. It has been estimated that EVA preparation time can be reduced to one-half hour by adopting these approaches.

ACKNOWLEDGMENT

The author wishes to thank Mr. L.R. Alton of NASA's Ames Research Center, and Mr. J.W. Patrick and Mr. E.F. Kraly of Rockwell International for their contributions to this paper.

AAS 75-250

THE ADVANCED TECHNOLOGY LABORATORY
C. Llewellyn and R. Milliken*

The Advanced Technology Laboratory (ATL) is the program of NASA Langley Research Center designed to fly space research missions employing the Space Shuttle and the European Spacelab. This paper discusses the evolution and current status of the ATL and examines some conclusions derived from integrated payload/Spacelab system point designs conducted by the Rockwell International Space Division.

Several studies were conducted by Langley which examined the fundamental concepts and systematically identified and described experiments suited to ATL missions. Specific principal investigators have participated in the continuing definition and development of experiments. A recent contractual effort was aimed at defining and developing the most effective checkout and integration concepts. A contract effort is just starting which will provide a complete ATL system definition. Langley coordination with payload planning activities, such as the Shuttle Payload Working Groups and the Spacelab and Shuttle Program Offices, has continued from late 1971 through the present.

SPACELAB PAYLOAD POINT DESIGN STUDIES

The background derived from the Langley ATL studies provided an opportunity for Rockwell International Space Division to conduct two integrated payload system point design studies. The two system designs and areas of investigation are illustrated in Figure 1. One payload is a pallet-only payload with a 5-section pallet. The other uses a long module Spacelab with two pallet sections.

Physical Layout

The pallet-only payload included 10 large sensors mounted on the pallet with controls and displays located in the Orbiter MSS/PSS area. The 5-section pallet easily accommodated the 10 sensors. The limiting factor for this configuration was space in the MSS/PSS. This was, in part, resolved by separating electronic equipment from the controls and

---
*C. Llewellyn, Shuttle Experiment Office, NASA-Langley Research Center; R. Milliken, Space Division, Rockwell International Corporation.

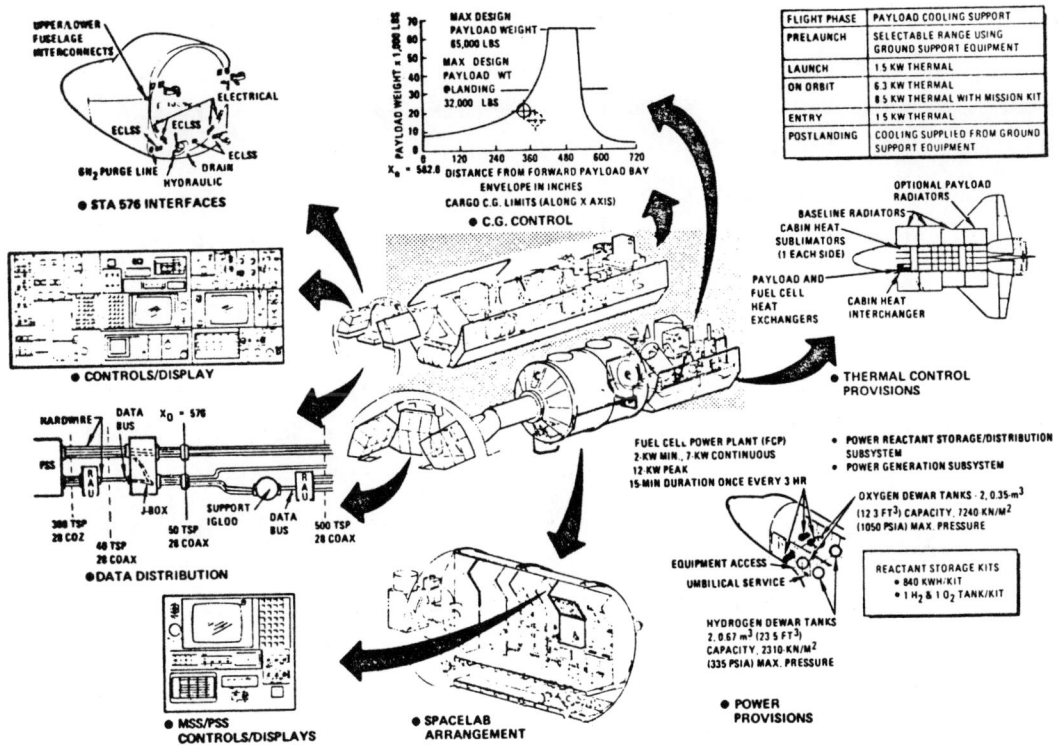

EXPERIMENTS IN ATL PAYLOAD POINT DESIGNS

Pallet-Only Payload		Module-Pallet Payload	
NV-1	Microwave Interferometer	NV-3	Multipath Measurements
NV-2	Autonomous Navigation	EO-2	Tunable Lasers for High-Resolution Studies of Atmospheric Constituents and Pollutants
EO-1	Lidar Measurements of Cirrus Clouds and Stratospheric Aerosols	EO-5	Laser Ranging
EO-4	Microwave Radiometer	EO-9	RF Noise Measurements
EO-7/8	Imaging Radar and Search and Rescue Aids	MB-3	Electric Field Opacity in Biological Cells
PH-2	Barium Cloud Release on Sunward Side of the Earth	PH-2	Barium Cloud Release on Sunward Side of Earth
PH-4	Molecular Beam	PH-3	Optical Properties of Aerosols
PH-6	Meteor Spectroscopy	PH-4	Molecular Beams
EN-1	Sampling Airborne Particles and Micro-Organisms in Space Cabin	MB-1	Colony Growth in Zero Gravity
EN-3	Environmental Effects on Non-Metallic Materials	EN-1	Sampling of Airborne Particles and Micro-Organisms in Space Cabin
CS-X	Contamination Monitor	CS-X	Contamination Monitor

Fig. 1 Preliminary Design Activities

displays and mounting it in two equipment canisters (igloos) on the pallet. The module/pallet payload included 11 experiments; 7 with sensors on the pallet, and 4 entirely in the pressurized module. The 7 sensors on the 2-section pallet were crowded and it was necessary to design both equipment and mission procedures for sequential deployment and retraction of certain instruments to eliminate interferences. The long module easily accommodated all equipment requiring environmental protection.

The pallet-only and module/pallet payloads weighed 24,332 and 27,174 lb, respectively, including growth allowances; both well within the Orbiter nominal landing weight of 32,000 lb. Center-of-gravity margins were more critical. The c.g.'s fell 15 in. and 4 in. aft of the forward edge of the allowable envelope for the module/pallet and pallet-only payloads, respectively. The pallet-only payload was adversely affected by the additional equipment forward in the Orbiter.

Mission Analysis

Seven-day mission plans were developed for each payload. Two active ground target locations, Mojave Desert and Chesapeake Bay areas (Langley), were designated. Crews of 2 and 4 were selected for the pallet and module/pallet payloads, respectively. Timelines were developed. It was concluded from these analyses that experiment operating times and cycles were satisfactory with the two active sites and available (inactive) target opportunities. Opportunities to view one or the other of the two active ground targets varied from 12 to 14 by adjustments of altitude or inclination. Careful selection of parameters could, however, maximize comparable viewing angles of a target(s) on a mission.

Analysis of crew activities for the module/pallet and pallet-only payloads indicated 75% and 78% active time out of total non-sleep time, which was considered reasonable. With a decrease from 4-man to 3-man crews, the active time for the module/pallet payload increased to 86%.

An analysis was made of extended mission durations. Kits for electrical power, RCS fuel and life support consumables were added as required. On this basis the module/pallet payload could have flown for a period of 12 days, limited by the 32,000-lb landing weight and by c.g. migration.

Controls, Displays and Data Management

Three design approaches were considered for controls and displays: (1) dedicated hardwired; (2) computer aided; and (3) fully automatic. The dedicated hardwired approach appears to permit design and development of independent self-contained experiments which minimize complexity and interfaces. This approach would provide the experimenter more complete responsibility for the design and control of his experiment.

Two factors forced the design of the pallet-only payload to the computer-aided approach. These were the limitation of control panel area (38% short of the dedicated requirement) and a desire to minimize the wire penetrations across the Orbiter/payload interface. This approach routes all control signals through computer remote access units (RAU's) and uses a keyboard and CRT extensively for control and display functions. Forty-seven percent of the panel area in an integrated control panel designed for the pallet-only payload was devoted to common controls and displays. The module/pallet payload used the dedicated control and display hardwired approach. A completely automatic system was rejected on the belief that it would be more complicated and costly.

In the case of the pallet-only payload it was found that the wiring across the Orbiter X_o 576 bulkhead between the MSS/PSS and cargo bay could be reduced from 500 to 50 twisted shielded pairs (TSP) by use of the computer-aided approach; 28 coaxial lines would be required in either case. In the pallet/module payload, approximately 50 TSP and 5 coaxial lines were required across the X_o 576 bulkhead.

Resource Requirements

Electrical power requirements were well within levels available from the Orbiter for both payloads, averaging approximately 70% of the available. Energy consumed by the two payloads was 829 kWh and 802 kWh, respectively, which can be provided by one standard energy kit (840 kWh). The heat load for the pallet-only payload was within 6.3 kW, which could be handled by the Orbiter without a kit. A standard Orbiter radiator panel kit was required for the module/pallet payload. Attitude profiles were prepared for both missions. RCS consumption was well within available capacities--2291 lb and 1970 lb for the pallet and pallet/module missions--compared with an available 6040 lb.

CONCLUSION

The ATL program will offer a dynamic potential for space research. Missions of 7 to 14 days with 8 to 15 experiments can provide operating times and numbers of experiment cycles appropriate to the development of advanced sensors and space research. Support requirements are well within the Spacelab/Orbiter capabilities. The larger published paper on this study provides additional detail.

AAS 75-251

SHUTTLE ENTRY TECHNOLOGY PAYLOADS[*]

Paul M. Siemers III[†]

The flight frequency of the Space Transportation System (STS) coupled with its large payload-carrying capability will provide an unprecedented opportunity for conducting aerothermodynamic/entry technology research. This STS research opportunity can be characterized into two distinct categories: (1) that research which will utilize the STS orbiter as the test vehicle, and (2) that research which will utilize a vehicle (or vehicles) launched from the orbiter for entry.

To date, on-going Langley Research Center studies have defined experiments as well as the support systems required for the shuttle launched research program.

The proposed Entry Technology Program will provide a flight data base from which accurate correlations can be performed relative to ground test and analysis data. These correlations will result in optimized designs for future flight systems.

INTRODUCTION

NASA's rapid and continued advance in aeronautics and space research has been due in large measure to the aggressive pursuit of flight research programs. The significant contributions of research aircraft, from the X-1 through the X-15, in expanding the speed envelope from transonic to low hypersonic speeds are well documented. Suborbital rocket probes such as Fire, RAM, Reentry F, and so forth, have increased the knowledge of heating and boundary-layer phenomena. The feasibility of suborbital flight and landing of lifting entry systems has been proved with the HL-10, M2-F2, and X-24 vehicles. Additional flight data in all disciplines are needed across the speed range, but flight programs have been severely curtailed for lack of funds. Many illusive questions relating

---

[*]Prepared for the American Astronautical Society's Twenty-First Annual Meeting — Space Shuttle Missions of the 80's.

[†]Aerospace Technologist, Aerothermal Branch, Space Systems Division, NASA Langley Research Center, Hampton, Virginia 23665.

to entry technology (flow-field behavior, heating, performance, materials verification, etc.) remain unaddressed from the standpoint of flight data. These unanswered questions result in system designs based on unverified ground data, which are conservative, and this design conservatism results in minimized scientific payloads and compromised system performance.

The mission and operational flexibility of the Space Transportation System (STS) can provide the agency with the means to address a number of these questions by facilitating access to full-scale free-flight data across the speed range. Data obtained on the orbiter itself will be pertinent to basic research and technology development tasks and invaluable in the design and development of advanced Space Transportation Systems. In areas where data are needed outside the orbiter flight envelope, the orbiter would act as the launch vehicle for payloads designed to address specific problem areas.

In recognition of this unprecedented opportunity to conduct entry research as an adjunct to STS orbital operations, Langley Research Center initiated an in-house study to define the specific LRC research that would be applicable to the STS. This activity has continued at LRC and been broadened within NASA as a result of the formation, by OAST, of the Entry Technology Study Team. The object of this study team is to determine the feasibility of using the STS in an entry technology program and then to define the technical and programmatic requirements to support such an effort. The data presented here are the results, to date, of studies conducted at and by the Langley Research Center in support of this activity.

DISCUSSION

The Entry Technology experiments exist in three distinct levels of sophistication and cost:[1]

- o The first level or category relates to those experiments which can be conducted on a minimal or no-impact basis during each STS mission as desired.

o The second category of experiments will capitalize on available excess volume within the payload bay. This volume can be utilized either for housing instrumentation or for transport of small research entry vehicles. In the case of the entry vehicle, this additional payload would be carried "piggyback" without impacting upon the primary mission goals.

o The third category of complexity is represented by a major advanced entry vehicle which would fill completely the available cargo volume and the launching of which would become the primary mission.

Category 1 Experiments

The primary emphasis of these experiments will be to provide the full-scale flight data required to resolve the many unanswered technology problems associated with entry and to provide a data base from which accurate extrapolation techniques of ground-based data may be developed.

Aerodynamic. The simulation of flight parameters with present ground facilities is often impossible, therefore, verification of theoretical approaches and experimental methods is needed. The utilization of the orbiter GN&C subsystem plus the determination of α, β, and free-stream density via an operational air data system would provide the data necessary to verify the force and moment characteristics of the orbiter across the speed range. Special emphasis would be placed on the hypersonic real gas and viscous flows, RCS/aero interaction and control effectiveness including flow separation effects. In addition, guidance and control system sensitivity to dispersions and off-nominal conditions can be predicted and evaluated.

Thermodynamic. As with aerodynamic parameter simulation, the simulation of all flow-field parameters simultaneously with present wind-tunnel facilities is often impossible, therefore, flight data are necessary for verification of theoretical approaches and experimental methods. The thermodynamic experiments will require the measurement of stagnation and surface pressure and temperature as well as heat flux and surface pressure fluctuations. Experiments will also be conducted which will define flow parameters, gas conditions, and chemistry in the shock layer through the use of direct measurement and sampling instrumentation (i.e., traversing probes, sampling mass spectrometers, etc.). These data will also provide for verification of viscous and real-gas effects on the thermodynamic phenomena.

Specific experiments will be conducted to verify boundary-layer transition prediction methodologies and criteria and to establish the effect of surface roughness, flight conditions, and acoustics on the onset and stability of transition. Additional experiments will be conducted to define and verify the effects of jet plume interaction, shock impingement, structural interference (e.g., corner flow), tile gaps, flow separation, and lee side flows on loads and surface temperatures.

Gap Heating. The results of the ground test experimental effort[2-4] relative to the effect on TPS heating of the thermal expansion gap on the orbiter reusable surface insulation (RSI) contributed to minimization of weight and cost for the TPS. However, more efficient design of advanced TPS requires the definition of ground-to-flight extrapolation techniques.

Data obtained on the STS orbiter will contribute to the development of the required extrapolation techniques and an understanding of the following orbiter TPS heating phenomena:

(1) Effect of tile edge radius on surface and gap heat transfer; integrated heat load to tile; and tile internal temperature distribution as affected by radiation relief resulting from edge radius and geometry.

(2) Heat transfer to intersections of streamwise and transverse gaps effects of gap width, running length, and boundary-layer parameters.

(3) Effects of spanwise pressure gradients on heating within transverse gaps.

(4) Promotion of boundary-layer transition by gaps transverse and parallel to the local streamline.

Flight experimental data will be obtained by replacing operational RSI tiles with appropriately instrumented tiles. The location and number of tiles replaced will depend on the specific data required. Figure 1 represents typical tile arrangements and parameters for gap heating experiments.

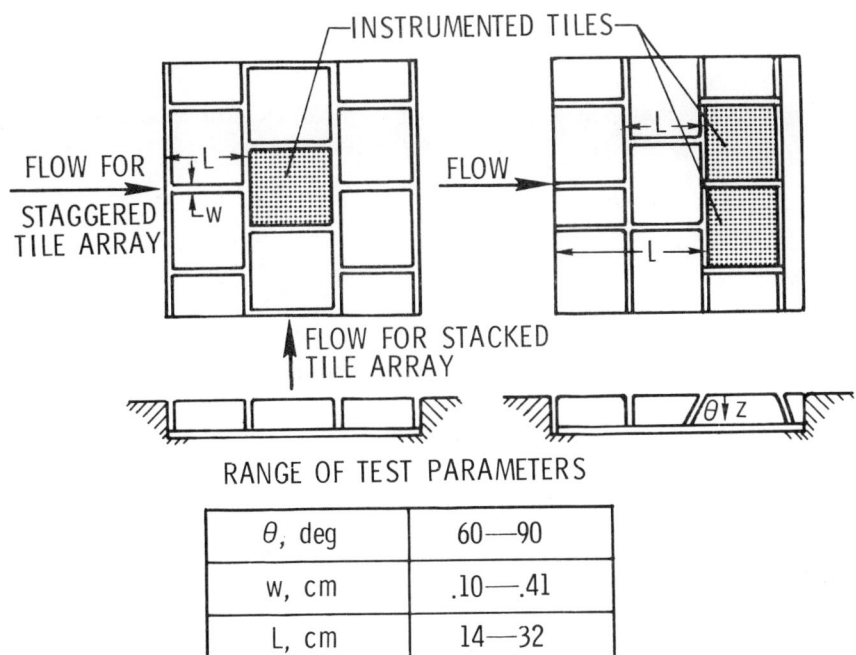

Fig. 1 RSI Tile Gap Heating Models

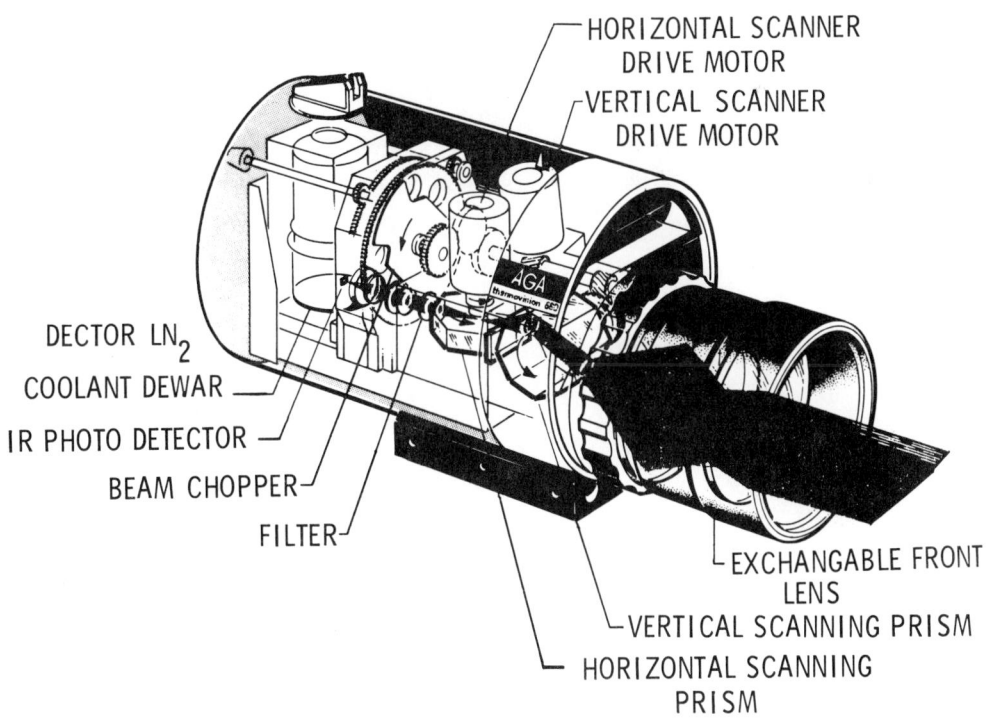

Fig. 2 Infrared Camera Electro-Optical Components

Lee Side Heating. Lee surface heating on STS orbiter configurations has sometimes been regarded as insignificant since the lee surfaces are generally sheltered from the main stream. However, experimental studies[5-7] have shown the strong influence of vortices on the heating. Some significant conclusions from these studies are: (1) peak vortex-induced heating at an angle of attack can exceed that for zero angle of attack, (2) peak heating is extremely sensitive to Reynolds number and Mach number, and (3) the geometry of the lee surface greatly influences the vortical heating. Although these studies have provided information which has been used in the design of the orbiter lee side TPS, flight verification of the vortical flow is required to establish ground-to-flight extrapolation techniques, verify heating rates, Reynolds number and Mach number effects, and to establish guidelines for vortex alleviation. The accomplishment of these objectives will result in the reduction of the weight and cost of the TPS on the lee side of future space vehicles including retrofitted orbiters.

The flight lee side heating data required can best be obtained from infrared observation of the surface temperature distributions. The infrared observation can be accomplished by utilizing a scanning system mounted in the tip of the vertical tail. Such a system will provide better coverage of the lee side than a large number of individual surface sensors, will cost less, and will not result in local disturbance in the measurement area. A proposed IR camera is shown in Figure 2 while mounting concepts being considered are shown in Figures 3 and 4. The approximate IR coverage map of the side-mounted tail camera is shown in Figure 5. Some design problems presently being addressed are solar radiation, effects of vibration and high noise levels during lift-off, scanner window heating during entry, and STS orbiter interfaces.

Materials/Thermal Protection System. Flight verification of the performance of the carbon-carbon leading edges as well as the RSI is of importance to personnel who have developed the design methodologies and the ground test techniques utilized on the qualification of the orbiter TPS materials. This verification will utilize pressure and temperature data obtained by techniques discussed previously. In addition, the

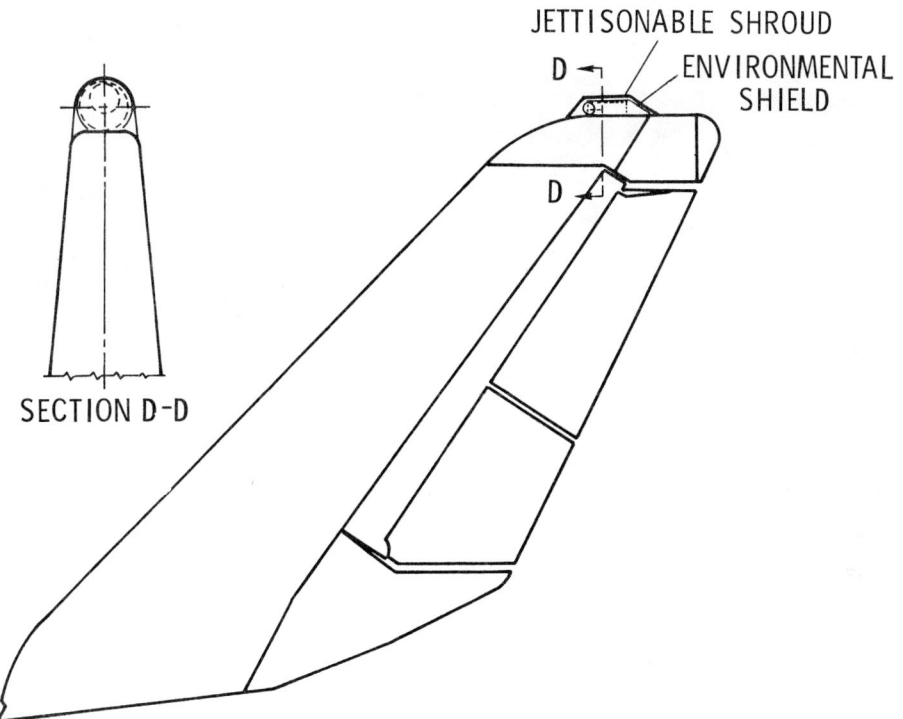

Fig. 3 External Mounting of IR Scanner

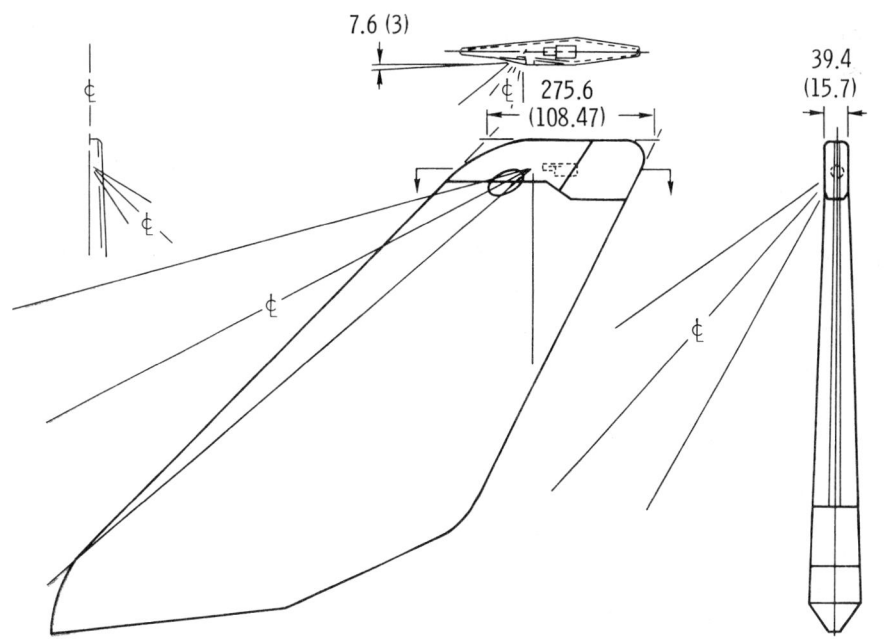

Fig. 4 Internal IR Scanner Mounting Concept

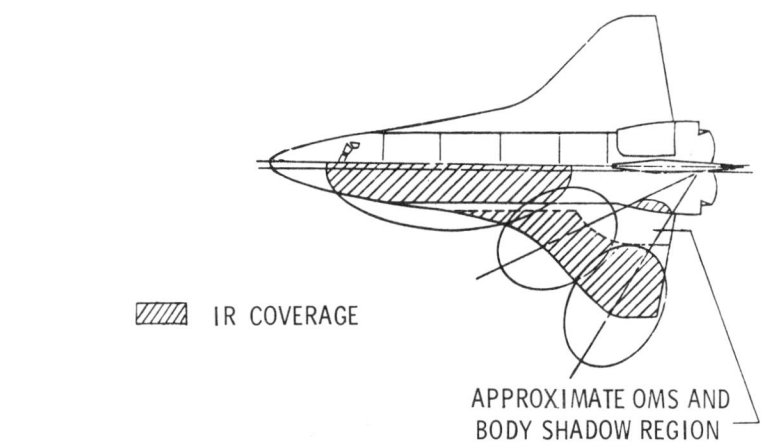

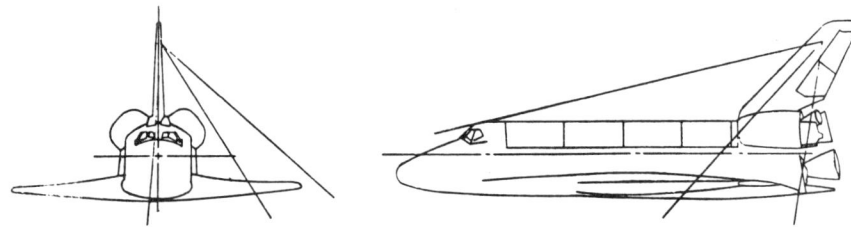

Fig. 5 Approximate IR Coverage for Tail-Mounted Camera

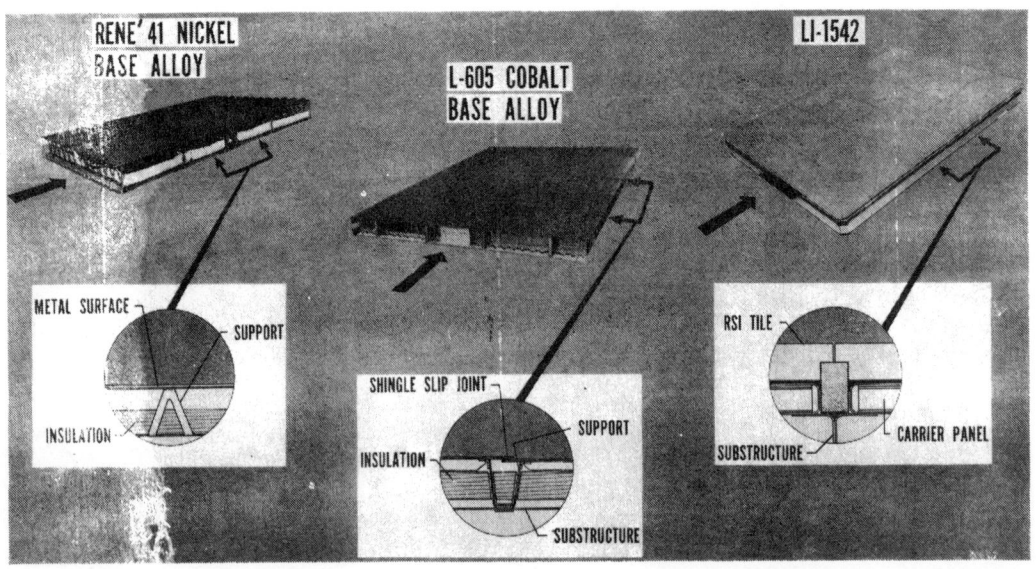

Fig. 6 Radiative Thermal Protection System

thermal map generated from these same data will be utilized to locate environmental regions for experiments relative to various thermal protection system material such as high-temperature polyimide foams, a lightweight low-density material proposed for orbiter lee surfaces,[8] and radiative metallic and RSI thermal protection systems.

Presently, three basic radiative TPS have been proposed and tested.[9] They are shown in Figure 6 and described in Table 1.

Table 1

CANDIDATE RADIATIVE THERMAL PROTECTION SYSTEMS

Material	Description	Ground test environment
1. René 41 (nickel based alloy)	Single corregated skin structure	26 Thermal 11 Aerothermal Cycles to T_{sur} = 1144K
2. L-605 (cobalt based alloy)	Corrugation stiffened	19 Thermal 13 Aerothermal Cycles to T_{sur} = 1200K
3. LI - 1542	Silica tiles	23 Thermal 12 Aerothermal Cycles to T_{sur} = 1200K GAP temp - 1370K

Instrumentation. The instrumentation of the orbiter carbon-carbon leading edge and RSI tiles has been studied[10] and surface pressure and temperature instrumentation is being defined for use on the orbiter during the development flights. This instrumentation will provide the surface pressure and temperature data required for the general thermodynamic/flow-field experiments, as well as provide the thermal data for materials and TPS evaluation and experiment location. To conduct a comprehensive entry technology program, this and additional instrumentation will be required during the operational flights. Langley Research Center is presently studying the concept of removable instrumented RSI tiles, Figure 7, which would replace uninstrumented operational tiles in the regions of experimental interest. These tiles would contain the required instrumentation (surface pressure instrumentation is illustrated in Fig. 7), power supply, and data recorder, thereby eliminating interfaces

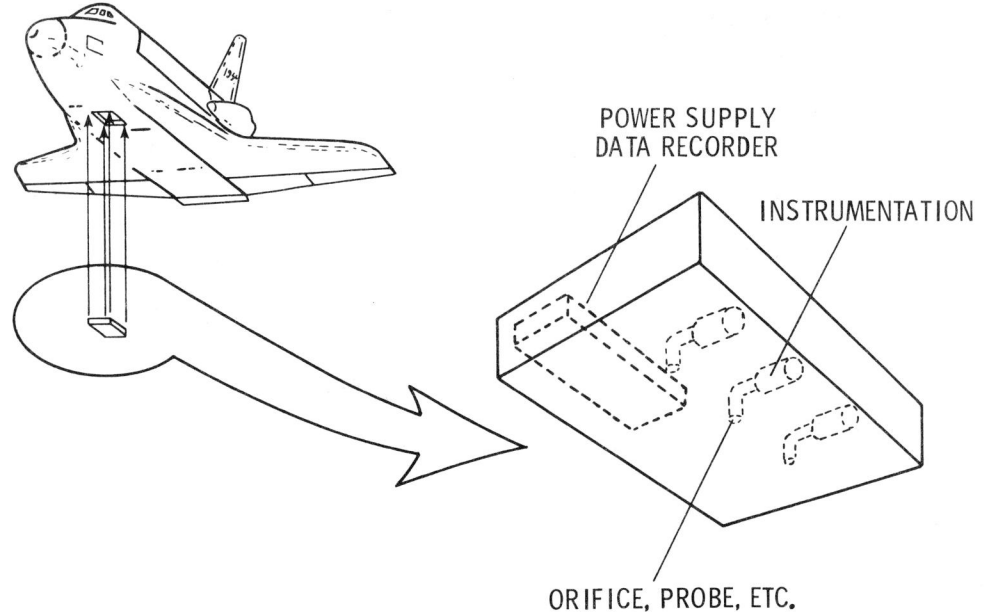

Fig. 7 Reusable Instrumented RSI Tile Concept

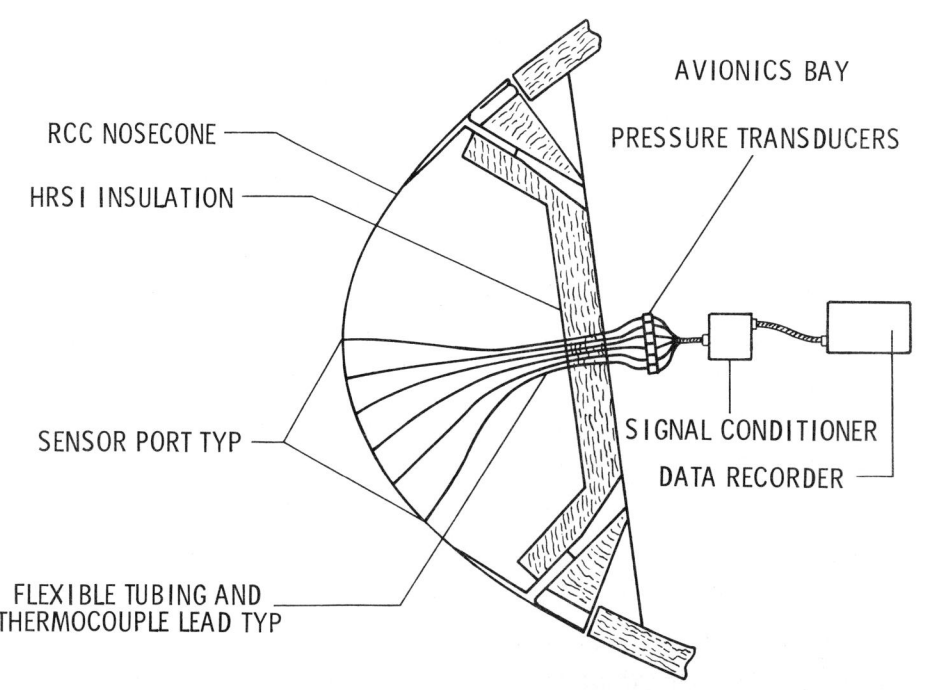

(a) Instrumentation Configuration

Fig. 8 Candidate Air Data System

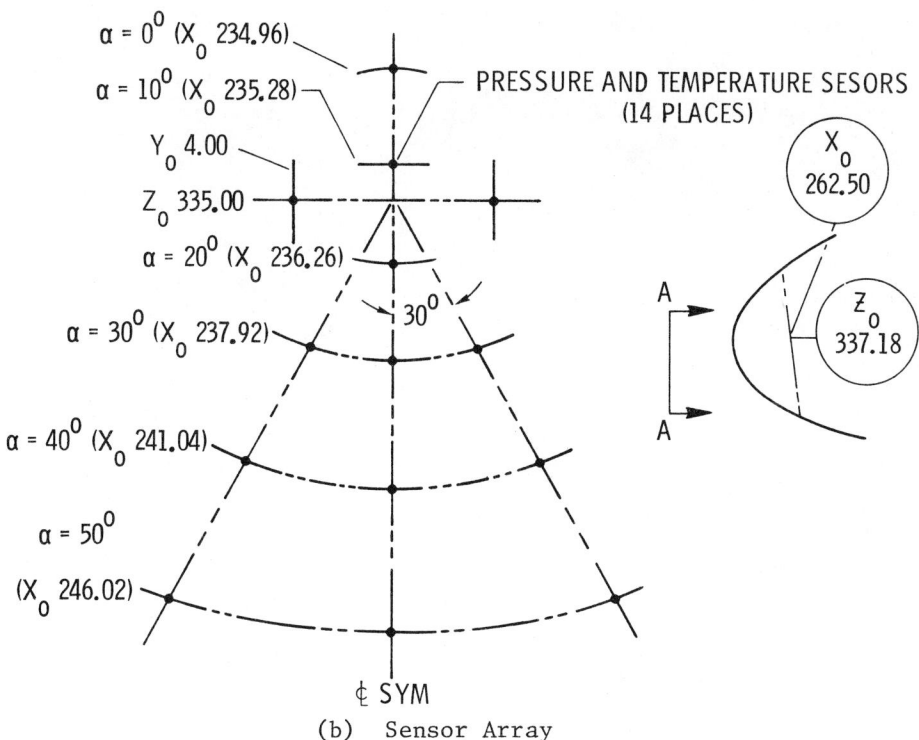

(b) Sensor Array

Fig. 8 Continued

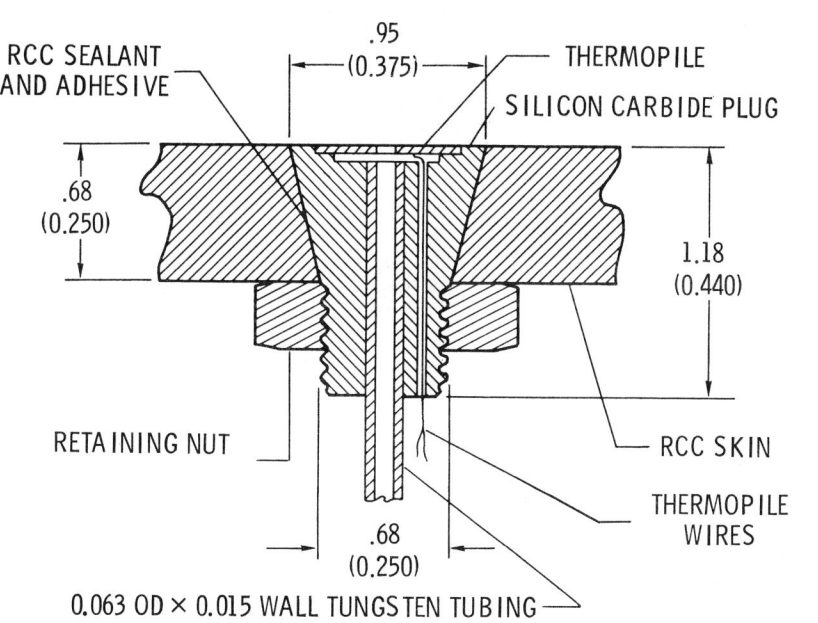

(c) Port Installation With Thermopile

Fig. 8 Concluded

with orbiter subsystems. To obtain stagnation region data, an air data system (ADS) is being proposed and studied by Langley Research Center. This will permit free-stream density and vehicle attitude to be determined at any time during entry thereby eliminating errors introduced by using the vehicle state data and an assumed atmospheric model. A proposed ADS is shown in Figure 8. A design, development, test, and evaluation program is underway to establish the feasibility of such a system.

Category II Experiments

Category II experiments are subdivided into two groups (A and B).

Group A. Shuttle borne experiments which require payload bay volume or require considerable orbiter subsystem interfaces. The experiments in this group are similar to Category I experiments in objective in that they are designed to resolve questions relative to aerodynamic performance and materials/structures.

Aerodynamic. Performance increases can be accomplished through the use of upper surface spanwise blowing.[11] The incorporation of such systems into the present orbiter or future STS would result in lift coefficient, C_ℓ, increases as indicated in Figure 9. This lift coefficient improvement would increase the allowable payload entry weight or extend the forward center-of-gravity (c.g.) range to provide a significant increase in payload capability. Flight tests for verification and qualification of such a flight system for possible use on a retrofitted STS are being evaluated.

Material/Structure. The development of advanced lightweight materials and structural concepts is a prerequisite for future lightweight high-speed aircraft and lifting entry vehicles, such as the heavy lift advanced Space Transportation System. Composite materials (boron-epoxy and graphic-epoxy) are presently being used in a limited manner on the STS orbiter. More extensive application of composites can result in a potential 20% weight reduction. Testing of these advanced concepts in the flight environment is highly desirable. Additionally advanced structural concepts such as bead stiffened panels[12] with fluted single

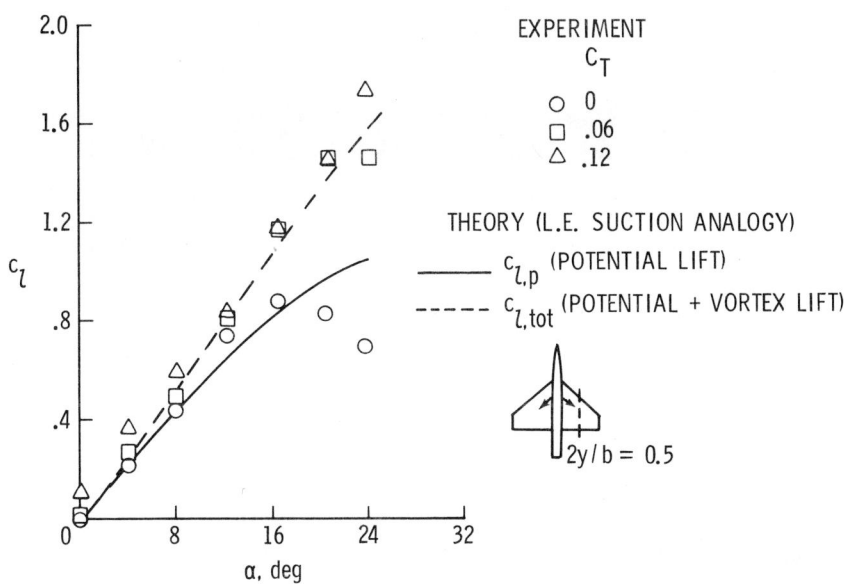

Fig. 9 Blowing Effects on Section Lift Curves

CONFIGURATION	CROSS SECTION SHAPE	CONFIGURATION	CROSS SECTION SHAPE
FLUTED SINGLE SHEET	t = 0.05(0.020), 103.5°, 7.1 (2.81)	CIRCULAR TUBULAR	2.6 (1.04), 90°, 3.4 (1.34), t = 0.06 (0.025)
FLUTED TUBULAR	t = 0.05 (0.020), 61°, 5.0 (1.97), 10.7 (4.21)	FLUTED TUBULAR	104°, 10.2 (4.02), 3.4 (1.33), t = 0.05 (0.020)
DESIGN LOAD 105 kN/m COMP. 35 kN/m SHEAR 6.9 kN/m² PRESS.	(600 lbf/in. COMP. 200 lbf/in. SHEAR 1 psi/PRESS.)	DESIGN LOAD 350 kN/m COMP. 70 kN/m SHEAR 13.8 kN/m² PRESS.	(2000 lbf/in. COMP. 400 lbf/in. SHEAR 2 psi PRESS.)

Fig. 10 Candidate Advanced Structural Concepts

sheet, circular and fluted tubular configurations (Fig. 10) utilizing aluminum and high-temperature alloys, that is, René 41, can result in system weight savings of up to 40%. Test panels should be flight tested as part of wing, fuselage, and control surface structures to demonstrate their applicability to an elevated temperature flight environment. Since the STS orbiter provides a unique capability for testing in this flight environment, the development of proper test techniques and procedures is under study. The results of such a test program would not only impact future systems designs but could be applied to the STS itself through retrofit.

SHUTTLE LAUNCHED EXPERIMENTS

To this point, this overview of entry technology flight test requirements has dealt with typical experiments which utilize the STS orbiter as a research vehicle. A large class of experiments exists which will utilize vehicles launched from the orbiter for reentry. These vehicles and associated experiments fall into one of two of the following categories:

Category II Experiments

Group B. Those experimental packages which require payload bay volume but are not the primary payloads for the mission and do not require a dedicated orbiter flight (piggyback payloads).

Category III. Those experimental packages which are the primary payloads for the mission and/or require a dedicated orbiter flight.

These payloads and experiments are not well defined at the present time but studies being conducted both at ARC and LRC will provide, within the next few months, a comprehensive evaluation of the STS orbiter launched entry technology research potential and capabilities.

The Category II Group B shuttle launched entry technology experiments are based on ballistic entry vehicle requirements of Earth entry, planetary, and DOD configurations. Experiments would be conducted to verify vehicle aerodynamic performance, thermodynamic/flow fields (e.g., verification of three-dimensional laminar and turbulent heating models), and TPS performance as it applies to the particular configuration. A simple recoverable vehicle (Fig. 11) adaptable to various forebody and afterbody geometries to evaluate the effects of cone angle, afterbody geometry, and so forth, is being studied. In addition to the performance testing, the vehicle would be a test bed for new concepts in entry systems such as propulsion, deceleration, avionics, and so forth. A secondary design goal is that the vehicle be capable of data canister delivery and payload deployment.

Also under consideration is the flight testing of an emergency astronaut reentry parachute system.[13] This system would include a suitable space suit, parachute, and life-support system to enable the astronaut to sustain life for about 2 hours during his reentry. Figure 12 shows a basic concept and typical performance requirements.

The Category III experiments are those experiments utilizing large or high-performance vehicles such as large to full-scale models of advanced space vehicle and hypersonic aircraft concepts. The experiments proposed are to verify, in a flight environment, vehicle performance, control effectiveness, real-gas effects, viscous effects, and the effect of canards, strakes, upper surface blowing, and so forth, on performance. The details of experimental requirements in this category will develop as future flight systems are defined.

Planetary Probe Program. The simulation of entry into Jupiter, Uranus, Saturn (the outer planets) requires the simulation of a heating environment controlled by radiation which cannot be fully achieved in ground facilities.[14] Simulation of design predictions and verification of flight performance relative to such heating mechanisms and levels can, however, be achieved from flight tests in the Earth's atmosphere.[15] A typical Earth simulation mission based on an entry angle of $-40°$, which provides a mean minimum ΔV requirement, and providing a $\dot{q}$ radiative of between 5.7 and 28.5 kw/cm^2 (Fig. 13) requires a ΔV of between 7 and 10 km/sec for a vehicle with a ballistic coefficient $M/C_D A = 120.3$ kg/m^2 and a nose radius $r_n = 0.3$ m. This ΔV range can be achieved with the Interim Upper Stages (IUS) candidates presently

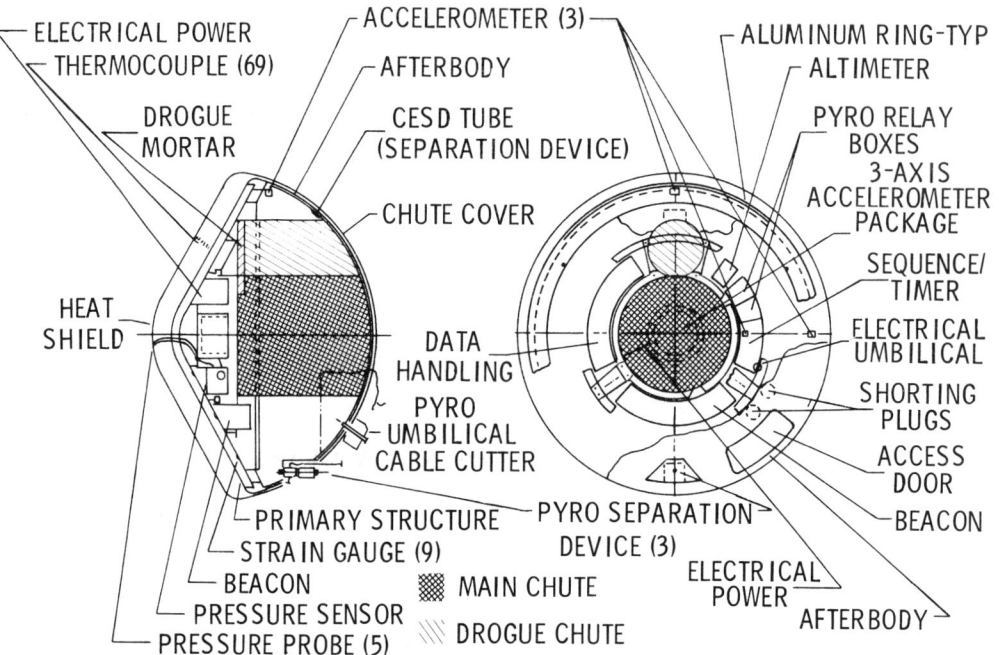

Fig. 11 Ballistic Entry Vehicle General Arrangement
(Planetary Configuration)

(a) Concept

Fig. 12 Emergency Astronaut Parachute Reentry System

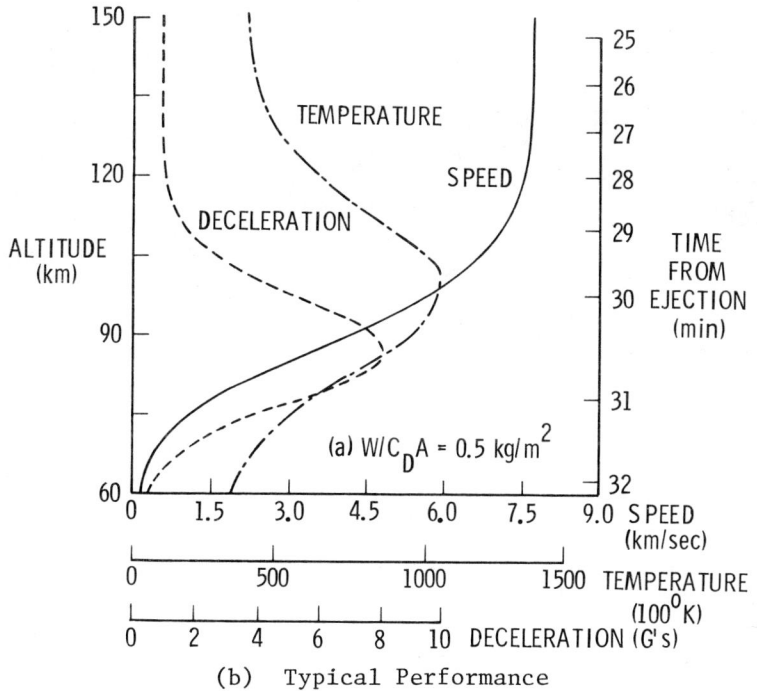

(b) Typical Performance

Fig. 12 Continued

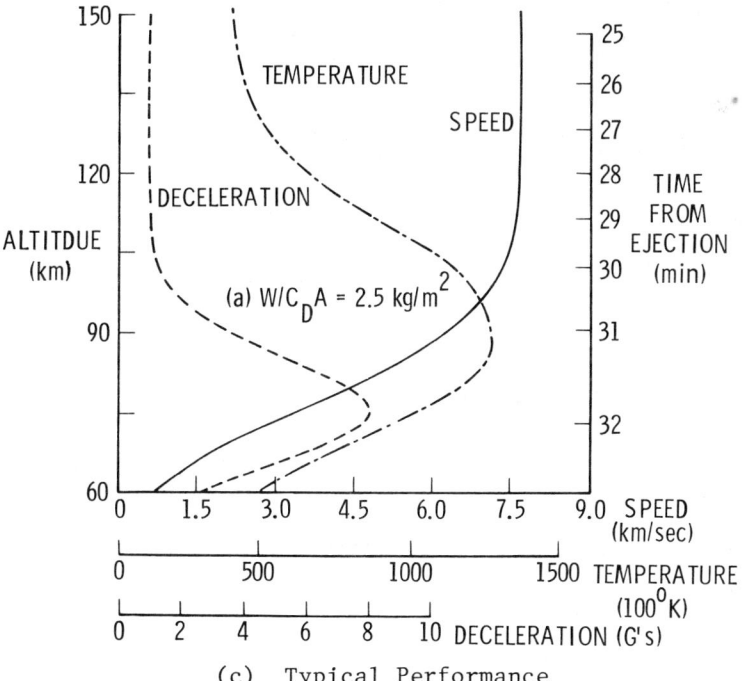

(c) Typical Performance

Fig. 12 Concluded

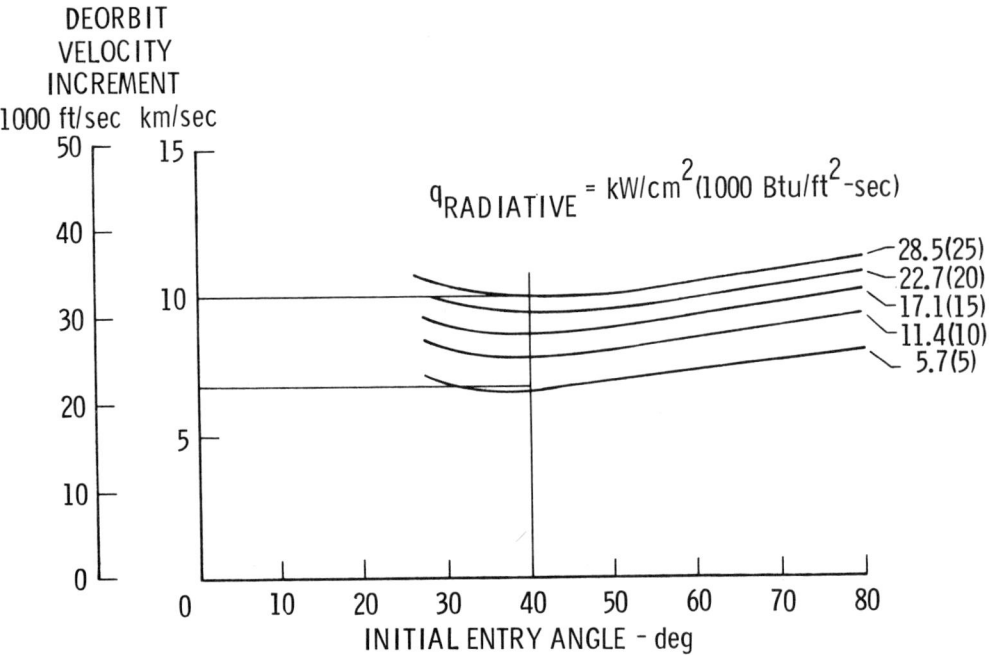

Fig. 13 Planetary Entry Simulation Performance Requirements

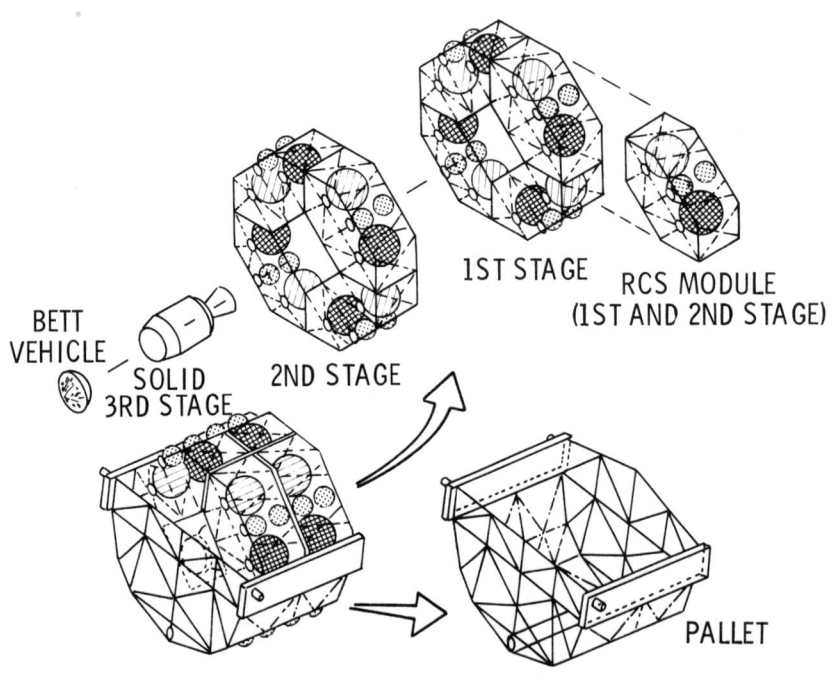

Fig. 14 Entry Technology System — Bett Vehicle

being studied by the DOD and other propulsion systems to be available in the 1980 period.

Alternate Upper Stage. Each of the upper stages presently being considered will require a large portion of the orbiter payload volume. In an attempt to reduce the large cost of dedicated (due to upper stage size) launched entry technology payloads, a study was performed to identify small alternate propulsion systems which could provide the high ΔV required for radiative flow-field simulations and other high-energy missions such as ICBM entry simulation. The system defined is compact, requiring less than 3 meters of payload bay length, high in performance, providing up to 9 km/sec ΔV to 100 kg, low in cost, utilizing orbiter ACS and RCS components thereby eliminating DDT and E, and modular, creating a multimission capability. The velocity package, Figure 14, is a three-stage system; the first two stages consist of four orbiter component propulsion modules.[16] These modules, as shown in Figure 15, can be grouped alone or in sets of two, three, or four; sets of four are then staged. The third stage is a small solid rocket motor. This propulsion system concept has been recommended by Langley Research Center for further study as part of the agency's advanced propulsion system studies.

CONCLUSIONS

As part of an overall NASA program, the Langley Research Center has undertaken a research and development effort aimed at acquiring flight data for substantiation of research and design methodologies for entry and high-speed flight technologies. The flights of the Space Shuttle Transportation System and the reentry of the orbiter will provide the opportunity to obtain these flight data.

Technology experiments have been identified in the areas of aerodynamics, thermodynamics, material/structures, thermal protection systems, and flight systems. These experiments will utilize both the orbiter and vehicles launched from the orbiter (ballistic and lifting).

The STS orbiter and launched payload experiments will provide verification and correlation between ground analyses and test methodologies

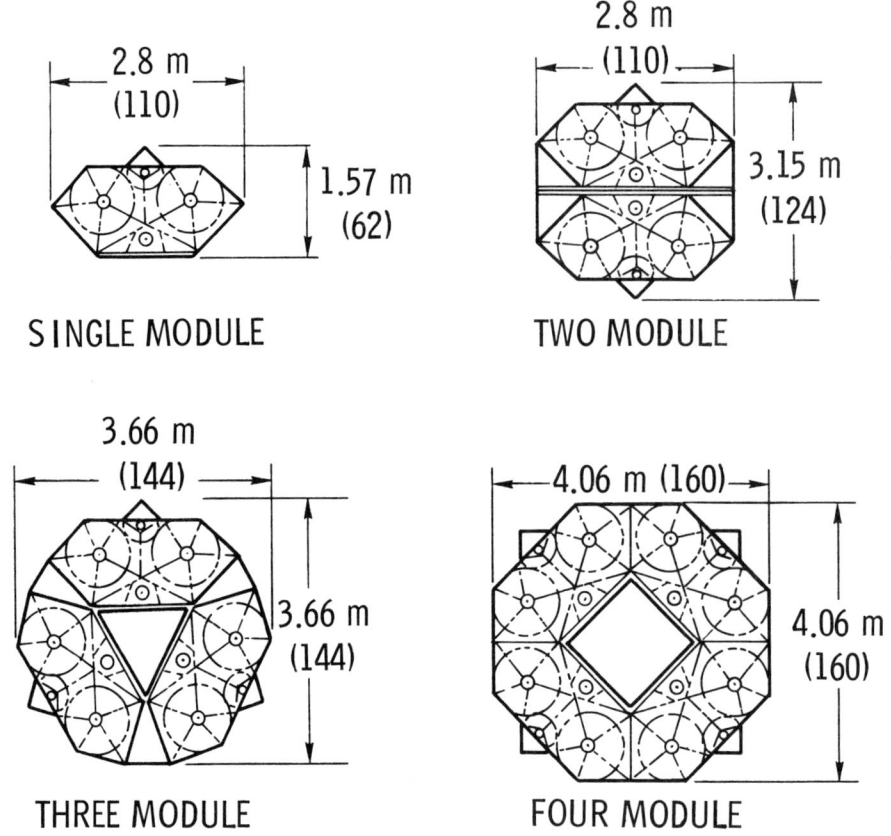

Fig. 15 Proposed Velocity Package: Grouping Arrangement

and provide for the establishment of extrapolation techniques from ground to flight, thereby reducing design uncertainties generally associated with flight vehicles. Specific orbiter experiments will provide data for the optimization and enlargement of the orbiter performance envelope through improved systems and decreased design uncertainties.

The feasibility of an orbiter launched payload experiment program including a multimission ballistic entry technology test vehicle has been demonstrated. These launched payload programs will provide payloads for experiments requiring flight environments out of the orbiter envelope and resolve geometry peculiar phenomena.

The proper implementation of the Shuttle Entry Technology Experiments Program will result in a cost-effective research program that will eliminate design uncertainties relative to the multifaceted problems of entry (i.e., structures, aerodynamics, and thermodynamics). It will also result in the flight qualification of advanced materials, structures, flight system concepts that will create lighter, more reliable engineering systems and thereby optimize payloads.

The incorporation of such a program into the Shuttle Payload Program will provide NASA the flight test capability it requires to continue to move forward in aeronautics and space research.

REFERENCES

1. Final Report of the Space Shuttle Payload Planning Working Group, Space Technology, Vol. 10, NASA GSFC, May 1973.

2. James C. Dunavant and David A. Throckmorton, "Aerodynamic Heat Transfer to RSI Tile Surfaces and Gap Intersections," Journal of Spacecraft and Rockets, Vol. 11, No. 6, June 1973, pp. 437-440.

3. David A. Throckmorton, "Pressure Gradient Effects on Heat Transfer to Reusable Surface Insulation Tile-Array Gaps," NASA TN D-7939.

4. David A. Throckmorton, "Heat Transfer to Surface and Gaps of RSI Tile Arrays in Turbulent Flow at Mach 10.3," NASA TM X-71945.

5. Jerry N. Hefner and Allen H. Whitehead, Jr., "Lee-Side Heating Investigations. Part I - Experimental Lee-Side Heating Studies on a Delta-Wing Orbiter," NASA Space Shuttle Technology Conference, NASA TM X-2272, 1971, pp. 267-284.

6. Jerry N. Hefner and Allen H. Whitehead, Jr., "Lee-Side Flow Phenomena on Space Shuttle Configurations at Hypersonic Speeds. Part II - Studies of Lee-Surface Heating at Hypersonic Mach Numbers," Space Shuttle Aerothermodynamic Technology Conference, Vol. II, NASA TM X-2507, 1972, pp. 451-467.

7. Jerry N. Hefner, "Lee Surface Heating and Flow Phenomena on Space Shuttle Orbiters at Large Angles of Attack and Hypersonic Speeds," NASA TN D-7088.

8. George L. Ball II, James W. Leffingwell, and Dennis W. Werkmeister, "High Temperature Polyimide Foams for Shuttle Upper Surface Thermal Insulation," CR-132572.

9. Herman L. Bohon, Wayne J. Sawyer, Roane L. Hunt, and Irving Weinstein, "Performance of Full Size Metallic and RSI Thermal Protection Systems in a Mach 7 Environment," AIAA Paper No. 75-800, May 1975.

10. K. R. Carnahan, G. J. Hartman, and G. J. Neures, "Thermal and Heat Flow Instrumentation for the Space Shuttle Thermal Protection System," Presented at the 20th National Symposium of the Instrument Society of America, May 21-23, 1974, ISA ASI 74226 (123-132).

11. James F. Campbell, "Augmentation of Vortex Lift by Spanwise Blowing," AIAA Paper No. 75-993, August 1975.

12. John L. Shidler, Herman L. Bohon, and Bruce E. Greene, "Evaluation of Bead-Stiffened Metal Panels," AIAA Paper No. 75-815, May 1975.

13. James J. Murray and Fred R. DeJarnette, "An Emergency Astronaut Reentry Parachute System" International Astronautical Federation XXVX Congress, Amsterdam, September 30-October 5, 1974.

14. Walter B. Olstad, "Planetary Entry Aerothermodynamics. Part 1 - Technology Requirements and Experimental Facilities," Astronautics and Aeronautics, November 1974, pp. 58-69.

15. Walter B. Olstad, "Planetary Entry Aerothermodynamics. Part 2 - Computational Analyses and Flight Experience," Astronautics and Aeronautics, December 1974, pp. 55-61.

16. W. R. McNeilly, "Advanced Shuttle Payloads Sizing Study," CR-132-714.

AAS 75-278

LANDMARK TRACKING TECHNOLOGY[*]

J. D. Welch[†], W. E. Sivertson[†], and R. G. Wilson[†]

INTRODUCTION

This paper addresses the preliminary design of an experiment to demonstrate a technique for automatically recognizing and tracking known Earth landmarks. This experiment is designed for Shuttle flight on board the Advanced Technology Laboratory, (ATL).[1] The utility of such an experiment can be of key significance for performing these functions for future applications, especially those related to earth resources investigations:

1. Providing precision data for vehicle attitude determination and/or payload pointing control systems relative to earth surface references.

2. Providing "real time" interpretation of Ground Control Point (GCP) data.

3. Providing key fix data for an autonomous navigation system.

JUSTIFICATION OF EXPERIMENT

This experiment makes use of the technique of coherent optional correlation via matched filtering to do landmark recognition, pointing, and tracking without requirement of precise a priori knowledge of spacecraft attitude or orbital parameters, or precise knowledge of earth mapping. Presently, in earth resources missions semi-automated ground-processing procedures are typically required to transform image data into a known ground coordinate system or to register one image with another. The concepts of the presently proposed ATL experiment provide for pointing of instrumentation coupled to the correlator to specific landmarks, e.g., ones on which repeated orbit-to-orbit measurements are

---

[*] The work reported herein was supported, in part, by the National Aeronautics and Space Administration under Contract NAS1-12550.

[†] General Electric Space Division

[‡] NASA-Langley Research Center, Hampton, Va.

sought, permitting a real-time match of landmarks and comparison of measurements.

In addition to providing a tool for landmark pointing, tracking, and repeat-observation parallel optical data processing techniques can ultimately provide basic instrumentation needed to perform combined on-board real-time spatial-frequency analysis, spectral analysis and spatial frequency filtering of data for purposes of recognition and classification of types of earth surface materials. As a further long-term development, the proposed ATL experiment can serve as the basis for designing autonomous navigation systems for spacecraft, based on landmark and starfield recognition and tracking.[2]

RECOGNITION AND TRACKING BY COHERENT OPTICAL PROCESSING

Coherent optical data processing with matched spatial frequency filters is the basis of the proposed method for recognizing and tracking. The matched filters are pre-made transparencies on which images of the selected landmarks are stored in the Fourier Transform (spatial frequency) domain. Mathematically, a matched spatial frequency filter is a two-dimensional Fourier Transformation of a conventional optical image into the optical information domain of spatial frequencies. The physical mechanism by which this is accomplished is the phenomenon of optical diffraction in coherent illumination and the matched spatial frequency filters are a special type of recorded diffraction spectra. Like many pattern recognition processes, this process is a two step process: filter synthesis and application of the matched filters in the pattern recognition process. In the present experiment a file of matched filters will be made based on available satellite photography in an Earth located laboratory. That file of filters will be taken aboard the ATL for experiments intended to achieve a correlation between candidate images obtained by telescope and one of the stored filters.

SPECIFIC OBJECTIVES OF PLANNED EXPERIMENT

1. To establish feasibility of using pre-made matched spatial frequency filters for recognizing and obtaining the vector direction to landmarks.

2. To achieve Objective 1 for a range of landmark types ranging from high contrast ones to those with less well defined features.

3. To perform the recognition and tracking functions listed under Objective 2 for selected landmarks over a wide range of angular viewing conditions and over a range of illumination, obscuration and atmospheric conditions.

4. To evaluate the effectiveness of man's role in the experiment.

5. To evaluate automated coherent optical processing in a satellite.

6. To evaluate fabricating matched filters in a satellite.

7. To evaluate non-coherent to coherent optical conversion devices.

IDENTIFICATION OF FLIGHT EQUIPMENT

The principal flight components dedicated specifically to this experiment are: viewing window, input telescope optics and tracking mirror, optical-to-optical interface device, exposure-control device, laser, Fourier transform optics, matched spatial frequency filter selection and storage device, an electro-optical correlation readout device, a display and control console and angle readout encoders and control servos for the tracking mirror. The experiment package will include a rigid structural form within which all of the components will be mounted. The frame assembly will also include light-tight covers so the coherent optical experiment can proceed without necessity of darkening the interior of the ATL. A plan view of the experiment package is shown in Figure 1.

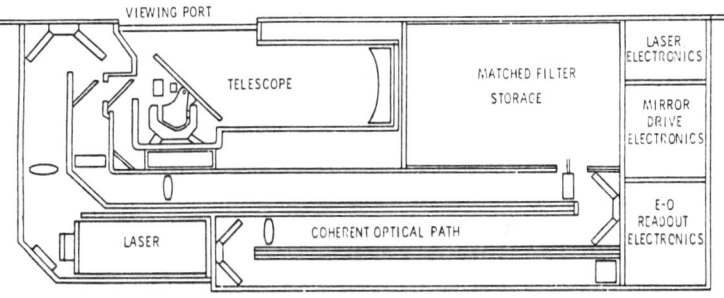

Fig. 1 A Plan View of Experiment Package

REFERENCE

1. Study of Shuttle - Compatible Advanced Technology Laboratory/ATL, NASA TMX2813, Sept 1973.

2. Holeman, J. M. and Welch, J. D., "The Application of Spatial Filter Techniques to Precision Autonomous Space Navigation," 17th International Astronautical Congress, Madrid, Oct 1966.

AAS 75-252

SHUTTLE COMMUNICATION EXPERIMENTS

John J. Woodruff* and Donald R. Peters+

Requirements exist for the development of Communication and Navigation systems during the Shuttle era, 1980-1991. These system requirements include links for inter space, space to ground, and ground to space communications; and are characterized by high data rates, multiple terminals/users, and long distances. Some of these requirements are directly associated with Shuttle, such as establishing communication links with the Tracking and Data Relay Satellite (TDRS) system; while others involve applications which will ultimately be implemented via dedicated satellite systems.

NASA is presently sponsoring Communication/Navigation Shuttle experiments in the areas of telecommunications systems and technology, data collection, and mobile systems utilizing radars, lasers, and interferometry to develop technology for commercial use. These experiments can be grouped into three categories: experiment development and demonstration to provide capability for other experiments (large antenna systems, laser communications, attitude determination systems), experiments operated to obtain data for users (search and rescue missions, electromagnetic mapping, and migratory studies) and synergistic payloads obtained by grouping together several experiments involving radar and radiometry to obtain simultaneous measurements which extend the individual data content and take advantage of common equipment.

A survey description of some of these activities presently being pursued by NASA is provided. Topics to be discussed include Adaptive Multibeam Antenna, Electromagnetic Environments, Microwave Multi-Applications Payload, Millimeter Wave Large Aperture Antenna, Bandwidth Compressive Modulation, and a Traveling Wave Tube Open Envelope Experiment.

---
*NASA Goddard Space Flight Center, Greenbelt, Md.
+General Electric Space Division, Valley Forge, Pa.

Requirements exist for the development of Communication and Navigation systems during the Shuttle era, 1980-1991. These system requirements include links for inter space, space to ground, and ground to space communications; and are characterized by high data rates, multiple terminals/users, and long distances. Some of these requirements are directly associated with Shuttle, such as establishing communication links with the Tracking and Data Relay Satellite (TDRS) system; while others involve applications which will ultimately be implemented via dedicated satellite systems.

The availability of Shuttle provides low cost/frequent access to space to develop the required Communication/Navigation technologies and exploit their use for operational applications. As a platform for space activities, Shuttle offers the advantage of large weight and volume, the availability of man, and the opportunity for reflight. These advantages, coupled with the global coverage and low attenuation factors associated with space experimentation, provide much of the justification for conducting Communication/Navigation experiments involving large antennas and lasers on Shuttle.

NASA is presently sponsoring Communication/Navigation Shuttle experiments in the areas of telecommunications systems and technology, data collection, and mobile systems utilizing radars, lasers, and interferometry to develop technology for commercial use. Many of these experiments utilize Spacelab to take further advantage of the availability and versatility of man. A listing of these Communication/Navigation experiments presently being pursued is shown below.

COMM/NAV EXPERIMENTS

Electromagnetic Environment Experiment (EEE)
CO_2 Laser Data Relay Link (Shuttle Laser Transceiver)
Large Reflector Deployment
Open Traveling Wave Tubes

Millimeter Wave Large Antenna Experiment (MWLAE)

Stars and Pads Experiment

Interferometric Navigation and Surveillance

Shuttle Navigation via Geosynchronous Satellite

Adaptive Multibeam Antenna

Precise and Accurate Time and Time Interval (PATTI)

ND: YAG Laser Transmitter

ND: YAG Laser Receiver

Bandwidth Compressive Modulation

Data Collection via Multibeam Array Antenna

Antenna Range Experiment

Coast Guard Mission

Interferometer Attitude Determination

Search and Rescue

NAVSTAR Global Positioning System

Microwave Multi-Application Payload (MMAP)

From the experiments listed above, five (5) have been selected for further discussion based on their state of development and priority.

Electromagnetic Environment Experiment

The objective of this experiment is to measure, map, and evaluate incident earth-emitted electromagnetic radiation at frequencies allocated for space use over the frequency range from 0.4 to 100 GHz. The experiment approach involves utilizing a set of four (4) different antennas and nine (9) receivers to cover the specified frequency band. Coverage is obtained by dividing the area to be mapped into a grid structure which is then swept by the antennas.

TWT Open Envelope Experiment

The objective of this experiment is to evaluate tube life by opening high power amplifier tubes to the space environment so that the collector is cooled by direct radiation into space, and the out-gassing products are removed from the immediate vicinity of the tube. Operation of the experiment is performed on an extendable boom which incorporates a protective shield against spacecraft contamination.

Millimeter Wave Large Aperture Antenna

This experiment is aimed at developing and demonstrating the feasibility of a large aperture (6-10 meters), wide bandwidth (1500 MHz), modularly deployed, electronically steerable millimeter wave antenna system. Design studies have investigated three applications areas: communications link, radar array, and radiometer array. General characteristics for these applications have been developed.

Bandwidth Compressive Modulation

The objective of this experiment is to measure the statistical and operational performance characteristics of experimental communication links that use coded or uncoded multiple phase and amplitude shift keyed digital modulation techniques to conserve RF bandwidth while increasing data rate. The experiment approach involves establishing one-way links from the Spacelab to ground station, and two-way data links between ground station via transponders in Spacelab. Specific types of test data will be transmitted for the purpose of measuring error characteristics. The links will use various modulation forms including uncoded or coded phase shift keying, multiple phase shift keying, and multiple phase and amplitude shift keying.

Microwave Multi-Applications Payload

The goal of MMAP is to develop an integrated set of specific microwave equipments into a synergistic payload capable of being utilized for many experiments in multiple disciplines such as Communication/Navigation, Weather and Climate, Earth Resources, and Earth and Ocean Dynamics. The key to the concept is the use of common equipment and time phasing the operation. A preliminary design employs three (3) antenna clusters to satisfy the requirements of twelve (12) different experiments on a single Shuttle mission.

AAS 75-253

LARGE DEPLOYABLE ANTENNA SHUTTLE EXPERIMENT[*]

R. E. Freeland[+], J. G. Smith[+],
J. C. Springett[+], and K. E. Woo[+]

INTRODUCTION

The Large Deployable Antenna Shuttle Experiment (LDASE) is a proposed program to evaluate the mechanical and electrical properties of a furlable 30 m diameter spaceborne antenna under zero-gravity conditions. The Shuttle will be used to transport the furled antenna into low earth orbit, where it will be deployed and tested. The integrity of the reflector surface will be measured using an amplitude-modulated laser technique. A spin-stabilized RF beacon, also carried by the Shuttle as part of the LDASE, will be ejected prior to antenna unfurlment and positioned in the same orbit to accommodate electrical performance testing of the antenna.

The LDASE has now reached the detailed planning stage. This paper assesses the future needs for large-scale spaceborne antenna systems, describes the proposed experiment, and explains the rationale on which it is based.

USER NEEDS FOR LARGE SPACEBORNE ANTENNAS

A number of space systems either require or could greatly benefit from the availability of large-aperture spacecraft-mounted antennas. Such large antennas will be needed in the fields of planetary probes, radiometry, communica-

---

[*] This paper presents the results of one phase of research carried out at the Jet Propulsion Laboratory, California Institute of Technology, under Contract NAS7-100, sponsored by the National Aeronautics and Space Administration.

[+] R. E. Freeland is a Member of the Technical Staff, Applied Mechanics Division; J. G. Smith is Group Supervisor, Applied Communications Research, Telecommunications Division; J. C. Springett is Manager for Research and Advanced Development, Telecommunications Division; K. E. Woo is Group Supervisor, Spacecraft Antennas, Telecommunications Division, all at the Jet Propulsion Laboratory, Pasadena, California.

tions, RFI detection, radio astronomy, and energy transmission. More generally, these classes of users can be categorized into (1) communications information transfer or (2) observation. Figure 1 shows how some of the present and projected space user requirements divide into these two categories as a function of user frequency and antenna aperture.

EXPERIMENT DESCRIPTION

The development of current structural concepts and designs with respect to performance, cost, weight, and packaging efficiency for large unfurlable antennas will result in hardware without sufficient stiffness or strength for a meaningful ground test program. As a result, the antenna system must be evaluated in the zero-gravity field for which it was designed to operate. Hardware tests and a demonstration will also be required for optimization of the final mechanical concept selected. The Shuttle-Spacelab system provides the means by which experimental verification can be economically undertaken.

The LDASE is intended to define, develop, and implement the mechanical and RF functional qualification testing of a large-aperture deployable reflector antenna using the Shuttle-Spacelab system. The recommended antenna size is 30 m for use at S-, X-, and K-band frequencies. An artist's conception of the mechanical configuration and deployment sequence for the LDASE is illustrated in Fig. 2.

The mechanical hardware demonstration will accomplish (1) unfurling and furling of a 30-m-diameter parabolic antenna, and (2) measurement of the reflector surface accuracy under varying orbital and thermal conditions. The reflector surface measurement, required to establish surface quality, is made by using an amplitude-modulated laser, mounted at the antenna base, in conjunction with a mechanically actuated scanning mirror located near the focus of the reflector and mounted on the deployable feed support structure.

The objectives of the RF tests are to evaluate the electrical performance of the LDASE antenna. The test utilizes a spin-stabilized RF beacon as an energy source. The beacon, positioned at a fixed distance from the Shuttle and in the same orbit, is a self-contained and independent system that is launched, prior to antenna deployment, from a spin table located in the Shuttle payload compartment.

The basic objective of a user demonstration is to verify in the space environment the applicability of a large reflector antenna for one or more specific RF experiments. It is not appropriate at this time to select the candidate for a user demonstration because (1) all potential users with current needs have not established firm enough requirements, (2) some potential users appear unaware of the current capability and the projected state of the art for large furlable antenna technology during the early 1980's, and thus requirements which could develop with a knowledge of such antenna capability have not been identified, and (3) potential users whose requirements will evolve between now and the late 1970's must be given serious consideration.

SUMMARY AND PROSPECTIVE

The need for development of large furlable antenna systems has been established. A Shuttle-Spacelab experiment to evaluate and demonstrate this technology has reached the detailed planning stage. The future objectives are (1) the complete implementation of the Spacelab experiment, (2) the solicitation, evaluation, and development of user requirements for an appropriate demonstration, and (3) the development of a mechanical and functional interface of the LDASE with the Shuttle-Spacelab system.

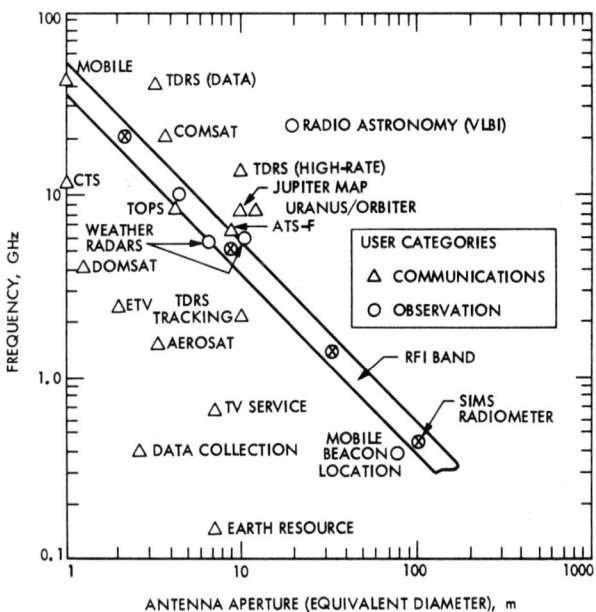

Fig. 1 Large Spaceborne Antenna Users as a Function of Frequency and Antenna Size

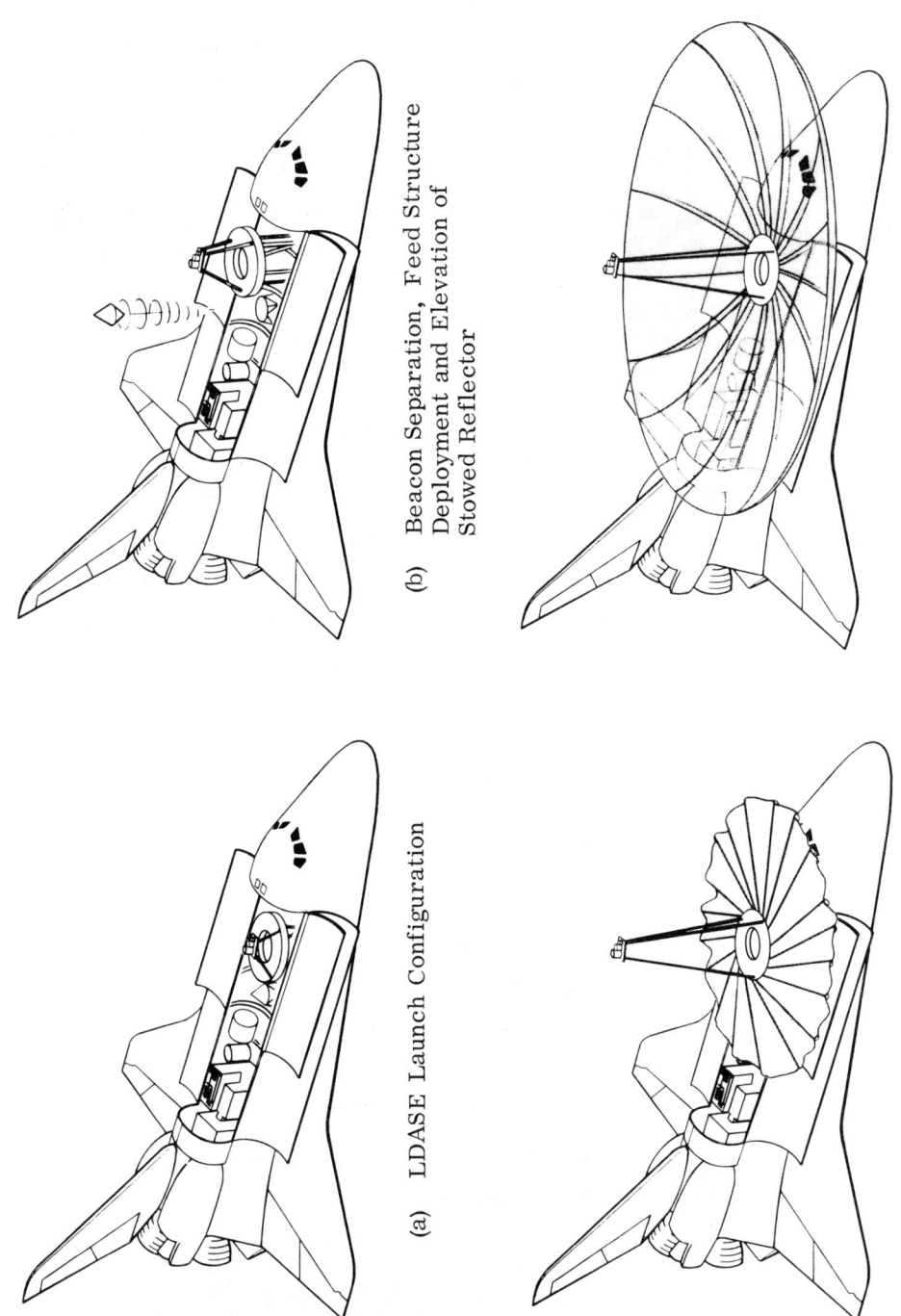

Fig. 2 Artist's Conception of the Mechanical Configuration and Deployment Sequence for the LDASE

(a) LDASE Launch Configuration

(b) Beacon Separation, Feed Structure Deployment and Elevation of Stowed Reflector

(c) Partial Reflector Deployment

(d) Fully Deployed LDASE Configuration

AAS 75-254

AVAILABILITY OF A COMMUNICATIONS SATELLITE, REQUIREMENT AND FEASIBILITY*

Gary D. Gordon†

INTRODUCTION

An estimated 45 communications satellites, including 9 international, 10 domestic U.S., 12 foreign, 3 for aviation, 3 for maritime carriers, 2 for disaster warning, 3 for TDRS, and 2 for R&D, will be in the geostationary orbit in the 1980's. This estimate does not include DOD satellites. Other estimates differ in detail, but yield approximately the same total. In addition, some earth observations satellites will have similar requirements.

Typical subsystem failures in communications satellites are classified in terms of reliability in Table 1. A striking feature is the large number of "design" failures, which occur early in the life of the satellite. In these cases a failure analysis shows that the initial reliability was less than planned, either due to an actual design error or due to quality control. On the basis of failure statistics, it is suggested that each new program can expect an average of two design failures, of which one may jeopardize the mission. On the average a design failure appears about one year after injection of the first satellite. An additional year or two is required to identify the cause and to procure necessary replacements.

AVAILABILITY REQUIREMENTS

There are no clear-cut requirements for availability, although past achievements may serve as a guide. From 1970–1974 the annual availabilities achieved from one U.S. earth station to another ranged from a low of 99.91 percent in 1970 to a high of 99.99 percent in 1972. An availability of 99.99 percent is equivalent to a 1-hour outage during the course of a year. Such an availability is a reasonable goal for many systems.

─────────────

*This paper is based on work performed at COMSAT Laboratories under Contract NAS 8-30849.

†Dr. Gordon is a Senior Staff Scientist with COMSAT Laboratories, 22300 COMSAT Drive, Clarksburg, Maryland 20734.

Table 1.

TYPICAL SUBSYSTEM FAILURES OF COMMUNICATIONS SATELLITES

Satellite	Component Failure	Type
COURIER	Decoder	Design
TELSTAR	Decoder	Design
	Battery	Random
RELAY	Power conditioning	Random
SYNCOM	Telemetry	Random
Early Bird	Fuel depletion	Wear-out
NIMBUS	Solar array bearings	Design
ATS-5	Attitude control	Design
TACSAT	Structural bearings	Design
DSCS-2	Deployable structures	Design
TELESAT	Power conditioning	Random
INTELSAT II	Battery	Random
	Propellant feed	Design
	Propellant relief values	Design
	Solar array degradation	Design
INTELSAT III	Structural bearings	Design
	Low orbit	Random
	Battery	Random
	Receiver	Design
	Transponder	Random
	Earth sensor	Design
INTELSAT IV	Receiver	Design
	Thruster	Design
	Earth sensor	Random
	Telemetry beacon	Random

It would be useful to know the financial value of increased availability. Although this value cannot be defined precisely, an order of magnitude estimate is $10M/year for one satellite. This order of magnitude estimate is based upon the willingness of various systems to pay for in-orbit spares. A similar estimate is obtained by considering that one country is willing to pay $3.5M/yr for "non-interruptible" service (as opposed to the $1M/yr which it would otherwise pay).

ACHIEVING AVAILABILITY

Many steps have been taken to ensure high availability in present communications satellite systems. These include quality control, testing, redundancy, and failure warning. When the satellite ultimately fails, the system is maintained by replacing satellites.

Four methods are suggested to increase the availability of a single satellite to 99.99 percent:

a. There continue to be a few problem areas in communications satellites. Notable examples are nickel-cadmium batteries, mechanical bearings, and traveling wave tubes. Research has been done on alternative techniques, such as nickel-hydrogen batteries, magnetic bearings, and solid-state amplifiers. While some advantages of these new developments have been publicized, their effect on satellite reliability has not been given sufficient weight.

b. To correct for design failures, the first satellite of a new system should be launched two or three years before the system is needed. Hence, any design defects can be detected and corrected in the remaining satellites. At this point the temptation to add other improvements must be resisted.

c. To achieve maximum improvement from redundancy, failures must be independent. When redundant subsystems are procured from the same manufacturer, they are based on the same design and usually have components from the same lot. If 10 percent of the failures are correlated so that they occur on all redundant components, the maximum increase in reliability is one "9" (unreliability decreases by a factor of 10). This is true regardless of the number of components in parallel. For components with a 0.99 or 0.999 reliability, redundancy with 10-percent correlation, even with four redundant boxes, is not as good as twofold redundancy with 0-percent correlation. Twofold redundancy from the same manufacturer is often justified on the basis of cost savings. However, for additional reliability, a second manufacturer should be used.

d. A fourth method of increasing availability is unmanned module exchange at geostationary orbit. A number of studies have proved that this is technically feasible and has cost advantages. Studies have shown that satellite availabilities of the order of 99.99 percent can be achieved. Costs can be minimized by servicing a number of different satellites with one free-flying servicer. Satellite reliability can be maximized by replacing design failures and failed redundant components.

While these methods are not new, they have not yet been implemented to full advantage.

AAS 75-268

MISSION CONSIDERATIONS FOR MULTIDISCIPLINE APPLICATIONS PAYLOADS*
John M. Macdonald†

INTRODUCTION

Space Shuttle sortie missions may combine a number of payload elements on a single mission of 7 to 30 days duration. Applications payloads on these missions could include those flown to acquire direct earth-related operational data, or to aid in the development of remote sensing and communications equipment and operating techniques for subsequent use with automated satellite systems. Such missions could also include payloads requiring a low-g environment such as Space Processing or Cloud Physics.

Combining several sets of these types of payload equipment on a single mission, such as the example group shown in Figure 1, allows sharing mission costs and more efficient use of mission resources. This combined payload set introduces some mission considerations of particular significance to such multidiscipline payloads.

PAYLOAD	MAJOR EQPT	PALLET–MOUNTED EQPT	ORENTATION POINTING TARGETS	OPERATING REQT		REMARKS
				DURATION HR/OPER	FREQUENCY OPER/DAY	
ELECTROMAGNETIC SURVEY	TWO CLUSTERS OF ANTENNA RECEIVERS, CONTROLS, BOOMS & GIMBALS		EARTH FIXED POINTED TO NADIR CONUS TARGET 24 HR COVERAGE	0.1	11 PER DAY	ONCE PER ORBIT OVER CONUS 7 HR/MISSION
EARTH RESOURCES IMAGING RADAR	PHASED ARRAY ANTENNA 3m x 10m TRANSMITTERS, RCVR CONTROLS, RECORDERS		EARTH FIXED SIDE-LOOKING OCEAN – LAND TARGETS	1.5	2 PER DAY	1/2 ORBIT TARGETS 2 PER 1 SHIFT 15 HR/MISSION
MICROWAVE RADIOMETER	ONE ANTENNA ARRAY 2m x 3m FOUR IR RADIOMETERS OBSERVATION TELESCOPE, RECEIVERS & CONTROLS		EARTH FIXED POINTED TO NADIR OCEAN TARGETS IN DAYLIGHT	1.0	7 PER DAY	1 OBSERVATION/ORBIT ON 7 CONSECUTIVE ORBITS 35 HR/MISSION
TWT OPEN ENVELOPE	OPEN ENVELOPE AMPLIFIERS 30m BOOM CONTROLS & DISPLAYS		ANY ORIENTATION POINTED TO COLD SPACE ON BOOM	4.0	6 PER DAY	ALTERNATELY OPERATE 2 EXPER CONTINUOUSLY 112 HR/MISSION
SPACE PROCESSING BIOLOGICAL & FURNACES	ELECTROPHORESIS UNITS FURNACES, FREEZERS, CONTROLS & DISPLAYS	(NONE – EQPT MOUNTED IN RACKS INSIDE MODULE)	ANY ORIENTATION	VARIES 1.4, 1.8, 4.2 & 24.0 CONTINUOUS	15 PER DAY	2 TO 4 OPERATIONS PER EXPIRIMENT 144 HR

Figure 1. Example payload for multidiscipline applications mission.

*This work was performed for the NASA Marshall Space Flight Center, Huntsville, Alabama, under Contract NAS8-29462.
†Advanced Systems Project Engineer, General Dynamics Convair Division, San Diego, California.

These considerations center on meeting the orbital requirements of each payload and providing the maximum possible return for each payload element, considering the competing demands for space and resources that are within the capabilities of the Shuttle/Spacelab systems. The degree of success in sharing resources may be greater than expected after a first review of requirements and incompatibilities of a group of applications payloads — largely because of the potential for sequentially scheduling the on-orbit operations of the various payloads to meet the requirements of each within the envelope of resources available. Sequential scheduling also helps to avoid operational conflicts between payloads.

An example of these mission considerations for a typical multidiscipline applications payload is shown in Figure 2. Mission considerations include the following.

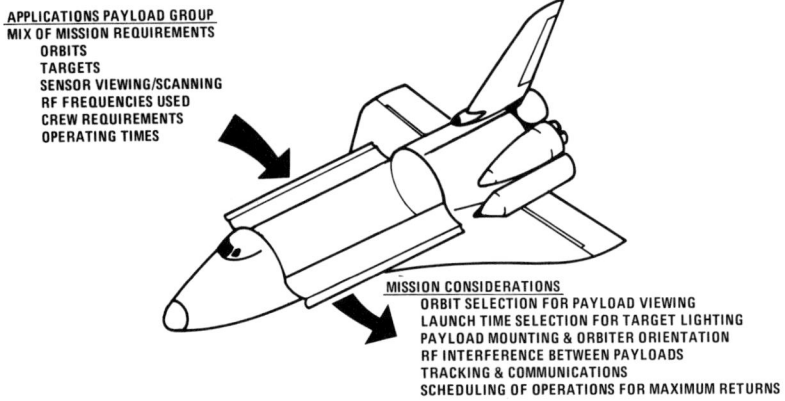

Figure 2. Multidiscipline applications payload missions.

1. Orbit selection to meet multiple payload viewing requirements.
2. Launch time selection for required target lighting and minimal impact of Orbiter thermal constraint.
3. Payload mounting and Orbiter orientation for viewing.
4. RF interference between payloads and with Shuttle RF systems.
5. Tracking and communications.
6. Scheduling of operations considering viewing opportunities, operational interferences and constraints, and availability of resources.

These mission considerations are discussed in the following paragraphs, along with an assessment of special solutions that provide for meeting the requirements of the example combined payload.

ORBIT SELECTION

The orbital parameters for applications payloads are selected to meet payload requirements for viewing earth surface areas and for providing the proper conditions for on-board experiments using the space environment. Selection of orbit altitude provides the proper balance between view swath width and sensor sensitivity or resolution. Altitude selection must also consider conditions such as acceleration levels and the ambient atmospheric (vacuum) pressure. Orbit inclination is selected to provide required global coverage, within the launch site range safety limits, for the viewing payloads. Altitude and inclination selection also considers providing the desired repeating or progression of the orbit track in relation to surface targets or areas.

Orbit selection may be constrained by Shuttle system payload weight capability versus orbit altitude and inclination. For earth-oriented payloads at high inclinations, the beta angle (orbit plane to sun-earth line angle) that will result from launch time selection must be taken into account, since Orbiter thermal constraints may impose limits to attitude hold periods. Other factors to be considered in orbit selection are natural background radiation (which varies with both altitude and inclination) and potential effects on radiation-sensitive films or electronic sensors. Contact times with ground/space communication and tracking networks for transmission of data, voice communications, and tracking for ephemeris determination should also be considered in orbit selection.

In the example case, a nominal orbit altitude of 400 km is readily selected from a review of payload requirements shown in Table 1. This is the only altitude within the minimum/maximum altitude ranges of all payloads and also is the desired altitude for the EM (Electromagnetic) Survey experiment for which altitude is particularly critical to antenna swath width and sensitivity. A nominal orbit inclination of 57 degrees is selected to provide maximum global coverage for the Radiometry and Radar payloads to include northern waters for high incidence of heavy seas and icebergs.

Table 1. Payload orbit requirements.

	Orbit Altitude (km)			
Payload	Minimum	Maximum	Desirable	Limiting Experiment Factor
EM Survey	**400**	463	402	Sensitivity vs Swath Width
Open TWT	360	N/A	550	Ambient Atmospheric Pressure
Radar	170	**400**	200	Resolution vs Swath Width
Radiometer	185	**400**	200	Resolution vs Swath Width
Bio/Furnace	Any	Any	Any	(accel due to aero drag $\leqslant 10^{-5}$ g)
Selected:	400 km (matches min/max requirements of payloads)			

	Orbit Inclination (deg)			
Payload	Minimum	Maximum	Desirable	Limiting Experiment Factor
EM Survey	**55**	65	60	Geographic Coverage
Open TWT	Any	Any	Any	None
Radar	30	90	60	Geographic Coverage
Radiometer	28	**57**	**57**	Geographic Coverage
Bio/Furnace	Any	Any	Any	(no relationship–
Selected:	57 Deg (maximum coverage within KSC range safety limits & min/max requirements of paylads)			

This nominal orbit of 400 km altitude by 57 degrees inclination provides good coverage of continental United States (CONUS) and global ocean and land areas, with an earth track that will progress westwardly each day to cover areas between the first-day orbit tracks shown in Figure 3. This coverage and track progression meets the viewing needs of this example set of payloads. However, this orbit has certain characteristics that warrant further study to improve viewing opportunities or crew activity scheduling. Basically, the orbit is largely repeating in nature, due primarily to orbit altitude, which determines the orbital period causing the repetition. As a result, the time of flight over specific targets will occur at about the same time each day. This is probably of no concern to some payloads such as the

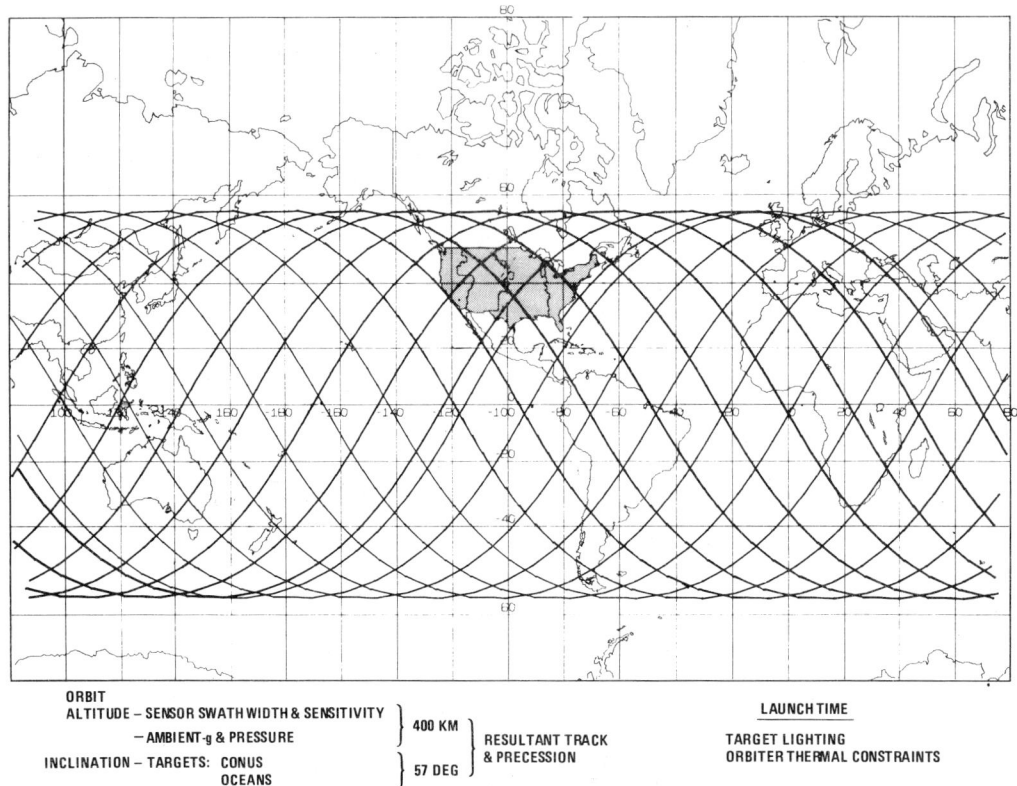

Figure 3. Orbit selection considerations – applications mission.

example Radar and Radiometry experiments, but does affect the capability for temporal coverage desirable for some payloads such as EM Survey equipment. Flight over CONUS will occur on about the first eight orbits each day, with no CONUS contacts on the latter half of each day's orbits. Thus, the coverage will be limited to the same 12-hour period each day. Variations in orbit altitude within limits considered potentially acceptable to payloads (350 km and 450 km) have little effect on the temporal coverage, shown in Figure 4. Variation in inclination will also have little effect, since this condition is primarily a function of orbit period.

LAUNCH TIME SELECTION

Launch time selection is primarily governed by lighting or local time requirements for measurements of the target areas, such as in the example case for providing capability for daylight viewing of ocean areas, desired for visual target selection/viewing and correlation photography. In this case, a launch time of 11:00 a.m. EST will provide daylight viewing of major ocean areas on the first day, which will be repeated each day during the mission. The 12-hour portion of the day to be covered for CONUS viewing on this mission is governed by this launch time selection.

The beta angle history resulting from an 11:00 a.m. EST launch on 21 March is shown in Figure 5. This angle is of importance on this mission because of the need to hold the Orbiter in an earth-oriented attitude for most or all of each mission day. When beta ≥ 60 degrees for this earth-oriented attitude, Orbiter thermal constraints may require periodic maneuvers for

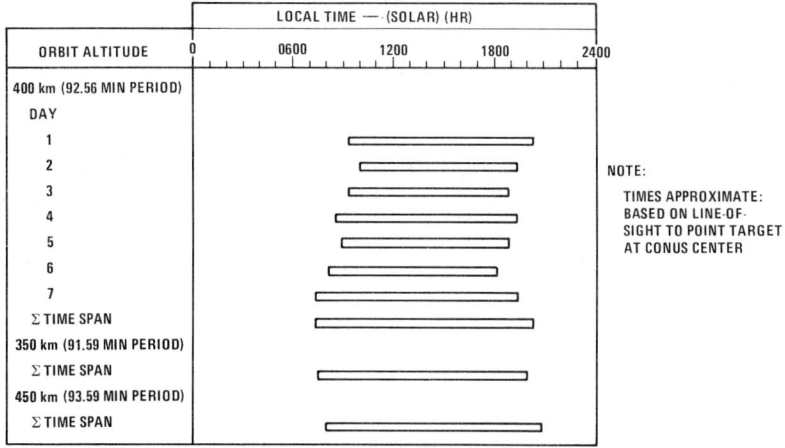

Figure 4. Temporal coverage of CONUS vs orbit altitude.

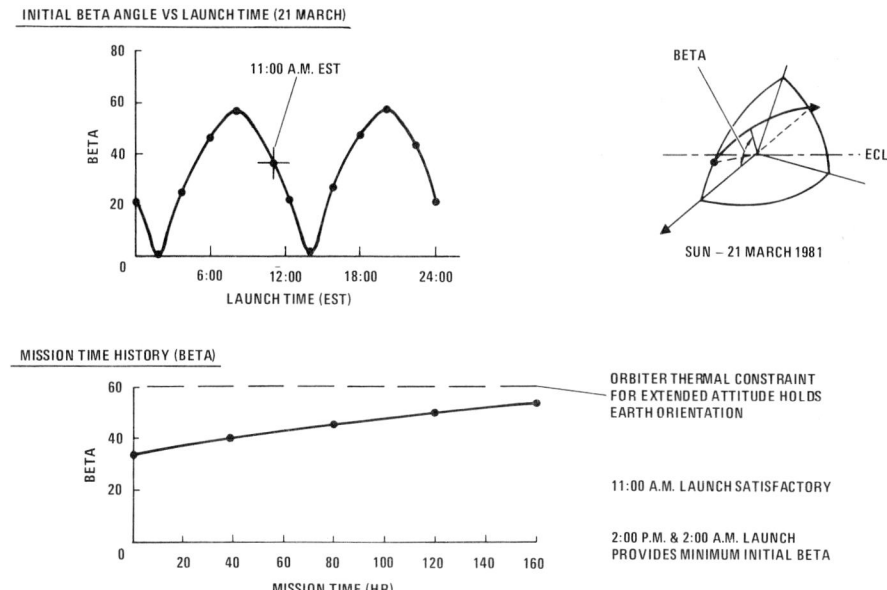

Figure 5. Consideration of Orbiter thermal constraint on launch time selection.

thermal conditioning. The 11:00 a.m. EST launch time results in maintaining beta at less than 60 degrees. The variation of beta for various launch times for a 21 March launch, 57 degree inclination orbit, is shown in Figure 6. These conditions of course vary with time of year of the launch. For this example payload case, further study of launch time appears warranted in order to consider the effect on target lighting of launch times that provide a greater margin with Orbiter thermal constraints.

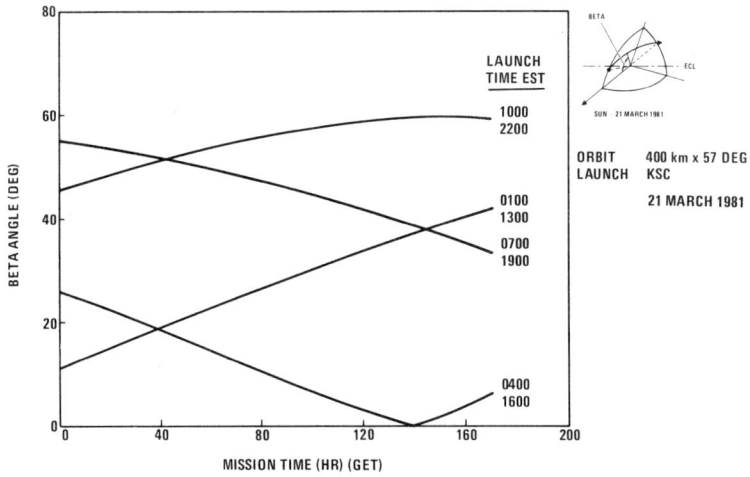

Figure 6. Beta angle history vs launch time.

PAYLOAD MOUNTING AND ORBITER ORIENTATION

Payload viewing requirements and physical size may result in mounting the antennas or sensors such that mission time must be allotted to sensor retracting and deploying, or reorientation of the Orbiter. A retraction/deployment requirement could be particularly severe on mission time if realignment or recalibration of a sensor is required following each deployment.

In the example case shown in Figures 7 and 8, a variety of sensor configurations are mounted on the Spacelab pallet. Each sensor is provided its required field of view and scanning space without the need for either retraction/redeployment or reorientation of the

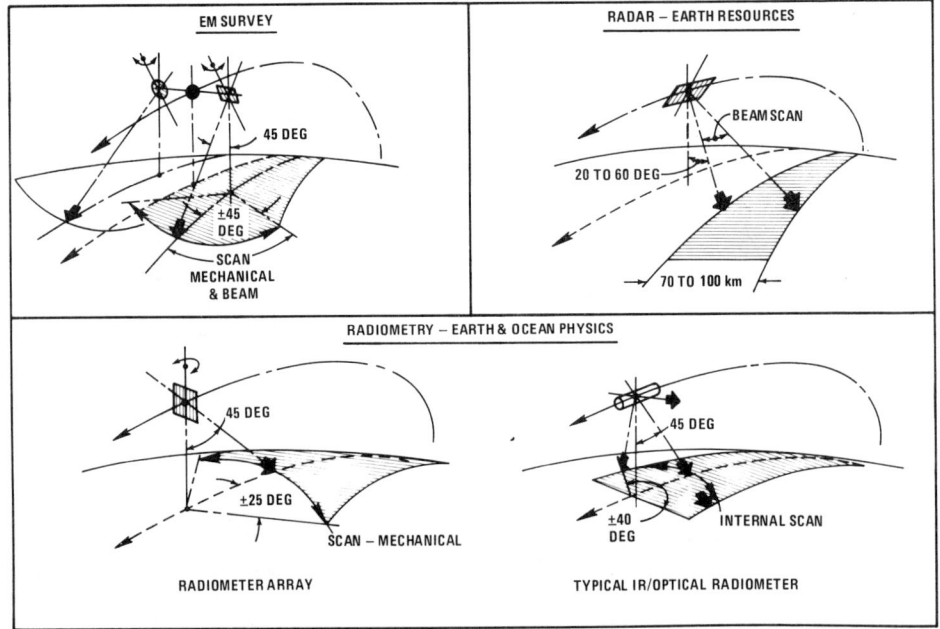

Figure 7. Payload sensor viewing patterns.

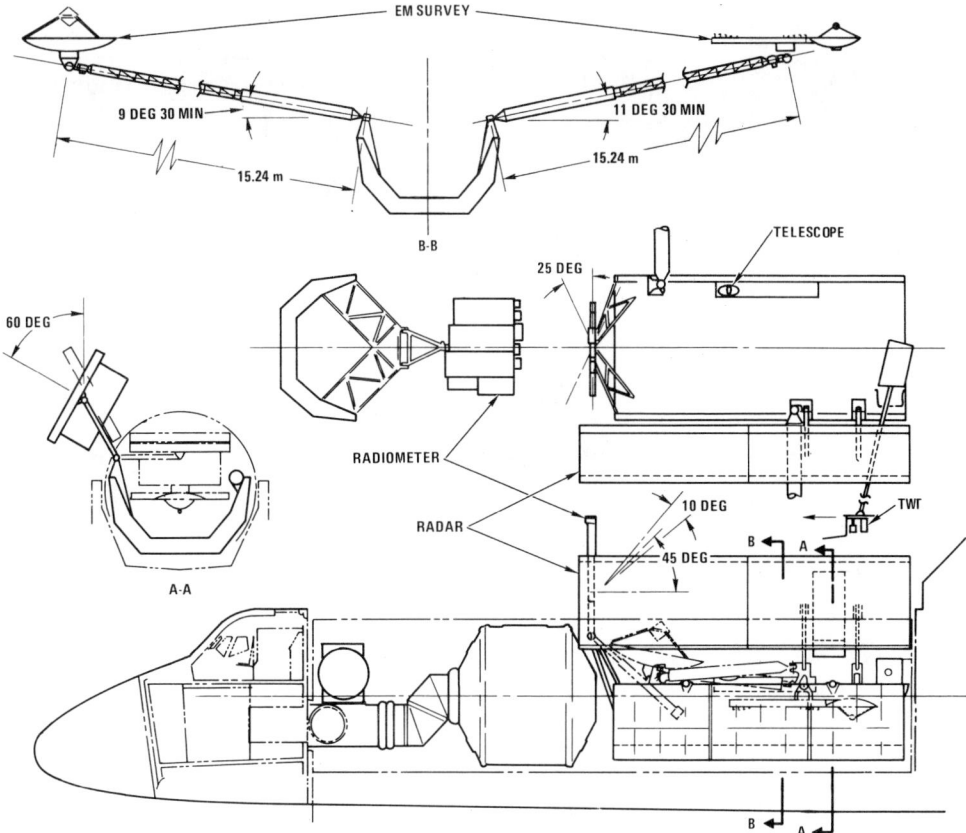

Figure 8. Payload sensor mounting.

Orbiter. One possible problem area lies in the potential that the mounts required for any of these sensors might be mission peculiar to match a certain set of accompanying payloads. Since reflight of the sensor may therefore require a new mount, avoidance of this need for new mounts on each mission may justify retraction/deployment operations.

In the configuration shown, the Orbiter would be oriented for viewing with the open bay pointed to nadir (Z-LV). The longitudinal axis is aligned with the orbit plane (X-IOP), to match the payload mounting selected in this case.

RF INTERFERENCE

The potential for RF interference between payload sensors or equipment is likely to be a significant consideration on many multidiscipline application missions, since many of the systems used for remote sensing and communications will transmit and/or receive RF energy. The Shuttle RF communications also enter into this consideration, particularly where real-time transmission of payload data is desired.

A possibly typical condition, illustrated in Figure 9, exists for the example set of payloads, where a significant conflict in RF transmission/reception is apparent. Possible solutions to these conflicts include sequential scheduling of payload operations, shielding or filtering, or postflight data processing. It may also be necessary to provide for buffer storage and a delay in the transmission of "real-time" payload data by the Orbiter communication system.

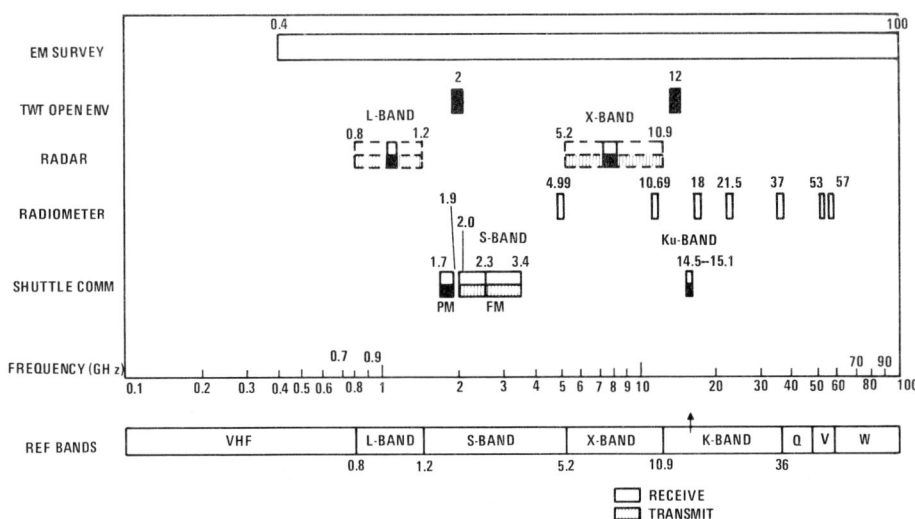

Figure 9. Mission RF system operating frequencies.

In the example set of payloads, target opportunities for the viewing payloads allow sequential scheduling of operations, as discussed later, and buffer storage, and delayed transmission is acceptable to avoid Shuttle RF interference.

TRACKING AND COMMUNICATION

Applications payload mission operations must consider communication — data/command/voice — and tracking requirements for the operational support of on-orbit experiment operations as a factor in orbit selection and operations scheduling.

Payload data may be recorded and, where required, dumped to ground at convenient operational times using efficient transmission rates via tracking data relay satellite (TDRS) or direct to space tracking and data network (STDN) ground stations. Uplink command for payload support or for remote ground control may include real-time voice during some experiment operations. Also, it may be necessary to use the STDN ground subnet for tracking to obtain the required positional accuracies of some payloads.

Figure 10 illustrates communication coverage with the basic Shuttle Ku-band antenna for the example orbit, with the Orbiter maintained in an earth-oriented attitude. The maximum communication gap of 3 hours and 24 minutes results from the occultation of the TDRS by the Orbiter due to the circumstance of Orbiter position and attitude. This coverage is adequate for the example payload since real-time data transmission is not demanded for any payload, although it may result in an impact to operations scheduling on some missions, in which case an additional antenna kit may be desirable.

The potential radio frequency interference (RFI) problem in the operation of the TDRS transmission link (in both S- and Ku-bands) caused by co-occupancy of the spectrum by the experiment and the TDRS transmission bands is dependent upon the level of transmitter power coupled to the experiment receivers and the receivers' sensitivity. The RFI problem is precluded for this example mission by recording the experiment data and avoiding transmission from the Orbiter during experiment operations.

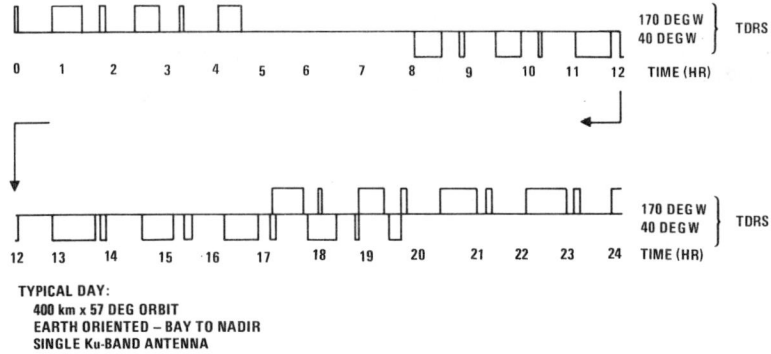

Figure 10. Ku-band TDRS communication opportunities.

OPERATIONS SCHEDULING

Payload operations scheduling should include consideration of viewing opportunities, avoidance of interpayload RF interference, resource availability, and efficient use of these resources. As discussed earlier, sequential operation of the payloads, where viewing opportunities permit, will avoid conflicts in RF interference and ease the competing demands for mission resources.

In the case of the example payload, viewing opportunities and requirements allow this sequential arrangement.

A preliminary schedule of experiment operations for a typical day on this mission is shown in Figure 11. This schedule will be repeated each day with some minor variation as discussed later. The schedule is designed to meet the payload objectives with the following major scheduling considerations.

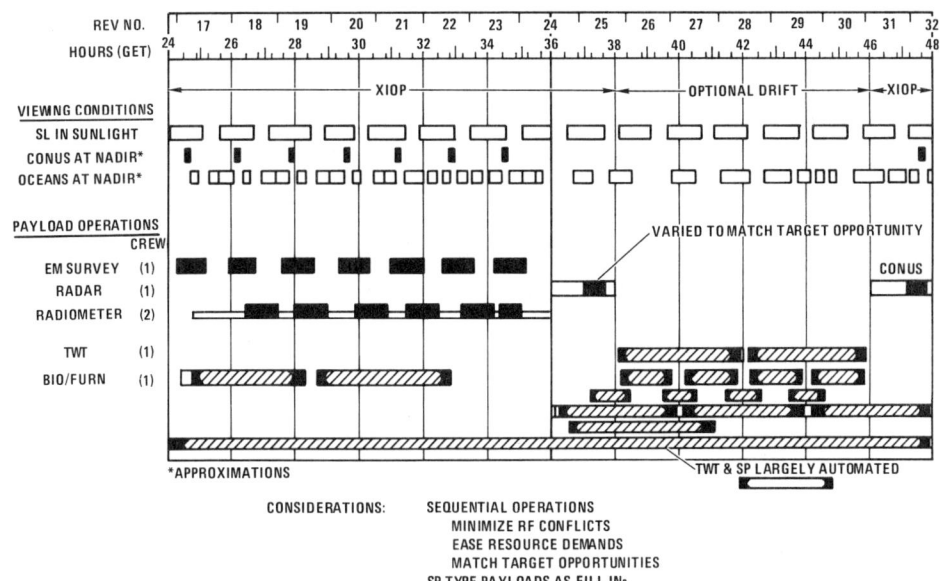

Figure 11. Generalized mission schedule — typical for days two through six.

- Opportunities for viewing CONUS for the EM Survey experiment will occur on each of the first seven to nine orbits each day.
- Operations of the passive/receiving RF experiments (EM Survey and Radiometery) should be scheduled only during periods when no RF transmissions are originating from other payloads or from the Orbiter.
- Operating times for the Radar should be variable throughout the mission to allow operation over a variety of land and ocean areas, plus at least several times over CONUS truth sites.
- Space Processing experiments can be operated at any time during the mission within crew and power/heat rejection constraints, and preferably during periods when accelerations are at or below 10^{-4}g; i.e., no RCS operation (which is estimated to result in about 4×10^{-4}g).

The overall operations schedule for a typical mission day depicted in Figure 11 provides for satisfying these conditions generally as follows.

- Both passive RF experiments are operated on the first shift, when CONUS opportunities exist for EM Survey, while both active RF experiments are operated on the second shift. This avoids the potential for RF emissions from the traveling wave tube (TWT) and Radar payloads interfering with the passive measurements.
- The Radar experiment can be scheduled for operation during any two periods on the second shift to coincide with ground target areas in view. The TWT experiment is operated for two- to four-hour periods alternately with the Radar to avoid creating RF interference with Radar measurements.
- The Space Processing operations are concentrated mainly on the second shift to coincide with crew availability and the potential for low-g periods where reaction control system (RCS) operation can be avoided, since earth orientation is required only during short periods for Radar measurements. Some of the primarily automated type Space Processing operations are scheduled for the first shift, when crew time is available for short intervals to support this type of operation.

Total mission operations are based on repetition of this typical day for all days except during the ascent activation and deactivation/descent phases at start and end of mission. The primary variation from the typical day occurs in the Radar operations and in the effects on other payload operations caused by variations in Radar operating times.

CONCLUSION

As in the above example case, several considerations apply to such multidiscipline applications missions:

- Orbit selection to meet payload requirements should consider variations that provide the best combination of viewing opportunities and operations schedules.
- Orbiter thermal constraints must be considered when selecting launch times to meet payload target lighting requirements during earth-oriented missions.
- Payload mounting design should consider the impact on mission time for any sensor retraction/deployment operations and the potential need for mission-peculiar mount design.
- Sequential scheduling of payload operations and delayed real-time data transmissions should be considered as a means to avoid RF interference between payloads and with

the Orbiter communication links, and to maintain resource demands within available levels.

- Payloads that are largely automated or whose operation is not orbit-time sensitive, such as Space Processing, should be considered for inclusion on a fill-in basis between opportunity-constrained operations, to provide maximum use of mission resources.

Consideration of these mission factors will aid in maximizing payload returns, mission sharing, and resource utilization.

AAS 75-269

THE LONG DURATION EXPOSURE FACILITY —
A SHUTTLE TRANSPORTED LOW-COST
TECHNOLOGY EXPERIMENT CARRIER

John D. DiBattista[*]

The Long Duration Exposure Facility (LDEF) is a passive spacecraft capable of remaining in space for extended periods. Its primary role is to accommodate advanced spacecraft technology experiments. The LDEF is space-shuttle delivered and retrieved. With retrieval, it offers unique opportunities to study in ground-based laboratories results from a wide variety of experiments after exposure in space.

Research and technology programs need to include space experiments and testing when —
- Ground facilities cannot provide adequate simulation.
- Space environment data are needed for design and operational criteria.
- Corroboration of data obtained in ground-based facilities is required.
- It is desired to verify a promising new concept so that it is space qualified prior to using it to supplant older, less advanced operational systems.
- The space environment itself can be exploited as a facility permitting new areas of investigation.

In the past, opportunities to perform technology experiments in space were limited. When the shuttle becomes operational, the opportunities for performing technology experiments in space can be greatly increased.

The Long Duration Exposure Facility is being developed by NASA's Office of Aeronautics and Space Technology as one means to accommodate many such technology experiments with the shuttle.

---
[*]Aerospace Technologist, LDEF Project Office, NASA Langley Research Center, Hampton, Virginia 23665. Member AIAA.

The LDEF, as shown in Fig. 1, will allow a unique mode of shuttle utilization which will extend and complement other planned shuttle utilization modes with, for example, Spacelab and/or pallets. The LDEF will be able to accommodate a wide variety of users who have simple technology and/or science experiments requiring a free-flying carrier. The LDEF is essentially a passive stabilized structure. Experiments, contained in trays, will be simply attached to the facility for transportation to and from space and for free-flying exposure in space.

Many experiments for LDEF will be completely passive with active data measurements made in the laboratory before and after the experiment exposure in space. Other experiments for LDEF which require active systems (power, data storage, programer, etc.) will include these as self-contained systems in their respective trays. This self-contained approach is to simplify the experiment integration and operation with the facility and shuttle.

The LDEF stay time in space will be limited only by its orbit decay. Since the LDEF will be passively stabilized, it will allow prolonged very low gravity levels and it will allow a relatively contamination-free environment. Being unmanned and low cost, LDEF can be used to perform experiments which present an unacceptable safety risk if performed on Spacelab or pallets in the shuttle payload bay.

The LDEF has been developed through an extensive dialogue with experienced space scientists and technologists as well as those who have had little or no contact with the Space Program. Because of this dialogue, every effort has been made to keep the LDEF simple from the users' point of view and thus encourage participation in using space for research purposes.

If any researchers think that an LDEF experiment may be of value to support his research program, or if additional information is desired on the LDEF accommodations for experiments and the approach to suggest an LDEF experiment, please contact the author.

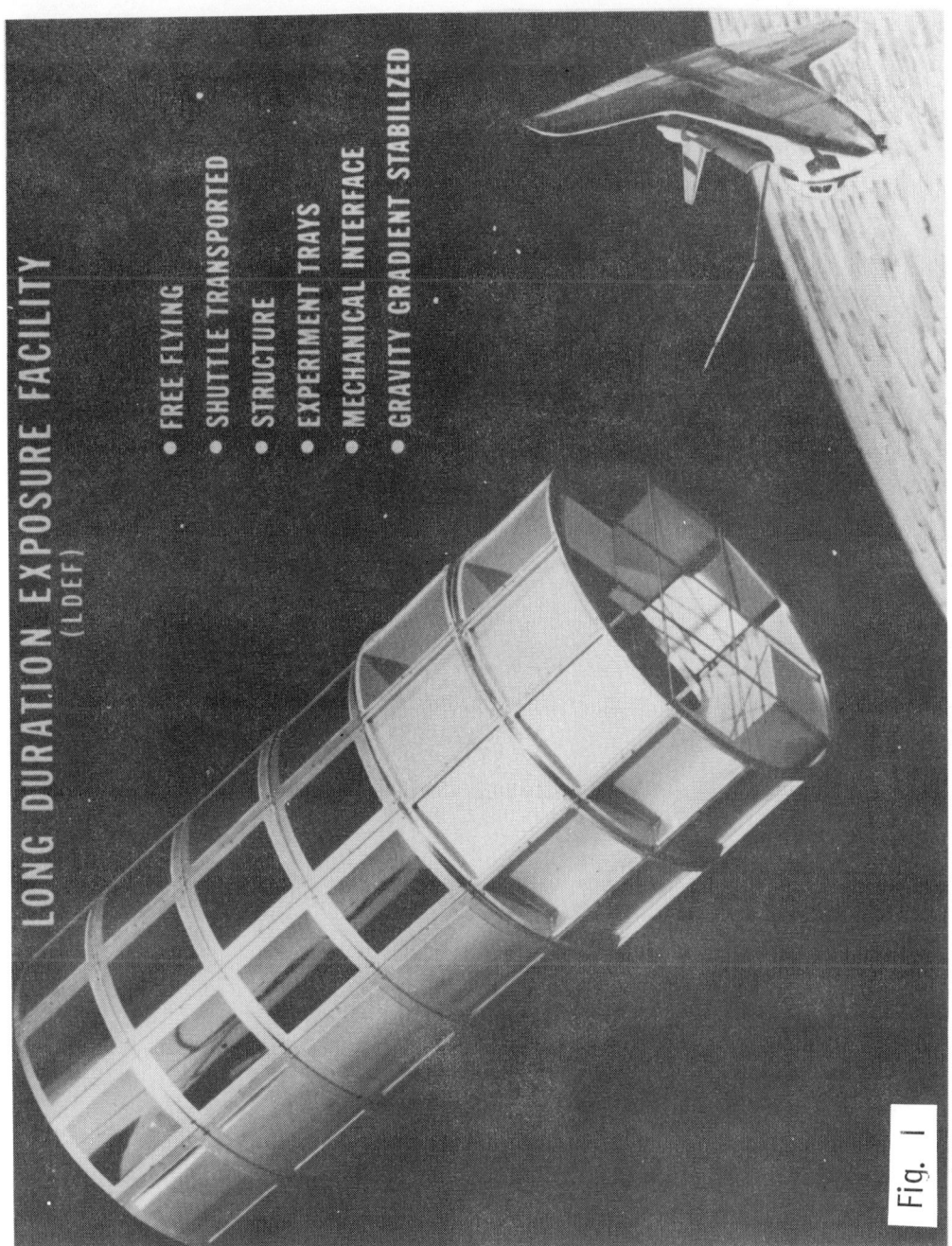

Fig. 1

AAS 75-270

PAYLOAD PLANNING FOR THE FIRST SPACELAB MISSION

- A EUROPEAN VIEW

A. v. Breitenstein[*] and
D. Davidts[*]

This paper gives a view of European planning activities in preparation for Shuttle/Spacelab Utilization Program. We attach special importance to the First Spacelab Mission and have attempted to highlight this aspect in terms of implications from our longer range planning activities.

[+] Dr. D. Davidts is Shuttle/Spacelab Utilization Program Manager and A. v. Breitenstein is Mission Project Definition Manager for Messerschmitt-Bölkow-Blohm Space Division

INTRODUCTION

It is our belief that the First Spacelab Mission has a special position in the overall Shuttle/Spacelab Utilisation program. We see this even though flight verification will impose constraints which will not be typical for subsequent missions. We feel the importance of the First Mission is in terms of its pilot function in the establishment of precedents that will influence follow-on activities. These will include those associated with equipment design, development and qualification as well as those for payload integration and operations.

Preparatory activities in Europe for this important multidiscipline First Mission Payload are presently concerned with formulating experiment equipment concepts and assessing system support activities required for their implementation. It is therefore our intention here to show by a few representative examples how we feel program and system aspects will influence First Mission Payload instrument performance requirements. In addition we wish to express our view of European integration activities as foreseen for the First and Follow-on Shuttle/Spacelab Missions.

As a European industrial firm we are interested in providing supporting services to our user community. The major objective of our planning studies is therefore to reach solutions which will lead to effective and low cost mission projects.

MISSION MODEL CONSIDERATIONS RELATIVE TO THE FIRST SPACELAB MISSION

The scientific objectives for a study of the European part of the First Spacelab Mission were selected out of approximately 250 proposals prepared by the European User Community. Criteria were therefore necessary to reduce the number of potential objectives to a manageable amount. The criteria utilized reflected both the unique objectives/constraints of the First Spacelab Mission and the relative priorities considering follow-on missions. As an example of the latter, the candidate German objectives were selected in accordance with a German National Mission Model. This model shows priorities in consecutive national and international flights and the desired mission opportunities for the instrumentation concerned.

Another mission model aspect is directly related to flight sequences and frequencies. In general, satisfaction of scientific and application oriented objectives dictate sequences and frequencies of desired flight opportunities which in turn leads to a dichotomy between general purpose facility type experiment equipment and experiment unique equipment. Consider for example the mission planning (Fig. 1) associated with the discipline Earth and Atmospheric Physics. Here the usefulness of Spacelab

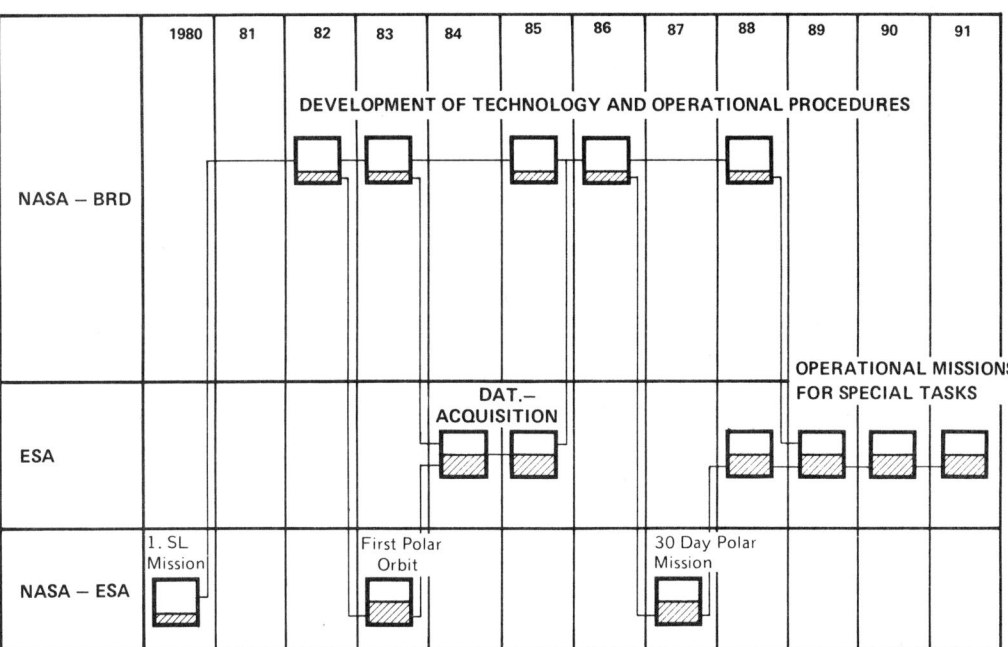

Fig. 1 Example Mission Planning For Earth And Atmospheric Physics

as an experimental test platform is demonstrated. Given an ultimate objective of having an operational earth monitoring satellite system it would be productive to make use of a general purpose facility type Microwave Radiometer/Scatterometer on the First Spacelab flight. The instrument would be used to investigate and narrow in on basic scientific and technological capabilities. Allowing time for further developments, subsequent flights would then be used to further optimize operational oriented sensing systems with emphasis on data management aspects. Later flights could then be used to verify the fully optimized equipment. The performance of payload hardware flown on the First Spacelab flight should therefore reflect such programmatic aspects.

OPERATIONAL IMPLICATIONS ON EXPERIMENT EQUIPMENT PERFORMANCE

Spacelab recoverability provides the possibility of experiment hardware reflight in case of experiment failure. Marginal flights can therefore be planned in advance much in same way as for cases where more than one successful mission is required to gather a complete set of data.

But thinking about "cost effective" experiments one should trade the costs of reaching alternative instrument reliability levels against the costs of the expected number of flights required to achieve experiment success with the related reliabilities. This means that total experiment cost includes not only instrument design, development, manufacture, and test costs, but also experiment operational costs (including integration) which is a function of both the instruments share in the total Spacelab Payload operations cost per mission and the number of missions flown.

We investigated this tradeoff in recent studies to see if we could identify reliability goals for First Spacelab Payload experiments. To do this we defined three different quality/reliability grades by detailing the activities, methods, procedures and control loops necessary to achieve each of these grades. Generally, the upper quality grade (UQG) corresponded to programs conducted for application or long life scientific satellites, the medium quality grade to nominal aerospace/defence industry practice while the lower quality grade reflected best commercial practice upgraded with some additional control loops. A low cost development approach characterized by a minimum number of development models was adopted for all cost assessments.

Then, considering factors such as equipment developement status, parts counts, part compositions, part quality/reliability levels, etc., total cost and reliability assessments for the alternative quality grades were developed for representative experiments. Recurring operational cost assessments per mission include the so-called launch ticket based on relative use of available Shuttle/Spacelab resources.

The results of one such study on a Free Flow Electrophoresis Experiment are depicted in Fig. 2. Here the result is clearly a desire for the medium to upper quality grade, with no great incentive to consider for example redundancy to further improve reliability.

	UQG	MQG	LQG
Instrument Reliability	.97	.86	.67
Minimum No. of Flights for Prob.Experiment Success=.96 (Nominal 20 Missions)	22	27	37
Design, Development Manufacture and Test Cost (MAU) +)	2.19	1.21	1.06
Operations Cost for Minimum No. of Flights (MAU) +)	6.47	7.61	10.21
Total Cost (MAU) +)	8.66	8.82	11.27
+) 1 Accounting Unit (AU) U.S. Dollar 1.27			

Fig. 2   Quality Grade/Reliability Tradeoff
For A Free Flow Electrophoresis Experiment

It is trivial to state the intuitive result that higher experiment operations costs (whatever the reason) tend to promote higher instrument reliability design goals. The interesting point is where the buckets and crossovers occur. To date we have identified no patent rules of thumb and therefore see the need to idendently analyse each experiment.

TEST IMPLICATIONS RELATIVE TO EQUIPMENT PERFORMANCE

European attention to space experimentation has primarily been focused on satellite programs with little man-in-the-loop space experience. Also, recognizing the unique capabilities of the Shuttle/Spacelab, we see further deviations from our traditional experiment programs. We have therefore tried to open our minds and re-look at the design and development areas and especially those concerned with qualification and acceptance testing and related quality assurance activities.

Our present approach features in the ideal case only one development model. This means we intend to go directly from design bread/brass boards to a proto/flight model. There will be exceptions but these will occur where safety considerations, design performance uncertainties, training requirements, etc. justify the addition of another development model.

Our test approach distinguishes between safety and other mandatory requirements, and an optional test program. Safety tests will demonstrate compliance with Shuttle/Spacelab safety requirements. Other mandatory tests will give some confidence in experiment function but will be primarily oriented toward ensuring compatibility with the Shuttle/Spacelab system and noninterference with other experiments. Experiments which can not successfully pass the safety and mandatory test program will be rejected.

Testing beyond mandatory requirements would be for purposes of increasing confidence that the experiment will function properly even under adverse conditions. As such, this testing should be largely the Users responsibility with the possible exception of where substantial investment of public funds warrant an independent authority.

The concept of an optional test program therefore arises from cost/
benefit considerations. In theory this means that the value of marginal
benefits associated with increased confidence in experiment functional
capabilities should be equated with marginal costs of increased testing
In practice, quantification problems will make this difficult. In any
case however the recommended amount of optional testing should reflect
the relative importance of an experiment in terms of its contribution
to mission objectives and/or the cost of experiment reflight.

We therefore see the need for flexibility in establishment of optional
test requirements as they should not apriori preclude the possibility
of a test flight.

PREINTEGRATION ACTIVITIES IN EUROPE

Spacelab Payload integration activities in Europe have both technical
and economic justification. We not only have the facilities and qualified personnel but also feel that proper support of our User community
requires proximity between experimenter and hardware. Furthermore if we
are to provide our Users with complete service we must understand problems and constraints associated with integration and resolve our own
conflicts. We recognize there are limitations with respect to what can
be accomplished below the complete payload but feel if we adopt a flexible
approach we can make a positive contribution.

For some experiments it may be most efficient to ship flight unit
assemblies directly to the U.S. for integration. For most equipment however we visualize the following level IV type activities being accomplished in Europe prior to shipment.
- o Integration of Flight Unit Assemblies
- o Testing of critical Spacelab System/Subsystem
  interfaces via Spacelab GSE
- o Safety and other mandatory tests
- o Variable amounts of optional testing for
  functional verification of hardware and software.

For other equipments we can foresee European integration activities extending into level III. This could include equipment which is functionally and operationally correlated such as earth and atmospheric packages, or equipment groupings which comprise experiment clusters such as space processing facilities and standard equipment. In such cases the European level III integration activities could include:

- o Integration of Experiment hardware with Spacelab hardware
- o Interface verification and compatibility testing
- o Combined environmental testing such as static, dynamic, modal survey, acoustic, EMC, etc.
- o Functional performance testing of flight hardware and software

As an example of existing European facility availability and capability consider those concentrated in Ottobrunn, Fed.Rep.of Germany (Fig. 3).

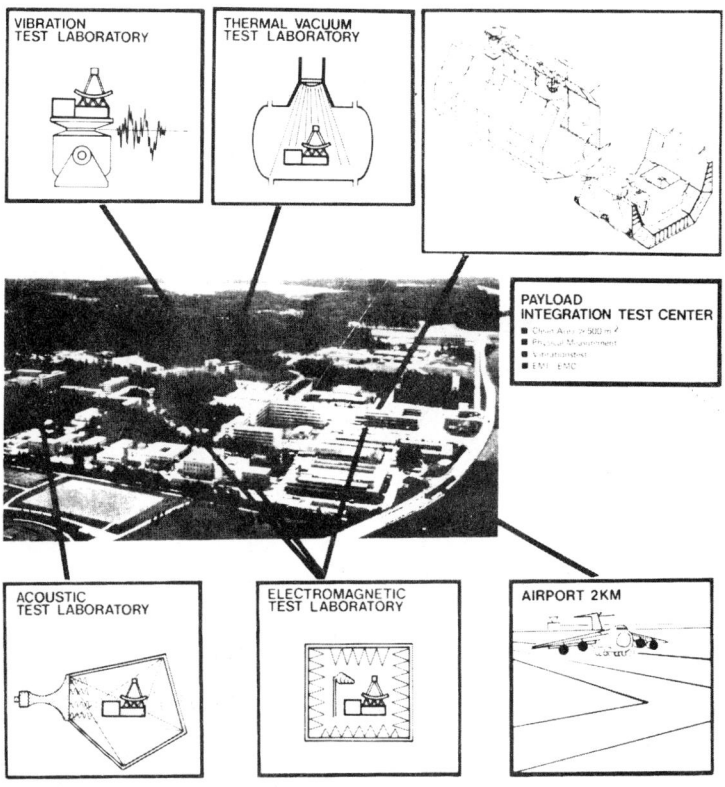

Fig. 3   Integration and Test Center in Ottobrunn

In this one location the combination of our own MBB facilities and the German National IABG test facilities can in principle satisfy all anticipated European payload element integration and test activities. This conclusion was the direct result of a nationally funded study to investigate facility requirements associated with the BMFT Mission Model.

EUROPEAN USER SUPPORT

We believe that European participation in the Shuttle/Spacelab program will be a continuous and growing activity. Furthermore, we see that each alternative European User will face the same if not identical problems during instrument design and development as well as during flight hardware manufacture and integration. If European participation is to reach its maximum potential we therefore see the need to remove possible duplications and redundancies and bring efficiencies into our activities. This we feel can best be accomplished by consolidating User support activities into a European User service organization. This organization would have the multilingual/multidisciplinary capabilities necessary to transcend problem areas and properly interface European activities.

Services performed by the support group would include analytical integration to examine for example, mission profiles, interface and software requirements, needs for mission dependent items and GSE, and integration and test planning requirements. In addition, questions related to interface compatibility, safety, thermal behavior, dynamic behavior, EMC etc. would be answered. Advisory assistance would also be given relative to the availability of standard components, interface hardware and GSE. Special emphasis would be placed on coordinating individual instrument test plans to ensure acceptability from the standpoint of safety and other mandatory test requirements.

MANAGEMENT APPROACH

The productivity of European participation in the Shuttle/Spacelab Utilization program will largely depend upon the chosen management approach. For this and other reasons this topic is currently the subject of much discussion in Europe.

We see program management being accomplished by a matrix type organisation such as that shown in Figure 4. Here integration, test and functional support would be provided to both individual experiment developers and small project teams that would be assembled to monitor and coordinate individual mission projects. This type of organization would foster assembly and dissemination of accumulated information on a program wide basis. It would thus promote efficiencies and preclude costly consequences associated with assembling and subsequently disbanding large autonomous mission project teams.

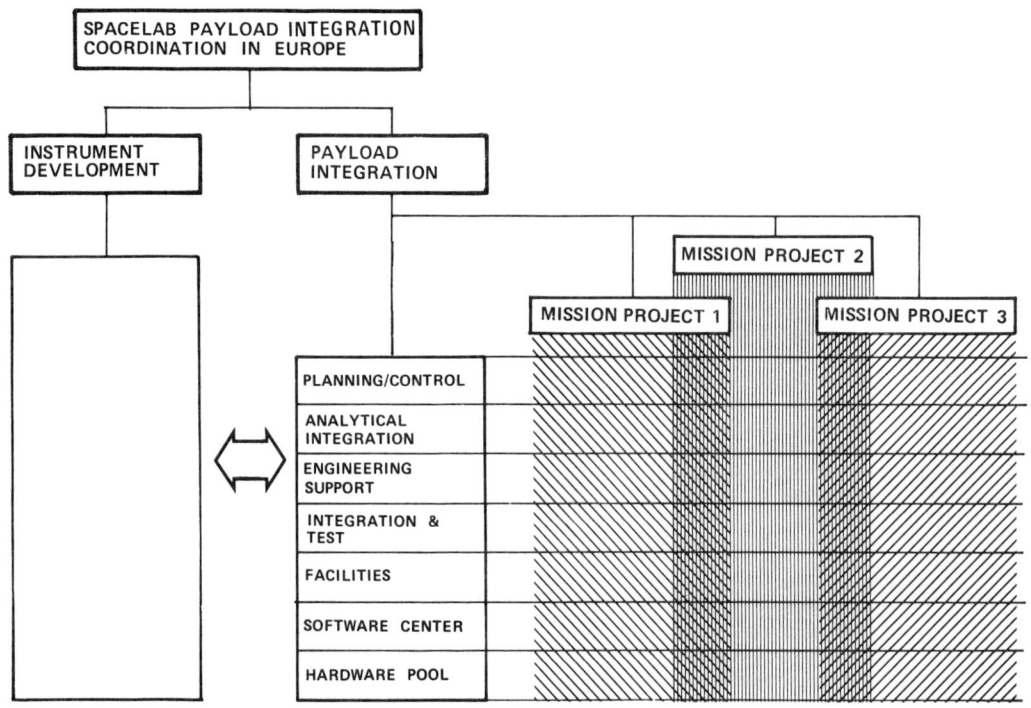

Fig. 4 Organisation Scheme For European Mission Projects

CONCLUSION

We have concluded that how Europe prepares for the First Spacelab Mission Project will be indicative of follow-on activities. It is this pilot function which we therefore emphasize in our study activities. Our goal is to help lay the foundation for a productive and viable European contribution to the Spacelab Utilization Program.

BIBLIOGRAPHY

The following is a list of MBB planning studies related to the Shuttle/ Spacelab Utilization Program in Europe:

- Definition of the Technical Requirements for an Earth Resources Payload for the Spacelab.
  ESRO Contract, December 73

- Follow-on Study for the Definition of the Technical Requirements for an Earth Resources Payload for the Spacelab.
  ESRO Contract, July 74

- Supporting Study for the Definition of Experimental Objectives of the First Spacelab Mission.
  ESRO Contract, December 74

- Follow-on Study to the Supporting Study for the Definition of Experimental Objectives of the First Spacelab Mission.
  ESRO Contract, April 75

- Definition of Model Missions for Spacelab
  GFW Contract, July 74 to present

- Preparation of Spacelab Mission Projects
  GFW Contract, July 74 to present

- Test and Integration Facilities and Equipment for Spacelab Utilization Program.
  GFW Contract, October 74 to present

- Data Management Systems for Spacelab
  GFW Contract, November 74 to present

- Principal Mission Project Flow Alternatives, Model and Test Philosophy
  GFW Contract, July 75 to present

AAS 75-271

SPACELAB UTILISATION IN DIFFERENT FIELDS OF
SCIENCE AND APPLICATIONS

Jacques Collet*

The European Space Agency (ESA) has investigated the
opportunities provided by Spacelab in the different
fields of scientific research and application
disciplines, including the new fields which are
opened up by the use of the manned laboratory. This
paper describes the studies to be carried out by
ESA in the Spacelab payload era. As an illustration
of the European interest in the utilisation of Spacelab,
the first planned payload experiments are indicated.
The paper explains how this particular payload fits
into the projected planning schedule for utilisation
of Spacelab.

INTRODUCTION

The European Space Research Organisation (ESRO) has been
superseded by the European Space Agency (ESA). ESRO is well
known for its excellent record in the field of automatic
satellites. Now ESA is broadening its activities to include
in its programme a launcher (Ariane) and a manned space
laboratory (Spacelab).

---

* Acting Assistant Director, Spacelab Utilisation,
  European Space Agency, Neuilly-sur-Seine, France

The aim of this paper is to define the significance of the latter for the programmes of the Agency.

In 1971-72 ESRO began to identify those areas where Spacelab could play a significant role in developing ESRO's traditional activity in astrophysics and geophysics in the scientific sector, and telecommunications/navigation and earth observations in the field of applications. At the same time ESRO investigated the new opportunities for utilisation of space which Spacelab could offer. Two new areas, as far as Europe is concerned, have been identified as very promising for space research; life sciences and space processing and manufacturing, the latter receiving very strong support from the user community.

There now follows a description of those Spacelab missions which are under study by ESA.

SCIENCES

Astrophysics

The astrophysicist realises immediately what advantages Spacelab could offer - the manned aspect, weight carrying capabilities, and the availability of an Instrument Pointing System (IPS), among others. In accordance with the wishes expressed by the scientific community in 1974, ESA selected the following missions for the phase A study.

Infra-red astronomy

The peak emission from many important classes of object lies in the $20\ \mu - 300\ \mu$ region which is inaccessible to ground-based observations due to the opacity of the Earth's atmosphere. Information in this region of the spectrum has been limited so far to mainly photometric data obtained with relatively small telescopes flown on aircraft, balloons and rockets.

The future development of this very exciting field requires high sensitivity measurements, with high spatial and spectral resolution, and thus the use of large telescopes operating in space and free from the restrictions imposed by selective atmospheric absorption. The advent of Spacelab in the next decade offers precisely such an opportunity for I.R. astronomy.

One of the major I.R. facilities required on Spacelab is conceived to be a large uncooled telescope which could be used for a wide range of I.R. observations. Its unique advantages, however, lie in its capability of carrying out high sensitivity photometry with high spatial resolution and of studying a wide range of astrophysical problems through measurements of atomic and molecular lines in the far infra-red.

The telescope studied by ESA (Fig. 1) is a 2 - 3 metre classical Cassegrain-type configuration with uncooled optics. Modulation is provided by rocking the secondary mirror. Its suggested focal plane instrumentation consists of a multi-band photometer, a polarimeter, a Michelson interferometer, and a heterodyne receiver allowing for multi-band mapping and polarimetry, and high and very high resolution spectroscopy. Compared to existing telescopes on the ground, or carried by balloons and aeroplanes, this telescope will provide more than one order of magnitude, greater sensitivity and, most important, a higher spatial resolution in the far infra-red.

The system requirements are consistent with the Spacelab capabilities. An IPS is required with a pointing precision of about 5 arc seconds and stability of a few arc seconds. The computer will be used for data reduction and to control the operational sequence. The major problems connected with a proper utilisation of this facility are the thermal control of the optics, the cryogenic system for the focal plane instrumentation, and the contamination induced by Spacelab.

This mission should command the support of the I.R. groups in Europe (about 20) dedicated to I.R. astronomy experimentation.

This mission, which features the highest spatial and spectral resolution planned at present, is complementary to NASA's cryogenic telescope currently under study. In order to ensure an overall coordination of these projects, the two Agencies have arranged an exchange of representatives between the NASA and ESA groups.

Ultra-Violet astronomy

ESA recognises that the NASA Large Space Telescope is the cornerstone of space astronomy. The Shuttle/Spacelab system will offer, in addition, the possibility of flying in space smaller telescopes to conduct various observational programmes of short duration, as well as covering sky areas larger than can be imaged by the LST; it also enables the use of sophisticated instruments and detectors for specialised work on particular objects. The scientific objectives of such a Spacelab facility would encompass virtually all classes of astronomical objects. The good imaging properties of the facility, its high spectral resolution and the fainter sky background will allow the making of meteorological surveys and atmospheric analyses of planets, the study of isotope ratios and chemical reaction in comets; it will also be possible to derive mass losses in stars, the abundance of rare elements and metals, and to study the evolution of hot stars; other uses are the study of the properties of interstellar medium, the observation of star formation sites, faint extensions, and the distribution of galaxies and quasars, as well as the detection of the (hypothetical) ionised component of the intergalactic medium.

To achieve these aims, a 1 metre diameter Ritchey-Chrétien telescope is being studied mounted on the IPS on the pallet (Fig. 2). The image size is between 0.3" and 0.5" over a field of about 5'. The stability of the guiding is foreseen to be 0.1" over 0.5 hours' exposure.

A large variety of instruments can be accommodated at the Cassegrain focus. A typical set comprises a high resolution spectrograph ($R = 10^5$; $m = 8.5$), a low resolution spectrograph ($R = 10^3$; $m = 13.5$), a photometer objective grating ($R = 10^2$; $m = 17$), and a high resolution camera ($R = 1$; $m = 26.5$).

Alternatively, more specialised instruments could be accommodated, thus permitting a large number of groups to perform experiments.

The level of scientific support in Europe is very high, when one considers that about 30 groups have submitted proposals for observation on TD-1 and IUE, of which about 10 have experience in balloon or space experimentation in this discipline. The project is eminently suited to meet the requirements of this large community and is within the capabilities of Europe to carry out.

X-Ray astronomy

Let us close our discussion of the astrophysic discipline by mentioning a somewhat more modest instrument which is also under study and which should provide very valuable data in high resolution spectroscopy and polarimetry X-ray astronomy. The proposed instrument consists of a number of large area Bragg crystal spectrometers for the study of important spectral features in the 2 - 10 keV energy range. The doubly bent focussing spectrometers are designed to study the spectra of variable sources with good time resolution and high sensitivity. These instruments can also be used to provide a polarimeter of exceptional sensitivity. The conical focussing spectrometer combines 1 to 5 arc minute spatial resolution with high spectral resolution. At energies below 3 keV a grazing incidence reflector array will be used to collect radiation for a number of Bragg crystals arranged around the mirror axis behind the focal plane. For broad band spectroscopy, either a solid state detector or a proportional scintillation counter can be used at the focus of the light collector. These devices would be employed to study the spectra of compact objects and of the low energy diffuse background radiation.

X-ray missions to be launched during the remainder of this decade will probably increase the number of known X-ray sources by a factor 10 to 100. Many of these objects will have been associated with optical and radio companions, but a detailed understanding of the X-ray emission process will not yet have been obtained. Such an understanding can be achieved by the payload briefly described here, which takes advantage of the reflight capabilities of the Spacelab to alter the emphasis of the work between missions, so that eventually optimised studies may be undertaken for a substantial number of X-ray emitting objects.

The payload proposed places only light demands on the Spacelab resources and does not contain any high precision moving parts. These characteristics combined make it very well suited to be included in a great number of Spacelab sortie payloads and to withstand the stresses of a high number of flights.

It is estimated that about 8 sortie missions of 7 days would allow for the performance of the desired observations on all relevant X-ray sources of the UHURU catalogue.

The interest in X-ray astronomy has been growing continuously in Europe, and an active community now exists, as evidenced by the high number of proposals for observations from EXOSAT which have been received from about 30 groups in Europe. Of these about 12 have experience with space hardware.

Geophysics

In the atmospheric field, a very strong interest exists in Europe for observing the disastrous effect of industrial pollution on the atmosphere, particularly the ejection of freon. It is considered essential to study the minor constituents acting on the stratosphere, and it is recognised in Europe that the active remote sensing technique using a laser is the most efficient way to do so. Very little has been done in the 30 - 120 km regions of the atmosphere which are not attainable by balloons or satellites. ESA has studied intensively the use of Spacelab-borne laser/radar (LIDAR).

The LIDAR will be designed to observe resonance absorption
or fluorescence of gaseous constituents and to measure
aerosol concentrations at different heights from the frequency
analysis of backscattered signals. Two processes are
considered :

- the backscattered process where pulses of collimated,
  monochromatic radiation are transmitted and echoes
  from atmospheric constituents are detected and analysed
  (time delay, intensity, Doppler shift).
- the forward scattering and absorption process using a
  reflector on a subsatellite where the absorption of
  light in selected spectral regions is measured.

The Spacelab is a perfect carrier for such an instrument
which has large volume, weight and energy requirements.
The availability of a payload specialist will permit the
performance of calibration, alignment, change of laser head,
etc.

Fig. 3 gives a possible configuration of the LIDAR with the
use of the Spacelab longitudinal airlock to permit man's
access to the laser.

Life sciences

This is a "new" discipline for ESA. The Agency has established
a list of objectives in order of priority for the first
Spacelab mission and has approached the large life scientist
community for recommendations on a Spacelab utilisation
programme in this field. As the vestibular field has been
given high priority, ESA is studying the design of a test
stand for vestibular research, which can be rectilinearly
accelerated and decelerated as well as oscillated along the
longitudinal axis of Spacelab. Such a facility, called a
Space SLED, will enable the study of man's animal and
vestibular adaptation to weightlessness, problems of visuo
vestibular adaptation, the cooperation between the semi-
circular canal and the otolithic system as well as the

problem of space sickness using rectilinear acceleration pulses as stimuli on the basis of weightlessness.

SPACE PROCESSING AND MANUFACTURING

The space processing and manufacturing discipline has received increasing support from the European user community since the first survey of potential interest started in ESRO a few years ago. As an illustration of this enthusiasm, more than half of the proposals received by ESA for the first Spacelab payload concern this discipline and emanate from government research institutes as well as from industry. ESA is studying Spacelab payload equipment which could serve as a basic facility for the different users in the field of metallurgy, crystal, glasses, etc. Different types of furnaces and sample positioning and stirring equipment are under study. Separation technique whose capabilities would increase tremendously under 0-g conditions, has been given very high priority within the Agency and ESA is starting a study on a free flow electrophoresis experiment which is proposed for the first Spacelab flight.

A vigorous space processing programme is proposed by ESA to its Member States, which is best demonstrated by the importance given by the Europeans to this discipline in the contemplated European complement to the first Spacelab payload. The space processing and manufacturing part of the first Spacelab payload would constitute the nucleus of future European space processing payloads.

APPLICATIONS

In the applications field (communications/navigation and earth observations) most of the missions in 1980s require geosynchronous orbit, with long duration for permanent services, both of which are beyond the capabilities of Spacelab. Nevertheless, a predominant role is foreseen by ESA in testing and evaluating systems in real space conditions.

THE FIRST SPACELAB MISSION

The objectives of the first Spacelab mission are primarily to test out the Spacelab system, but NASA and ESA have decided to conduct significant scientific and technological research during this mission. The verification of the Spacelab system imposes some constraints on the mission (sequence of orbiter orientation mode, configuration, reduced power, energy and crew time, etc.) which have been taken into consideration by a Joint NASA/ESA Planning Group who had the task of submitting to the Heads of ESA and NASA a list of experimental objectives for the mission. This task has been completed and the implementation period has begun.

ESA, in its exercise to define the European complement of the first Spacelab payload, performed in 1974-75 a series of iterations with the user community and the different ESA delegate bodies. This activity started in July 1974 with a wide consultation of the European user community. A "Call for Ideas" was initiated and 260 proposals were received. These proposals were analysed, and checked as far as compatibility with Spacelab, constraints and the other payload elements were concerned.

7 different options were proposed for selection by the ESA bodies and a typical payload was chosen.

The European complement of the first Spacelab payload

The proposed European complement of the payload is characterised by two major packages :

- in atmospheric physics, by a LIDAR system for active sounding of the atmosphere, complemented by passive sounding;

- in space processing and manufacturing, by the specific field of separation techniques, crystal growth, metallurgy and fluid physics.

In addition, in the scientific field there are proposals for a small astronomical package and the vestibular test band.

The applications disciplines are represented by a navigation experiment (one way navigation using an atomic clock) and in earth observation, where it is proposed to test microwave technique performances by using a scatterometer-altimeter-radiometer. Small technological experiments (e.g. heat pipes) will complement the payload.

The payload is a good example of ESA's future utilisation programme. It consists of a mixture of basic elements to be part of dedicated Spacelab payloads (e.g. the LIDAR which is contemplated as being a part of the future AMPS payload, space processing experiments which would constitute the nucleus of future European space processing payloads) and of experiments to be tested on Spacelab for ultimate use on operational satellites (e.g. microwave experiments).

CONCLUDING REMARKS

All activities in the ESA traditional disciplines are planned to take advantage, to different degrees, of Spacelab's capabilities and the European scientific community is anxious to use this new system. Spacelab is attracting, within Europe, new customers to space and, particularly in the field of space processing and manufacturing, a very great interest has been built up. The applications mission planners are very enthusiastic to test, in real conditions, the systems they envisage for future operational missions. The European Space Agency considers the first Spacelab payload to be an accurate reflection of its interests, and is planning its mission model for the early 80s around the basic element of this payload. After having started the development work on the Spacelab system, ESA is developing a vigorous programme for its subsequent utilisation.

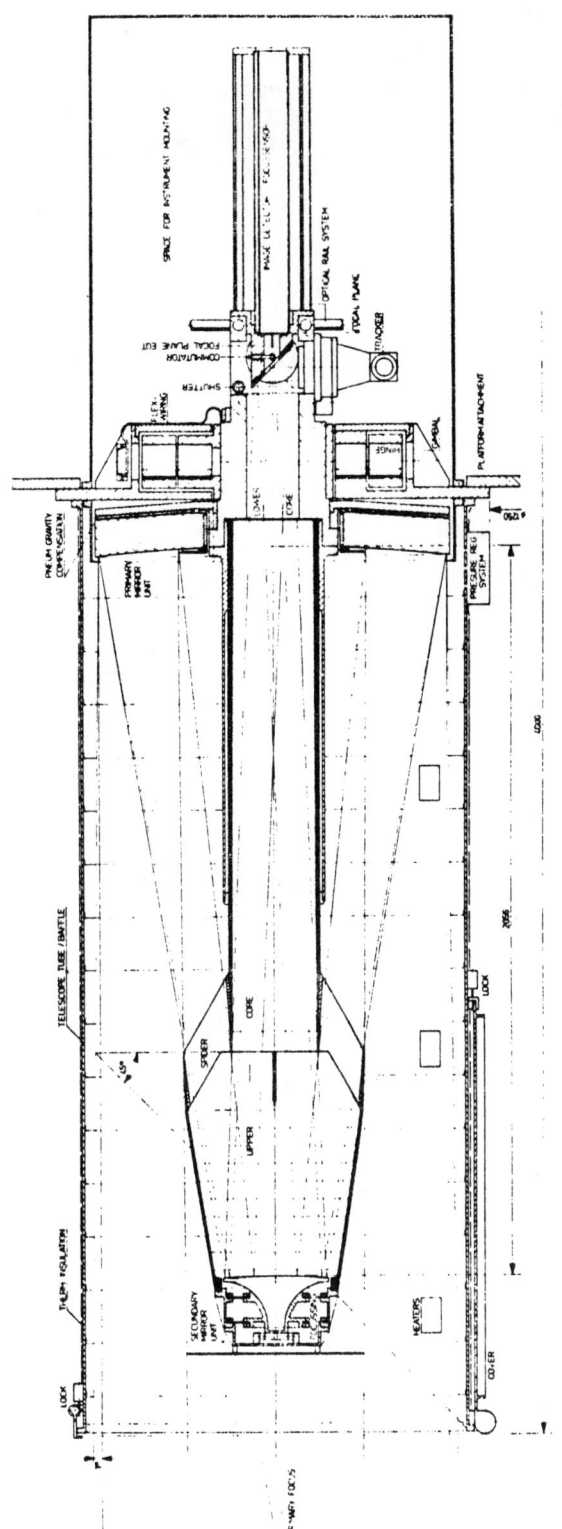

FIG.1

1 METER UV TELESCOPE

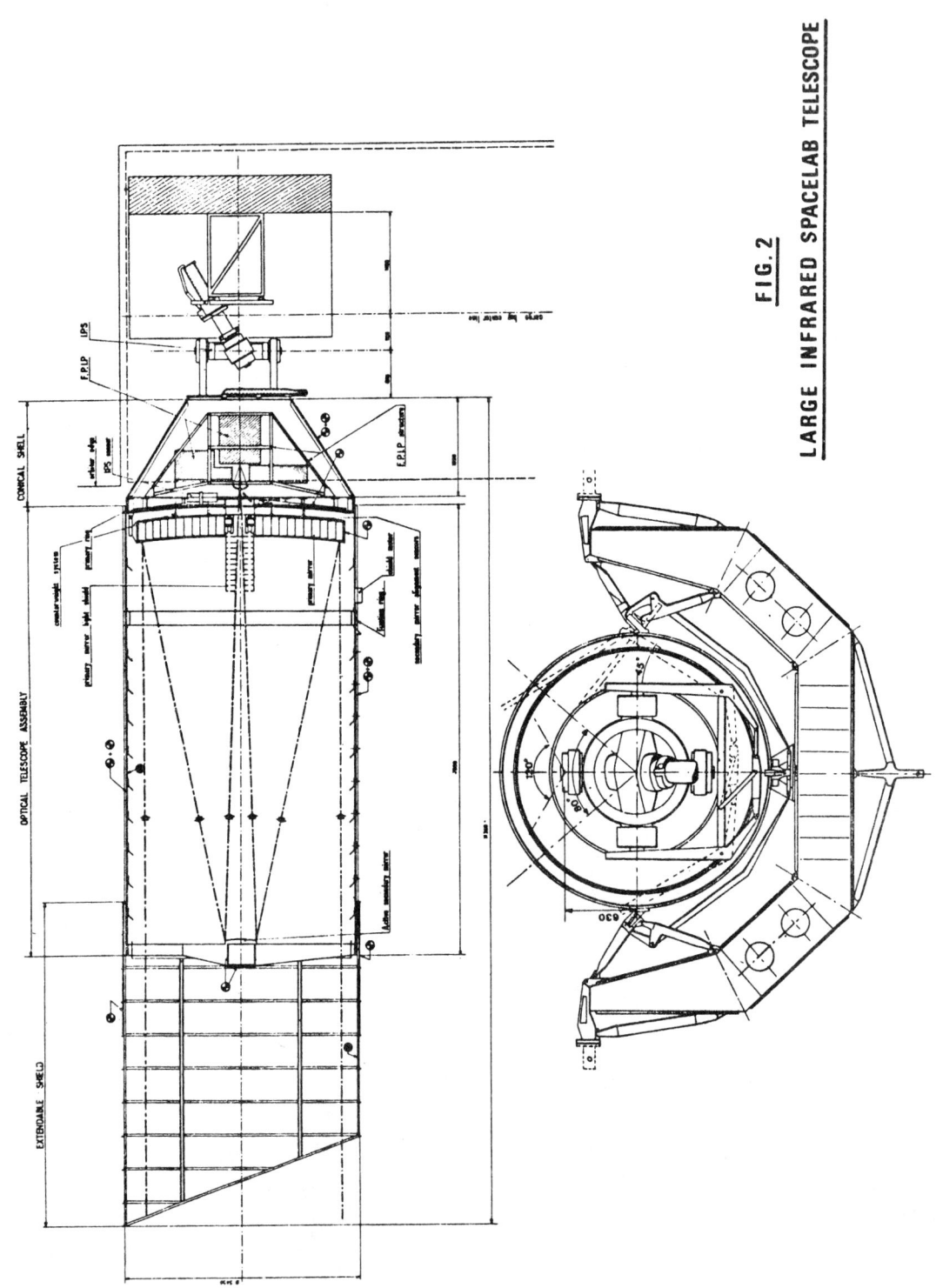

FIG. 2 LARGE INFRARED SPACELAB TELESCOPE

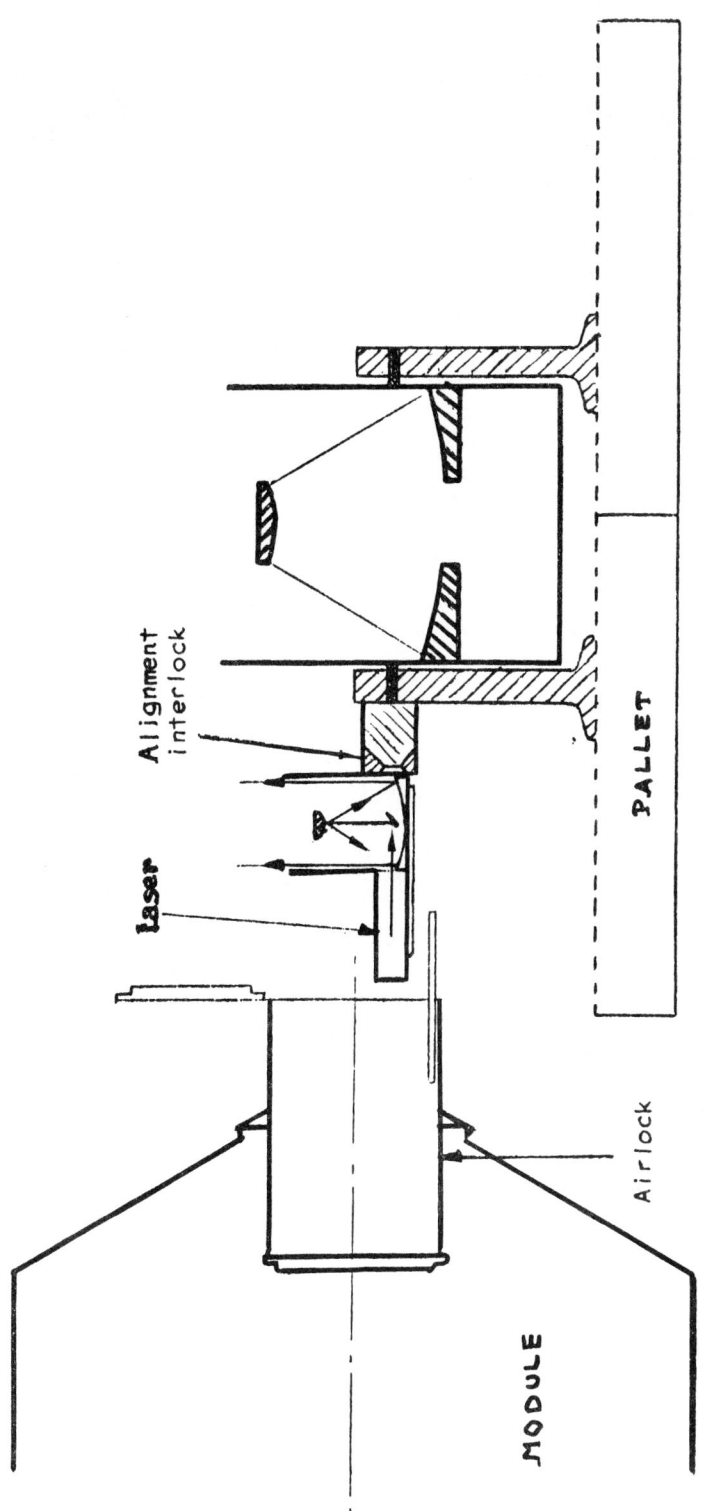

FIG.3
LIDAR SYSTEM CONFIGURATION

AAS 75-291

EARLY SPACE STATION USER ACCOMMODATIONS

Donald R. Saxton
NASA Marshall Space Flight Center

and

Harry L. Wolbers
McDonnell Douglas Astronautics Company

With the reduction in cost and complexity of delivering scientific and technical personnel and payloads to space provided by the Space Transportation System (STS) currently under development, fascinating new opportunities for research and development programs present themselves. To fully exploit this capability, McDonnell Douglas Astronautics Company working recently under contract to NASA's Marshall Space Flight Center* determined that a free-flying manned facility — space station — placed in permanent earth orbit is a viable and cost-effective adjunct to the basic Shuttle system. Together, the Shuttle and the free-flying manned orbital facility offer unprecedented opportunities for the pursuit of knowledge and the application of space technology to the benefit of all mankind (Fig. 1).

The free-flying manned orbital facility could be supported at regular intervals by the STS and would provide living and working quarters for a team of scientists and technicians. Such a facility would enable the scientific community to pursue programs directly related to the improvement of life on earth—notably, earth resources management, pollution control, global communications, weather forecasting for agri-

---

* Contract No. NAS8-31014, Manned Orbital Systems Concepts (MOSC) Study

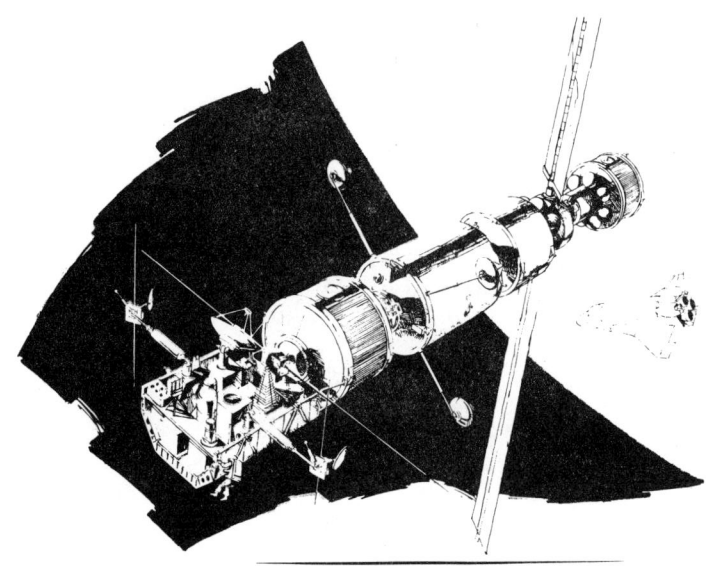

Fig. 1 A Free-Flying Manned Facility (Early Space Station)

culture and disaster warning, manufacture of critical materials and medicines, and new energy sources for the world's growing needs.

In developing the most cost-effective approach to extended-duration missions, available hardware and technology from the Skylab, Orbiter, and Spacelab programs should be used insofar as practical and the lessons learned from Apollo and Skylab experience should provide the basis for the many critical design decisions. Since the payload weight constraints for each Orbiter flight are 65,000 pounds at launch and 32,000 pounds for planned landing, a modular approach to facility buildup appears to offer the best means of providing growth and flexibility to meet future needs while remaining within the basic weight and volume limits of the Orbiter.

An investigation of a broad spectrum of research requirements suggested that a minimum crew size of four would be required to provide a sufficiently broad reservoir of skills and a sufficient number of on-orbit manhours to provide a flexible base for accommodating projected workloads. As demands increase this core system could be expanded with additional facility modules and additional scientific and technical personnel.

The four-man baseline configuration derived in our analysis is shown in Fig. 2. The core vehicle consists of a subsystem module and a habitability module. This core vehicle would be delivered in one Shuttle launch and normally would be left on orbital station for the

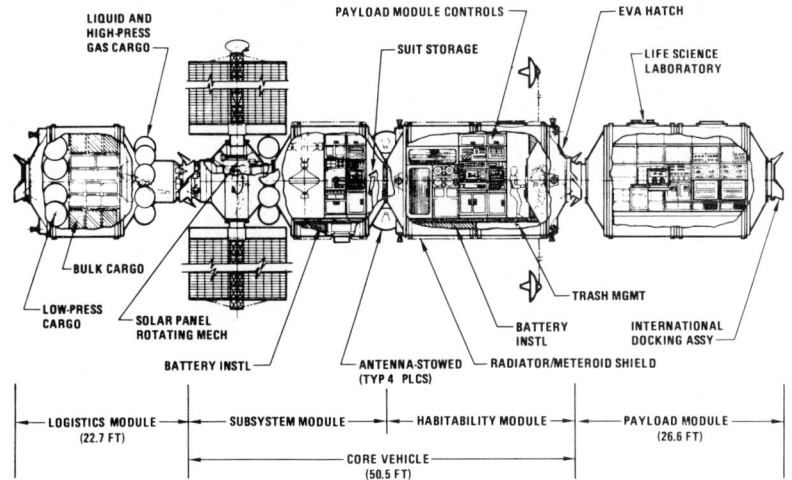

Fig. 2 Four-Man Baseline Configuration

life of the facility. The logistics modules and the payload modules would be launched on supplementary Shuttle flights as required, although nominally we would propose that they be delivered and replaced at 90-day intervals.

A summary of the principal subsystems is presented in Table 1.

Table 1
SUBSYSTEM SUMMARY

SUBSYSTEM*	SELECTION	SOURCE
• CREW ACCOMMODATIONS		
WASTE MANAGEMENT	CENTRIFUGAL SEPARATOR	ORBITER
CREW EQUIPMENT	RESTRAINTS, PERS GEAR, ET AL	ORBITER/SKYLAB
• ENVIRONMENTAL	1 ATM	SPACELAB EXPERIMENT
CONTROL/LIFE SUPPORT	CLOSED H_2O (VAPOR COMPRESSION)	ORBITER
	OPEN O_2 (LiOH FOR CO_2 REMOVAL)	
• ELECTRIC POWER	25 KW SOLAR ARRAYS (12 KW AT BUS)	SEPS
	36 KWH BATTERIES (12)	ORBITER
• DATA MANAGEMENT		
– EXPERIMENT	DISTRIBUTED (1 MBPS SERIAL DATA –	ORBITER/SPACELAB
– VEHICLE	40 K WORD MEMORY)	ORBITER
	CENTRALIZED	
• COMMUNICATIONS	S-BAND	ORBITER
	Ku BAND	ORBITER
• STABILITY/CONTROL	CMGs (3) (18,000 FT LB SEC EA)	SKYLAB – IMPROVED
	SENSORS (EDGE TRACKER,	ORBITER/SKYLAB
	GIMBALLED STAR TRACKER, SOLAR)	
• REACTION CONTROL/	COLD GAS – GN_2	SKYLAB
PROPULSION	60 K LB SEC TOTAL IMPULSE	
	14 THRUSTERS AT 200 LB EA	
• STRUCTURAL/	MODULAR – PRIMARY STRUCTURE	SPACELAB
MECHANICAL	DOCKING ASSEMBLY	ASTP

*APPROXIMATELY 75 PERCENT OF MOSC COMPONENTS IDENTIFIED AS AVAILABLE HARDWARE OR TECHNOLOGY

The basic concept provides a number of growth options (Fig. 3) leading to geosynchronous facilities and expanded operations in low earth orbit. The basic design can provide a long-duration manned facility for the support of satellite servicing and the assembly of large space structures. These missions may involve space structure assembly projects in which large assemblies such as radio telescopes are assembled manually and then moved to the desired operational orbit by unmanned Tugs. The maintenance/checkout and crew-supported functions necessary to achieve these capabilities are inherent in the design concept. Interface capabilities could be designed into the initial modules with a minimum impact on system weight and cost. By using the baseline configuration in low earth orbit to perform these types of support tasks, Shuttle launches would be reduced and the on-orbit stay time of serviced equipment would be increased. In summary:

- A continuously manned orbital facility provides a platform for research, applications, and implementation including the assembly of large structures, on-orbit space manufacturing, etc.

- A free-flying facility as defined in this paper is the most cost-effective way to provide a continuing manned presence in space.

- The anticipated world problems of the 1990's must be addressed and solved in the 1980's. An extended-duration manned orbital facility can make significant contributions to their solution.

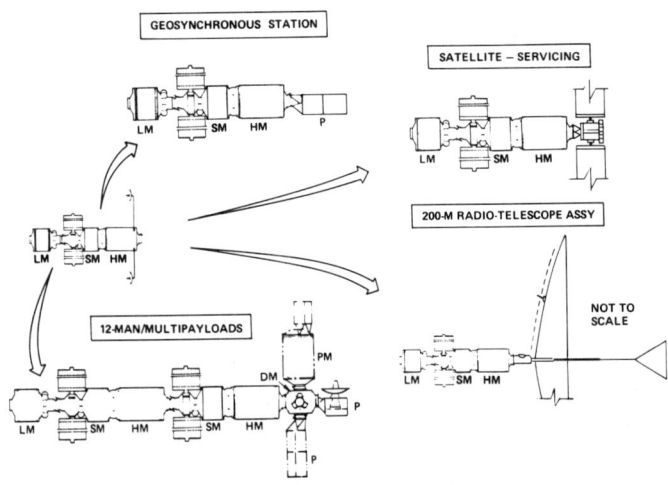

Fig. 3 Future Mission Aspects, Growth Options

AAS 75-264

AUTOMATION OF
SPACE PROCESSING APPLICATIONS SHUTTLE PAYLOADS*

Walter E. Crosmer[+]
Oakley T. Neau[+]
James Poe[‡]

The NASA Space Processing Applications (SPA) Program is examining the effects of weightlessness on key industrial material processes, such as crystal growth, fine-grain casting of metals, and production of unique and ultra-pure glasses. Work has been performed and is underway to define an inventory of equipment to conduct a series of experiments to obtain data on these effects. Some of the processes lend themselves to automation for reasons of safety and optimum performance. Process temperatures in excess of 2000°C require elaborate safety precautions and sample treatment if manual operations are undertaken. Automation of space processing payload equipment will also increase the number of potential Space Shuttle flight opportunities and increase the overall productivity of the program.

The SPA Program has identified six categories of material science research and technology that can be exploited in a weightless or low-gravity environment; metallurgical processes, processing of electronic materials, glass and ceramics processing, biological preparations, and physical and chemical processes in fluids.

Twelve different automated facility concepts were developed, encompassing the experiment requirements and representing variations in the hardware implementation. The characteristics of these concepts are

---

*This work was supported under Contract NAS 8-30741 from NASA Marshall Space Flight Center

[+] Bendix Aerospace Systems Division, Ann Arbor, Michigan

[‡] NASA Marshall Space Flight Center, Huntsville, Alabama

summarized on Table 1. The facilities were designed to accommodate "research size" material samples.

Table 1
AUTOMATED FACILITY CONCEPTS SUMMARY

Concept	Description	Facility/Experiment Capability
L-1	Electromagnetic Levitation (35 kHz). Vacuum (10^{-5} to 10^{-7} Torr). Electron-Beam Heating. Multiple Samples (6).	Spherical samples, 1 to 4 cm diameter. Low resistivity, high secondary emission. Melting, purification, and homogeneous solidification of metals, their alloys and compounds, and chalcogenide glasses.
L-2	Electromagnetic Levitation (100 kHz). Vacuum or Atmosphere (10^{-7} Torr to several atmospheres). Induction Heating. Multiple Samples (6).	Spherical samples, about 2 cm diameter. Low resistivity, low secondary emission, e.g., beryllia dispersion in beryllium.
L-3	Electromagnetic Levitation (15 MHz). Vacuum or Atmosphere (10^{-7} Torr to several atmospheres). Induction Heating. Pre-Heating (imaging or resistance). Multiple Samples (6).	Spherical samples, 1 cm diameter. High resistivity materials requiring pre-heating to improve electromagnetic efficiency, e.g., high melting point oxide glasses. Controlled cooling.
L-4 L-4A L-4B	Acoustic Levitation. Inert or Active Gas (5 Torr to several Atmospheres). (a) Resistance Heater. (b) Arc Imaging Heater. Multiple Samples (6).	Spherical samples, 1 to 4 cm diameter. Ultra-high resistivity, glasses and crystals, controlled cooling. (a) Low absorptivity materials. (b) High absorptivity materials.
F-1 F-1A F-1B	High-Temperature Resistance Heater Tube Furnace. (a) Multiple Furnace Units (6). (b) Sample Handling (6).	Self-contained cartridge samples. Equivalent diameter 4 to 10 cm. Glasses and crystals to 2,200°C. Controlled cooling.
F-2 F-2A F-2B	Low-Temperature Resistance Heater Tube Furnace. (a) Multiple Furnace Units (6). (b) Sample Handling (6).	Self-contained cartridge samples. Equivalent diameter 1 to 2 cm. Immiscibles, composites, and low-temperature (less than 1,200°C) crystals.
F-3A F-3B	(a) Moving-Zone Image Heater. (b) Fixed-Zone, moving sample; multiple samples or large rod.	Zone refining and directional solidification. Rod-shaped samples. (a) 1 cm diameter by 58 cm long, to 1,100°C. (b) 2 cm diameter by 58 cm long, to 1,900°C.
E-1	Continuous Flow Electrophoresis Unit.	Three biological specimens. Collect up to 50 separated fractions.

Five of the concepts were selected for preliminary design and analysis. Special consideration was devoted to mechanical, thermal, and electrical interfaces with the spacecraft. The functions of core equipment, which provides the central processor unit and data collection and control functions, were investigated with a view toward equipment standardization. Characteristics of the selected facilities are summarized in Table 2. Power requirements are dependent on the material and sample size, as indicated by the range of requirements on the table.

Table 2

PROCESSING FACILITIES DESIGN SUMMARY

Processing Facility	Weight (Kg)	Volume (m³)	Sample Material	Sample Diameter (cm)	Peak Power (KW)	Average Power (KW)
L-1 Electromagnetic Levitation	528	3	Tungsten Tungsten GeTe	1 2 2	4.5 7.6 4.5	2.0 4.8 1.0
L-4A Acoustic Levitation	369	3	Silicon Silicon Alumina	4 11.5 1	3.3 5.8 6.0	2.7 4.5 3.6
F-1A High-Temperature Furnaces (6)	291	1.4	TiO$_2$ YIG	4 (equiv.) 10 (equiv.)	5.1 3.0	3.4 1.9
F-2A Low-Temperature Furnaces (6)	303	1.4	CdSe Al$_2$O$_3$ in Al	1 (equiv.) 2 (equiv.)	2.1 2.0	1.8 1.9
F-3A Zone Refining (Image Heater)	212	0.6	CuAl	1 (equiv.)	3.7	3.3
F-3B Zone Refining (Resistance Heater)	406	1.3	Silicon	2 (equiv.)	4.1	3.0
E-1 Electrophoresis	114	0.2	NA	NA	1.4	0.9

Automated payloads made up of one or more facilities have been defined for conducting equipment and technique verification tests and SPA experiments on early STS flights. It is expected that most of these missions will be shared with other disciplinary areas so that the entire payload will not be devoted to the SPA equipment.

Three payload concepts for accommodation in the STS were derived; (1) a cargo bay mounted payload dependent on the Shuttle Orbiter for power and heat rejection, (2) an independent payload with a single fuel cell arrangement and its own heat rejection unit, and (3) an independent payload with two fuel cells and its own heat rejection capability.

Fig. 1 shows the two fuel cell concept highlighting some of the payload equipment. This payload can be accommodated in 2.4 to 3m (8 to 10 ft) of cargo bay length with a total weight of 3600 Kg. A typical five-day on-orbit mission consists of performance of a molten zone crystal growth experiment, material purification by zone refining, six low temperature (less than 1200°C) vapor crystal growth processes in a closed container, and containerless processing of six samples in an acoustic levitation facility. The support unit provides peak power of 15.2 KW and total energy of about 1500 KW hr to the payload equipment.

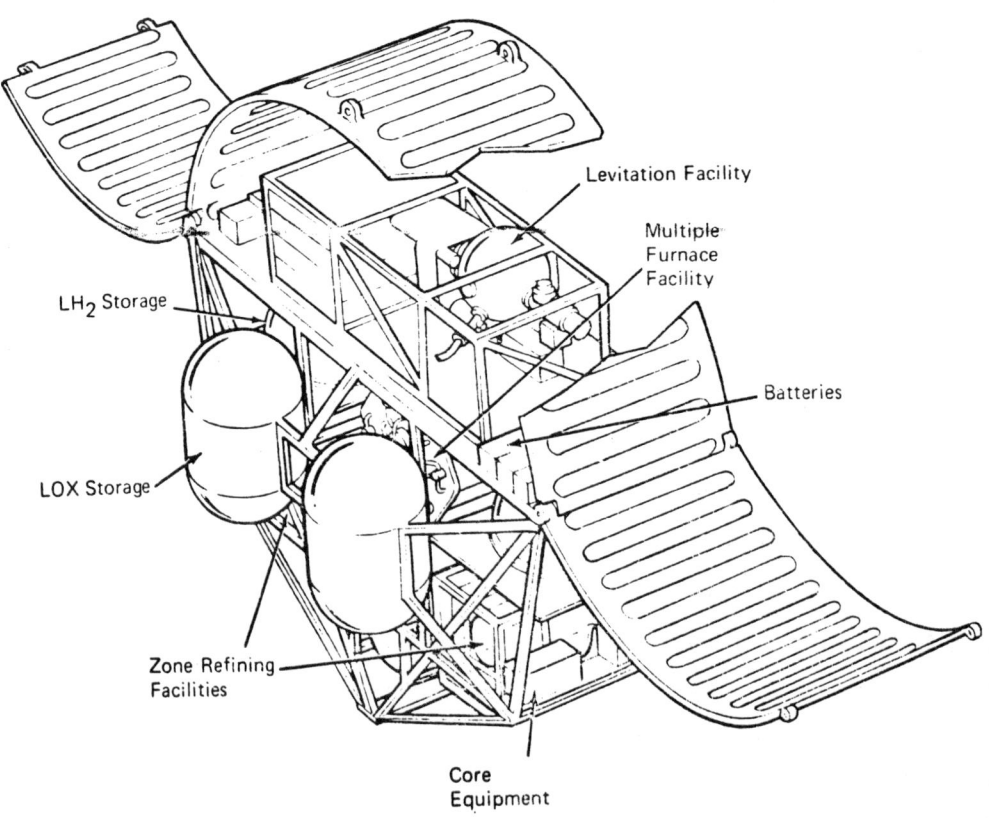

Fig. 1 Two Fuel Cell Payload Concept

AAS 75-265

SHUTTLE BENEFITS FOR AUTOMATED RETRIEVABLE
CRYOGENICALLY COOLED PAYLOADS

John O. Simpson
Thomas M. Spencer
William H. Follett, Jr. *

Requirements for planned and contemplated missions for infrared astronomy, relativity, and cosmic ray physics are used to develop dewar size requirements and to demonstrate the capability offered by planned liquid helium cryogenic systems for other experimentation. Cryogenic system sizes for the missions reviewed are grouped around two cryogen quantities, 150 kg and 500 kg. This suggests two "standard" cryogenic automated payload sizes for the coming era of Shuttle applications, physics and astronomy investigations. Present booster capability restricts experiment size, weight, operating life, and cost latitudes. Use of Shuttle for weight/cost design latitude, retrieval, servicing, and maneuvering for precise orbits and injection attitudes provides higher payload capability at lower cost, and allows serious consideration of "standard cryosats" for cryogenic temperature experimentation of the future. Approaches to the standardized satellites are straightforward. The conclusions reached by this summary survey are that the shuttle capability for supporting cryogenic automated payloads allows planned and contemplated requirements to be met with two standard shuttle-launched payload designs, and that the present and planned technology development status for cryogenic payloads is compatible with the shuttle operational schedule.

---

* Ball Brothers Research Corporation, Boulder, Colorado 80302

INTRODUCTION

The desirability of measurements requiring long-term operation of equipment in earth orbit at temperatures below 2° K is well established. Experience with cryogenic containers on unmanned boosters and on Apollo missions has demonstrated the practicality of handling and managing cryogenic fluids through ground preparation, boost, and zero-g environments and operating phases. Passive low-pressure dewar systems appear to be the simplest, lowest cost, most reliable and safest systems for attaining long term orbital operating temperatures in the 1 to 4° K range. This paper explores briefly the benefits offered by the Space Transportation System for sizable liquid helium dewar systems planned and contemplated for future automated payload missions requiring long life and low temperatures.

AUTOMATED PAYLOAD CRYOGENIC REQUIREMENTS

A summary discussion of mission requirements relating to cryogenic needs for several planned and contemplated missions follows. The information is tabulated (Table 1) following the discussion.

Infrared Astronomy

The planned joint Netherlands/United States infrared astronomy mission scheduled for Delta launch prior to shuttle availability is designed around a 0.6 meter Casegrain telescope which will do an all-sky scan survey and will point to specific targets for extended periods. One year mission operation duration is planned. A sun synchronous polar orbit at about 600 km controls the earth/sun/telescope axis geometry for thermal purposes. The telescope is pointed generally up nadir. The mission satellite is not specifically designed for retrieval since it will fly several years before the shuttle polar orbit retrieval capability exists. Although Delta restraints limit the dewar size, and thus the operating life and/or the number of detectors which can be used, mission hardware of the approximate size planned would be appropriate for a first mission even if shuttle were available for launch and retrieval. This is so because the cryogenic technology development programs which were begun several years ago are based on a dewar sized for Delta launch. It is appropriate that this general hardware size be flown before larger systems are produced. Thus, planned shuttle flight scheduling is timely for follow-on missions using the same, and larger, hardware.

The astronomy mission satellite carries about 100 kg of liquid helium in a dewar with outer shell diameter of about 1.5 meters and length of about 1.7 meters. Heat load is about 70 milliwatts of which about 30 milliwatts is detector dissipation. The dewar has a cylindrical experiment cavity, which houses the telescope and detectors, of about 0.7 meter diameter and 1.5 meters length.

Contemplated long duration future IR astronomy missions will probably be based on the one-meter aperture telescope size now planned for sortie mode operation in the shuttle bay. Free-flying telescopes of this size require a much larger dewar system since heat loads will be appreciably higher due to longer mission operation times, supports being larger and more detectors being of interest. An estimate of dewar sizing for a two-year mission carrying a one-meter diameter by 2-meter long telescope in the experiment cavity is 2.3 meters in diameter by 2.3 meters in length with about 450 kg of liquid helium. Clearly, this must be an STS payload, and must be a free flyer due to duration of missions.

Fig. 1 shows the general sizes for planned and future astronomical support dewars. Fig. 2 shows the Netherlands IR Astronomy Satellite.

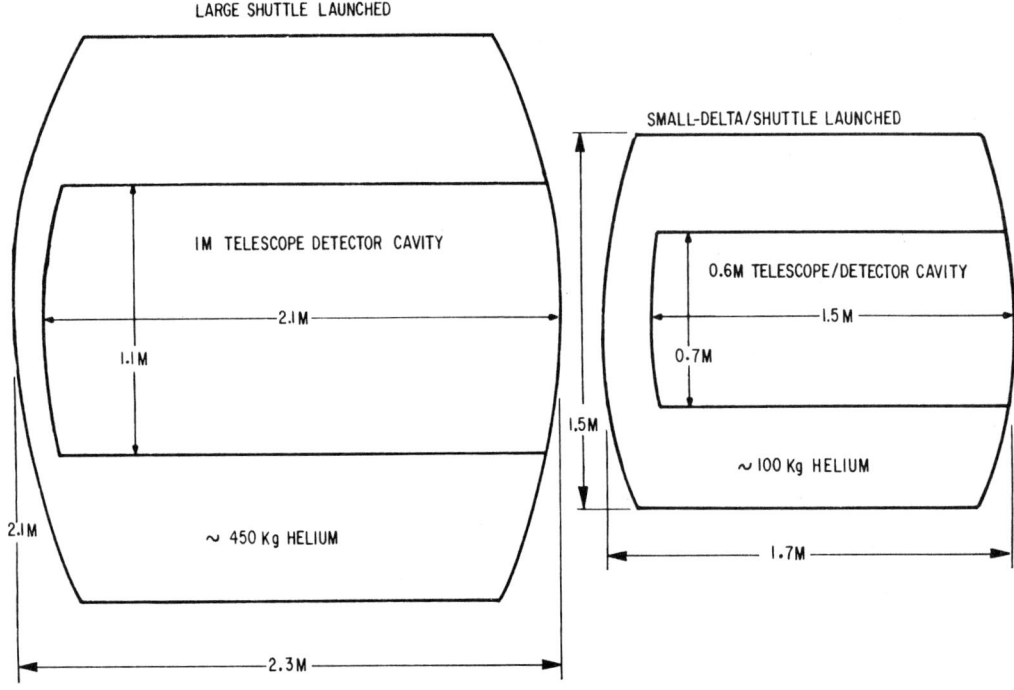

Fig. 1 Astronomy Mission Dewar Sizes

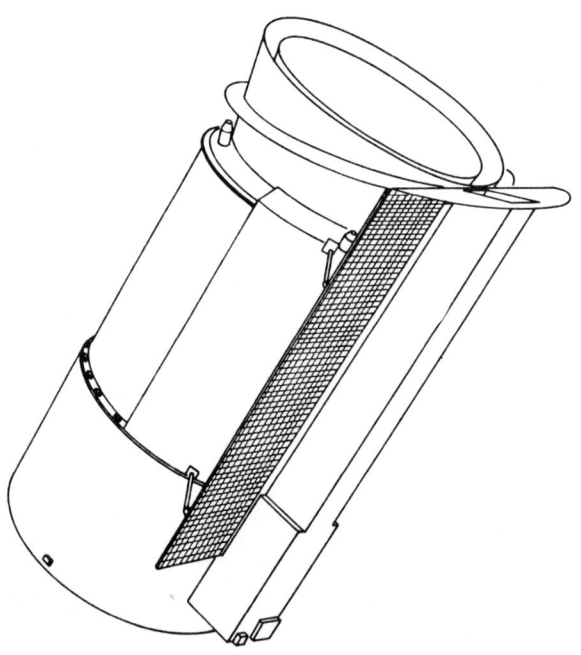

Fig. 2 IR Astronomy Satellite

Relativity Experiment

The relativity experiment in the planning calls for a Shuttle launch in 1981 (Mission No. 10) in a 460 km, 57° orbit, retrieval after six months, and reflight for one year in a polar orbit early in 1983. Considerable work has been done on the technology development and design definition for this mission by Stanford University (the "inventors"), Marshall Space Flight Center, and Ball Brothers Research Corporation (BBRC). Since a number of technological capabilities of interest beyond the relativity missions have been developed, these are described to show potential usefulness for other experiments.

Gyroscopes. The heart of the relativity experiment is the cryogenic gyroscopes. Einstein's general theory of relativity predicts that a gyroscope in orbit around the earth will undergo about 7 arc seconds per year relativistic precessional drift caused by the motion through earth's gravity field. In order to measure this drift with the precision needed to distinguish between Einstein's theory and competing theories, accuracy of drift measurement must be about 0.001 arc second per year. To do this, the gyroscopes must be operated at superfluid helium temperatures for thermal/mechanical stability, use a superconductor London moment readout, and use superconductor magnetic field shielding. Accelerations from air drag, solar pressure, satellite attitude control thrusting, and gravity gradients must be no more than about 10^{-9} g's to keep Newtonian precessional drifts suitably low. These several capabilities are discussed in the following paragraphs.

Zero-G Environment. The application of the Stanford DISCOS-type drag free control system, required to keep air drag accelerations for the gyros below 10^{-9} g's, has application for other experiments requiring "zero g" environment. Thus, the capability required for the relativity experiment produces a capability for combining extremely low g-fields and extremely low temperatures.

Stable Thermal/Mechanical Environment. The relativity experiment gyros and reference star tracker telescope are all made from high quality, very homogeneous quartz parts. They are all optically contacted together and all parts are at the superfluid helium temperature. This produces an extremely stable thermal/mechanical environment virtually free of temperature gradients and mechanical creep. Thus, the environment needed by the relativity experiment hardware is an environment of utility for other experiments requiring extremely stable conditions and extremely low temperatures.

Superconductor Environment. The position of the gyro spin axes cannot be determined by conventional means. Even light beam readout photon accelerations cannot be tolerated. Thus, readout by London moment detection is used (the gyro ball rotor is coated with niobium and this rotating superconductor produces a very small magnetic dipole moment). A Josephson junction (superconducting ring with a "weak link") is used to detect the London moment field coupled into the ring. This capability combined with the superconductor temperatures should have application to other superconductor experimentation at these extremely low temperatures.

Low Magnetic Field Environment. The gyro ball and the readout loop, being superconductors, will trap magnetic flux as the critical temperatures are reached. Trapped flux will completely overwhelm the small London moment flux, and must be reduced to near zero levels. This requires two techniques. First, the magnetic field at the gyros must be less than 10^{-7} gauss as the cool-down occurs, and second, superconducting shields must completely surround the gyros and their readout loops to assure that no field penetrates to the gyros after removal from the low field environment after cool-down. Thus, the relativity experiment causes a capability to exist for producing a combined environment of extremely low magnetic fields and extremely low temperatures.

If these environments are all combined, as must be so for the relativity experiment, the capability for an experiment chamber at 2° K, 10^{-9} g's, and 10^{-7} gauss is achieved. In combination with the space vacuum, it is difficult to imagine a more passive environment for experimentation.

Autonomous Space Navigation. These highly drift-free gyroscopes form the basis for an autonomous space navigation system. After initial inertial referencing to stellar sources, a three-axis inertial reference can be maintained with 1 arc second per year accuracy without any reference updating required and without gyro torquing (by use of an appropriately modified readout system). Thus the relativity experiment technology offers a dramatic improvement in attitude reference technology of a highly counter-measure resistant nature.

Attitude Control. The relativity experiment satellite uses the boil-off helium gas for attitude control and drag-free control with proportional control gas valves. This valve and control technology is useful for other cryogenic missions for maximum utility of the boil-off gas. Some missions, depending upon external torques, can use boil-off gas exclusively; others may require supplemental control gas or magnetic control.

The relativity satellite is built around a one-year life dewar carrying 150 kg of liquid helium. The dewar is 1.5 meters in diameter and 1.5 meters long with an experiment cavity 0.5 meter in diameter and 1.25 meters long. Expected heat load is 125 miliwatts. Fig. 3 shows the dewar and

Fig. 4 shows the satellite. The relativity experiment missions must use the STS to realize the savings planned by retrieval and reflight using one satellite for two missions.

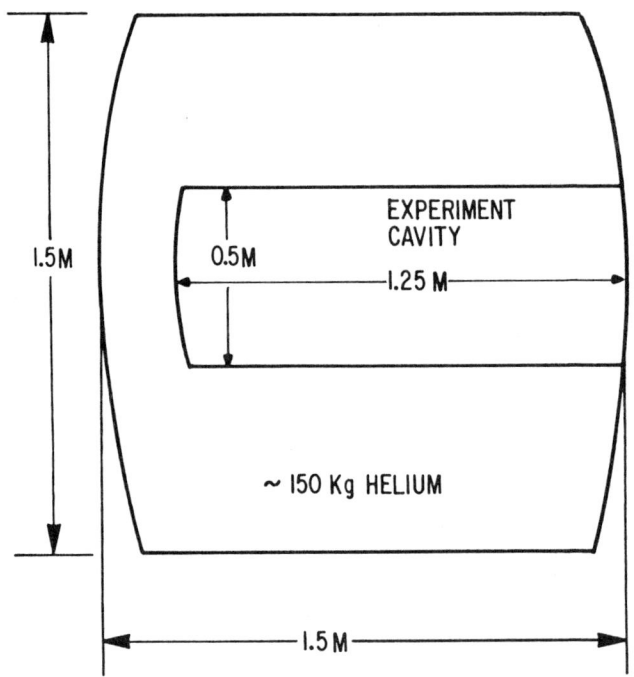

Fig. 3 Relativity Explorer Dewar Size

Cosmic Ray Physics

High energy cosmic ray spectrometers requiring extremely high magnetic fields are practical with the use of superconducting magnets. Such a spectrometer contemplated for shuttle launch contains the superconducting magnets in a two-meter diameter by two-meter long dewar holding 320 kg of superfluid helium for one-year mission operation lifetime. Superfluid helium is required to allow energy absorption from magnet discharge without depletion of the cryogen as would occur with normal helium. The magnets and support structure are contained in a 0.6 meter diameter by 1.75 meters long experiment cavity. Fig. 5 shows the dewar size.

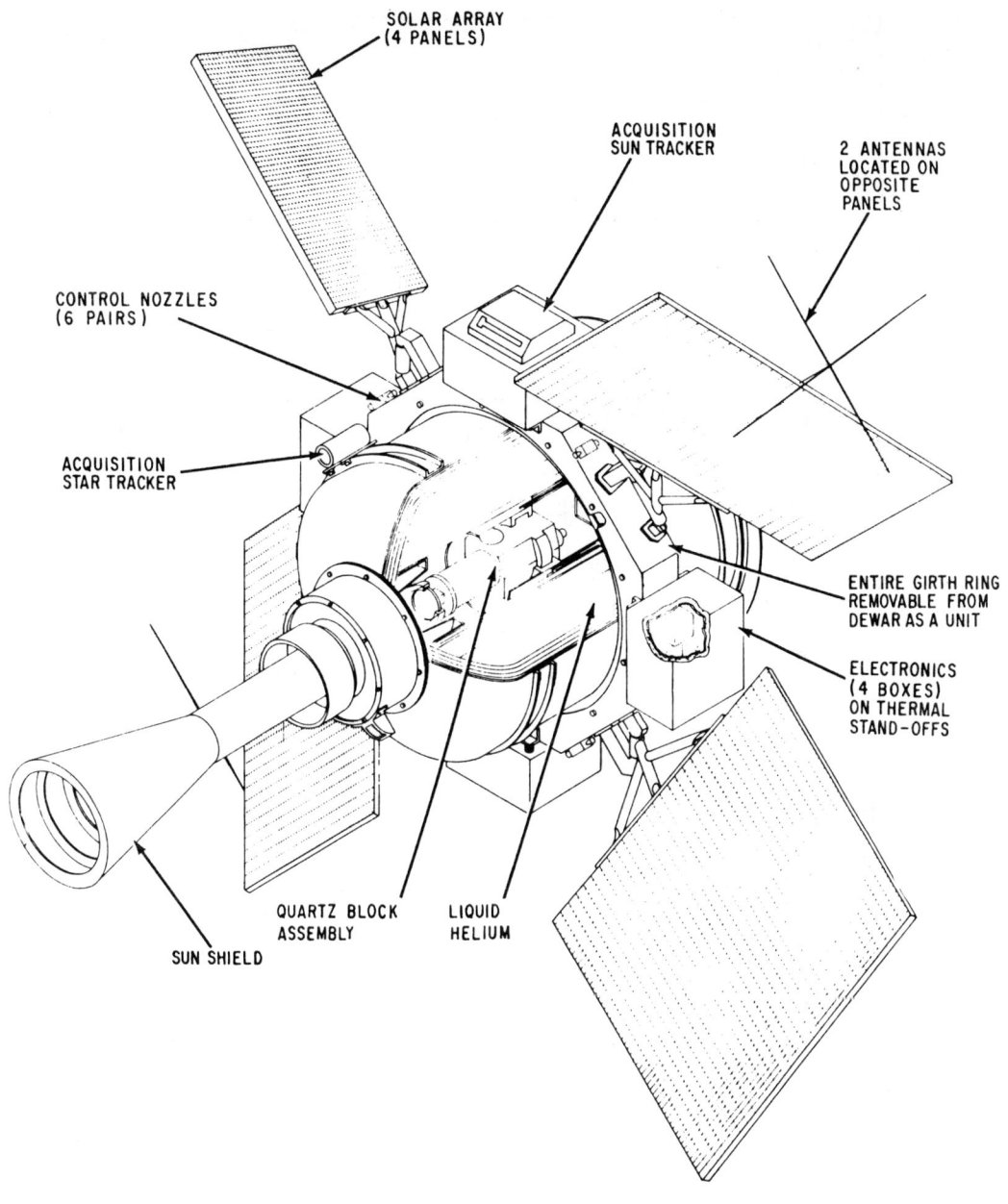

Fig. 4   Relativity Explorer Satellite

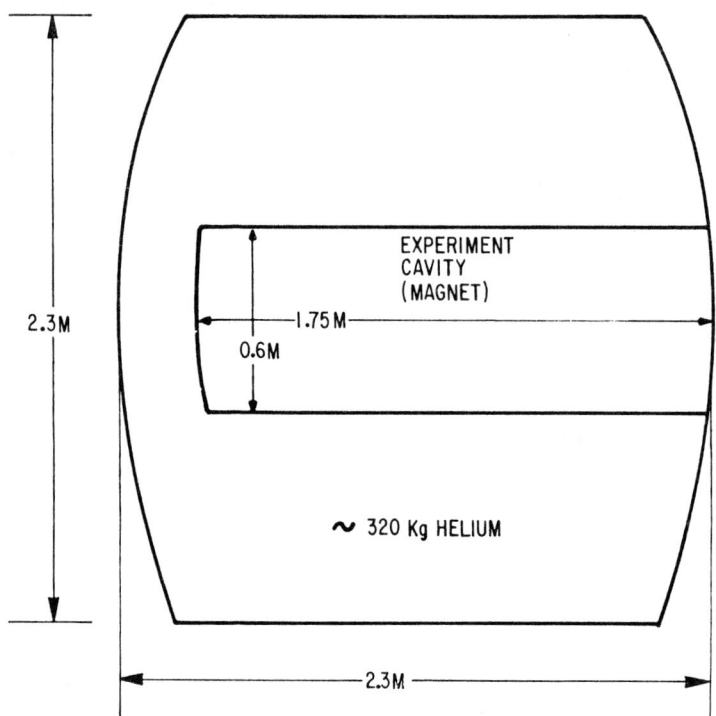

Fig. 5 Cosmic Ray Physics Dewar Size

Summary of Mission Requirements for Dewars

Table 1 - DEWAR REQUIREMENTS

Mission	Dewar Size		Experiment Cavity Size		Weight of Cryogen	Mission Duration
	DIA (M)	Length (M)	DIA (M)	Length (M)	(kg)	(Yrs)
IR Astronomy (Delta Shuttle)	1.5	1.7	0.7	1.5	100	1
IR Astronomy (Shuttle)	2.3	2.3	1.1	2.1	450	2
Relativity	1.5	1.5	0.5	1.25	150	1
Cosmic Ray	2.0	2.0	0.6	1.75	320	1

CHARACTERISTICS OF REQUIRED CRYOGENIC SYSTEMS

Table 1 shows that two distinct system sizes are appropriate; a small system with 100-150 kg helium storage, and a large system with 320-450 kg helium storage.

The small system can use a 1.7 meters diameter by 1.7 meters long external dimension dewar (6 by 7 feet) with an experiment cavity 0.7 meter diameter by 1.5 meters long with about 150 kg helium capacity.

The large system can use a 2.4 meters diameter by 2.4 meters long external dimension dewar (8 by 8 feet) with an experiment cavity 1.1 meters diameter by 2.1 meters long with about 500 kg helium capacity.

As a quite conservative rule of thumb, the dewar weight will be about four times the cryogen weight. System (with cryogen) weights for the two "standard" sized systems then are about 750 kg (1650 lbs) and 2500 kg (5500 lbs). Fig. 6 shows "standard" sizes.

The flexibility of use of the two systems is clear. The larger system can be used for very long missions with nominal heat loads, for shorter missions with higher heat loads, and for missions requiring the larger experiment cavity dimensions. The smaller system can be used for experiments for which its characteristics are more appropriate.

One design characteristic common to both dewar systems is the experiment cavity. This design allows telescope open apertures, star telescope windowed apertures and sealed cavity experiments. With this design feature, the experiment hardware may be separately prepared, calibrated and tested in ground equipment dewars and then installed in the flight dewar just prior to system testing before delivery to the Shuttle launch facility. The experiment can even have its own outer shell (cavity interface) configured as a dewar to maintain low temperature during transfer from a ground dewar to the flight dewar. This "dewar within a dewar" concept is planned for the relativity experiment.

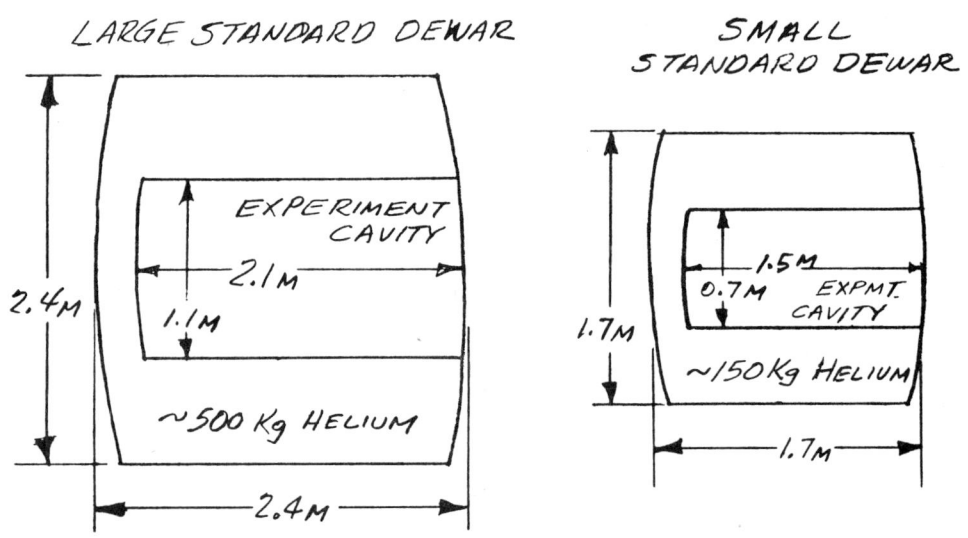

Fig. 6 Standard Cryogenic Spacecraft
Standard Dewar Sizes

PRE-STS RESTRAINTS

Experiment Size and Operating Life

As a practical matter, present booster size and weight capability restrains dewar size to about 1.5 meters diameter. Larger dewars could be physically enclosed in the heat shields, but practical satellite design for automated cryogenic payloads dictate either a "bellyband" of spacecraft equipment around the dewar center of mass or an "end cap"

for support equipment. The dimensions of equipment mounted
around the dewar use up the remaining shroud diameter in
general and solar array must be accommodated. Fig. 7 illustrates this situation.

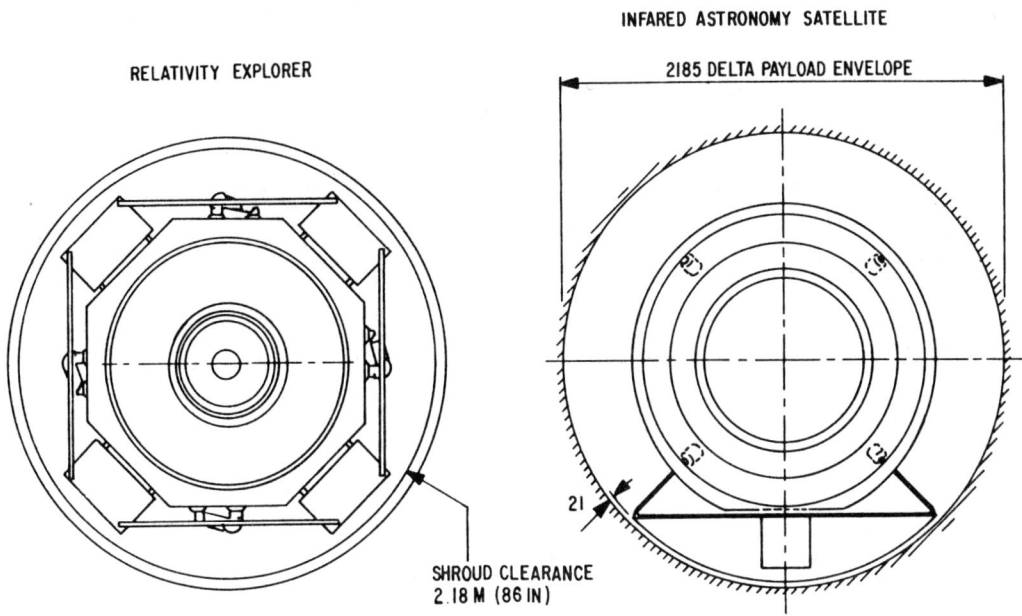

Fig. 7 Unmanned Booster Shroud Restraints

Dewar mechanical and thermal design dictates cylindrical
configuration with length-to-diameter ratios as near one as
is practical. Spherical containers are too expensive to in-
sulate. L-over-D ratios of as high as two are reasonable,
but longer thinner dewars have much higher heat leaks than
do short fat ones.

Thus present boosters can handle the smaller size of the two
"standard" sizes considered above, but have no hope of
handling the larger size. The present launch capability
then restricts experiment size to that for the smaller

standard dewar and restricts mission operation time to one year or so for experiments which present any significant heat load to the dewar.

Cost

An automated payload designed around a helium dewar system represents perhaps the ultimate example of mission limitation by quantity of an expendable. The designer is usually forced to complex optimizations in order to get the biggest "fuel load" he can on-board. This means preserving weight for fuel and looking for all the dimension he can get to use for his "tankage". Thermal constraints have to be placed on the experiment being accommodated, causing the experiment design to be more expensive and the measurements to be restricted. Present boosters, with high thrust accelerations, require a heavy duty dewar pressure vessel to outer shell suspension system. If this system is sized for the launch loads, it is an appreciable heat short. If the system uses retractable supports, removed when in orbit, it is complex, expensive, and represents a significant critical single point mission failure mode.

Thus designing and building a cryogenically cooled system under the restrictions of present boosters requires a relatively expensive approach.

SHUTTLE BENEFITS

Retrieval Capability Feature

Probably the single most important benefit for cryogenic automated payloads is the capability of Shuttle for retrieval. For payloads with predictable life of expendables, routine schedules for retrieval and reflight can be precise since there is no question of performance beyond predicts. Many reflights of the same payload with different experiments are practical.

Orbital Servicing Feature

With the shuttle payload weight and volume capability, any single mission cryogenic automated payload should be able to carry enough cryogen to satisfy mission objectives with one flight. Thus, for these payloads, replenishment of the cryogen on-orbit is probably not necessary. However, for a facility type payload, for example, an astronomical telescope, dewar refill is a real benefit. It seems reasonable that, after a two-year observing time, it would be desirable to replace the instrumentation before continuing the mission. This brings up the potential of instrument change out on-orbit and dewar refill. By placing the payload in the bay, change-out of instruments should be straight-forward since the dewar instrument cavity design makes the instrument accessible. Dewar refill in the bay from a storage dewar source should also be straight-forward. All support system components are readily accessible for change-out as well.

If standard subsystem modules are used, for example those conceived by Goddard for SMM, EOS, GRE, and ERTS, subsystem change-out could be easily done.

Payload Weight Feature

Most of the optimization which must be done now for unmanned booster payloads to control weight can be eliminated in the design for shuttle boost. Dewar size is not restricted, so payload operating life is lengthened. Pressure vessels can be made from stainless steel, which is usually cheaper and easier to weld than aluminum. Stress margins can be much larger where cheaper, but heavier designs are appropriate. Lightweighting machining and chem-milling is not necessary. Generally the design can be approached from the utility and value viewpoint rather than the lightweight optimum viewpoint. Hardware reliability will be increased.

Payload Size Feature

Benefits here are in the dewar size lattitude and the potential for eliminating deployments (solar array, sunshades) which have been necessary to get all the hardware under the unmanned booster shrouds. The larger size standard cryogen system, which could not be flown on an unmanned booster, fits easily in the orbiter bay. Dewar sizing is not restricted, so payload operating life is extended. Sensor heatload can be increased significantly and compensated by just carrying more cryogen.

Orbiter Maneuver Feature

Payloads flown by unmanned boosters must do their own initial acquisition of targets. Usually backup systems just for acquisition are needed. The payloads either work or all is lost. If precision orbits are needed, entire dependence is placed on booster accuracy, or else a ΔV system must be carried by the payload.

With the orbiter, initial acquisition can be done prior to release by orbiter maneuver. Precision orbits can be obtained prior to release. Orbiter can fly formation with the payload to be sure the payload is healthy; if it's not, it can be returned for repair.

A typical orbiter-aided injection sequence could be as follows. Assume a particular guide star as the acquisition target. The orbiter star tracker is gimbaled to pick up the guide star. The star tracker orients the orbiter to a position where the yaw axis is pointing to the guide star. The payload star tracker reads out payload look axis error to the guide star. The orbiter star tracker is gimbaled to the necessary offset angle to null the payload pointing error. Orbiter is now limit cycling about the true payload look axis. The payload is functionally checked out. The payload is released with small initial acquisition error. Orbiter flys formation for one or two orbits while all payload functions are checked, either from on-board orbiter or via ground link. If the payload is healthy, orbiter proceeds with its mission work. If the payload is sick, orbiter recovers the payload for repair (module replacement) or return to earth.

Orbiter Payload Attachment Feature

The orbiter provides at least three potential payload attach methods which could be used for a cryogenic automated payload. Payloads may be located horizontally above the tunnel, vertically in the bay on an attach fitting mounted to a pallet section, or horizontally in the bay using attach hardware as conceived by Goddard for the SSM or GRE missions or on a support ring. Figs. 8, 9, and 10 illustrate attachments.

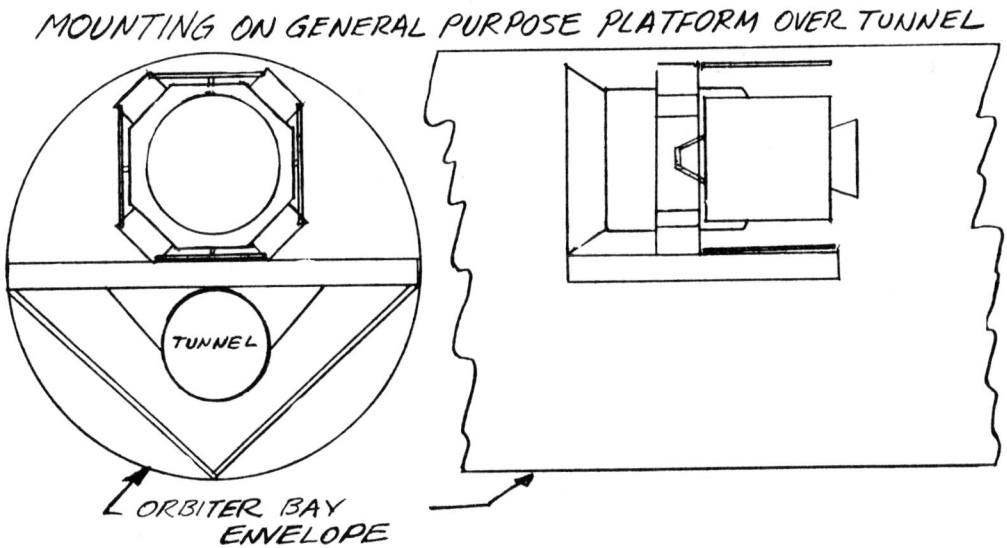

Fig. 8 Shuttle Operation Orbiter Payload Attachments

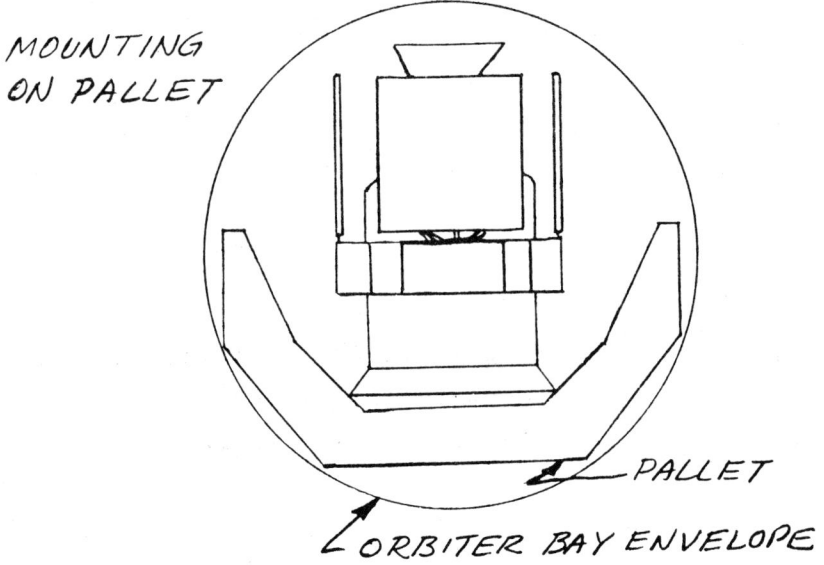

Fig. 9 Shuttle Operation Orbiter Payload Attachments

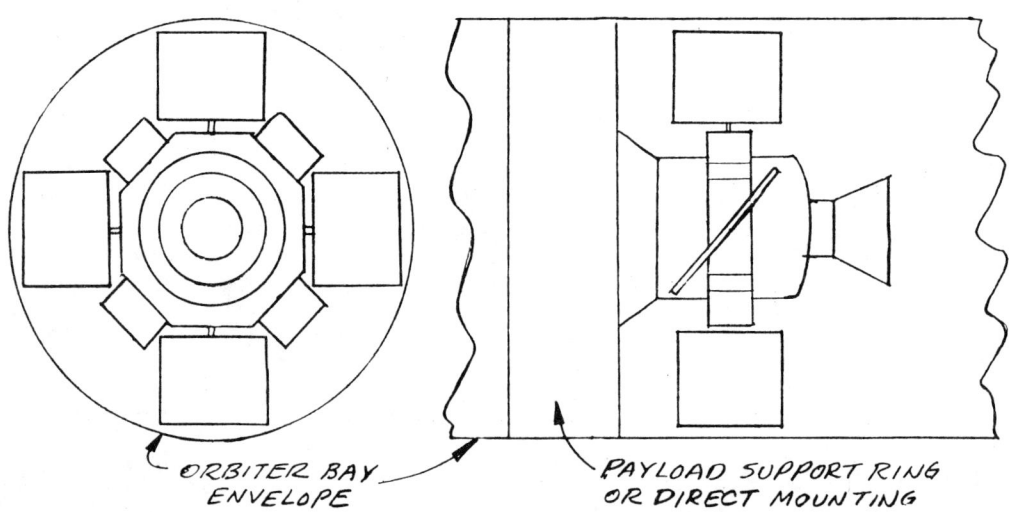

Fig. 10 Shuttle Operation Orbiter Payload Attachment

The benefits realized from features offered by the STS are highly significant, as reviewed above, for the technical advantages and operational lattitudes offered. Large payloads, which could never be flown by the usual unmanned boosters, can be easily handled by the orbiter. The benefits offered from the cost viewpoint, in comparison to unmanned boosters are also highly significant. Recent studies for the relativity mission show 15 percent of program costs saved for just a two-mission program even without recovery of the satellite after the second mission. To realize the maximum cost saving offered by STS for cryogenic payloads requires that a complete follow-through with standardized hardware be accomplished. The next paragraphs describe how this may be done.

STANDARD HARDWARE APPROACH

Since the dewar sizing requirements predict two standard sizes for dewars, it is reasonable to suggest two standard spacecraft for shuttle launched cryogenics work. The experiment support equipment can be the same for both spacecraft, whether the standard subsystem modular approach is taken or the standard component (black box) approach is taken. If the standard subsystem approach work now being done comes to fruition, then use of this equipment would be the most reasonable approach for the "cryosats". Figs. 11 and 12 show these two approaches to standardization. A third standard satellite approach, shown in recent studies for the relativity mission as feasible for the dewar size required for those missions, is to use an existing satellite such as the HEAO Spacecraft Equipment Module (SEM). With this approach, components would be standard and would be replaced during refurbishment on earth between missions.

Standard satellites can fly in any orbit from shuttle by using a flat solar array for sun synchronous orbits and an "omni" array for non-sun synchronous orbits as shown in Fig. 13.

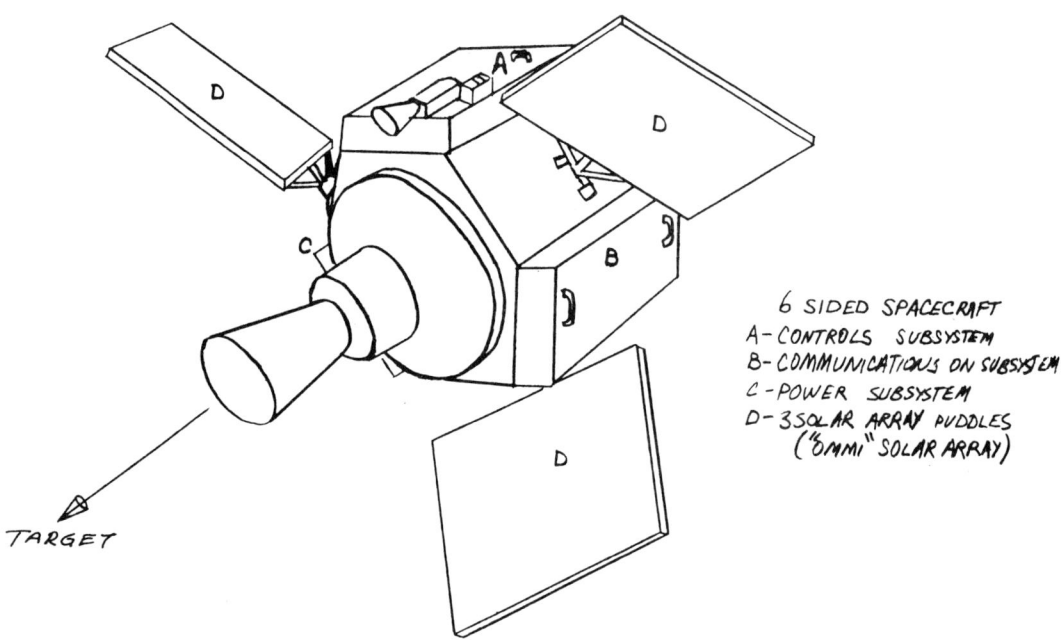

Fig. 11 Standard Spacecraft, Three Replaceable Modular Subsystem

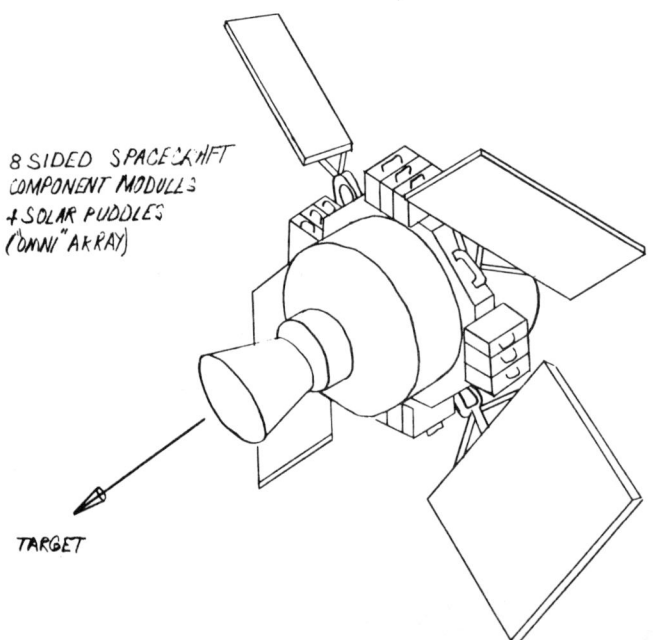

Fig. 12 Standard Spacecraft, Replaceable Modular Components

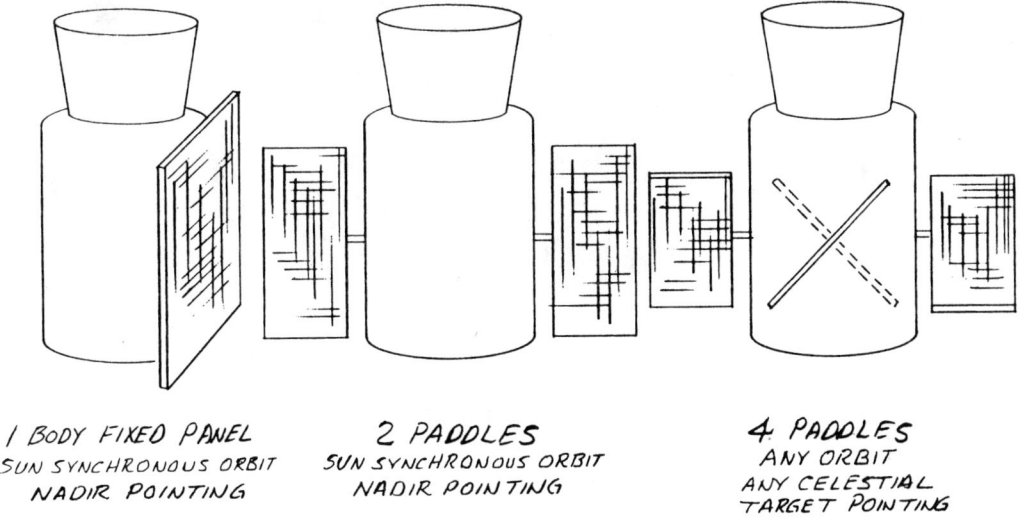

Fig. 13 Standard Solar Arrays

CONCLUSIONS

The shuttle capability for supporting cryogenic automated payload flight allows planned and contemplated requirements to be met readily with two "standard" sized payloads. Shuttle operational status scheduling fits well with technology development schedules and planned mission launch timing.

REFERENCES

1. Everitt, C. W. F., W. M. Fairbank and L. I. Schiff, Theoretical Background and Present Status of the Stanford Relativity Gyroscope Experiment, ESRO Colloquium on the Significance of Space Research for Fundamental Physics, Interlaken, Switzerland, September, 1969.

2. Bull, J. S., Precise Altitude Control of the Stanford Relativity Satellite, SUDAAR No. 452, March, 1973.

3. Ball Brothers Research Corporation, Mission Definition for the Stanford Relativity Satellite, Final Report F71-07, December, 1971.

4. Ball Brothers Research Corporation, Mission Feasibility for the Stanford Relativity Experiment on Scout, Final Report F73-03, June, 1973.

AAS 75-267

RETRIEVAL OF THE HEAO-C SPACECRAFT WITH SPACE SHUTTLE

David H. Mitchell[†]

INTRODUCTION

The High Energy Astronomy Observatories (HEAO) will be launched into low-earth orbit for cataloging of the universe in the X-ray and gamma-ray wave lengths and the high energy charged particle cosmic ray component. The HEAO program is managed by the NASA George C. Marshall Space Flight Center (MSFC) and the spacecraft is built by TRW Systems. One launch is planned each year in 1977, 1978 and 1979. The Observatories are 18 to 20 ft long, 92 in. diameter, weigh 6,000 to 7,000 lb and are launched by an Atlas-Centaur.

HEAO-C will complete its mission in 1980. Since this is the same time frame as the Shuttle Orbiter development flights, HEAO-C could be used as a testbed to develop and demonstrate its satellite retrieval capabilities. Therefore, MSFC contracted with TRW Systems to define a preliminary plan for a phased development program to determine requirements and implement Shuttle recovery capability in the HEAO-C Observatory.

DESIGN MODIFICATIONS FOR RETRIEVAL

The study identified two major design conditions for successful recovery of the HEAO-C by the Shuttle. First is the requirement to survive for up to three years until an Orbiter is available. To do this, the initial HEAO-C orbital altitude must be increased and the Observatory must be put into a passive storage mode that utilizes gravity-gradient and aerodynamic torques. The HEAO-C Attitude Control Subsystem (ACS) must be reactivated before Shuttle rendezvous since Observatory motion in the passive mode exceeds the current capture specifications for the Orbiter Remote Manipulation System (RMS).

[†] Space Vehicle Division, TRW Systems Group, Redondo Beach, California

The second design condition for successful recovery is compatibility with the Orbiter during capture and stowage in the payload bay. The design modifications for this condition fall in two areas - manned safety requirements, and structural mounting.

For Orbiter flight crew safety, there is a positive disconnect of all Observatory electrical power before it enters the payload bay. Pressure vessels that could be hazardous, such as batteries and tanks for propellant and experiment fluids, will be continuously monitored for abnormal pressures and/or temperatures. To minimize any hazard, residual propellant and experiment fluids are dumped before retrieval. Temperatures and pressures must still be monitored - the Orbiter will not rendezvous with HEAO-C until SAFE signals are given. All Observatory subsystems are disconnected by switching of a Safe/Arm device by the Orbiter RMS at capture. Once in the payload bay, SAFE signals are transmitted to the Orbiter flight crew by hardlines, using Orbiter power. Safing commands from the flight crew go through the same path.

Structural attachment of HEAO-C in the Orbiter payload bay is accomplished through the Observatory separation ring at its base and special upper-end fittings, as shown in Figure 1. Mechanical support equipment is required in the Orbiter for mating with HEAO-C, positioning it in the bay, and holding it securely during deorbit. Equipment to do this has been designed, using two retention frames.

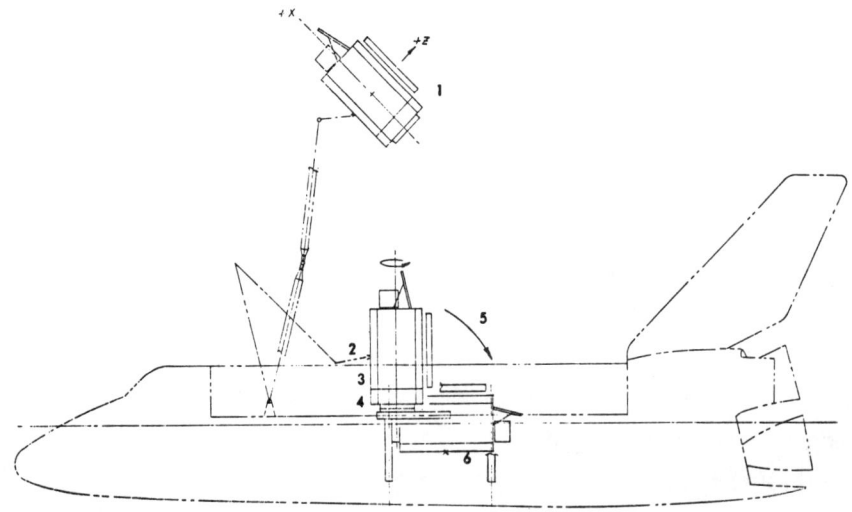

Figure 1. HEAO-C Retrieval by Space Shuttle Orbiter

OPERATIONAL SEQUENCE FOR HEAO-C RETRIEVAL

The operational sequence developed for retrieval of HEAO-C is:

1. HEAO-C is put into an initial orbital altitude that will provide extra lifetime. The specific altitude will depend upon a detailed tradeoff between orbit life and increased experiment radiation background at higher altitudes.

2. At the end of the regular HEAO-C mission, residual experiment fluids are dumped and the ACS is put into a passive storage mode. Tape recorders and experiments are turned off, but all other subsystems are kept on.

3. Gravity gradient and aerodynamic torques provide earth-oriented passive stabilization until just before retrieval (Figure 2). Observatory power and thermal control are degraded, but are sufficient to allow reacquisition of the ACS before retrieval.

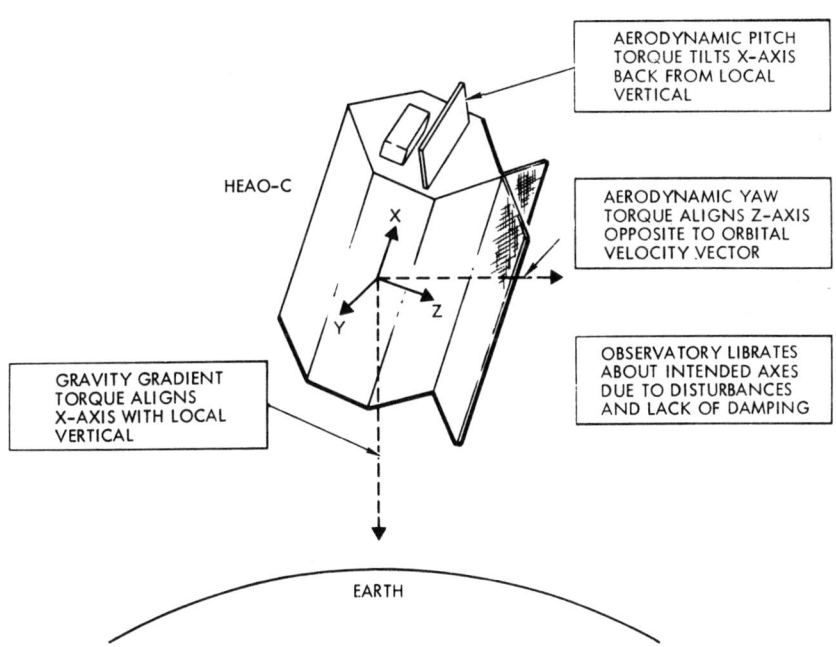

Figure 2. HEAO-C Long-Term Passive Storage Mode
(Earth Oriented)

4. The Observatory state-of-health is checked periodically (one or two times a year).

5. Reactivation of the Attitude Control Subsystem and subsequent Shuttle rendezvous are planned for the end of a maximum solar array illumination period, when the necessary Observatory equipment is warm.

6. Observatory SAFE signals (status of tanks and batteries) are transmitted to the Orbiter by RF before retrieval is initiated.

7. The HEAO-C Attitude Control Subsystem is reactivated. This can be done with or without the batteries.

8. Rendezvous and docking are accomplished, with the Orbiter as the active member. Upon engagement by the Orbiter Remote Manipulator System (see Step 1 in Figure 1), the remaining propellant is depleted and the Safe/Arm device deactivates the Observatory.

9. The Remote Manipulator System captures the HEAO-C, places it on the Primary Retention Frame (Step 2), and orients it until the spacecraft index pins engage holes in the support spider (Step 3). The hardline for Observatory safety data and commands is connected by motor.

10. After the RMS detaches, the Observatory is rotated 180° about the Orbiter yaw axis for proper solar array orientation. The Observatory separation ring is then locked to the spider (Step 4).

11. The Primary Retention Frame rotates the Observatory 90° about the Orbiter pitch axis (Step 5). The upper-end fittings engage the Secondary Retention Frame and are locked in place (Step 6).

12. The Observatory is passive in the Orbiter bay, except for safety signals and commands, during de-orbit, re-entry, landing and ground maneuvers.

AAS 75-235

THE MULTIMISSION MODULAR SPACECRAFT
FOR THE 80's

Robert O. Bartlett[*]

Frank J. Cepollina[*]

The challenge to NASA is to prepare for the missions of tomorrow with manpower and funding constraints of today. The Goddard Space Flight Center has met this challenge with the Multimission Modular Spacecraft (MMS). This spacecraft design has evolved over the past six years while studying various potential missions. The key to the concept of the MMS is modularity and flexibility to accept mission unique hardware with minimum impact on the basic spacecraft bus. Beyond this, it was imperative that this multiple mission bus be cost effective even though it would not be of an optimum design for many missions having minimum performance requirements.

The MMS performance and cost will capture 34 of the 43 potential spacecraft missions which have been initially studied. Gamma Ray Explorer (GRE) will be the first mission to utilize the MMS. Requests for proposals from industry to supply the modular subsystems for GRE will be released in the fall of 1975 if the mission is approved.

1. INTRODUCTION

The design of the Multimission Modular Spacecraft has evolved over the past six years. It was born out of frustration in attempting to use an existing spacecraft having a fixed structure for additional missions. From its genesis, the corner

---

[*] NASA Goddard Space Flight Center

performing a multitude of applications, space science and bio science missions, has been cost effectiveness. Faced with a situation of fiscally constrained budgets, an increasing rate of inflation, and the desire to maintain the same rate of technological and scientific progress which has been achieved in the last 15 years, Goddard Space Flight Center has embarked on a course of using a single spacecraft design to support a significant number of low earth orbit and geosynchronous missions.

The important cost saving aspects of the MMS design lies in the following:
- maximum utilization of standard components;
- design of a set of standard subsystem modules such that these subsystems can provide the services required by earth pointing, stellar pointing, solar pointing, and geosynchronous missions;
- design of a spacecraft system such that it can fully exploit the space shuttle's two way capability so as to extend the usefulness of the mission by either in-orbit resupply and/or retrieval of modular subsystems and/or instruments.

These features are the key to the capability of the MMS to capturing missions whose performance requirements are minimal. Beyond this, flexibility has been a design ground rule and potential users will find that they can:
- select the level of redundancy from non-redundant to fully redundant based upon their own cost/weight/reliability trade-offs;
- alter the solar array size and orientation along with battery capacity;
- alter the attitude control system performance by changing the on-board computer software to meet the unique requirements of their mission;
- add an Actuation Module with large reaction wheels or CMG's and magnetic torquer bars to handle very large payloads without impacting the spacecraft design;
- add propulsion systems for reaction control, orbit adjust and orbit transfer as required;
- accept fine error signals from mission unique sensors and/or instruments;
- add additional banks of computer memory up to 64 K words;
- program the telemetry format;

- interface with TDRSS;
- operate in any orbit and orientation without thermal constraints;
- capability for on-orbit servicing with "kits" provided by design;
- be launched by Shuttle, Delta or Titan vehicle.

The following sections will present an overview of the design of the four modular subsystems and the structural, thermal, and electrical systems which support them. The four subsystems are power, attitude control, communication and data handling, and propulsion. A more detailed presentation of this material can be found in Ref. 1.

2. STRUCTURAL SYSTEM

The baseline MMS configuration contains three subsystem modules supported by the thrust axis load carrying structure. The three modular subsystems are attitude control, communication and data handling, and power. A transition adapter and a base adapter complete the structural elements of the system. A propulsion/actuation module may be added at the base adapter if required by a specific mission. Other mission unique items include the solar array, the solar array drive, the solar array deployment mechanism, a set of communication antennae and a mission adapter. The MMS configuration is shown in Fig. 2-1.

The structure above the modules shown in Fig. 2-1 is the transition adapter which accommodates many different interfaces. The mission unique adapter, which supports the payload, bolts onto the transition adapter with essentially a flat clean interface. This allows the payload to be developed, integrated, and tested independently of the spacecraft. Other hardware such as the solar array drive motors are attached to this structure, as well as the array drive shafts.

The bottom of this adapter has a universal bolting pattern which will accommodate a non-resupply structural attachment member, or by the same bolting pattern, a load carrying resupply mechanism.

The launch configuration for the spacecraft differs only in the manner in which it is restrained. A mission which is orbited by a conventional vehicle such as the

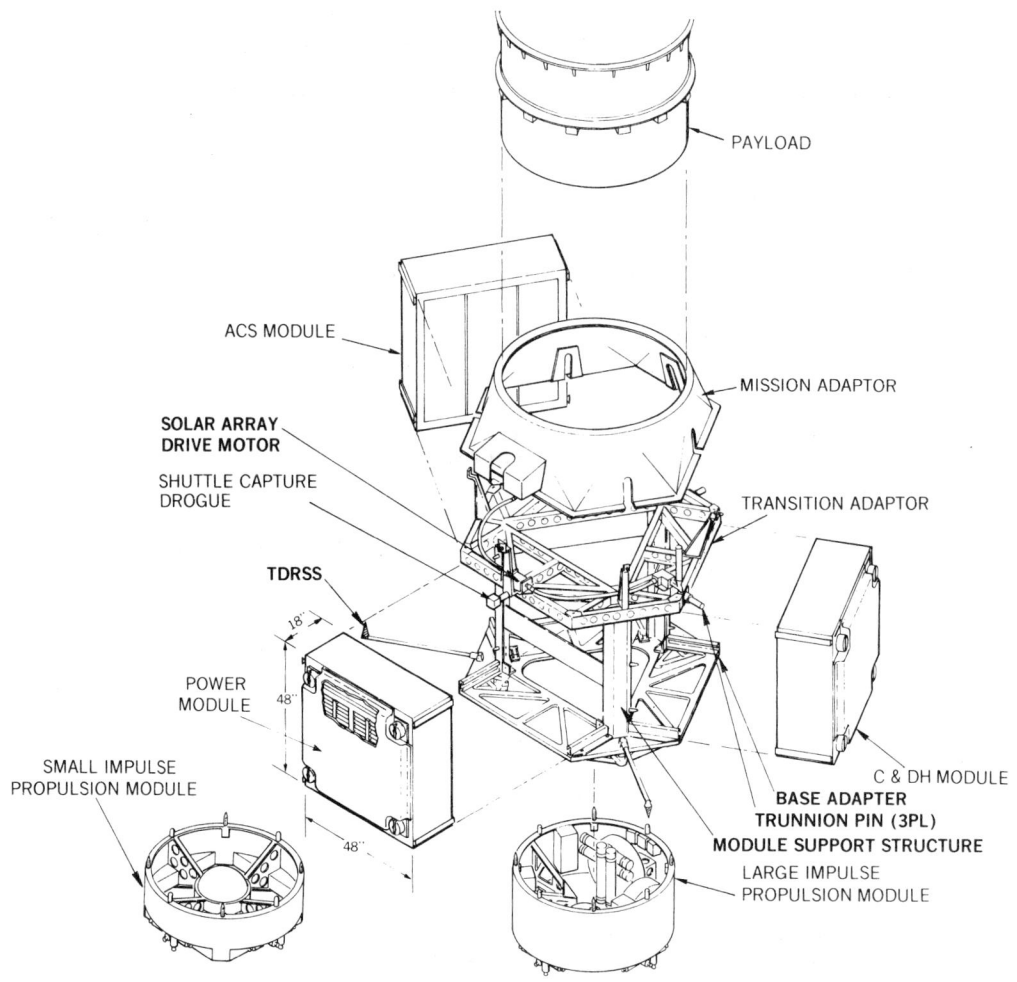

Fig. 2-1 Multimission Modular Spacecraft (Exploded View)

Delta will utilize an aft end mount separation system. Separation will be accomplished with a proven redundant stud and explosively actuated collapsible nut system. Reliable electrical umbilicals across the separation interfaces are also attached. After separation, a deployment sequence will be initiated by the spacecraft for all of the stowed appendages. The Delta 2910 launch vehicle configuration can carry a 2700 pound payload instrument on top of a 1300 pound spacecraft, for a total of 4000 pounds, to an altitude of 270 nautical miles at 28.5° inclination. The diameter of the payload is limited to 86 inches and the height is limited to 179 inches. A limited amount of conical volume is available above the cylindrical volume.

The transition adapter is used for a shuttle orbiter launch and for retrieval. Three trunion pins located on the periphery of this adapter fit into a clamping arrangement of the Flight Support System (FSS). In orbit the shuttle manipulator attaches to a drogue point on the side of the transition adapter and raises the spacecraft free of the launch restraints trunion. From this point the spacecraft can undergo a final electrical checkout and be allowed to be a free-flying spacecraft. The size and weight class of these payloads is not restricted by the modular spacecraft design since launch/retrieval loads induced by the adapter do not pass through the spacecraft. Also the propulsion/actuation module can accommodate large reaction wheels or CMG's and large magnetic torques and/or RCS thrusting. In addition the modular spacecraft design allows an Interim Upper Stage (IUS) or Transtage to be employed when extreme orbit transfer is required for a given mission.

The photograph in Fig. 2-2 shows a full scale earth observation type payload model being tested in the shuttle mockup at a North American-Rockwell facility. This would be the payload configuration just prior to orbital delivery.

Fig. 2-2 Testing of MMS Model in Shuttle Mockup

3. ELECTRICAL SYSTEM

The spacecraft electrical system provides the following functions:

1. power distribution;
2. signal distribution;
3. central grounding point;
4. signal conditioning and control utility.

Unregulated +28 volt power will be distributed to all spacecraft/system loads. The bus characteristics will be as discussed in Section 6. Unfused power buses will distribute power to each subsystem module and each instrument will be redundantly fused to protect the spacecraft bus. Power line voltage drop will be limited to approximately one percent of the supply voltage exclusive of the voltage drop across the fuses.

Distribution of command and telemetry data between the Communications and Data Handling module (C&DH) and other subsystem modules and the instruments will be handled via remote decoders and mutliplexers as described in Section 5. This party line method of signal distribution will minimize the need for numerous interfaces and provide significant immunity to noise.

A Central Ground Point (CGP) will be provided on the module support structure in the vicinity of the power subsystem and C&DH subsystem interface connectors. The CGP will be the busing point to the structure for signal returns and AC circuit returns for carrier or servo systems employed in the subsystems or instruments. Power returns will be bused in the power module and then returned to the CGP to minimize common impedance in the power distribution circuitry.

A Signal Conditioning and Control Unit (SCCU) will be designed for installation on the module support structure. The purpose of this unit is to house the control, conditioning and telemetry circuitry for those spacecraft functions which are not directly related to operation of the primary subsystems modules. It also provides structural housing for mission unique equipment and serves as a "catch-all." The addition of this module will minimize the number of interface connections required through connectors which must be demated and mated during resupply operations.

The circuitry within the SCCU will accommodate the following functions:

a. command decoders and associated circuitry;

b. arming, firing and bus protection circuitry for pyrotechnic functions required for appendage release, dust cover release, etc.;

c. power circuitry for control of electro-mechanical actuators such as may be required for restowable appendages;

d. power control circuitry for structure heaters;

e. signal conditioning circuitry and remote multiplexers for telemetry data such as appendage position, structural temperatures, pyro bus arming, etc.;

f. control of special functions which may be required for resupply operations.

All major system harnesses will be capable of being installed as integral assemblies. Harnesses that are internal to the spacecraft subsystem modules, will be designed so that assemblies can be replaced without removing harness. Installation or removal of the harness from the module should be possible without removing the assemblies.

The main structure harness will be divided into two major segments so that changes to either the spacecraft or instrument sections can be made without seriously delaying work on the other subsystems. This feature would permit, for example, a major change in the instrument wiring without significantly delaying integration or test of the spacecraft subsystems. Fig. 3-1 shows the proposed harness configuration.

Access for ground test of the spacecraft subsystem modules will be provided for by test connectors on the modules at positions to be determined as system design progresses. These connectors will afford access during all phases of module or system test, possibly including this capability during on-stand launch operations and orbital resupply. By the discreet use of proper isolation circuitry and stimulus inputs it will be possible to perform major tests on system performance without any chance of damaging the system or breaking connectors affecting flight circuits.

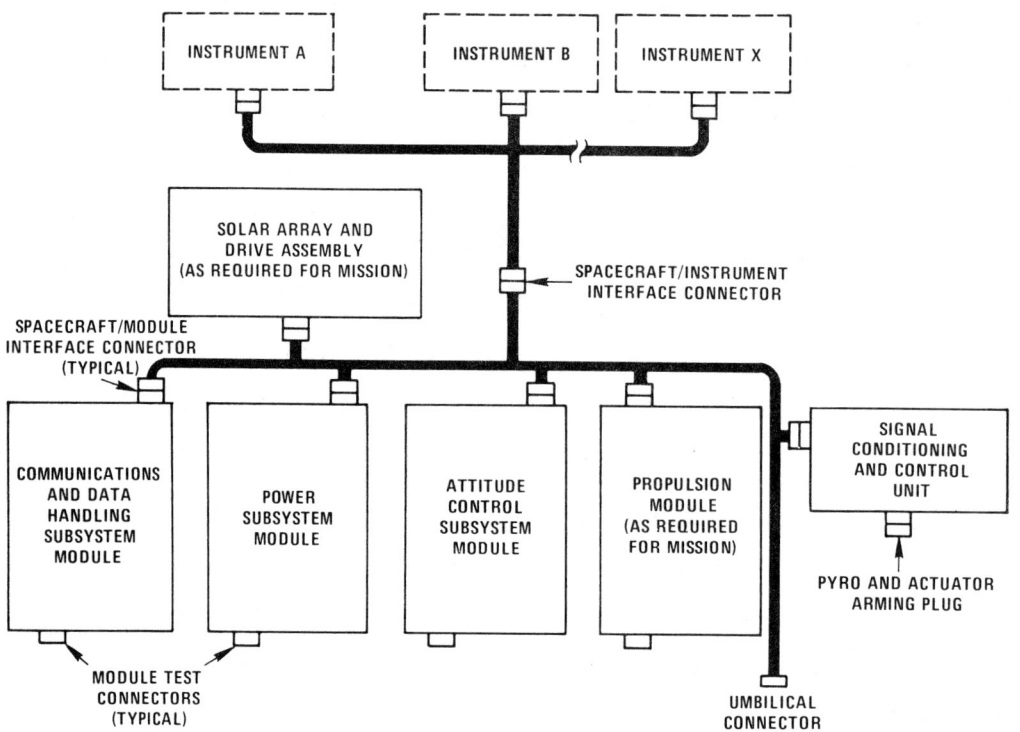

Fig. 3-1 System Harness Configuration

4. THERMAL CONTROL SYSTEM

The thermal design of the modular spacecraft satisfies two major objectives: (1) maintains all spacecraft components within acceptable limits for all phases of flight, and (2) accommodates all conceivable missions with a common design concept such that little or no change to the thermal configuration is necessary. Satisfying the latter objective will result in a minimum of thermal engineering manpower and thermal testing for the entire profile of missions. All thermal hardware employed in the design has been flight qualified on previous missions such that no developmental costs will be incurred. The concept of using passive thermal control supplemented by heater power is a sound technical approach which has been employed on many GSFC programs, and has been thoroughly evaluated for specific missions of the modular spacecraft. However, a louver design for the three subsystem modules has been selected as the baseline in order to more fully satisfy the overall objective of having a standard spacecraft design for all missions.

The orientations of the spacecraft for earth observation, solar and stellar missions have been optimized to achieve minimum variation in thermal environment for all modules. This effectively has minimized the extent of the thermal design requirements to satisfy all conceivable missions. Calculations to date indicate that a single design will satisfy virtually all missions.

Each module is configured with two sets of louvers mounted directly to the outboard side of the equipment baseplate or heat sink. The heat sink is covered with a high emittance coating to enhance the regulation capability of the louvers. A separate cover (outer skin) acts as a radiator as well as a protection, for the louvers, from solar impingement. The outside surface of the cover will be coated with a stable low α/ϵ material, Teflon for most missions and second surface quartz mirrors for synchronous missions or those missions that expose the radiator to a high concentration of charged particles (that may degrade the Ag Teflon). Modules that always face away from the sun have the Ag Teflon or second surface quartz mirrors mounted directly to the heat sink and no cover is provided.

All other surfaces of the modules are covered with multi-layer insulation blankets to isolate the surfaces radiatively from both the space environment and the remainder of the spacecraft. Satisfactory thermal isolation of the modules from the support structure is achieved with the use of fiberglass inserts, sleeves, and titanium bolts where applicable.

The module support structure is insulated and provided with heaters which can be activated, if necessary, to minimize structural temperature gradients when very fine pointing stability is desired.

A separate heater system on the structure and modules is included for use during shuttle resupply operations. Shuttle power will be used during these periods to maintain the required structure temperatures; thus avoiding any chance of mechanical misalignment and/or binding between the module and module support structure.

The propulsion/actuation module is passively controlled, supplemented by heaters on the hydrazine lines and tanks to maintain temperatures above the 5°C freezing point. Radiators are provided on the module side surfaces for those missions which require reaction wheels and associated drive electronics.

The orientation of the spacecraft for an earth pointing mission is such that the power module receives little or no direct sunlight. This is achieved with one-sided solar illumination, (Fig. 4-1), regardless of the variation in orbit inclination, from noon to twilight. The ACS module, oriented in the anti-earth direction, receives extremes in sunlight from edge on, or zero illumination of the radiator in a twilight orbit, to a $1/\pi$ cyclical variation in a noon orbit. The C&DH module radiator receives nearly full solar intensity for twilight orbits to nearly zero solar input during noon orbits. Fig. 4-1(b) represents a possible alternate orientation for earth pointing missions where the spacecraft is rotated 90°. For this

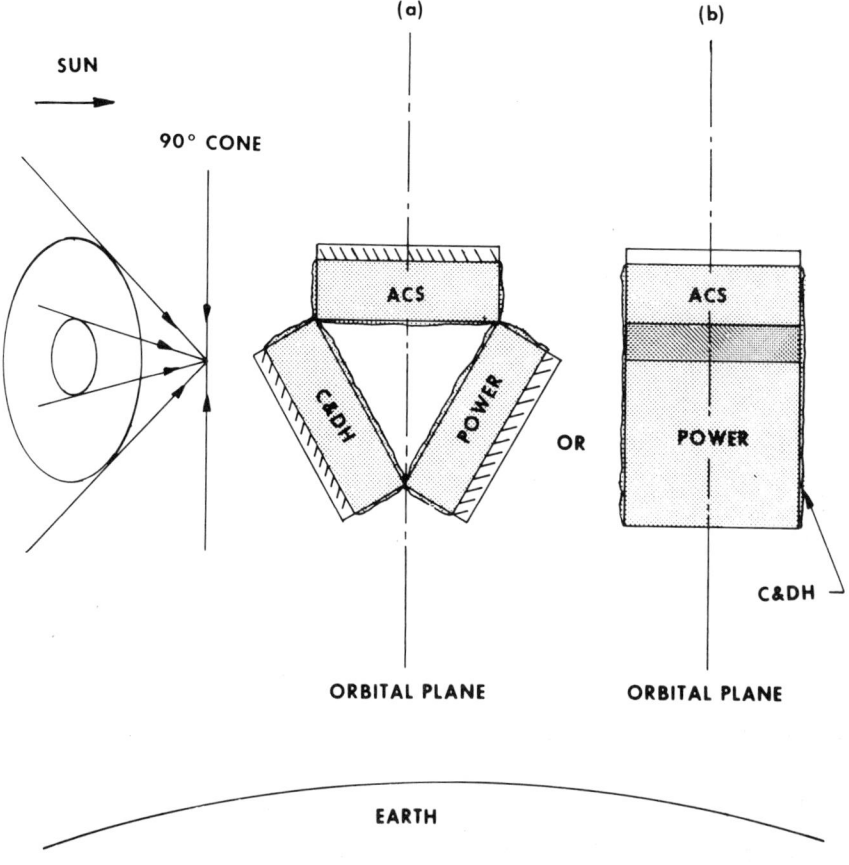

Fig. 4-1 Spacecraft Orientation for Earth Observation Missions with Sun Angle Variation

alternate orientation, the ACS module environment flux variation is the same as before. The environmental inputs to the Power and C&DH modules are equal to each other but different than they were for orientation 1(a). Both modules must now accommodate extremes from no sunlight (edge on for twilight orbits) to approximately 0.25 suns orbital average intensity for noon orbits.

The spacecraft orientation for stellar pointing missions (Fig. 4-2) has been chosen so only the C&DH module radiator is in sunlight. This is possible since the sun need only be in one place and limited to a ±90° angle variation. Therefore, the C&DH module radiator will be illuminated by the sun at an angle which

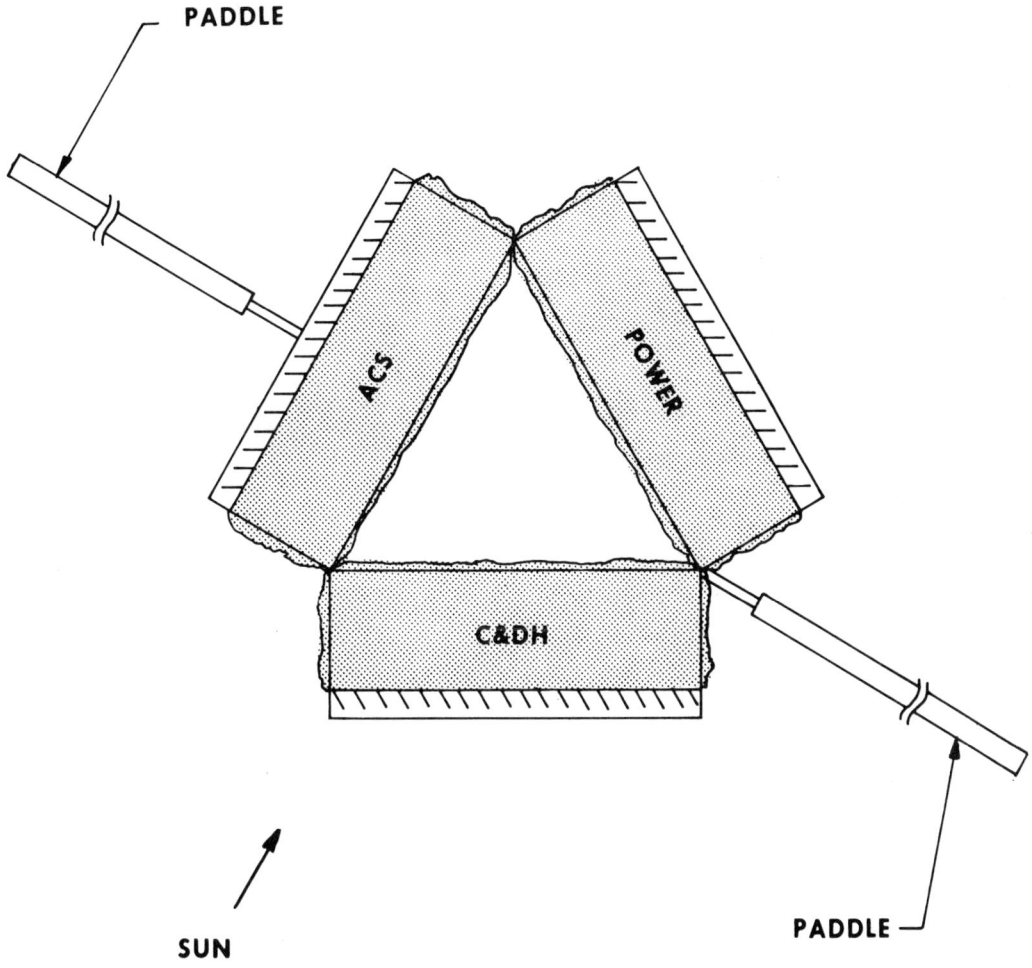

Fig. 4-2 Spacecraft Orientation for Stellar Missions

depends upon what source the instrument is observing. The Power and ACS modules, on the other hand, are never exposed to sunlight during any normal pointing operation.

5. COMMUNICATIONS AND DATA HANDLING (C&DH)

The Communications and Data Handling Subsystem provides a means for ground and on-board control of all spacecraft and sensor functions and for retrieval of observatory low to medium rate data. This subsystem consists of the Communication Equipment (i.e., RF transmitters and receivers) and Data Handling Equipment which is composed of a command group, a telemetry group, and an on-board computer. A subsystem block diagram is shown in Fig. 5-1 and a layout of the module is shown in Fig. 5-2.

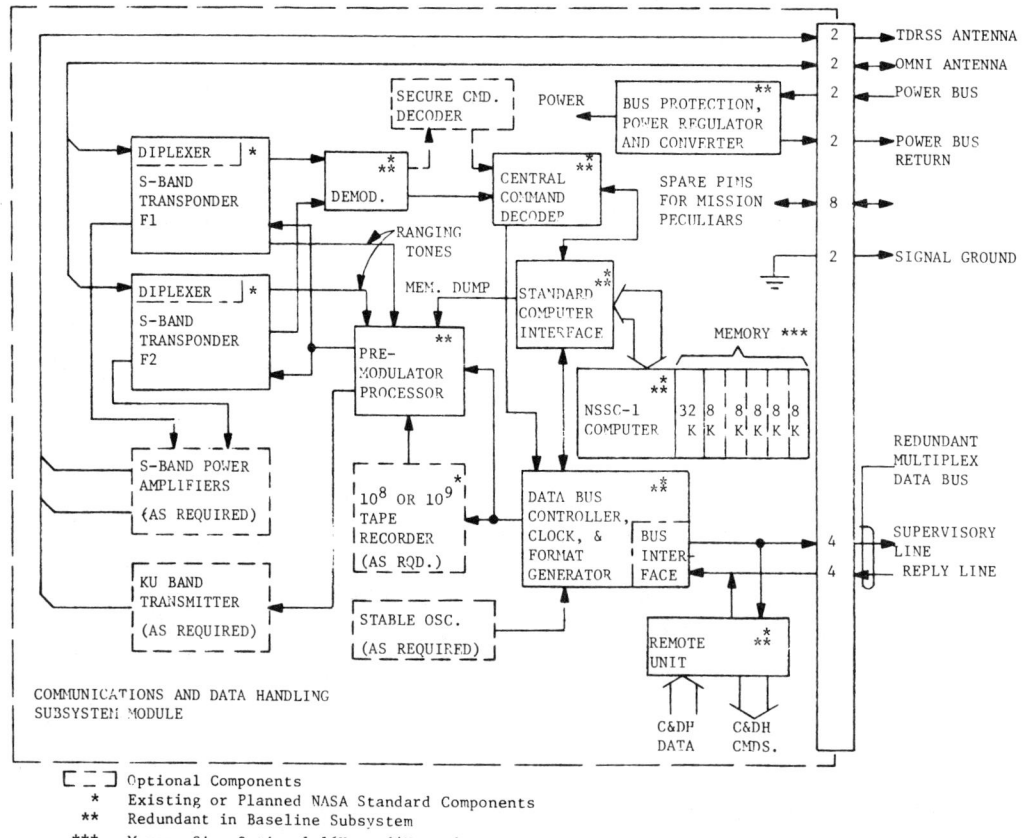

Fig. 5-1 Communications and Data Handling Subsystem Module

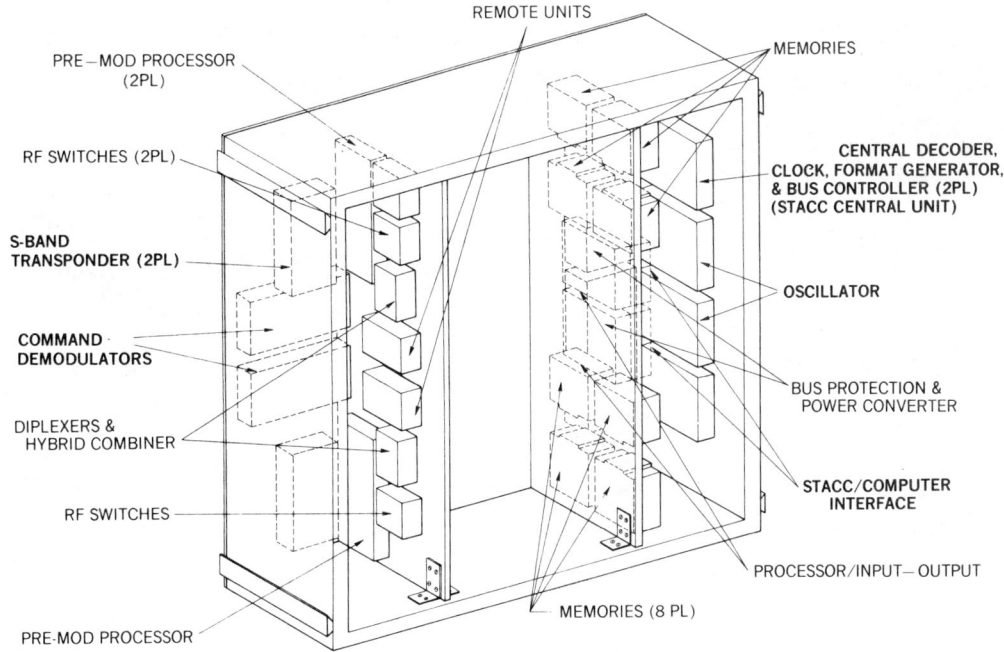

Fig. 5-2 C&DH Module

Key features of this subsystem are:

- all major subsystem components are either existing or planned NASA standard units;
- an S-Band communications subsystem with omni-directional antenna coverage is used for receiving command, transmitting housekeeping and narrowband sensor data, and precise tracking;
- with the addition of S-Band power amplifiers or Ku-Band transmitters, communications is compatible with TDRSS service;
- use of a centralized on-board computer will provide low cost implementation of many on-board functions and will permit autonomous operations;
- the command and data handling hardware design is based on the concept of remote multiplexing of telemetry data and remote distribution of commands. Command and telemetry data are routed to and from other spacecraft and instrument subsystems via a serial digital multiplex data bus to minimize interconnect problems and to allow sizing the system to actual requirements.

- the on-board computer communicates with all observatory subsystems by time sharing the multiplex data bus. This composite hardware interconnection forms an integrated electronic system which provides very flexible, yet standard, hardware;
- signal conditioning of passive transducers is provided, when needed, by a constant current source which is applied to the devices at the time of sampling;
- the telemetry format is controlled by either the computer or a read-only memory; therefore, parameter selection and sample rates can be defined after user design is complete and modified from mission to mission at minimum cost;
- all telemetry and command formats fully comply with GSFC Aerospace Data System Standards;
- the command rate is a shuttle compatible 2000 bits per second with 48 bits per word;
- the telemetry bit rates are command selectable 1 kbps, 2 kbps, 4 kbps, 8 kbps, 16 kbps, 32 kbps, or 64 kbps and can accommodate the shuttle automated payloads mission model prepared by MSFC (July 1974 revision). The 16 kbps data rate is compatible with the Shuttle/Payload Interface Requirements Document.

With reference to the subsystem block diagram (Fig. 5-1), a functional description shows the S-Band receiver passing commands received from the ground through a demodulator to a central decoder unit which distributes the commands for real time execution through a bus controller to remote units or feeds the commands to the on-board computer for delayed execution. Delayed commands are processed as a function of spacecraft time or event and are passed from the computer back to the central decoder which distributes the commands to using subsystems through the remote units in the same manner as real time commands. Imbedded in the command format is information identifying the command as (1) real time, (2) delayed, or (3) a computer program word. In operation, the central decoder places commands received from the ground and the computer into unique time slots on the data bus supervisory line, so that a conflict of simultaneous command execution is avoided. Commands may be executed from the ground and the computer at maximum rates of 42 and 1000 per second, respectively.

As for command, the concept of remote, distributed functional elements is employed for data acquisition and is integrated into the telemetry subsystem as shown in the C&DH block diagram. The master unit for data acquisition is the format generator. It contains a read-only memory (ROM) for producing a baseline telemetry format. In operation, the format generator sequentially takes instruction words out of memory and sends them to remote units via the multiplex data bus supervisory line. When addressed, the remote unit sends the selected data back to the format generator via the response line and a composite bit stream of up to 64 kbps is fed to the computer and transmitter or tape recorder as desired. If necessary, data are converted from analog to digital in the remote unit. As a back-up, the computer, by command, can assume the role of the read-only memory and consequently control the transmitted format. This two-way interface also allows the computer to sample data sources for its exclusive use at a rate up to 64 kbps in addition to receiving the normal telemetry stream. Time sharing of the data bus supervisory line for both command and telemetry functions is achieved by the bus controller which places remote unit addresses and commands into unique time slots.

6. POWER

Several candidates were considered during the studies to develop a standard power subsystem. Many factors were considered. Foremost among these were cost, efficiency, simplicity, and flexibility to capture a large number of missions without modification. The number of candidate systems were reduced to two: (1) an unregulated direct energy transfer system that uses a partial shunt for bus voltage control, and (2) an unregulated system that includes direct energy transfer as an option and a series power regulator unit that controls bus voltage and battery charging. The later was selected.

A simplified block diagram of the power module is presented in Fig. 6-1. The solar array is divided into two electrical sections. The "auxiliary" section is

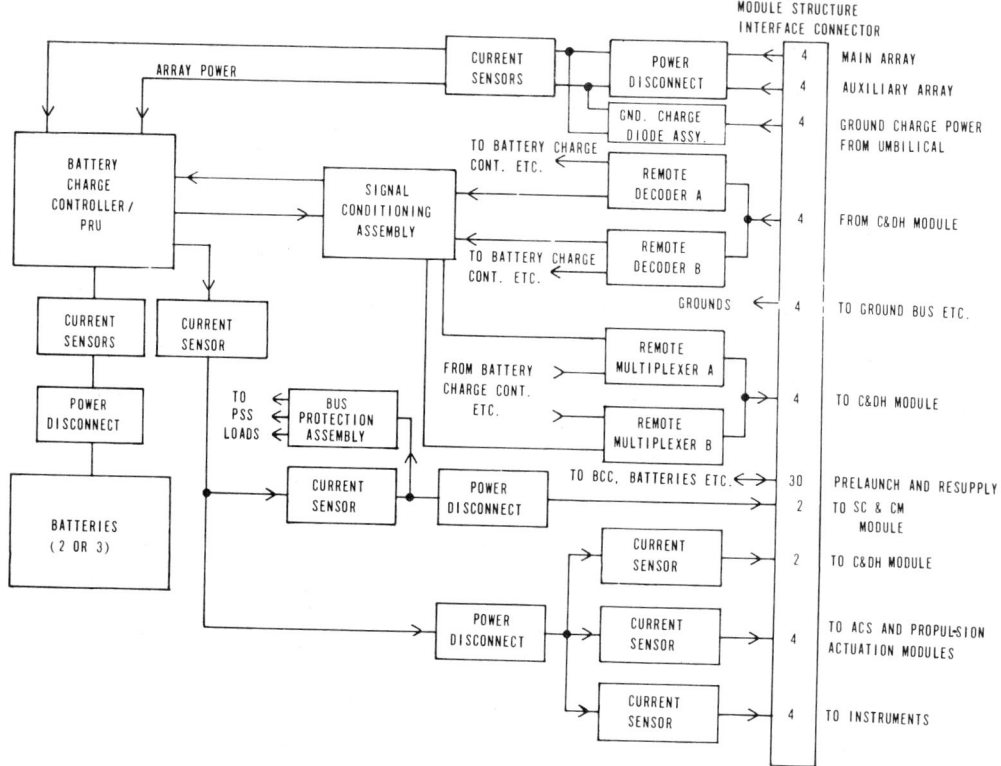

Fig. 6-1 Block Diagram — Power Subsystem Module

connected directly to the load bus and is sized to produce the spacecraft minimum load requirement. The balance of the array power is provided by the "main" section which is connected to the load bus through the power regulator unit (PRU). The "auxiliary" array concept is an option for the user.

The PRU accepts unregulated DC power from the solar array "main" section, transforms this energy from one voltage level to another, supplies the load bus requirements in excess of the "auxiliary" array section, and controls battery charge currents.

The PRU also contains a by-pass relay whose main function is to apply the "main" array section to the load bus when high peak loads are planned. The relay is normally open and must be commanded closed. Relay closure is inhibited by a signal from the array bus so that the relay will not close when the array is illuminated.

This prevents the relay from closing when high currents are present. It also can be used at the beginning of light and as a back up to the PRU if desired.

The subsystem design will accept one to three batteries without changing the number of battery chargers. The design of two batteries is the proposed minimum configuration. The standard 20 a.h. or the standard 50 a.h. battery will be used for all missions. Thus a maximum of 150 amp-hours of energy storage are available.

Battery charging and discharging will be accomplished in the parallel mode for all cases. This is the least complex approach of the two proposed, although it does reduce somewhat the flexibility in battery utilization. Battery cells should all be from the same plate lot and care should be exercised to ensure that the thermal design will maintain temperatures as required for proper sharing of charge currents.

Fig. 6-1 also shows the interfaces within the power module and to the other subsystems and the payload. The power subsystem provides unregulated, +28 ±7 volt power to all users. Regulated power for each subsystem and the payload is to be provided by the user and it is intended that this circuitry also provide power ground to signal ground circuit isolation in addition to the required output voltages, regulation, filtering, etc.

Remote decoders and remote multiplexers provide all command and telemetry interfaces between the power and the communications and data handling (C&DH) subsystems via a few interface wires. These interfaces also provide input/output access to the on-board computer which is in the C&DH module. Subsystem design is based on the use of redundant remote decoders and multiplexers. In applications where the redundancy is not required, the redundant units may be omitted without any changes to the harness.

The Signal Conditioning Assembly provides all interface circuitry required to match the input/output interfaces of the power control circuitry to those of the remote decoders and multiplexers. It also contains heater power control circuitry, temperature monitoring circuitry, and DC-DC converters which supply all regulated power required within the power subsystem module. Redundancy is incorporated for all critical circuitry.

The Bus Protection Assembly provides redundant fusing for all non-critical subsystem loads. The fuses are operated well below their normal ratings to avoid failures other than those due to circuit overload. The primary and redundant fuses are on separate plug-in cards to permit verification of interface connections and testing of individual fuses.

Power disconnect circuitry is provided at all power input/output interfaces to the module. The circuitry consists of a group of redundant power relays which may be used to deactivate the spacecraft during ground test and prelaunch operations or during resupply via the shuttle. Control power for these relays and relay status indication will be provided via the spacecraft umbilical.

The current sensors and redundant excitation supplies are housed in a common enclosure with the power disconnect relays. The sensors provide current monitoring of all major input/output buses. Regulated DC power for the excitation supplies is provided by the Signal Conditioning Assembly.

All power circuitry for the Power System, with the exception of the array and optional array drive, are housed in a standard subsystem module. The power input/output and umbilical functions are routed to the spacecraft structure via heavy duty rack and panel type connectors which mate automatically as the module is inserted into the spacecraft structure. Test connectors for subsystem monitoring and battery conditioning and mounted at a convenient position on the side of the module.

7. ATTITUDE CONTROL

The primary function of the Attitude Control Subsystem (ACS) is to orient and stabilize the spacecraft relative to a desired target. The basic design approach applied to the ACS for the modular spacecraft has been to adapt proven techniques and, where possible, utilize flight qualified and/or Low Cost Standards Office equipments. The basic control system element is an on-board computer that processes all sensor derived information and, in conjunction with various types of stored information, generates the appropriate control signals to operate the reaction control devices. The on-board computer, which is shared with other spacecraft subsystems, is physically located in the Communication and Data Handling (C&DH) Module. The degree of flexibility provided by the on-board computer allows the use of a single ACS Module design to meet the requirements of a diverse group of payloads with regard to both mission and configuration. The ACS Module can support earth-oriented, solar physics and astronomy payloads in both near earth and geosynchronous orbits. An analog backup control loop is provided to insure the capability for shuttle retrieval or resupply in the event of anomalous behavior in either the computer or ACS equipments.

With the exception of coarse sun sensors on the solar array and mission unique fine error sensors which may be incorporated in payload instruments, all control sensors are housed within the ACS Module. Payload instrument fine error sensors, which would be primarily related to solar physics and optical astronomy payloads, can be utilized to enhance control performance. The degree of enhancement is primarily defined by the characteristics of the particular fine error sensor. The ACS includes the reaction control devices that would be utilized during normal operations. These are the reaction wheels and magnetic torquers. Mass expulsion reaction control equipments are housed within the Propulsion/Actuation Module which is discussed in detail in Section 8. The Propulsion Module would also be capable of accommodating large reaction wheels, large torquer bars and/or control moment gyros as may be necessary to complement the reaction control equipments in the ACS Module for very large spacecraft or spacecraft with very unique requirements.

The basic ACS configuration is fixed for all mission types except geosynchronous orbits where the magnetometer and magnetic torquers can be deleted. The balance of the configuration options relate to reliability improvements through redundancy. A block diagram of the basic ACS Module is defined in Fig. 7-1. The shaded components are those that may be added as a mission option to improve reliability through redundancy. For normal modes of operation, all sensor information is transferred to the computer via a remote multiplexer in the ACS

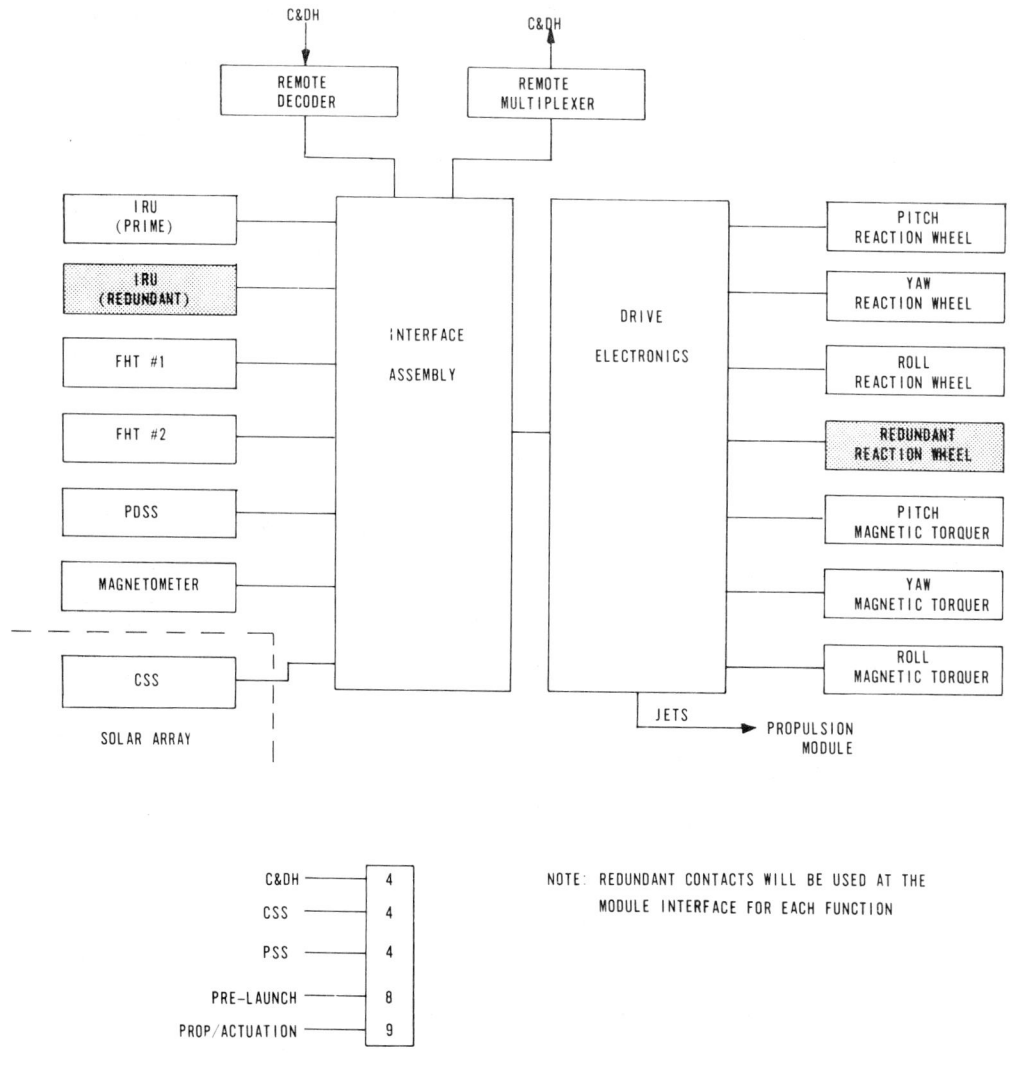

Fig. 7-1 ACS Module Block Diagram

Module. This information is downlinked as telemetry data and/or made available to the on-board computer for processing the control algorithms. The reaction control equipments respond to commands generated by the computer and transferred to the reaction control equipment drivers by the means of the data bus and a remote decoder located in the ACS Module in a manner similar to the method of introducing ground commands.

The performance of the control system will vary with respect to pointing accuracy and stability as a function of various mission unique factors. Therefore, the pointing accuracy and stability performance levels will be categorized per mission type. The performance characteristics are defined in Table 7-1. Error budgets are used as a basis for estimating the pointing accuracy and stability characteristics. The error budget does not include mechanical alignment errors. The earth pointing accuracy characteristics are based on the ephemeris knowledge being accurate to 100 meters and an orbital altitude of 700 kilometers. The accuracy of ephemeris determination varies considerably with orbit altitude. The projected accuracy for the 1980 time frame is approximately 100 meters for 500 kilometer orbits. The accuracy improvements as a function of increase in altitude

Table 7-1

ACS PERFORMANCE

	Stellar & Solar Mission		Earth Pointing Mission[4]
	w/Payload Sensor[1]	w/o Payload Sensor	
Pointing Accuracy (Degrees - 1σ)	$< \pm 10^{-5}$	$< \pm 0.01$	$< \pm 0.01$
Pointing Stability[2] Average Rate Deviation (Deg/Sec)	$< \pm 10^{-6}$	$< \pm 10^{-6}$	$< \pm 10^{-6}$
Pointing Stability[3] Attitude Jitter (Degrees)	$< \pm 10^{-6}$ [5]	$< \pm 0.0006$ [6] ($\leqslant$ 20 min)	$< \pm 0.0003$ [6] ($\leqslant$ 30 sec) $< \pm 0.0006$ [6] ($\leqslant$ 20 min)
Slew Scale Factor Stability		0.01% (30 days)	
Slew Rate (Deg/Sec)		0 to 0.2	

Notes:

(1) ACS performance limit with ideal sensor (Applies only to axes normal to line-of-sight of payload sensor)
(2) Based on $\geqslant$ 30 minute intervals
(3) Relative to average rate deviation as baseline
(4) Based on 700 kilometer orbit altitude
(5) Based on no disturbance torques internal to spacecraft.
(6) Based on the effect of internal disturbance torques limited to $\pm 0.2 \widehat{\sec}$

above 500 km is significant. The degree of accuracy with which the ephemeris can be predicted is also a function of orbit altitude and diverges rapidly in a 500 kilometer orbit. For orbit altitudes of 700 kilometers or greater, the constants necessary to compute the ephemeris with the on-board computer are expected to be sufficiently accurate to provide ephemeris prediction to 100 meters for a period of a few days (one day at 700 kilometers — three days at 1000 kilometers).

8. PROPULSION/ACTUATION MODULE

The purpose of the Spacecraft Propulsion Subsystem (SPS) on the Modular Spacecraft is to provide the impulse necessary to (1) correct launch vehicle injection errors, (2) offset the effect of external disturbances to maintain the correct mission altitude, (3) unload the momentum wheels, (4) serve as a backup to the momentum wheels in the event of a failure, and (5) accommodate larger reaction wheels/control moment gyros (CMG) and magnetic torquers.

An additional function is to provide an orbit transfer capability on certain missions. Studies conducted at GSFC and by the GSFC study contractors have shown that a cost savings of several million dollars per shuttle launch can be realized if the shuttle launches the spacecraft into a low earth orbit (nominally 250 nautical miles) and an on-board propulsion system is added to the spacecraft to transfer the spacecraft to its mission altitude (nominally 450 nautical miles) and then back to 250 nautical miles for shuttle retrieval or resupply. Mission and spacecraft parameters were used to derive the requirements for the propulsion subsystem functions contained in Table 8-1.

Two propulsion subsystems have been designed to meet the requirements of the various missions being considered for the Multimission Spacecraft (MMS). The first one, SPS-I, meets the reaction control and orbit adjustment requirements for spacecraft in the 2,500 pound class that would be launched by a Delta 2910. The second one, SPS-II, meets the requirements of reaction control, orbit adjust and orbit transfer for spacecraft in the 4,000 to 10,000 pound class that would be launched by the Shuttle. SPS-II is a direct extension of SPS-I that is achieved by changing to a larger propellant tank and adding orbit transfer thrusters. A summary of the capability of these two subsystems is given in Table 8-2.

Table 8-1

PROPULSION SUBSYSTEM REQUIREMENTS

Reaction Control Functions	Delta	Shuttle
Initial Stabilization and Restab.	880 lb-sec	880 lb-sec
Control During Orbit Adjust Maneuvers	65 lb-sec	65 lb-sec
Momentum Wheel Unloading	1,455 lb-sec	3,000 lb-sec
Orbit Adjust Functions		
Injection Error	45 fps	45 fps
Orbit Maintenance	5 fps	5 fps
Control During Orbit Transfer Maneuvers	-	3,150 lb-sec
Orbit Transfer Function		
Maintenance Orbit Establishment	-	650 fps
Retrieval at 250 nm circular	-	650 fps

Table 8-2

MMS PROPULSION CAPABILITY

	SPS-I	SPS-II
Dry Weight (lbs)	80	220
Propellant Weight (lbs)	55	1,050
Propellant Growth (lbs)	165	None
ΔV (ft/sec) (Ws/c = 6,000 lbs)	65	1,350
Total Impulse (lbs-sec)	12,100	231,000
Reaction Control	Yes	Yes
Orbit Adjust	Yes	Yes
Orbit Transfer	None	Yes

The reaction control thrusters have a nominal thrust rating of 0.2 lbf and provide a torque to the spacecraft of 0.8 ft-lb minimum about each of three orthogonal axes. These thrusters operate in a pulse mode with an on time of 0.1 seconds and an off time of 3.9 seconds. The thrusters can be fired via the remote interface unit or by hard line direct from the attitude control module. The reaction control thrusters perform the following functions:

1. initial stabilization of the spacecraft;
2. reaction torque to counteract the torque produced during reaction wheel unloading;
3. restabilization of the spacecraft;
4. removal of disturbance torques during orbit adjust firings;
5. coarse (±1°) attitude control of spacecraft while in safe hold mode.

The orbit adjust thrusters have a nominal thrust rating of 5 lbf and are used to perform the following functions:

1. removal of launch vehicle injection errors;
2. maintenance or change of orbital parameters.

The thrusters are located outboard with the 0.2 lbf thrusters to form four rocket engine modules (REM). In this location the orbit adjust thrusters can be used for pitch and yaw control during orbit transfer maneuvers.

Thruster orientation combined with the aft mounting arrangement of the propulsion/actuation module avoids the undesirable effects of heating, contamination, extraneous torques and translations due to plume impingement.

The orbit transfer system is sized to transfer a 6,000 lb spacecraft from a nominal shuttle delivery altitude of 250 nautical miles (nm) to an operational altitude of 450 nm and then back to 250 nm for retrieval or resupply. Redundant nominal 100 lbf thrusters are provided for the orbit transfer function. These thrusters are canted to fire through the spacecraft center-of-mass. This system has a total impulse capability of 231,000 lbs-sec and can deliver a ΔV of 1350 ft/sec to a spacecraft with a gross weight at liftoff of 6,000 lbs.

The basic SPS-I has a single 16-1/2 inch diameter tank that can hold 55 lbs of hydrazine when operating in a 3:1 blowdown mode. Sufficient room is available in this module to add three additional tanks that will extend the propellent storage capability to 220 lbs of hydrazine.

SPS-II is sized to hold 1050 lbs of hydrazine when operating in a 3:1 blowdown mode. The baseline design uses a single cylindrical tank with an elastomeric bladder for propellant expulsion.

9. MISSION CONSIDERATIONS

The performance capabilities of the MMS envelopes the requirements of 23 generic missions which represent 43 separate flight articles. These missions and their requirements as they are now known are shown in Table 9-1. The data presented here were developed in Ref. 2 with the exception of two missions, AIRSAT and IXE, which were subsequently added.

The results of the mission performance requirement and cost benefit study are presented in Table 9-2.

Table 9-1

MISSION CHARACTERISTICS

Mission	Orbit (All Are Circular)	Experiment Payload Weight kg (Net)	Experiment Power Solar Array Considerations	Attitude Control	Design Life	Payload Configuration Size	Experiment Data Storage Communication Data, Comm Band	Special Considerations and Comments
				Space Science				
EGRET	28° Inc. 500 km	1140 kg 2500 lb	100 W Deployable Fixed Array	5° Pointing 0.1° Stability Map Celestial Sphere Every 6 Mo. Spin About Sun Vector 5° A/C 1° Determ.	1 Yr.	Single Large Experiment 66" Dia x 98' Long	10^8 Cap, 4 kbps & S-Band	Shuttle - 10^8 or TDRSS Real Time
Solar Maximum (SMM)	29-33° Inc. 288 nm	500 kg 1100 lb	200 W Deployable Fixed Array	Sun Pointing 1-5 Sec RMS Pointing 5 Min Stability 1 Sec Accuracy	6 Yrs.	Several Instruments	6 kp/s 10^8 Cap. (S-Band)	TDRSS Compatibility or Storage Contamination Critical Optics STS
Infrared	Polar 810 nm	320-900 kg 700-2000 lb	100 W	1 Arc-Sec Pointing 0.5 Arc-Sec Stab.	1 Yr.	S/C 70' x 10' TELE 50"	10^8 Cap, 4 kbps & S-Band	
Electrodynamics (EE)	Polar >4 RE/ECC	A. 225 kg 500 lb B. 450 kg 1000 lb	100 W	1° Pointing	1 Yr.	60" Dia x 48" H	10^8 Cap. 4-40 kbps & S-Band	
IUE B	Geosynch	115 kg 250 lb	50 W	1 Arc-Sec Pointing	3 Yrs.	Central Hole 60" Dia x 60" H 30" Dia x 72" H in S/C	40 kbs & S-Band	TDRSS Real Time
X-Ray & Gamma Ray Flare	A. Heliocentric B. 28° Inc. 270 nm	A. 140 kg 300 lb B. 450 kg 1000 lb	100 W	1° Pointing	1 Yr.	70" Dia x 80" H	10^8 Cap. 4-40 kbps & S-Band	
HEAO - B1 II	28.5° Inc. 250 nm	3600 kg 8000 lb	500 W Fixed Array	1° Coning Rotating About Sunline	2 Yrs.	Shuttle Cargo Bay	25 kb/s S-Band	TDRSS, Shuttle
LST	300 nm 28.5° Inc.	7100 kg 20,000 lb	1,500 W Movable Probably	Pointing Acc. 0.05 Sec Pointing Stab. 0.005 Sec	1 Yr. 10-15 Yrs. Refurb.	3 Instr. 13' Dia x 43'	1200 Command for 24 Hrs. 2×10^9 bps, S-Band	Shuttle Launched-Recoverable. May Be TDRSS or STDN

Table 9-1 (Continued)

Mission	Orbit (All Are Circular)	Experiment Payload Weight kg (Net)	Experiment Power Solar Array Considerations	Attitude Control	Design Life	Payload Configuration Size	Experiment Data Storage Communication Data, Comm Band	Special Considerations and Comments
IXE	28° Inc. 250 nm	150 kg	40 W	2 Arc-Sec Pointing	2 Yrs.	2' Dia 11' H	20 kb/s	
Soft X-Ray Telescope	28° Inc. 270 nm	450 kg 1000 lb	100 W	1 Arc-Min	1 Yr.	70" Dia x 50" H 30' x 10' Instr.	10^8 Cap. 4-40 kbps & S-Band	
Radio Lg. Baseline Interferom	Sync	115 kg 250 lb	50 W	3° Pointing	1 Yr.	70" Dia x 50" H	10^8 Cap. 4-40 kbps & S-Band	
Relativity	Polar 540 nm	640 kg 1400 lb	100 W	0.05° Arc-Sec Pointing 0.0005 Arc-Sec Stab.	1 Yr.	4' Dia x 6' H	10^8 Cap. 3 kbps & 84 kbps Real Time & S-Band	
Applications								
EOS A & A' Thematic Mapper (TM) & S-Band MSS or LANDSAT	Polar Sun-Sync 99° Inc. 380 nm	250 kg 550 lb	220 W One Degree of Freedom Array Drive	Nadir Point 36 Sec Pointing 0.0086 Sec/Sec Rate 2 Sec for 20 Min	2 Yrs.	Two Large Boxes (3 x 3 x 6 ft) 1-1/2 x 1-1/2 x 3	120 mb/s X & K Band	Delta or T-III B (WTR) Critical Thermal and Contamination Requirements. Radiative Cooler for TM Mission Peculiar Wide Band Communication Link. ΔV Req. TDRSS
GRAVSAT	90° 162 nm	635 kg 1400 lb	325 W	3° Pointing Accuracy	2 Yrs.	0.225×10^6 cm^3	1 kbps & S-Band	TDRSS Compat. ΔV Req.
Geopause	90° 16,200 nm	139 kg 305 lb	N/A	1° Pointing Accuracy	5 Yrs.	0.4×10^6 cm^3	1 mbps & S-Band, K-Band	TDRSS Compat. Shuttle Retrival
Severe Storm Observatory (SSOS) or STORMSAT	Geostationary 19,323 nmi	91 kg 200 lb	60 W One Axis Array	Points to Selected Earth Targets 0.1° Determine to 0.08° Stability 2 Sec/ 20 Min	3 Yrs.	Single, Medium Sized Sensor 2-1/2' Dia x 6'	6 mb/s S-Band	Radiative Cooler ΔV Req. Shuttle & IUS (ETR) AASIR Instrument
Environmental Monitoring Satellite (EMS)	103° 920 nm	320–545 kg 700–1200 lb	220 W	1° Pointing Accuracy	2 Yrs.	986×10^3 cm^3	4 kbps & X-Band Coverage Continuous	TDRSS Compat. Shuttle Retrival

Table 9-1 (Continued)

Mission	Orbit (All Are Circular)	Experiment Payload Weight kg (Net)	Experiment Power Solar Array Considerations	Attitude Control	Design Life	Payload Configuration Size	Experiment Data Storage Communications Data, Comm Band	Special Considerations and Comments
AIRSAT	565 km 57°	200 kg	125 W	0.1° Pointing Accuracy	9 Mo.	2' H x 4' Wide 4' Deep	18 kb/s	LSIR Lem Scan Inversion Radiometer
Grav Gradiometer	90° 160 nm	680 kg 1500 lb	100 W	1° Pointing Accuracy	1 Yr.	16×10^6 cm³	4 kbps & S-Band	TDRSS Compat.
EOS B - B' TM & HRPI	705 (380 nmi) Polar Sun-Sync 99° Inc.	385 kg 850 lb	220 W Array	Similar to EOS A	2 Yrs.	Two Large Boxes 3.6×10^6 cm³	Up to 250 mb/s X & K Band	Delta or T-III B (WTR) Similar to EOS A ΔV Req. TDRSS
EOS - C & E	270 nm 50° Inc.	365 kg 810 lb	400 W 2 Axis Array	0.01° Pointing	2 Yrs.	2.7×10^6 cm³	150 mb/s X & K Band	ΔV Req. - C Shuttle Launch - ETR TDRSS
SEASAT-B Radar Altimeter Radar Scatterometer Multifrequency Microwave Radiometer V&IR: SAR	486 nm Polar Non Sun-Sync 108° Inc.	500 kg 1100 lb	500 W Ave Up to 800 W For Short Periods With SAR Drive	Nadir Pointing 0.5 to 0.25° Control: 0.2 to 0.1° Determine	2-3 Yrs.	Several Large Antennas. Size and Configuration Depends on Sensor Design. Approach: SAR 3 x 5 m	60 kbps 60 mb/s for 10 Min S-Band	Does 180° Yaw Turn Midyear ΔV for Drag. Make-Up TDRSS. Limited Jitter
SEOS (Synchronous EOS) LEST (1-1/2 m Dia)	Geostationary 19,323 nmi	1085 kg Including Telescope Structure (2300 lb)	160 W Fixed Array	Points to Selected Earth Targets 5 Sec: Stability for 12 Min 0.1 Sec	2 Yrs.	2.5 m O.D. x 3.5 m Long Fold Up Shade	60 mb/s S-Band	55 Meters/Sec/M ΔV for Stationkeeping Shuttle IUS
TIROS - O	500-1000 km 103°	345 kg 770 lb	270 W Fixed Array	40 Sec Pointing 10^{-6} Deg/Sec	2 Yrs.	30×10^3 cm³	10 kbps S & X Band	TDRSS, Shuttle
EOS D & D'	EOS B - B' TM & HRPI	EOS B - B' TM & HRPI	EOS B - B' TM & HRPI	EOS B - B' TM & HRPI	EOS B - B' TM & HRPI	EOS B - B' TM & HRPI	EOS B - B' TM & HRPI	EOS B - B' TM & HRPI
Life Science								
Biomedical Experiment Scientific Satellite (BESS)	555 km 300 nmi Circular 37.7°	1140 kg 2500 lb	300 W	3 Axis Stab. 10^{-3} g Max	1-6 Mos.	Biological Specimens	5-10 kbps S-Band	(Shuttle Launch & Retrived)

Table 9-2

SPACECRAFT MISSION MODEL MMS CAPTURE CAPABILITY

	MMS CONFIGURATIONS			STRUCTURAL RECONFIGURATION + LARGE TORQUERS
BASELINE	RCS/OAS	LARGE TORQUERS/OAS		
GRE	LANDSAT	BESS		LST
SMM	STORMSAT	GRAVSAT		
IXE	AIRSAT			
INFRARED	SEOS			
X-RAY/GAMMA RAY	IUE-B			
EMS	RADIO INTER.			
RELATIVITY	SEASAT-B			
TIROS-O				
SOFT X-RAY				

NOT CAPTURED
- GEOPAUSE
- GRAVITY GRADIOMETER
- EE
- HEAO BLK II

10. CONCLUSION

The final system configuration presented in this paper is a synthesis of 6 years of study at GSFC of which one year represents a detailed review and study by 3 systems contractors and 3 years of development and testing of subsystems and Shuttle interfaces. The chronology of studies applicable to the MMS is shown in Fig. 10-1.

Present plans call for the Energetic Gamma Ray Experimental Telescope (EGRET) mission to be launched as the Gamma Ray Explorer (GRE). This mission, if approved, would be the precursor for the MMS. Because of GRE funding constraint, several components, which are available, will be used in the GRE configuration. Although adequate for the GRE mission performance requirements, these

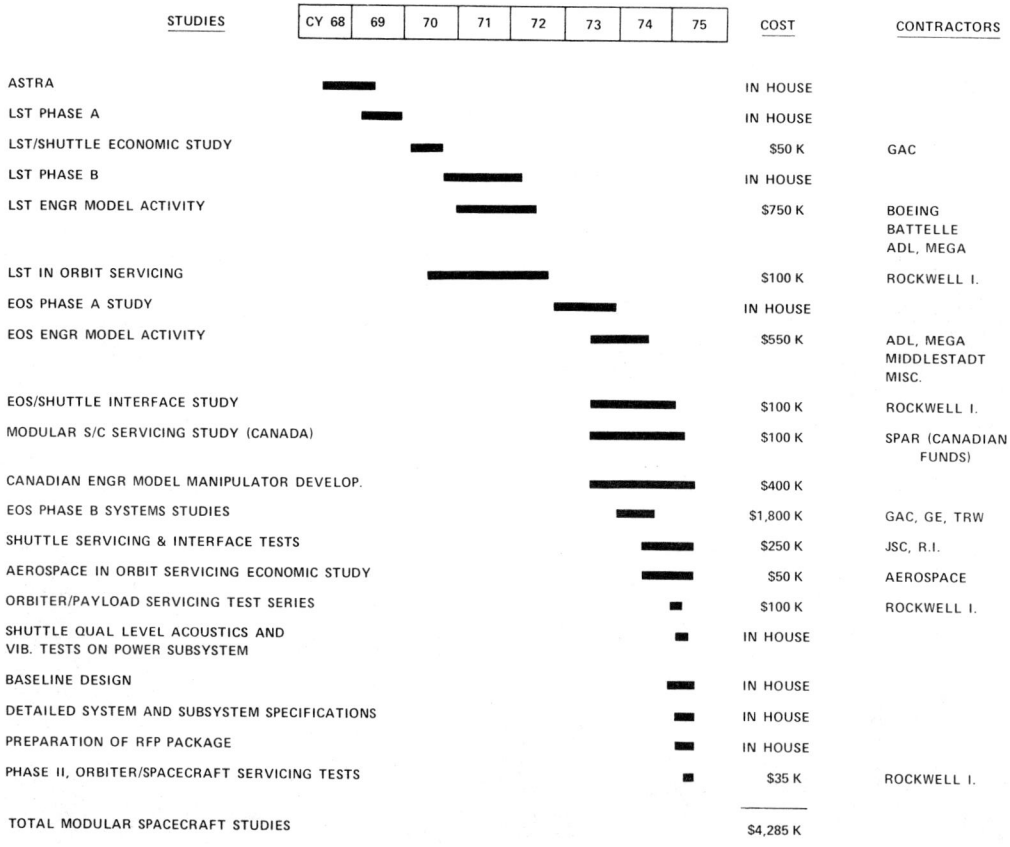

STUDIES	CY 68	69	70	71	72	73	74	75	COST	CONTRACTORS
ASTRA									IN HOUSE	
LST PHASE A									IN HOUSE	
LST/SHUTTLE ECONOMIC STUDY									$50 K	GAC
LST PHASE B									IN HOUSE	
LST ENGR MODEL ACTIVITY									$750 K	BOEING BATTELLE ADL, MEGA
LST IN ORBIT SERVICING									$100 K	ROCKWELL I.
EOS PHASE A STUDY									IN HOUSE	
EOS ENGR MODEL ACTIVITY									$550 K	ADL, MEGA MIDDLESTADT MISC.
EOS/SHUTTLE INTERFACE STUDY									$100 K	ROCKWELL I.
MODULAR S/C SERVICING STUDY (CANADA)									$100 K	SPAR (CANADIAN FUNDS)
CANADIAN ENGR MODEL MANIPULATOR DEVELOP.									$400 K	
EOS PHASE B SYSTEMS STUDIES									$1,800 K	GAC, GE, TRW
SHUTTLE SERVICING & INTERFACE TESTS									$250 K	JSC, R.I.
AEROSPACE IN ORBIT SERVICING ECONOMIC STUDY									$50 K	AEROSPACE
ORBITER/PAYLOAD SERVICING TEST SERIES									$100 K	ROCKWELL I.
SHUTTLE QUAL LEVEL ACOUSTICS AND VIB. TESTS ON POWER SUBSYSTEM									IN HOUSE	
BASELINE DESIGN									IN HOUSE	
DETAILED SYSTEM AND SUBSYSTEM SPECIFICATIONS									IN HOUSE	
PREPARATION OF RFP PACKAGE									IN HOUSE	
PHASE II, ORBITER/SPACECRAFT SERVICING TESTS									$35 K	ROCKWELL I.
TOTAL MODULAR SPACECRAFT STUDIES									$4,285 K	

Fig. 10-1 Chronology of Study Activity

components do not have the ultimate performance needed for full MSS capability. In addition, GRE funding will only support the procurement of the modular subsystems from industry with their integration into GRE being accomplished in house at GSFC. Fig. 10-2 depicts GRE being delivered into orbit.

Therefore, the proposed development of MMS will be a two phase activity. The principal difference between the GRE prototype configuration and the MMS is the incorporation of four new components and the development of a propulsion module not needed for GRE. These components, under development by the NASA Low Cost Systems Office, are the standard gyro, momentum wheels, TDRSS/STDN transponder, and power regulation unit. Coincident with the introduction of these four standard components, it is planned that industry will assume the role of integrating the spacecraft. The MMS would then be available to the mission project or payload integration contractor to complete the development and test of the flight

Fig. 10-2 GRE Being Delivered into Orbit

system. The first five missions identified as candidates for the MMS are:

- Solar Maximum Mission
- AIRSAT
- LANDSAT-D
- STORMSAT
- Biomedical Experimental Scientific Satellite

Of these missions, only LANDSAT-D would be launched by a conventional vehicle since if requires a polar orbit prior to the Shuttle being operational from the WTR.

In conclusion, there are two key points to be made in support of the timely development of the Multimission Modular Spacecraft.

- The limited national resources available for aerospace research, development and application must be concentrated on payloads and their evaluation with minimum investment in the fundamental service system of the spacecraft.

- It is essential that the development of the MMS occur during the same time frame as the development of the Space Shuttle so that the shuttle/payload interface can be based on actual user needs rather than projected user requirements.

The Multimission Modular Spacecraft is a concept consistent with the cost conscious exploration of space for the benefit of mankind.

ACKNOWLEDGMENT

The authors wish to acknowledge the support and assistance of the MMS Cadre in the definition and design of the Multimission Modular Spacecraft. This study group was established under the cognizance of the GSFC Engineering Directorate, J. Purcell, Director. Each Cadre Manager draws heavily on his/her line organization and/or contractor's support. Thus a complete listing of all personnel contributing material to this paper is not feasible. The MMS Cadre is comprised of the following system, subsystem, or functional managers:

Structural	J. Webb
Thermal	S. Willis

Electrical	C. Hoffman
Attitude Control	J. Kull
Communications and Data Handling	C. Trevathan
Power	D. Harris
Propulsion	W. Wooddruff
Environmental Test	J. Greenwell
Reliability and Quality Assurance	H. Doyle
Shuttle Interface	R. Mattson
Mission Operations	R. Des Jardins
Operations Control Center	T. Moore
Network Support	T. Spencer
System Configuration and Documentation	D. Witters
Ground and Flight Software	A. Merwarth
Ground Support Hardware	R. Rhodes
Administrative Support	W. Woodyear
Resources	R. Martin
General Business	L. Lopatin
Procurement	W. Pohl
Financial Analyst	F. Meyerle

REFERENCES

1. "Low Cost Modular Spacecraft Description," X-700-75-140, GSFC, May 1975.

2. Official Correspondence from: Ron Crawford, NASA HQ/KC, to: All NASA Field Centers. Subject: Spacecraft Common Usage and Standard Bus Assessment, December 10, 1974.

AAS 75-273

SHUTTLE/IUS PERFORMANCE FOR PLANETARY MISSIONS[*]

M.J. Cork[+], J.M. Driver[+], J.L. Wright[±]

INTRODUCTION

This condensed paper summarizes a study conducted during the past year by the Jet Propulsion Laboratory, Advanced Technical Studies Office, and the Battelle Columbus Laboratories, NASA Launch Vehicle Project, for NASA. The study was performed to obtain a better understanding of the planetary missions proposed for the 1980s, the Interim Upper Stage (IUS) system capabilities as derived from the Air Force-funded IUS studies, and Shuttle/IUS mission profile options for performance enhancement.

KICK-STAGE SIZE CONSIDERATIONS

Selection of kick-stage size for high-energy missions can have a significant impact on IUS performance. The Shuttle system's 65,000-lb (29,500-kg) guaranteed performance to low Earth orbit causes the kick-stage size optimization to have different characteristics than the kick-stage optimization of expendable vehicles. If the Shuttle cargo capability is not filled, additional kick-stage mass

---

[*] This paper presents the results of one phase of research carried out at the Jet Propulsion Laboratory, California Institute of Technology, under Contract No. NAS7-100, and the Battelle Columbus Laboratories, under Contract NASw-2018, sponsored by the National Aeronautics and Space Administration.

[+] Jet Propulsion Laboratory, Pasadena, California.

[±] Battelle Columbus Laboratories, Columbus, Ohio.

gets a free ride to Earth orbit. For larger kick stages, however, propellant off-loading from a lower stage may be required. The optimum size kick stage for several IUS configurations, at C_3 values of 80 to 140 km^2/s^2 (where C_3 is the square of hyperbolic excess velocity with respect to the Earth), appears to be in the range of 2000 to 2300 kg.

Several IUS contractors proposed the TEM 364-4 solid motor, with a stage mass of about 1200 kg for high-energy missions, because it is developed and would be less expensive than a new motor. However, the value of the performance to be gained from a more optimally-sized kick stage should be factored into the cost comparison when kick-stage sizes are ultimately selected for IUS missions.

SHUTTLE ELLIPTIC ORBITS

Excess Orbital Maneuvering System (OMS) propellant in the Shuttle Orbiter may be used to place the Orbiter and its cargo in a 100-nmi perigee elliptic orbit with apogee as high as 450 nmi. This maximum altitude is a thermal protection system constraint for entry conditions associated with single-burn deorbit. However, most missions cannot achieve that large an orbit because apogee altitude obtainable with the fixed ΔV OMS is dependent on the true anomaly of the ellipse. The true anomaly is, in turn, dependent on mission energy (C_3) and declination of the escape asymptote. Most missions should be able to obtain an apogee altitude of 300 nmi with a performance gain of 4 to 7% compared to circular orbits. This return is sufficient to warrant use of the elliptic orbit profile for the most demanding planetary missions.

MISSION PERFORMANCE

Potential requirements for 14 planetary missions of interest in the 1980s, based on the February 1975 planetary mission model, were evaluated in the reference study. The four most demanding are the Pioneer Saturn/Uranus/Titan Probe and the Mariner-class orbiters of Mercury, Jupiter, and Saturn. Although the IUS candidates were evaluated against all missions in the latest model, it should be recognized that the IUS configurations proposed for planetary missions were based on the pre-1985 missions of the 1973 NASA mission model. This model did not include the more severe requirements for the Mercury, Jupiter, and Saturn orbiters discussed herein.

The Pioneer Saturn/Uranus/Titan probe mission consists of three spacecraft launched on different years with the probe destined for Saturn, Uranus, or Titan. The Uranus mission requires a $C_3 = 142$ km^2 to reach Uranus in 7 years from Earth launch. The Saturn or Titan probe missions, however, can use a reduced launch energy at the expense of a somewhat lengthened time to Saturn. All IUS candidates capture the Uranus mission, but some require use of a two-burn injection sequence to reduce gravity losses. Single-burn sequences may be used for Saturn or Titan missions with increased flight time.

All of the IUS candidates capture some version of the 1980, 1983, or 1988 Mercury Orbiter missions using multiple Venus-swingby techniques evaluated by Hollenbeck, et al. Some IUS candidates require that space-storable (e.g., fluorine/hydrazine) spacecraft propellants be used to reduce injected mass. Direct launch opportunities are much more demanding of IUS performance.

Requirements for orbiter missions of Jupiter show wide variations over the 1983 to 1987 time period, with 1985 being the most demanding opportunity. Ultimate injection requirements of the Shuttle/IUS will depend on the launch year, type of spacecraft propulsion, Jupiter orbit characteristics, and total spacecraft mass in orbit. Only the most capable IUS candidates with mission-optimized kick stages capture the most demanding mission options in 1985.

Mariner Saturn Orbiter missions using direct trajectory launches in 1986 through 1989 are beyond the capabilities of expendable IUS candidates. Use of a Venus-Earth-gravity assist mode for this mission allows capture by all IUS candidates at the expense of an additional two years flight time and some increase in spacecraft complexity.

CONCLUSIONS

The IUS candidates offer a variety of configuration options for planetary missions. Performance can be significantly enhanced for the high-energy missions by making maximum use of the Shuttle payload capability and providing proper staging ratios for the various upper stages. Shuttle elliptic parking orbits provide significant performance gains, and their use should be promoted for planetary missions. However, the performance gain obtainable is mission dependent, and care must be taken with generalized predictions. Planetary missions are the most demanding, performance-wise, of the missions in the NASA program. All IUS candidates capture most of the planetary missions; however, the options available for the most demanding planetary missions will depend on the specific IUS configuration selected for development and the programmatic phasing of the IUS and the NASA Tug.

AAS 75-279

POSSIBILITIES FOR REDUCING
HIGH-ENERGY PERFORMANCE REQUIREMENTS

G. R. Hollenbeck*

In recent years, techniques have been developed for utilizing gravity-assist effects from intermediate planets to improve performance and/or reduce flight time to difficult planetary targets. The first significant application of these benefits was accomplished by the highly successful MVM program which employed gravity swingby of Venus to reduce launch energy (and costs) for the initial flyby of Mercury. The same basic methods have been used to deflect the Pioneer 11 Jupiter flyby spacecraft to a Saturn encounter trajectory.

The current Planetary Mission Model is predicated on further exploitation of gravity-assist techniques. Remnants of the ambitious Grand Tour concent are evident in the Mariner Jupiter Saturn, Mariner Jupiter Uranus, and Pioneer Saturn Uranus mission designs. Also, the difficult Mercury orbiter mission will be possible with multiple gravity swingbys of Venus. In this case, the mission flight time of 31 months represents an operational price to be paid for the performance improvement over direct ballistic flight.

NEW FLIGHT TECHNIQUES FOR OUTER PLANET MISSIONS

A recent development has been identification of methods to utilize the gravity-assist potential of Venus and/or Earth to improve performance for missions to the outer planets. Two basic techniques have been verified, both of which offer the prospect of approximately doubling launch vehicle delivery capabilities over direct ballistic flight values for the price of increasing flight time by about 2 years (Ref. 1). The first of these techniques is predicated on successive Venus and Earth swingbys (dubbed VEGA for Venus-Earth Gravity-Assist) followed by a spacecraft velocity maneuver to achieve the final desired aphelion radius. The performance benefits originate in the low launch energy requirements ($C_3 \sim 20$ km^2/sec^2) to initiate the planet encounter sequence.

---

* Senior Staff Engineer, Martin Marietta Corporation, Denver, Colorado.

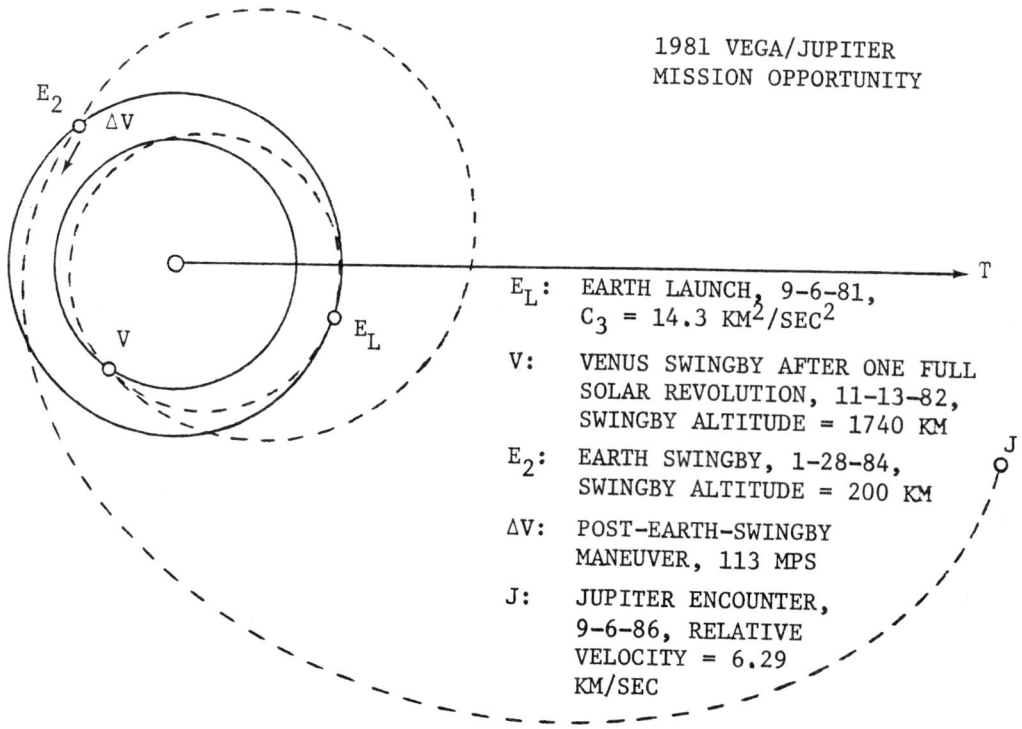

Fig. 1   Typical VEGA Flight Profile

The VEGA flight technique, illustrated on Figure 1, is characterized by initial perihelion radius inside the orbit of Venus. While this class of thermal environment should not be a major cost factor for a totally new spacecraft, modifications of existing Pioneer and Mariner designs could be quite expensive. For this reason, an alternate flight technique was developed which does not depend on Venus swingby.

Dubbed ΔVEGA (for ΔV-Earth-Gravity-Assist), this latter flight technique, depicted on Figure 2, involves initial launch outward from Earth orbit, a retrograde velocity maneuver at aphelion to produce an Earth-crossing orbit, and gravity swingby of Earth at either of two optional locations. Performance capabilities and phasing time requirements for this technique are comparable to the VEGA method. However, the minimum perihelion which must be experienced by the spacecraft is about 0.85 AU. A secondary characteristic of ΔVEGA missions is the existence of two distinct launch periods for each launch opportunity to any specific outer planet. The two launch periods are separated between centers by about 12 weeks and could prove of significance to operations in the Shuttle era.

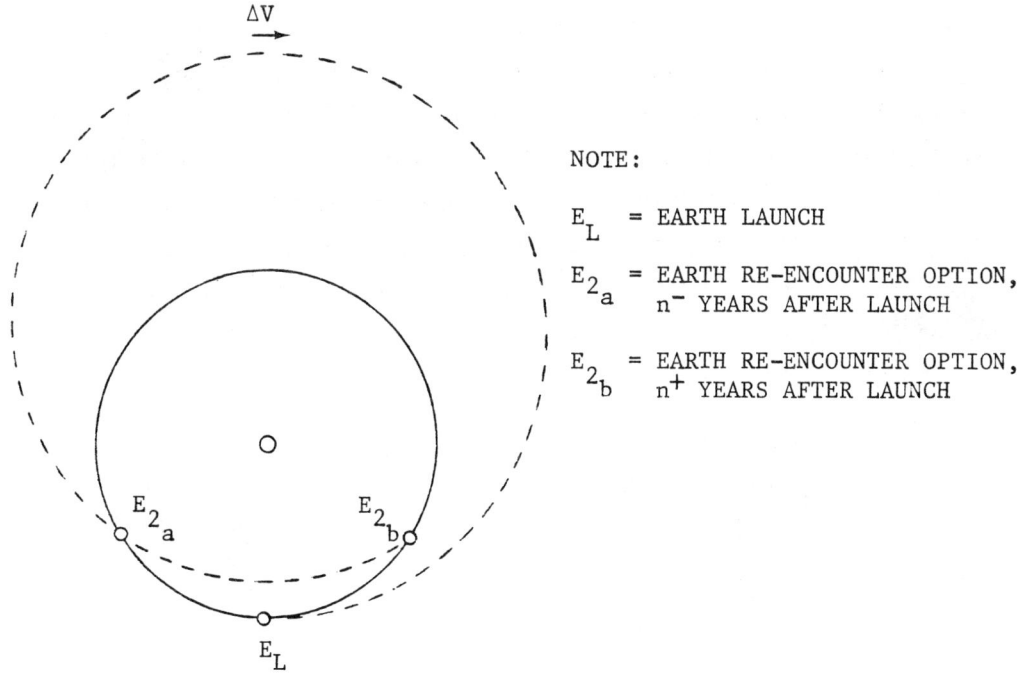

Fig. 2   ΔVEGA Heliocentric Geometry Options

LAUNCH VEHICLE CAPABILITIES

The performance advantage factors offered by the new flight techniques apply to all planetary launch vehicles. For example, the difficult Mariner Jupiter Orbiter and Mariner Saturn Orbiter missions, which substantially exceed the direct launch capabilities of Titan IIIE/Centaur/TE364-4, can both be performed by this launch vehicle if the VEGA or ΔVEGA techniques are employed. Of course, the mission flight times are increased by the necessary planet phasing interval of about 2 years. This latter consideration must be weighed in context with the value of increased spacecraft mass and/or reduced launch vehicle requirements.

Current mission plans for Mariner Jupiter Orbiter and Mariner Saturn Orbiter are based on launch in 1985 and 1987 respectively. Therefore, capabilities of the Titan IIIE/Centaur launch vehicle are probably academic. Of more significance, the new flight techniques affect the requirements imposed on the Interim Upper Stage operating in conjunction with Shuttle. As shown on Figure 3, only a few of the competing IUS candidates can deliver the Jupiter Orbiter spacecraft with conventional ballistic flight. The Saturn Orbiter mission, which is currently planned for launch after availability of the Space Tug, is within the capabilities of a single IUS configuration. If the VEGA or ΔVEGA flight techniques were employed, both of these demanding missions fall within the capabilities of even the lowest-performing IUS candidates. Combining the new flight techniques with a high-performance stage such as Space Tug would result in delivery capabilities to the outer planets sufficient to support consideration of totally new types of massive spacecraft and a wide range of new options for scientific exploration.

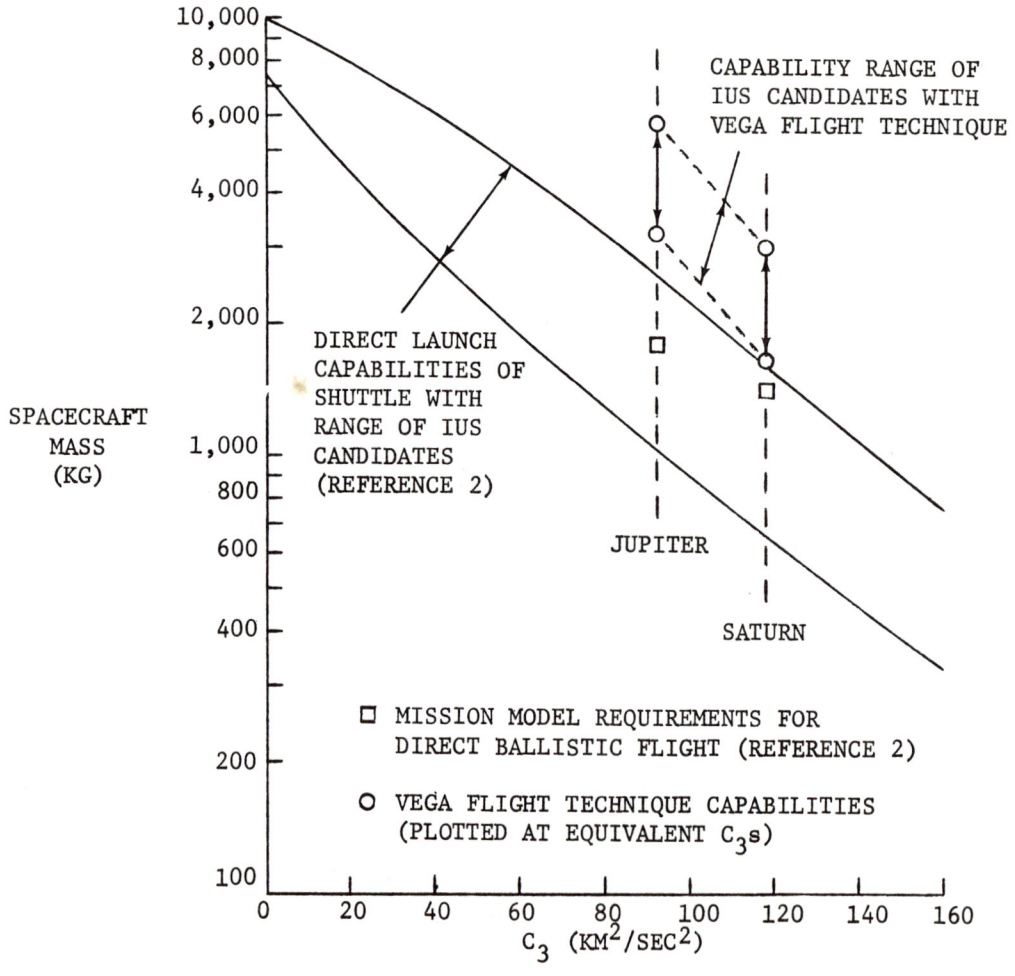

Fig. 3 Performance Potential of Space Shuttle

CONCLUSIONS

The most demanding performance requirements in the current Planetary Mission Model are those associated with the Mariner-class Jupiter Orbiter and Saturn Orbiter missions. The high launch energies required for conventional ballistic flight, in combination with the large spacecraft masses desired, exceed the capabilities of most of the competing candidates for the Interim Upper Stage planned for use in conjunction with the Space Shuttle.

New flight techniques have been developed which redistribute performance requirements between launch vehicle and spacecraft. With the low launch energies involved, net delivery capabilities to the outer planets can be approximately doubled for the operational prices of increased complexity and extended flight time.

With the new flight techniques, the difficult Jupiter and Saturn Orbiter missions fall within the capabilities of even the lowest-performance IUS candidates. For high-performance upper stage configurations, the current spacecraft mass requirements could be tripled and doubled for Jupiter and Saturn missions respectively.

The new flight techniques offer program options for reconciling conflicting IUS requirements as well as providing growth potential for future planning. In conjunction with orderly evolution of upper stages, the Space Shuttle could support the objectives of outer planet exploration through the remainder of the century.

REFERENCES

1. <u>New Flight Techniques for Outer Planet Missions</u>, Paper No. AAS 75-087, July 1975, G. R. Hollenbeck, Martin Marietta Corporation.

2. Shuttle/IUS Performance for Planetary Missions, Paper No. AAS 75-273, August 1975, M. J. Cork and J. M. Driver, Jet Propulsion Laboratory and J. L. Wright, Battelle Columbus Laboratories.

AAS 75-274

IMPACT OF SPACE TRANSPORTATION SYSTEM ON PLANETARY
SPACECRAFT AND MISSIONS DESIGN*

Philip M. Barnett[+]

GENERAL

At JPL we have made studies to define and understand the alternatives for planetary spacecraft operations with the Space Transportation System (STS).[1] The STS presents a new set of interfaces, operational alternatives, and constraints in the prelaunch, launch, and near-earth flight phases of the mission, when compared with today's expendable launch vehicles.

Although this study concentrated on planetary missions, our conclusions are applicable to other mission classes launched with the Space Shuttle. This is an interesting conclusion in itself, and it says something about the requirements that planetary missions will place on the Shuttle. Based on the results of this study, it can be concluded that there are no planetary spacecraft-unique requirements imposed on any of the elements of the Space Transportation System Kennedy Space Center (KSC) launch site facilities, or the operational support system elements. Interim Upper Stage (IUS) performance, which is possibly critical for certain planetary missions, was not examined specifically in this study.

To define and understand the operations concepts and alternatives for the missions discussed, it is necessary to consider all elements of the Space Transportation System — payload and all other elements which provide operations

---

* This paper presents the results of one phase of research carried out at the Jet Propulsion Laboratory, California Institute of Technology, under Contract NAS7-100, sponsored by the National Aeronautics and Space Administration.

[+] Member of the Technical Staff, Jet Propulsion Laboratory, Pasadena, California.

support for the Shuttle/payload mission. Therefore, when considering the operations concepts and alternatives for Shuttle-launched missions, it is necessary to include at least the STS, spacecraft, KSC ground support facilities, tracking networks, launch and flight operations control centers, and the payload or spacecraft operations center.

MAJOR FACTORS/CONCLUSIONS

During the course of this study, several major factors were identified and conclusions drawn as first-order drivers in defining, scoping, and understanding planetary spacecraft mission operations concepts for the Shuttle era. These factors are:

1. The Shuttle Orbiter turnaround requirement (landing to launch) is 160 working hours (within a 2-week calendar time). All spacecraft/IUS/Orbiter integration options will require first mating of the flight spacecraft with the flight Orbiter within this 160-hour allocation — thereby precluding off-line "pathfinder" operations with a flight Shuttle Orbiter.

2. Planetary launch opportunities and launch periods are fixed for a given planetary mission. The requirement for launching two planetary spacecraft within a given launch opportunity, coupled with the launch period/opportunities constraints and the 160-hour Shuttle turnaround time, will affect prelaunch ground operations and the number of Orbiters, Mobile Launch Platforms, and launch pads needed to support a planetary launch. In general, however, planetary launch periods impose no additional requirements above those required to support the STS-level 160-hour turnaround.

3. The Shuttle/planetary spacecraft launch through injection time will be longer than in current expendable launch vehicle missions and is characterized by a real-time interaction with the Orbiter crew and Shuttle Flight Control Facility. Spacecraft power requirements and spacecraft power interfaces with the Orbiter and IUS, for the time from launch through Sun acquisition, will be affected. During the earth-orbit phase, an option exists to perform some spacecraft checkout prior to committing the payload to either deployment from the Orbiter or (after deployment) ignition of the Interim Upper Stage. This checkout can be performed either attached to the Orbiter or after Payload/Orbiter separation. In any case, a spacecraft status evaluation and launch-commit decision analogous to prelaunch on-pad

system verification will be made after the spacecraft has been subjected to the Shuttle launch environment. The choice of on-orbit repair or spacecraft return is available. The effects on spacecraft design and operations of an in-orbit spacecraft repair or return, in case of failure in orbit, are topics not treated in this study but are worthy of future study.

4. Communications alternatives for one- or two-way communication between the spacecraft and the JPL Mission Control and Computing Center during the prelaunch, launch, and on-orbit phases (up to the time of Deep Space Network acquisition) will tend to be more complex for Shuttle-launched missions relative to expendable-launched missions. The in-flight spacecraft/IUS/Orbiter and ground data path alternatives are currently being defined and will include most (if not all) of the following major communication system elements:

   a. Planetary spacecraft
   b. Shuttle Orbiter
   c. Interim Upper Stage
   d. Tracking and Data Relay Satellite System (TDRSS) including TDRSS ground station
   e. STDN/MIL-71
   f. Johnson Space Center Mission Control Room
   g. JPL Mission Control and Computing Center

5. Definition of the Interim Upper Stage is considered critical to defining/understanding planetary missions operations concepts and alternatives. This study activity has essentially treated the IUS parametrically by analyzing various ranges of spacecraft/IUS/Orbiter interface alternatives and describing the implications on spacecraft operations alternatives accordingly. As mentioned earlier, IUS performance was not treated.

   The IUS/spacecraft interface definition will affect spacecraft communications from earth orbit, data handling, power, and overall operations alternatives.

6. Spacecraft protective shroud requirements are primarily driven by the current Shuttle contamination estimates. Some attention is also being given to deriving acoustic isolation during the powered-flight phase by

means of this type of shroud. Inclusion of a protective shroud requires operational consideration both for the prelaunch spacecraft/IUS/Orbiter integration phase and for the on-orbit phase at the time of shroud deployment. Shrouds will then, in addition to weight and volume penalties, add complexity to the prelaunch and on-orbit operations phases.

REFERENCE

1. STS Planetary Mission Operations Concept Study, Final Report, Document 760-122, April 15, 1975 (JPL internal document).

AAS 75-275

USING THE SHUTTLE FOR FUTURE ADVANCED
PLANETARY MISSIONS

L. D. Friedman[*]
W. Scofield[+]

This presentation is a collation of several future mission possibilities that the authors have been involved in examining. The ideas represent early consideration of future missions for the Shuttle era that have received study in the planetary program. Other more broad-brush surveys and "quickie" mission descriptions have been examined in Advanced Propulsion, Future Technology, and Shuttle Planning studies. The topics discussed here go one (albeit small) slice deeper into the subject.

The first consideration is what characteristics will the STS provide that provide positive use on future planetary missions -- either in planning or in execution? A listing is given in Fig. 1.

The next consideration is what technology is being pursued in the planetary program that relates to use of the STS characteristics? Some are listed in Fig. 2. The relevance of these to Shuttle should be obvious: SEP, NEP, Cryogenic Propellants and Gravity Assist all are enabling ideas for high-energy mission achievement -- launching large masses on the STS and enabling them to reach and orbit the distant planets. The in-orbit checkout and assembly capability to be supplied by the STS will enable the test and eventual usage of large structures for antennas, solar sailing, large telescopes, etc. and for employing new instruments first on Earth, later on other planets.

---

[*] Jet Propulsion Laboratory, California Institute of Technology

[+] Martin Marietta Corporation

Given the characteristics and exploitation of them for future planetary technology, what missions might we consider for the fully developed Space Transportation System? Figure 3 lists planetary missions under study or potentially available for consideration in the 1988-2000 time period. Characteristics (Fig. 1) of the STS and technology (Fig. 2) to be developed for the planetary program will make these missions possible at reasonable cost. The first three, outer-planet observatories, MSSR, and Titan Lander, are briefly discussed in this presentation. The remaining will await further study.

The first mission is the outer-planet observatory. To consider a specific example we use a mission concept called the Jupiter System Exploration Facility (JSEF). The concept is summarized in Fig. 4 with objectives and component vehicles matched to yield a mission concept. Note that about 8000 Kg in Jovian orbit is called for (compared to 750 Kg Mariners now in consideration). The masses are not important, indeed the economics of STS usage may dictate many launches of individual components rather than one launch of many components. The important idea is the system devised to perform the required Jovian exploration. A possible mass-velocity history (hence mission profile) is given in Fig. 5, with the previous comment meaning that the reader should concentrate on the part of the diagram past Jovian orbit insertion.

Delivery of the JSEF by the STS is considered in Fig. 6. The performance of STS for outer planet missions is the subject of Fig. 7. There the injected mass vs. launch energy for Shuttle Interim Upper Stage launches as well as Tug launches are plotted. The injected mass for direct missions and for VEGA-type gravity-assist missions (Hollenbeck, this conference) is given. The latter are given by the mass remaining after Earth swingby and after ΔV has been expended in the amounts shown. Note the top labeling of the abcissa, showing regions of mission achievement.

The initial mass - orbited mass tradeoffs for the NEP System mentioned in Fig. 6, is shown in Fig. 8. Stearns (this conference) has discussed NEP System readiness. Clearly, both the Shuttle-launched NEP System and the dual Shuttle-launched advanced chemical system on a VEGA-type trajectory are relevant to this mission.

With injection masses of 1500-2000 Kg suggested in Fig. 7 for the Shuttle/IUS, a satellite orbiter at Ganymede becomes feasible. This is possible with other gravity-assist techniques in Jovian orbit, such as powered swingby orbit insertion and orbit pumping (reduction of orbital energy by use of satellite encounters) to reduce the spececraft's relative speed at the satellite. Potential for a Ganymede orbiter is shown in Fig. 9. The ΔV needed for Ganymede orbit insertion (GOI) (from a Jovian orbit) is 900 m/s. For navigation, 400 m/s is assumed. Jupiter orbit insertion (JOI) of 675 to 1025 m/s is assumed depending on launch year (these values are low due to aforementioned gravity assist employment). The final mass - initial mass tradeoffs are then shown. Whether a satellite orbiter is desirable in light of the potential for repeated encounters from Jovian orbit is problematical and awaits further study.

Having the Shuttle system, with the IUS and Tug stages, available to the Mars Surface Sample Return (MSSR) mission planning process has opened up mission options that could not be considered in the past. Figure 10 shows the mission options that can be mixed and matched to produce some 1920 simple combinations of mission profiles. Most of these options, for example quick return flights requiring less than 3 years or so to perform, are usually impossible with expendable launch vehicles of the Titan IIIE/Centaur class.

Martin Marietta has developed an MSSR mission search computer program that not only calculates the performance requirements for all of the 1920 mission profile combinations but also searches through launch and encounter periods of 400 days or more duration on a day by day basis to select the optimum launch periods and interplanetary trajectory shaping. This work was performed under contract to the Jet Propulsion Laboratory.

Figure 11 is a typical output of this program showing the performance potential for one mission mode over a number of years. This search has produced a number of interesting mission possibilities such as the short

(462-day) mission that can be flown on the Tug in 1986 using the Type I or opposition class profile. Also, on the right-hand side of the table where maximum excess payload (MAX ERV margin) is identified, opportunities such as the 1988 Type I profile using the Tug, stand out.

Another area of MSSR mission planning where the availability of the Shuttle system has been an asset is in the retrieval of the sample at Earth. A great deal of controversy has arisen over the potential dangers to the Earth's biosphere of bringing a Mars sample back that could contain alien organisms.

Figure 12 shows two options for capturing the sample: direct entry of a heavily protected capsule for retrieval on the surface; and, capture in Earth orbit for retrieval by the Shuttle. The Shuttle retrieval offers the possibility of examining the sample in orbit for potentially dangerous biological activity before bringing it into Earth-based laboratories.

Thus, the Shuttle has contributed to both the performance potential and the safety and feasibility of a mission that is considered by many planetary scientists as the most exciting step ahead in planetary exploration.

Titan, the seventh satellite out from Saturn has stimulated a great deal of interest among scientists because of its relatively dense atmosphere, elevated temperatures, and abundance of methane. This combination points to the possibility that life could originate and develop in the atmosphere or on the surface.

Martin Marietta is now engaged in a study of Titan exploration missions for the NASA Ames Research Center. The study covers observations of Titan from a Saturn orbiter, atmospheric probes to Titan, surface penetrators, and Titan soft landers.

Figure 13 shows how the performance requirements of these missions match the launch performance capabilities of the Shuttle/Tug, Shuttle/IUS, and Titan IIIE/Centaur systems. It is evident that the availability of the Shuttle systems will open up the mission planning options for Titan exploration considerably.

CHARACTERISTICS FOR POSITIVE USE

- HIGH PERFORMANCE OF THE STS
- IN-ORBIT CHECKOUT OF SCIENCE AND ENGINEERING SYSTEMS
- IN-ORBIT ASSEMBLY
- HIGH LAUNCH RELIABILITY
- LARGE PAYLOAD VOLUME
- RENDEZVOUS AND RETRIEVAL
- POTENTIAL LOWER COST
- POTENTIAL COMMONALITY AMONG MISSIONS

Figure 1

RELATED PLANETARY TECHNOLOGY DEVELOPMENTS

SEP, NEP

LARGE STRUCTURES (COMMUNICATIONS, PROPULSION, INSTRUMENTS, SAILS)

RENDEZVOUS AND DOCKING

VEGA AND OTHER GRAVITY ASSIST

CRYOGENIC PROPELLANTS

NEW SCIENCE INSTRUMENTS FOR COMPARATIVE PLANETOLOGY

Figure 2

ADVANCED MISSIONS FOR THE "DEVELOPED" STS

- OUTER PLANET OBSERVATORIES
- MARS SURFACE SAMPLE RETURN
- TITAN LANDER
- VENUS LANDER
- ASTEROID ORBITING / LANDING / SAMPLE RETURN
- SOLAR OBSERVATORIES
- MERCURY SAMPLE RETURN

Figure 3

JUPITER SYSTEM EXPLORATION FACILITY

COMPONENTS	OBJECTIVES		
	PARTICLES AND FIELDS	ATMOSPHERE AND IONOSPHERE	PLANETOLOGY
ORBITER	IN-SITU EXP'TS	SUPPORT	REMOTE SENSING EXP'TS SUPPORT TO LANDERS POSSIBLE RETURN ANALYSIS
SUBSATELLITE	IN-SITU EXP'TS	-	-
DEEP PROBES	IONOSPHERE LOW ALTITUDE PARTICLES	IN-SITU EXPERIMENTS	GRAVITY EXPERIMENTS
CALLISTO LANDER	IN-SITU EXP'TS	-	IN-SITU EXP'TS
Io LANDER (LATER LAUNCH)	IN-SITU EXP'TS	-	IN-SITU EXP'TS SAMPLE RETURN
ROUGH LANDERS	-	-	IN-SITU EXP'TS

e.g. 2000 kg ORBITER   (2) 500 kg PROBES
      1500 kg PROPELLANT   1500 kg LANDER
       400 kg SUBSATELLITE   1000 kg ROUGH LANDERS

Figure 4

JUPITER SYSTEM EXPLORATION FACILITY
A POSSIBLE MASS-VELOCITY HISTORY

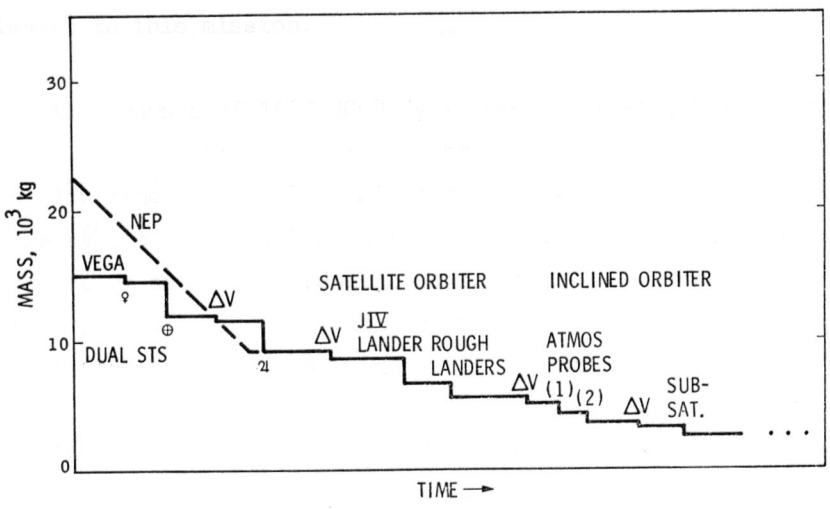

Figure 5

 JUPITER SYSTEM EXPLORATION FACILITY DELIVERY

DUAL SHUTTLE/TUG-VEGA-CRYOGENIC CHEMICAL

    LAUNCH   15,000 kg

    ORBIT   7500 kg

    T_{FL} 1800 days

SHUTTLE/NEP (240-400 kWe, I_{SP} = 6000-10,000 sec)

    LAUNCH   20,000-30,000 kg

    ORBIT   6000-10,000 kg

    T_{FL} 1000-2000 days

LONG LIFE FACILITY

- MANY COMPONENTS
- LATER LAUNCHES
- ADAPTIVE EXPERIMENTS

Figure 6

STS INJECTION MASS POSSIBILITIES

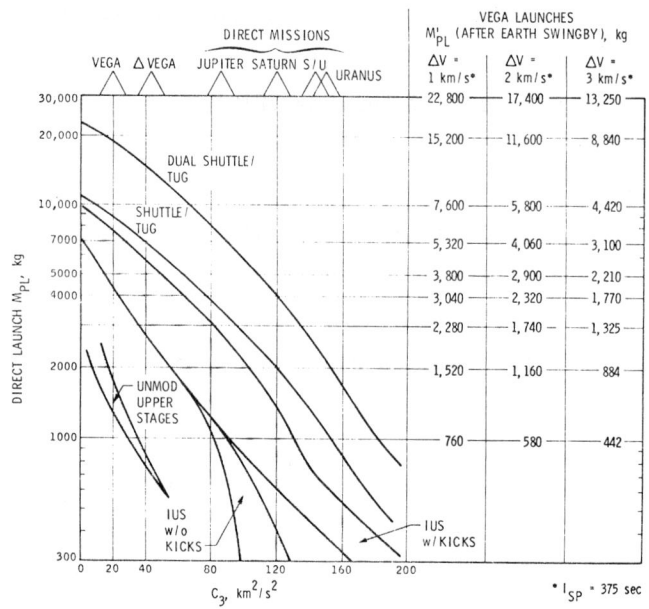

NEP JUPITER ORBITER PERFORMANCE
15 R_J CIRCULAR ORBIT

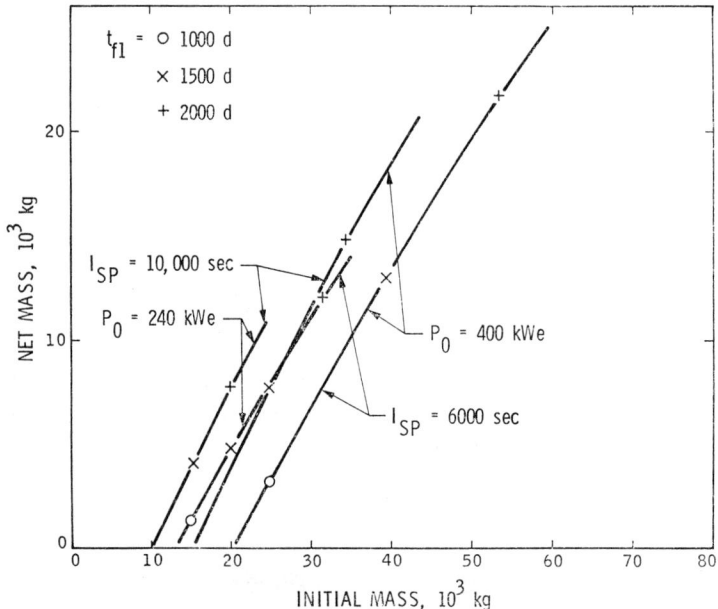

Figure 8

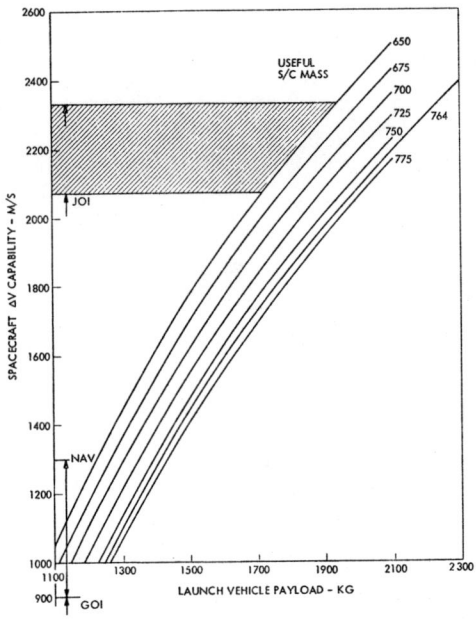

Figure 9

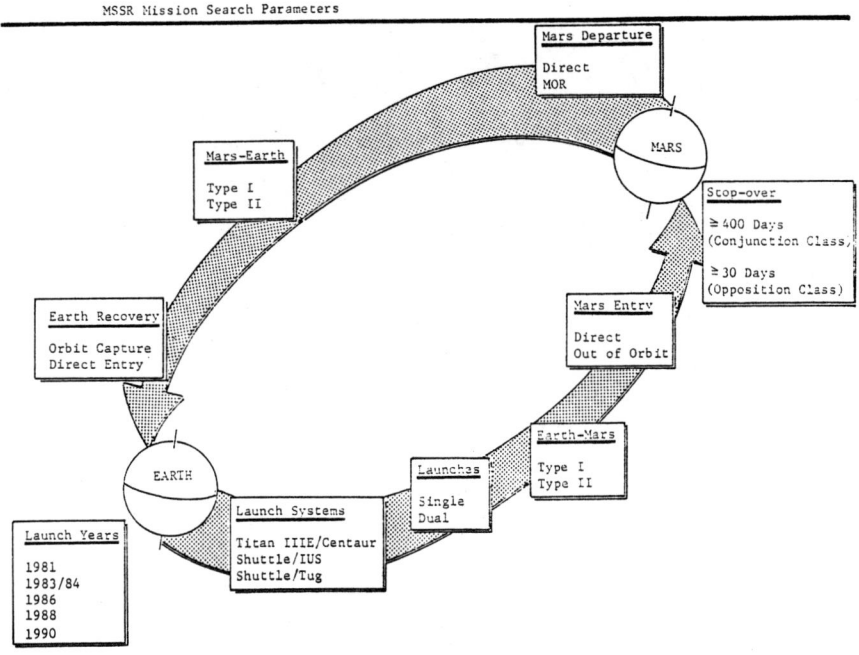

Figure 10

MSSR MISSIONS: DUAL LAUNCH, OUT OF ORBIT LANDING, MOR MODE

YEAR	TYPE	L/V	MINIMUM MISSION TIME (DAYS)	ERV MARGIN (KGMS)	STOPOVER TIME (DAYS)	MAX ERV MARGIN (KGMS)	MISSION TIME (DAYS)	STOPOVER TIME (DAYS)
1981	I	IUS	852	31	240	279	982	420
		T3E	-	-	-	-	-	-
	II	IUS	870	36	150	783	1020	380
		T3E	900	20	190	314	1010	360
1983/ 1984	I	IUS	863	31	230	348	945	380
		T3E	-	-	-	-	-	-
	II	IUS	898	22	190	496	1008	380
		T3E	928	20	250	123	1004	480
1986	I	TUG	462	59	30	1332	902	380
		T3E	882	0	320	14	938	480
	II	TUG	761	8	50	872	921	310
		T3E	947	13	480	48	957	480
1988	I	TUG	731	63	100	1351	921	410
		T3E	811	27	200	313	976	480
	II	TUG	751	59	30	890	986	330
		T3E	-	-	-	-	-	-
1990	I	TUG	780	0	140	884	960	450
		T3E	-	-	-	-	-	-
	II	TUG	756	102	30	1311	966	300
		T3E	906	2	200	198	976	320

MARTIN MARIETTA

Figure 11

Figure 12

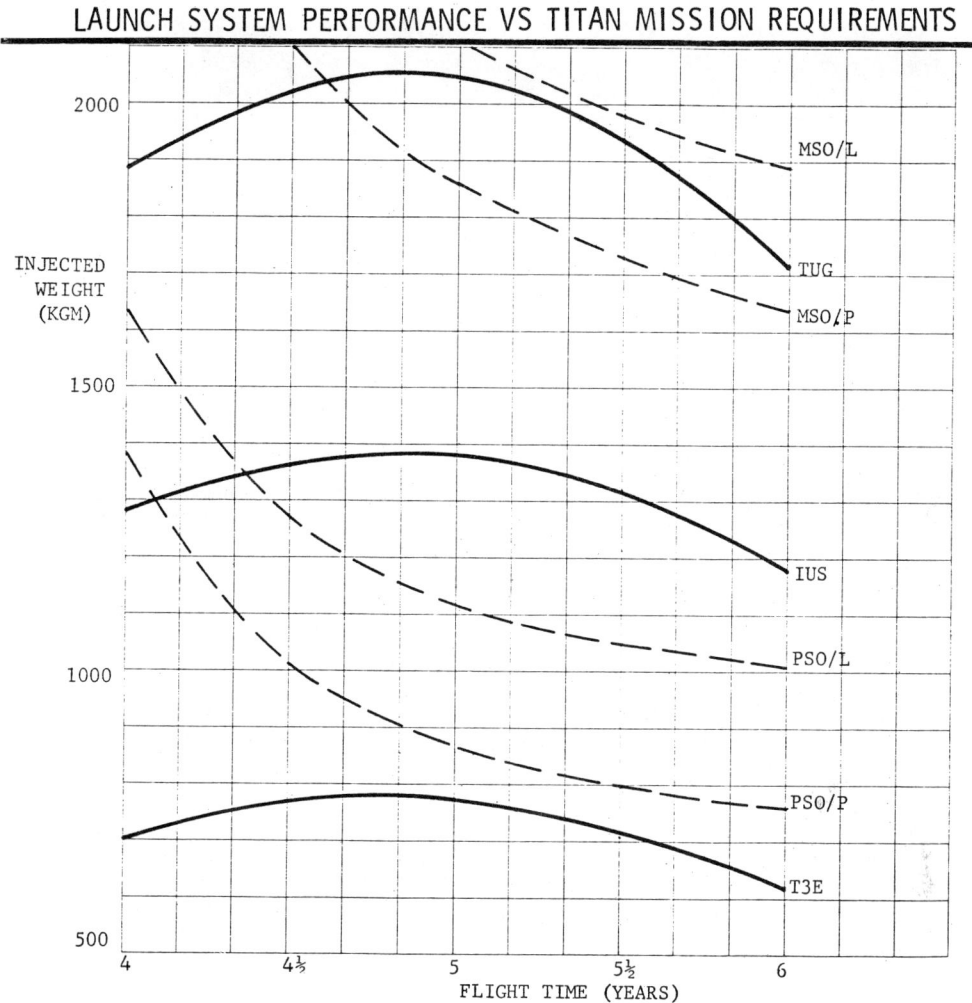

Figure 13

CAN WE USE THE STS TO IMPROVE THE PLANETARY PROGRAM?
PANEL DISCUSSION

MODERATOR

John Niehoff, Science Applications, Inc.

PANEL

Jack Wild, NASA Headquarters, OMSF

Daniel Herman, NASA Headquarters, OSS

Phil Culbertson, NASA Headquarters, OMSF

Robert Parks, JPL, California Institute of Technology

Niehoff - I have a number of comments I am going to start out with, and then we will have some overview statements by the various members of the panel, and then we would like to have some fairly free discussion. I think we are particularly fortunate in having the individuals on the panel that we have today. They are all very much involved at the higher echelons of NASA's planning efforts in regard to Shuttle, and they are probably much more knowledgeable than any of the rest of us here on many of the facets around which decisions are being made relative to the Shuttle. We should take advantage of their presence here, and I think we can have a very good discussion on some of the issues that were raised earlier in the day from various papers. Naturally, the use of the Space Transportation System, where planetary is concerned, evolves very much around the upper stage. Without it, the planetary missions really don't exist. That is going to be the center of the panel discussion, and we will then branch out to other subjects -- particularly interface problems and interface benefits between stages, payloads, and the Shuttle. To start, I would like to make some remarks about the economics of the Shuttle. These remarks will be made from a planetary missions viewpoint rather than from an overall NASA viewpoint or from an entire traffic model viewpoint. Then, I have asked Jack Wild to give us a presentation of the NASA viewpoint relative to upper stage performance and operations schedule. I have asked Dan Herman of the Planetary Programs Division of OSS to give the planetary programs viewpoint of upper stage requirements, performance, and what various options will be available for the planetary programs. Following these two overviews, I would like to have some discussion and questions on this general subject of upper stage performance and operations. Then we will have two similar overview comments, one by Phil Culbertson of OMSF on the Shuttle viewpoint of interface requirements with upper stages and payloads. Finally, Bob Parks of JPL will give us a brief statement of the payloads viewpoint of the interface.

I would like to state that the planetary mission model as it presently stands is very well balanced in that it includes all targets of the solar system with very few exceptions. It proceeds at a rate of about two launches a year from now through 1987 when we have four launches -- there are several years where we have three launches. I won't dwell on the individual missions for we are all rather familiar with them. Let's take a look instead at some of the statistics of this model. In all there are 15 projects, and in those 15 projects there are 21 launches. Starting in 1981, there are 18 potential Shuttle launches. For Tug missions, I've chosen an IOC date of 1985, which is very conservative. There are eight launches which would be potential Tug missions. There are an additional two more missions if we take an earlier IOC of 1984. If you take present estimates in FY75 dollars of all these 15 projects, the total bill would be over 2 billion dollars. Cost range of the missions of these 15 projects is about an order of magnitude scaled from 30 to 300 million dollars. The median project cost is $142 million and the average project cost is very close to that, about $150 million per shot. This is just to show you by comparison what the cost of launching these projects would be. Those costs I mentioned before were exclusive of all launch costs. If indeed, which is not the plan, but if we were to try to implement this plan with expendable launch vehicles -- particularly the Atlas/Centaur and the Titan/Centaur the present workhorses in the planetary program -- and augment that with Solar Electric Propulsion (SEP) as necessary, the cost, the accumulated cost of the transportation system is about a billion dollars or about 50% of the project cost. A very expensive bill just to get out of earth orbit. This is the prime motivation for getting something like the Shuttle which promises reuseability and cost savings.

Now, what I will do is suggest three different scenarios by which we would implement an upper stage with the Shuttle to give the planetary program a Space Transportation System capability that they need. I've tried to select three scenarios which really bracket all possibilities. I'm not suggesting that any one is of any particular interest to anybody at NASA at this time. These are just typical examples that span the range. The first one, scenario A, says the planetary program accepts the IUS, as selected, but recognizes the fact that it must provide some additional development funds in order to get to escape-mission configurations needed. This could be by stacked IUS's or putting on a kick stage, or what have you. The cost of that additional increment to the planetary programs I set, as an example, at $25 million. Scenario B is one in which the planetary program rejects the IUS and decides to go with its present workhorse, the Centaur D1T, as the stage for the Shuttle. And it does so independent of the NASA, or the Air Force, or anyone else. And that's an area where I set an RDT&E of $50 million. The third, scenario C, says that NASA rejects the IUS, and the planetary program works with NASA as a whole to develop a low-technology Tug at a very early time frame of 1981 so that they can implement their missions with that system. NASA's planetary programs contribution, but not the total RDT&E, would be $50 million as well. These are just round figures in order to give us something to work with to see how some of the comparisons come out. The low-technology Tug is recoverable as is the full technology Tug and I've set the recovery costs at 2 and 4 million. The 4 million here was to reflect a much hotter upper stage to make recoverability more available. Now, the Shuttle launch-cost I set at $12M per flight in FY75 dollars, and the high-technology Tug is assumed to be developed under Shuttle program funding -- with no RDT&E assessment against the planetary program budget. Unit costs for the upper-stage launch options are: IUS ($5M), modified Centaur ($10M), low-technology

Tug ($15M), and high-technology Tug ($20M).

Let's take a look at some comparisons then in terms of accumulated launch costs. We see that with scenario A, which is using the IUS with an additional $25M in RDT&E from planetary programs, there is an accumulated cost by the end of 1987 which is only about 50% of the expended launch vehicle cost. Certainly its a goodly savings. If we implemented the Tug in 1985 to do the last eight missions, the costs would be slightly less but very slightly less. Let's take a look at scenario B, which would be the cost of planetary programs going alone as it were, paying $50 million to modify the Centaur and then using SEP where necessary, and that cost is about 40% less than the total. Here the Tug makes a somewhat larger difference because of its recoverability. Finally, scenario C, which is the low-technology Tug, falls between scenarios A and B, and Tug does not make any further difference at all because both concepts would be recoverable there. We see that the cost differences between the scenarios are certainly within the noise-level, which was very significant with this kind of analysis that I made. One would not want to draw any conclusions between the various options, but one could certainly draw conclusions between the options and expendable vehicles. That would be that the options using the Shuttle system with a wide range of possibilities, that's the upper stage, saves 50% of the cost of expendable vehicles which is 50% of the project cost. So we have reduced the transportations costs to a factor of about 25% by going to Shuttle. If you take this as a cost benefit, the benefit-cost ratio, where the benefit is the difference between the expendable and the Shuttle launches and you take the cost as being the launch cost and the project cost, going to the Shuttle saves you on the order of 20%. This is another way of looking at that, over the expendable vehicle cost.

In summary, the cost benefit of an early and complete transition to the Shuttle is considerably more significant than cost benefits identified in this little analysis between various upper-stage alternatives and, strictly from a planetary viewpoint, an early introduction of the Tug does not seem to be an essential feature of the STS system for the planetary program. If we get an upper stage in 1981 which is capable of performing the missions, as defined, we do not gain any significant new economies by going with the Tug in 1985.

Herman - I would like to comment on just one thing. I don't think I disagree, but I would like to clarify something. In your definition of the low-technology Tug concept, let me just translate those words to mean a recoverable IUS. If you do have a recoverable IUS, and if you plan your traffic in such a manner that you do several recoverable missions to geosynch, and plan to expend the IUS for planetary mission thus amoritizing the cost of that IUS over 4, 5, or 6 missions, I think you come up with an answer that says that the most cost beneficial approach, for planetary missions, is to get a recoverable IUS going in orbit.

Niehoff - Well, I've already shown in the samples I've had here that we have used recoverability in the planetary missions wherever possible.

Herman - What you have not done, what you have committed is the recurring cost of an IUS against the planetary mission whenever you've expended.

Niehoff - That's correct, yes.

Herman - OK, what I'm saying is that's not necessarily valid. Let's assume that we do five geosynch missions with an IUS. Then, on the sixth mission, use it for a planetary mission and expend it.

Niehoff - It turns out that there aren't that many. That's probably true, but I would question whether its a significant saving or not. I don't want to argue at this point about whether the cost of that expended, recoverable IUS is $15 million or $10 million or $5 million. I don't think its going to alter the conclusion here that the early transition to a Shuttle is perhaps more important than what we transition to in terms of the upper stage.

Question - Have you looked at your cost model for different mission models? Past history indicates we don't fly everything we hope to. Generally its actually always on the low side; we don't fly as many missions as we initially plan. This appears to me to be a fairly optimistic plan for the future.

Niehoff - This plan is the most conservative plan that NASA's come up with in the last five years. If you cut this down further, certain benefits are decreased because the expendable vehicle costs go down. There comes a time in which there is a more important question asked. At what time is it no longer viable to mount a planetary program? I'm not going to say that this is the minimum, but I would certainly say we are getting close to that. I did not specifically answer your question. I did not look at anything less ambitious than this.

Question - I notice that in your three scenarios the unstated assumption in every case was that the IUS was not a Centaur. What was the reason for that assumption?

Niehoff - The assumption was that whatever the IUS was it would cost $5 million a shot. That was the only assumption that was made, and I was not going to make any assumption as to which particular IUS it would be. In fact, there was no use of any particular IUS in the capture analysis that was necessary to assign the vehicles for the cost. We assumed a general IUS performance curve which, in fact, all of these contractors are using with one configuration or another.

Herman - There is one other point that I want to emphasize before we go whole hog. Everyone, of course, recognizes that the planetary traffic is a very small percentage of the total traffic, and on an economics analysis basis the planetary missions cannot dictate, in a major way, the choice of the IUS. So, as John prefaced his remarks, he is taking the viewpoint from a planetary missions standpoint only. That should be recognized.

Niehoff - Let me say one additional thing. The fact I used Centaur in scenario B as the present planetary workhorse did not mean it was excluded as a candidate in the IUS scenario A. It was certainly included there as well.

Dan Herman has indiated that there are probably some cost differences in the upper stages, particularly with recoverability, and we have seen this in other studies that we have done that also stress the point that early transition to Shuttle is one of the big benefits of the STS system.

I would like to ask Jack Wild if he would make a few comments related to the NASA viewpoint on upper stages for the Shuttle and perhaps make a tie-in here with planetary programs. Then, I'll have Dan give the planetary programs viewpoint on the upper stage.

Wild - Before I start, I would like to make a comment or two on the session we had here just previously. There are a couple of nasty rumors going around. One of them is that manned space flight, in developing the STS system, are ignoring the planetary people and not taking their requirements into consideration. And then there is one that counters that, which Danny referred to in an oblique sort of way, that this planetary tail is wagging the dog and that those guys are just dead set on having everything their own way and they have no consideration for the other kinds of payloads. I can assure you that the former is not true, because the planetary missions have taken a large part of mine and a number of other people's time over the past year in the IUS excerise that we are just in the process of finishing up. I think, too, that the papers that were given here today certainly give a lie to the second one, because I think you are doing an excellent job of contingency planning, if you will, by looking at what you can do with the type of capability that may be available to you. I think that its a completely realistic look and I was very gratified by the papers, from my personal viewpoint. I think it shows a realistic outlook and a real cooperative spirit on the part of you people who are doing the work.

Getting to the subject end of the comments, I would like to spend a few minutes on three subjects. I would like to talk a little bit about the status of the IUS and the Tug at this time; and about the performance of the IUS and the Tug; and then say just a word about the operations complexities on the IUS. Very briefly on the IUS, most of you heard yesterday, I am sure, that the IUS is under evaluation now as a result of the studies that have been made and that the Air Force is expecting to make a decision about the middle of September as to which IUS will be selected. They are planning the initial operational capability of the IUS for mid-1980 to be coincident with the operational capability of the Shuttle. I don't think I have to say anything more about that.

Planning for the Space Tug is still proceeding. As you know, we are still in the study phase on that. Current plans are for phase B study to be initiated in fiscal year 1977 for an IOC in late 1983. Now, John asked me to make some comments about how firm that late 1983 date was. I think about all I can say is that we are not an approved program yet.

When you are in the non-approved status, you're subject to continual review, particularly at budget time. That's where we are right now -- we're being reviewed in the light of the fiscal year 77 budget. I anticipate that within the next month or two we will get either a confirmation that we will go into a phase B in fiscal 77 or we'll be told to hold off and target for a later date. I think that this decision, you appreciate, will be made primarily in budget terms. The decision to go to the IUS, instead of directly to a Tug, was made strictly on budgetary terms. Our studies have established that if we go to a Tug immediately the long range economics will pay off. The sooner we can get a Tug into the system, from a strictly economics viewpoint, the better off we are. But we're faced with the near-term problem of Shuttle funding and of payload funding. The payload people have been somewhat starved, as all of you know, the last few years because of the large amount of the budget going into the Shuttle. These are all considerations that have to go into making this decision in addition to the mission projections and the requirements for those missions, some of which we have talked about here today. I've talked about two or three minutes on this and I guess what I've told you is that the current plan is still for fiscal 77 but this is in review and we should have a decision within a month or two. Also, within that month or two we will have a decision on the IUS, which I'm sure you will appreciate may very well impact what will be the IOC date on the Tug -- how soon we feel we need to go to a high-capability, reusable vehicle.

On the subject of performance, I would like to show you what has turned out to be the most popular slide of the day. I think you have seen this about four times now. This is a mission model plotted with a Titan IIIE/Centaur/TE364 imposed on it, and let me talk the Tug first. The Tug performance comes in well over anything that you see on that chart, and so from a performance viewpoint in the getting of the planetary missions there is absolutely no problem if we go to the Space Tug. Now in the case of the IUS, each of the IUS candidates will deliver performance for planetary which is better than the Titan IIIE/Centaur/TE364. They will have trouble with Mariner Jupiter Orbiter and the Mariner Saturn Orbiter. Now, we have done some work in-house on these and as you heard today there are ways of getting those missions with any of the IUS candidates by putting SEPS on it, by going to some of these more exotic trajectories-- such as the VEGA, and there are ways of capturing those two missions with any IUS that might be selected. I think I can say, from a performance viewpoint, there are ways of getting the missions that you have projected in the mission models.

On the operations, I think Joe Cork pretty well illustrated what the differences in the operations problems are going to be. As he indicated, the candidates that we have for planetary missions range all the way from single stages, single stages with a kick stage, single stages with perhaps more than one kick stage, drop tanks, solid auxiliary stages, various configurations. These, obviously, are going to have some impact on the operational complexity -- the number of firings, the way to fly your missions, the trajectories, and so forth. For obvious reasons, I don't want to get into a rating of 1 to 5 for the complexity. I think what I can say is that both NASA and the Air Force have evaluated the IUS candidates, and we feel that any of them are operationally acceptable. There are obvious differences in complexity, but the mission can be formed from an operational viewpoint as well as performance viewpoint with any one of the IUS's that are in the running.

Question - You said that some of the IUS candidates do have trouble with MJO/MSO. Since the earliest anticipated launch date for either of those is '85, that is a couple of years beyond when you hope to get the Tug built.

Wild - I'll clarify that. The mission model that was set up for the IUS candidates ran through 1984. So, all of the IUS candidates as they were presented were able to get the total mission model through 1984. Now, I was extending that. I was saying suppose we have a contingency and the Tug doesn't appear, then we could get those missions with the IUS. They would have some trouble. There would probably have to be some changes, but there are ways of doing it.

Herman - I want to dwell on a couple of points. One is to really emphasize John Niehoff's suggestion that it would behoove the planetary missions to move to the Shuttle/IUS as rapidly as possible. I would like to state one concrete action that was actually taken which, incidently, would be of interest to any ESA, MBB, or European people in the audience. The original mission model shows that the Jupiter swingby out-of-the-ecliptic mission would be in 1980. Our current plan is to do that in '81 and to do that in a dual spacecraft mode with the Shuttle/IUS. Now, there are three reasons for that. One is that, based on the near-term constraints that we envision, we just cannot see how we can procure a Titan/Centaur launch vehicle for that mission. Noel Hinners informed ESA of that at a meeting that occurred in February. That would mean that if we do that mission with an expendable launch vehicle it would have to be done with an Atlas/Centaur. If it is done with an Atlas/Centaur, to get any degree of performance at all even with a single spacecraft, the optimum launch year is '83, even though the minimum energy launch year to Jupiter is '81. The reason is that in '81 you are plagued by a launch asymptote constraint which doesn't allow you to get the kind of performance that the energy requirement of Jupiter would appear to allow you. The most effective way to accomplish that mission from both a fiscal standpoint and a performance standpoint in a dual spacecraft mode is to do it in '81 on a Shuttle/IUS. That's our current plan. The fallback position is still to do it in a single spacecraft mode on the Atlas/Centaur. That's the one point I want to make.

Another point is this. Jack Wild has indicated that he has looked at a host of backup scenarios on how we can accomplish the mission model, no matter what the IUS story is, and no matter when the Tug comes into being. Depending upon what happens, we pay a penalty for it, but there is one thing that is of concern. It touches the mission that we are hoping to get a start on in fiscal '77. It has become evident to us that NASA management, (and rightfully so, I'm not quarelling with their assessment of it) in weighing the sceintific thrust we want to initiate, views a return-on-assets situation as a very strong factor in a yes or no decision. Now, as I mentioned before, given a mission that takes 7 years before you get a payoff, as opposed to a mission that takes two years to get a payoff -- other things being equal such as the same degree of support from the scientific community -- the pressures will very strongly go to the mission with the shortest flight time. Therefore, even though a guy like Jerry Hollenbeck has developed an extremely elegant and very powerful technique to effect high-energy missions in a low-energy mode, we still have to be concerned about the additional trip time that that concept demands you to pay. Trip time is very important to us. Hopefully the MJU mission will go, but if disaster strikes and it doesn't, my assessment is that one of the facets which will kill it would be the relatively long trip time which we could do nothing about. I have a further personal opinion (This is not a NASA position but is a personal opinion.) that if MJU does not go, in my estimation, we will never fly another Titan/Centaur vehicle after the MJS mission. We would await the Shuttle/IUS before we do any high-energy planetary program.

Question - How far are you willing to go on the short flight time? Would you be willing to put in your money into a SEP, for example, to augment the IUS in order to get short trip time?

Herman - Again, that's something that you have to look on a return-on-assets position. What do you pay for not only the RDT&E costs but what are the recurring costs of a SEP as compared to the recurring costs of a chemical stage? And that has to be weighed in looking at the total return on assets position.

Question - How does that situation look right now?

Herman - We have not done the study to date. You know, really assessing if there was a committment to a large number of SEP flights, what the recurring costs would be.

Question - Well, actually the problem occurs in the later time frame so we have time to do something about it. For example, 1985 or 1987 are when we will be looking at these difficult, high-weight, outer-planet missions that could take a long time if we are forced into a VEGA or Delta VEGA mode. Given that time of planning, if we couldn't have the Tug, wouldn't planetary be interested in SEP?

Herman - Let me backtrack by saying, without these specific issues you raised, we are counting on the development of SEP as reflected in our mission models. There are three missions that are slated as SEP missions, the out-of-the-ecliptic solar observatory, the comet rendezvous, and the asteroid rendezvous. We have had to slip those missions based on a restraint on the technology funding for SEP that occurred last year. As was mentioned in a session that Pres Layton chaired this morning, the current plan by the Office of Aeronautical and Space Technology is to bring SEP into a technology readiness state by 1979 or '80. We are counting on using it for the first mission for the out-of-the-ecliptic solar observatory '84, with a project new start at '81. We would like to move those missions closer to the present. We would hope that the funding is there to accelerate the SEP development program. But, forgetting about the issue that I've just raised, SEP is a necessary item in our arsenal in order to do the missions in our mission model. It has nothing to do with the aspect of IUS versus trip time.

<u>Question</u> - Jack, I would like to ask you a question on the Tug. Talking about phase B study work, how concrete is the Tug design? For example, cryogenic versus space storable, recoverable-only versus recoverable/retreival capability. Are these options still open in those studies, or have we pretty well decided what we want?

<u>Wild</u> - Well, I think that if we would get the go-ahead for a phase B start in '77, I think that our Baseline Tug which we have now would remain the baseline. I think we have done enough work that we can say, with a high degree of confidence, that we need the performance that a hydrogen/oxygen stage will give to us. I think that storables, in the long run, for a Tug which will operate in the late '80's will probably not give us the kind of performance we think we will need for a new development. Now, if the decision is made to slip the phase B perhaps as much as 2 years, then I think that we would want to take another look at the design of the Tug. The Baseline Tug, as most of you may or may not be aware, is pretty much based on current technology. We are using the RL10 engine, very much as it is, with some minor modification and with no real increases in the technology. I think if we get the kind of design that an additional two years would give us, we would probably want to go back and reassess this. We would look at what the requirements are going to be, what we see in the late 80's in the way, perhaps, of higher requirements, manned capability on Tug, and would want to look at the possibility of designing a Tug which would evolve more easily into satisfying those kinds of requirements.

Question - Del Tischler recently had an article in AERONAUTICS AND ASTRONAUTICS which made the point that a cryogenic Tug should only be planned if retrieveability was a strong element of the future NASA program. Do you concur with that conclusion? In other words, if you are just going to recover it without recovering payloads, you shouldn't do it.

Wild - We haven't done enough analysis, I believe, on Del's specific proposal for me to give you a firm answer on that. My opinion is, and we are looking at it, my opinion is that over the long pull, if we're talking 10 years of usage, that a cryogenic Tug, even if you never recovered a payload, would pay for itself on a reusability basis alone. I think it would cross over with Del's more modest arrangement somewhere in the '85-'87 time period. In Del's proposal, he talks about a small cryogenic Tug boosted by two solids. And I think that when we look we'll find that it doesn't cost much more to build a large cryogenic Tug than a small cryogenic Tug -- you simply make the tanks bigger. So, I think the development costs on Del's arrangement are probably comparable to, or maybe even perhaps more than, a single stage.

Question - Question for Dan. I imagine that you cannot comment on the relative viability of planetary programs versus astronomy programs, such as LST, but do you have any insight into whether the NASA budget can support new starts in fiscal '77 for both MJU and LST? Maybe it will be an either/or situation.

Herman - I hope so. If we were going out of the planetary arena, I would not be in my current job. I firmly believe that NASA does have a commitment to sustain the exploration of the solar system, and I further believe that in the Shuttle era it will enable us to do a more effective exploration program. As John Niehoff indicated, no matter what the IUS candidate is, we will be able to do more missions for less money. On the near term issue, the one you ask, you know, on whether MJU is mutually exclusive with LST I would hope that it isn't. The only point that I would want to make is that if we don't get a new start now, the MJU program is dead. That's not the same as LST. We have to make that '79 launch window or cancel MJU with the Titan/Centaur launch vehicle.

Niehoff - I'd like to now change the discussion and talk about some of the interface problems and some of the interface benefits that are going to be incurred by transition to the Shuttle transportation system. We will start out with Phil Culbertson of OMSF to give us an overview from the Shuttle viewpoint of interface requirements.

Culbertson - I'm not sure I'm going to specifically address the question in the way that you stated it -- or the issue, but I don't think that Jack or Danny did either. I don't feel all that bad about it, but I'll get to the point that you raised before I'm through. I wanted to comment a little bit on the process that we have been going through to try to determine what these interfaces should be before I talk about what the interfaces are. When I speak of interfaces, I'm speaking in a very broad sense in that I would like to include in my definition of interface for this particular meeting what the performance of the systems should be, what the physical and software interfaces should be, and what the operational interfaces should be. The process we have been going through over the last three years, in trying to figure out what these machines should be, has to me been a very intriguing process. It is one that I don't know has a parallel, certainly in the space program, and probably and very possibly not in the industry itself. For the first time, we have tried to develop a single machine which is destined to become almost a monopoly or to control the launch business as a monopoly -- with the possible exception that some of the small launch vehicles may very well stay in the inventory. But, for sizable payloads, the Shuttle is going to replace the presently existing systems, as we see it today. Now, even Coca-Cola, Henry Ford, and Ma Bell have competition, and so in trying to decide what the Shuttle should be, what the use, and what the Tug also should be, we have had to look at a very broad range of frequently conflicting requirements placed on it or requested of it from the different program offices which will intend to use it. It was in trying to decide what the system should be like that we could have gone to either of two extremes or stayed somewhat in the middle. We could have, in attempting to accommodate the requirements and desires of everybody, made the system so complex that it could have barely gotten off the ground all by itself, quite alone without any payloads in it at all. On the other hand, we could have made it so narrow in its performance range or its capabilities range that it would have been very difficult for anyone to have used it, because there was no way it could have done what was being asked of it by a particular organization. So, we have, in a very mysterious process which I can't fully define, and in which I participated for the last three years, worked with all the users and have tried to arrive at some kind of middle ground on this system. This probably does not completely satisfy any individual user. On the other hand, it probably comes close enough to the vast majority, if not all the users, that with some slight modification in the way of interface equipment or that sort of thing it can do the job its being asked for without a great deal of compromise.

Now, in getting to this point we found that we had to educate both ourselves and a lot of potential users of the system. We all had to understand that when more is asked of the Shuttle for an individual mission in a commodity which is not needed by other people, then that detracts from the utility of the Shuttle in performing those other missions. So, for every one user that is helped, there is the potential of half a dozen others could be penalized for many of the requirements that are being sought. In a similar fashion, we find ourselves, because we are budget-limited, in a position that anything any of us, as individual payload users of the system, would ask the Shuttle to invest in for increased capability would really be taking money out of one pocket and putting it into another. As a result we have tried to achieve a balance both in the capabilities of the system and in the resources that go into the system. Whether or not we have achieved the right balance I think we won't know for many years. I think we are fairly close to the point now, at least in the Shuttle, that we will agree that the major decisions have been made.

There are minor vernier corrections to be made in areas that are somewhat open. We are still fighting about the amount of space left for the payload operator in the aft wall of the upper flight deck. There are some communications problems that remain open. The extent to which the Shuttle will be able to relay spacecraft data down to the ground without increasing the error bit rate too much is being evaluated still. There are some problems in RFI that many of us feel are not yet fully solved and very possibly won't be solved until we fly. RFI has that unique property that's hard to analyze on paper, and it is frequently a surprise at the end, (Many of you know that better than I.) and so we will learn more about RFI in about three years when we begin to fly this machine. By and large, the principal decisions are being made, and have been made, and the job now is how best can we use the Shuttle and what kind of interfacing equipment will we have.

We have tried to design the Shuttle, to a degree, as a flat-bed truck or as a flat-bed railroad car, and there will be what I refer to all along as buffer systems which make the ride easier for the individual payload user. Some of these will be provided out of a standard inventory of the Space Transportation System. Others will be provided by individual users. The kinds of things I am referring to are those peculiarities that only some programs face. For instance, in the planetary program there is a requirement for sterilization that basically-speaking no other program has. One of the early requests from the planetary program was -- the cargo bay has to be sterile. The Shuttle came back -- it isn't going to be, you know nobody else wants it sterile. It would cost X number of pounds and X number of dollars. So, the planetary vehicles will have shrouds where they must be sterile. It will be the requirement and function of and the responsibility of the planetary program to provide its own shroud and do its own sterilization. It's not quite that simple for back-contamination, and we really haven't licked that problem. We haven't faced it squarely yet, but we will when the time comes. I use that just to illustrate the kind of situation where we find the Shuttle does not do all things for all people -- but it never will, never could have, and if we had tried it would have been a failure. Well, so much for the first thing that's not on the agenda, but I wanted to mention it.

With respect specifically now to the stage interface between the Shuttle and the Tug, I just want to make a couple of comments which are more on the viewpoint that we have for this interface than the actual interface. In the first place, the IUS is a separate vehicle from the Shuttle. It will not be used in all Shuttle missions; it will be used in a very high percentage of Shuttle missions. It does enter into the Shuttle from an

operational standpoint in kind of a peculiar way. If you were to set
yourself up a hypothetical model of various organizations who are trying
to operate a mission, one of them is the guy who is trying to operate
the Shuttle. His mission begins at liftoff and goes through a time when
the IUS and the planetary spacecraft are removed from it. Then, the
Shuttle continues its mission and comes back home. The IUS mission
starts at liftoff, goes through the time it is removed from the Shuttle,
then goes as long as it is required to burn, and then it separates from
the planetary spacecraft. If its a Tug, it may come back. If it isn't,
it goes off and does its own thing in solitary confinement, but it no
longer has any part of the real mission. The spacecraft mission starts
at liftoff, goes through the time that it leaves the Shuttle, goes
through the time that the IUS is burning, and then goes off and does its
own thing. So, if you are thinking of three people who are operating
the system, then the three of them are working together up to a certain
point. First the Shuttle guy goes away as he finishes his part of it.
The IUS and spacecraft men continue their part of it for a while, and
then the IUS guy goes away and the planetary man goes on for years thereafter. This situation raises some interesting kinds of jurisdictional
problems and planning problems which are complicated by one other dimension I didn't introduce. That is, the Shuttle may go off and do
something else after it furnishes its part of the planetary mission. It
may have some other payloads it wants to do something with, and so now
there is a new bunch of people who are participating with the Shuttle
who are also participating with the first part of the planetary mission,
too.

The IUS fits in the middle of this situation in a peculiar kind of way
in that the planetary people want the IUS to belong to it during those
planetary missions. The Shuttle people think its part of the Space
Transportation System and ought to belong to it. Each one has his own
justifiable reasons for thinking each of these two ways. Now, the IUS
people don't like either one, and they would like to go out and do their
own thing, as a matter of fact quite separately. I think that probably
won't happen -- that third option. For those times in which the IUS is
within or in close proximity to the Shuttle, the overall control of the

IUS will very likely be with the Space Transportation System organization and specifically the guys who are most worried about the Shuttle from the standpoint of safety. That does not mean that someone other than a Shuttle operator will be actually worried about the IUS in operating it, getting ready to fire, and actually firing it. But he will have to be pretty friendly with the guy who is operating the Shuttle during this time period. Once it is fired for the first time and goes away from the Shuttle, then that interface no longer needs to prevail. We haven't yet solved this problem entirely, but to a degree those kinds of problems are influencing the kinds of interfaces that are going to be set up not only from the management standpoint but, more specifically now, from the operations standpoint.

The IUS has another peculiar relationship with the overall program and that is from the space transportation standpoint. We have an overall responsibility to provide the delta-V that is being asked for by the planetary program, or by the communications organization that wants to get to geosynchronous, or anybody elase who wants to use the STS. From our standpoint of a transportation organization, we want to try to essentially guarantee the delivery of the specific mass to a given velocity in a particular direction. We will control the IUS as to its configuration and its quality and the materials that go into it with the same sense of rigor that we control the Shuttle. On the other hand, it is the policy of the transportation organization, in this case, the Office of Manned Space Flight, that we will not impose such controls on the spacecraft. That is someone else's responsibility, and we don't propose to try to influence the extent to which that spacecraft is going to work or not work because of materials, quality, and that sort of thing. We will, however, have certain safety requirements that we will impose and that will have to be met by the spacecraft. We are working hard to make those the minimum that we can. We have a draft of the Safety Requirements Document which has been about two years in the making and which, I think, sets the right tone. It is not complete in that, for instance, when we ask for hazard analysis it can be a very simple thing or a very complex thing. Until we really run a hazards analysis and make sure it satisfies everybody, we can't say how simple or complex that's going to be. From the standpoint of the requirements which we will impose (materials flammability, toxicity, that sort of thing) and even from the standpoint of hazards analysis, I believe we have gone a significant way toward trying to make life a little easier for the spacecraft people. So, those are some of the less obvious interfaces that I wanted to mention.

With respect to the specific requirements that the Shuttle will place on the spacecraft in addition to safety, there are major similarities between the planetary spacecraft and geosynchronous spacecraft as they interface the Shuttle. They are both kind of unique from other payloads that will drop off in low orbit, because they both involve a propulsive stage, and, therefore, the provision we would like to have the Shuttle afford is on-orbit checkout before you commit the spacecraft beyond the reach of the Shuttle. If there is a problem you can then bring the spacecraft back to earth. It is considerably more difficult with a planetary or geosynchronous spacecraft because they don't like to deploy antennas, or power supplies, or solar arrays, or other appendages since they would get a propulsive burn after that. To a degree, I think we can still provide for some rudimentary checkout on orbit even for the planetary spacecraft. I would hope that as we move from here to there we will find more effective ways of doing this. We haven't, I think, plowed all the ground that can be plowed. Another similarity between the two destinations is that both of them are weight critical, and that means that some of the other inherent capabilities of the Shuttle to make weight less of a penalty are not nearly so useful in the planetary program. You have much more difficult tradeoffs than for the spacecraft that wants to say in low-earth orbit. There is a major difference in the way you run a weight trade-off analysis.

Another factor to consider is that neither planetary spacecraft nor synchronous spacecraft want to use the TDRS, the Tracking and Data Relay Satellite; although synchronous missions can to a degree but its not very good for them. The Shuttle is tuned to the TDRS and that means that right off the bat we have some kind of communications problem between the planetary spacecraft and the Shuttle, particularly when they want to communicate with each other in low earth orbit before they get away. We are working that problem, and I think we are close to a solution. As a matter of fact, when we first started worrying about communications between spacecraft outside the Shuttle and the Shuttle itself, we said well its simple, we will just make the Shuttle compatible with everything else. As it turns out, that doesn't pay, because there are so many other demands that the Shuttle would turn into a communications system and that's all. We are trying to resolve that problem, but that's a tough interface which will take a little bit more work. I should also mention RTG's. In the NASA stable, the planetary spacecraft are the only ones with RTG's. They have some unique problems, and that's something we are working with the planetary program to make sure that we can handle. In normal operation, the RTG is no problem. In the case of an abort, you have to worry about the RTG a little bit more than you would about batteries and solar arrays. So, that's giving us a little bit of a headache. Other than that, it seems to me that there aren't too many things that are unique about planetary spacecraft.

<u>Parks</u> - We have had a very good and thorough set of both papers this morning and comments this afternoon, and in that sense there is not much more new to talk about. I would like to kind of summarize, or emphasize, some of the things that I think are important out of what we have heard today so far as it relates to planetary usage of the IUS and Shuttle. Along those lines, I think it should be quite obvious to everyone that the planetary program is able to dream up things to do far more than the money is available to do them. There is, therefore, a strong motivation to rush to anything that offers any reduction in the cost of doing these missions. We are really seriously motivated to really take advantage of that and so we are working on that very deligently.

Another very important characteristic that I think relates to Shuttle usage by the planetary program is the fact that the launch opportunities are infrequent and short. These range from once a year, in some cases, out to maybe once very 12 or 20 years, depending upon where you are going and what you are trying to do when you get there. We find cases where the launch period is required to be as short as 10 days or even less, if you try to extend the launch period beyond that, you start paying significant penalties to accommodate that longer period. So, one of the important things that we are looking for in the STS is what it can do to offer increased launch reliability, on the one hand, and to increase launch-on-time assurance, on the other hand. These are both very important factors. Right now we have to do a lot of backup launch planning, primarily, not solely, but primarily based upon the contingency of a launch failure. We have to, right now, pay some serious penalties in extending the launch periods over what they optimally otherwise should be because of either complications of getting two launches off from a single launch pad or of other considerations just to increase, to a reasonable level, the likelihood of launch. I'll talk about both of those considerations a little bit later.

We have a number of very complex and very important interfaces as you have seen today. I would like to talk first about these interfaces in the preparation-for-launch period. In that area, it appears to be completely practical to install and/or replace the spacecraft on the launch pad more or less in the fashion we have done it with the launch vehicles we have used in the past. The use of a container around the spacecraft will, in fact, give reasonable solutions to the cleanliness and quarantine issues that have been discussed and are important for many of these planetary missions. The timelines for these activities have indicated that you have a reasonable minimum for the time in which you have to involve the Shuttle itself. The time from the last spacecraft checkout to the time you actually launch is a reasonable period of time, and the turnaround time if you have to replace the spacecraft on the pad also appears to be reasonable. There are some special problems with things like the RTG's, which have been alluded to, that have to be resolved. All in all, it seems to me that many of these problems can be solved along the lines of what we have done in the past. We, therefore, have a lot of background experience and can have some reasonable assurance that they are going to be quite workable with the Shuttle.

In the launch-to-injection phase, which is the next region I would like to talk about, this reliability and launch-on-time assurance question is a very important one. I think the question of launch reliability does not boil down to just -- what's the reliability of getting that one particular launch off? If you have an abort, either from the standpoint of the Shuttle itself or if you bring a spacecraft back for one reason or another, how soon can you get another one off during that period of time? I suspect that, in many cases where time is very critical, you may want to have a backup system standing by with the ability to bring the second Shuttle and spacecraft into operation. Because, I doubt if you would have time to turnaround that first combination after one of those abort events. In time, that is, to hit your short launch period. If you could have either of those capabilities, either a high assurance that you are going to make it all right or a quick turnaround capability, there are many things we wouldn't have to do that we have had to do in the past about providing assurances and special features, like backup launch capabilities, and could materially reduce the cost.

We haven't talked about the guidance system and its interfaces here to any great degree. I think that's because it isn't a serious problem at the present time. We're quite satisfied with the guidance accuracy we are getting out of our present vehicles. In fact, if it were important to do so, we could probably degrade that accuracy to some extent. We are not inviting anybody to do that, but if it provided other opportunities, particularly if it increased the performance in some significant way, accuracy degradation would be possible because the main spacecraft penalty is just a little bit more fuel to correct the error.

The acoustic environment, and maybe the thermal environment -- particularly when you have an RTG, as well as the safety problems are ones that we are not really as comfortable about yet as we would like to be. This is simply because a lot of them aren't really that predictable from the decisions and actions that have gone on to date. We really need to see a little bit more about those, but we have no basic reason at the present time to sense that they are going to be serious problems relative to practices that we either have used in the past or would probably have to use.

The data and command question is an important one that is being worked on, and I don't think completely understood but yet no real problems are coming up there. I for one have been pretty strong in trying to maintain the capability of bringing the spacecraft back if it doesn't check out properly in orbit. That means the capability of properly checking it out at that time too. The case for that isn't actually completely solid here, because prior experience indicates that might not be a likely event to contend with. Given the fact that you can check out a spacecraft in vacuum, in zero G, which you can't do back here on earth, there are also a lot of things you can't check out on orbit because you are in the environment of the Shuttle itself or still attached to it. You cannot unfold solar panels and there are other things you can't do. So, on-orbit does not show a completely clear-cut advantage, but I think we should continue to understand what its potential is and, at least initially, be prepared to exercise it.

In this operations control question that was just discussed, I don't anticipate a serious problem. I think we can work it along the lines that we have worked with the standard launch vehicle people in the past. I agree with Dan's comment very much that everything else being about equal the shorter the flight time the better off you are. We would really love to take advantage of any excess performance along those lines that can be provided. That doesn't mean that there won't be individual tradeoffs when the best answer, in fact, does involve some extended flight times.

In conclusion we think that the STS potential for the planetary program is very good. We are eagerly awaiting the IUS decision so that we can really get at some of these problems which right now don't have a very specific definition.

Herman - We are convinced that the Space Transportation System can afford us a good deal of savings in effecting a planetary mission, but there is one area of concern which really shouldn't be a concern. And that's this, and I'll use Phil's example of a flatbed truck. If I go to buy a car that's a real bargain, but I have to buy windshield wipers as an accessory; a wheel as an accessory; a transmission as an accessory; a horn as an accessory, its no bargain. The thing that would concern us about the Shuttle is the aspect of the ancillary equipment we would need in order to fly a planetary mission. One aspect of it is from the standpoint of safety as Phil mentioned. The original concept of the Space Transportation System, the Shuttle, was that it was supposed to be more along the lines of airplane technology rather than Apollo-type technology. If that philosophy is adhered to and demands of assurance of safety, documented to the enth degree, are not imposed on us, then I think using the STS will truly effect cost savings. But on the other hand, if we are driven to a mode where we have to pay a heck of a penalty in the spacecraft itself, that additional cost might tend to overweigh the cost savings due to transportation alone. And that's something maybe Phil might want to comment on.

Culbertson - No, I agree with you completely. We can kill you with requirements. I don't think we will.

Herman - I don't think so either.

Culbertson - There is a fundamental problem, and that is that about 20,000 people within NASA and about 200,000 contractors got adjusted to a mode of operation which is called Apollo. Those 220,000 people have got to change. I can get up and preach it, and you can get up and preach it, Bob can get up and preach it, we can all do that. But, until it happens, its going to be -- you know. I've got my fingers crossed, I really have. I see a lot of good words and I see a lot of good work. But there are a lot of people in the system who figure the other way worked great and they're not going to change. I remains to be seen, and it isn't all NASA either. I tell you that we share the responsibility, I think, among the principle contractors we are working with as well as NASA. I've seen evidence of reluctance on the part of all of us to change. And that will kill the Shuttle. I don't disagree with you in any sense at all. And I hope we all remember that.

ENERGY IN THE SHUTTLE ERA

ENERGY IN THE SHUTTLE ERA

Program Chairman   Gary Glenn, Stearns Roger Corporation
Co-Chairman   Allan Anglund, Mountain Bell

AAS 75-280

ENERGY RESEARCH OVERVIEW/ALTERNATIVES

FOR ENERGY DEVELOPMENT

Thomas J. Vogenthaler[*]

R&D on new energy supplies will help shape the energy mix of the future. Both energy conservation and currently available technologies will determine whether and how the country's energy future is attained. Each probable energy source is examined to present an overview and the research needs.

It is the objective of the United States to reduce dependency on foreign energy sources by decreasing energy demand and by increasing domestic resource development. Both parts of this program involve the use of new technology to make maximum efficient use of America's energy resources. Research and development of these technologies, some of which are derived from the space program, will play a significant role in reducing our dependence on energy imports from foreign countries.

Research and development on new energy supplies will help to shape the energy mix of the future. This includes technologies which will provide new or lower cost supplies within the time frame of the next 5 to 10 years as well as some which will produce longer range benefits. In addition, conservation measures must be developed to reduce the need for scarce energy sources. It is both conservation and the impact of the currently available technologies which determine the country's energy future. The

---
[*] Director, Colorado Energy Research Institute

time is short. Let me give you a picture of research needs by different energy sources currently in use or likely to become available.

OIL AND GAS

Oil and gas now provide about 75% of the nation's energy and will continue to be the main source during the 1970's and 1980's. The oil and gas industry in the United States is already highly developed, and has in progress some of the research necessary to increase production.

Given the importance of this sector of our economy, it is essential to assure that this research and development is being pursued rapidly enough. The areas of most uncertainty, therefore, offering the biggest pay-off are in <u>exploration</u> and <u>enhanced recovery</u>. Application of the latest geophysical and computer technologies may be able to increase the success rate while better drilling technology could reduce both the costs and time of drilling. Examples of the latter will require the use of new materials which maintain their structural integrity at higher temperatures; the use of plasma torches to cut through resistant rock structure, etc. In addition, basic research and mapping of major U.S. source rock formations is needed to determine their significance to future exploration.

In the area of improved <u>recovery</u> from existing wells, <u>secondary</u> recovery is in widespread use but tertiary recovery methods are still in development. Secondary recovery involves the injection of water or gas into the well to stop the decline in subsurface pressure, thus pumping more oil to the surface.

More advanced methods of recovery, "tertiary" recovery, have been under laboratory development. Most promising techniques include the introduction of steam, in situ combustion, carbon dioxide, miscible flooding agents and surfactants. Suitable techniques used will vary from oil field to oil field, depending on the characteristics of the formation.

The importance of these efforts can be shown in the potential for additional petroleum recovery in the United States. U.S. oil reservoirs discovered to date contained approximately 430 billion barrels of oil. Primary and secondary operations get about 1/3rd or 140 billion barrels, leaving 2/3rds or 290 billion for other methods. Economic extraction of this resource would put America well on the road to self-sufficiency.

Oil and gas R&D have been primarily the province of private industry but the urgency implied by the current critical situation has led to a much more active Federal role in fiscal year 1975.

COAL TECHNOLOGY - MINING

Coal is a corner stone in the plans for energy self-sufficiency. It is America's most abundant fossil fuel resource, comprising 94% of the "identified" primary energy reserves. Despite this abundance, it supplies only 17% of our total energy requirements. In 1920, it provided 78% of the nation's energy.

The basic strategy for R&D is to accelerate its substitution for oil and gas in the domestic energy markets. Two main obstacles stand in the way however, the technology for mining the coal and the technology for using it.

First, the production of existing mines must be increased, and second, new mines must be opened and brought on line quickly. Efforts to meet these goals must concentrate on improving mine safety and health and reducing the environmental costs of mine operations.

Underground mining techniques have shifted in recent years from manual operations to mechanized room-and-pillar mining or to the use of the continuous miner. The continuous miner is now able to cut and load faster than the haulage or ventilation systems can accommodate. To increase the production of existing mines, the operating time of the continuous miner could be increased.

Currently, the machine operates, on the average, about 25-30% of the time. This could be increased to 50-60% by the implementation of several steps.

First, develop methane drainage in advance of mining since increased methane release increases the danger of explosion. Early drainage would also allow the operation of several mining faces concurrently.

Secondly, continuous haulage of coal from the miner to the surface belt transport would increase productivity. Continuous belt transport could carry the coal from mine fact to the surface.

In order to speed the opening of new mines, several research areas should be investigated. The adaptation of large diameter continuous boring equipment now available for metal mines could develop the shafts and entries of coal mines. Surface mining equipment could both strip and restore the overburden of the mined area instead of only removing it. In addition, research could develop geological methods such as more extensive use of remotely sensed data for more efficient evaluation and planning of new coal mine development.

Improved technology for reclaiming strip-mined land is also needed since current coal economics favor surface mining.

The other major area of coal R&D involves its use in compliance with air quality standards. When most coals are burned they produce unacceptable quantities of sulfur compounds, particles and nitrogen oxide. Stack gas scrubbing technology has been under development for neary a decade, but controversy rages over its effectiveness and reliability. Both federally and industrially funded R&D work is now underway to remedy this, but the problems are large and the urgency great with existing air quality laws.

Additional R&D is needed to identify and control other pollutants produced when burning coal.

COAL TECHNOLOGY - SYNTHETIC FUELS

Another way to overcome the problems of using coal is to convert it to liquids and gas which substitute for natural gas and petroleum. Coal, when converted to oil or gas, yields the equivalent of two or three barrels of oil per ton. With perhaps 500 billion tons of recoverable coal resources, the U.S. has the equivalent of more than a trillion barrels of oil in the form of coal. These synthetics could play a major role in reducing the domestic supply and demand gap, thus reducing the necessity for imported energy supplies.

Realistically, it is estimated that only 1.5 to 2.0 percent of the coal production will be converted into synthetics if the processes are developed to the point that they are commercially viable. Synthetics are an expensive form of energy to produce, but may be profitable if current world oil prices are maintained. Gasified and liquified coal products are highly compatible with the energy supply infrastructure which is already in place.

Coal is a carbon fuel, like oil and natural gas. The process for making gas from coal reflects the fact that coal does not have enough hydrogen relative to its carbon content. Consequently, to upgrade coal to other hydrocarbon fuel types, the conversion must include a means of combining hydrogen with the coal - THE MORE HYDROGEN IN THE PRODUCT, THE MORE GAS PRODUCED.

The source of hydrogen for all conversion processes is water. Hydrogen may be supplied through direct transfer from steam in contact with hot coal, or by a separate step in which hydrogen-rich gas is obtained by reacting steam with hot coal or iron oxide.

Coal gasification research is geared toward producing several types of synthetic gas. High-BTU gas is composed essentially of methane, is of pipeline quality, and has a heating value of about 100 Btu per cubic foot of gas. Medium or low-Btu gas would be used primarily for electrical generation and has a heating value of 100 to 500 Btu's per cubic foot. The production of these types of gas requires different adaptations of the gasifying process.

The primary commercial process for gasification is the Lurgi process, developed in Germany in the 1930's. The basic process currently in use abroad produces a low-Btu gas. In plants being designed for use in the

U.S., an additional step will be introduced -- methanation -- to create a pipeline quality, high-Btu gas. This step, now being tested is necessary to produce a product compatible with the existing natural gas pipeline network. The medium or high-Btu gas produced would be sent through the pipeline network due to its significantly higher delivery cost.

The only commercial liquefaction plant is in South Africa. The state-of-the-art is not so far advanced as gasification. Current experimentation indicates that liquefaction is possible, but would be a costly fuel compared to crude oil.

Research for the nation's needs must solve many problems in order to bring these potentially valuable synthetic products into production. The major constraint to any process will be the availability of water. Projected plants will require 10-17,000 acre-feet of water per year for a 250 million cubic foot per day gasification plant, or perhaps twice that for a 100,000 barrel per day liquefaction plant. This would place a heavy burden on the water-poor areas, particularly in the West, where much of the domestic coal is located.

As in coal mining, there is a shortage of skilled personnel for the conversion technologies. Projected limitations on steel making and fabricating facilities for the heavy-walled pressure vessels necessary for conversion processes appear restrictive.

As with many other energy-related problems, the environmental problems need to be defined and solved. The disposal of waste from the processed coal will present a problem of leaching of soluble pollutants from waste piles.

Technical problems abound in the "Conventional" above ground plants. These include methods of introducing coal into a high-pressure system; controlling reactions at high temperatures and pressures; defining construction materials which will retain their integrity under hostile condition; and equipment to satisfactorily separate solids from liquids and gas at high temperatures.

Currently, there is doubt that the technology will be deployed at all in the absence of any guarantees of the market for the products. The financial risks are high and there is a difference in opinion on the future availability of natural gas. R&D efforts in the near term will concentrate on these questions about the future of synthetic-fuels-from-coal.

OIL SHALE

One of the country's largest potential energy resources is oil shale, which yields up to 60 gallons of shale oil per ton of rock. This oil can be re-refined to yield the same range of products as natural crude. Deposits are located in Colorado, Utah, and Wyoming. Eighty percent of the rich formations (30 gallons per ton or more) are located in Colorado. The total amount of deposits of rich shale is about 600 billion barrels, but substantially less than that is recoverable under current technology and economic conditions.

Shale can be either strip mined or deep mined. The surface mines resemble copper pits rather than strip coal mines. The most common type of underground mining is the room-and-pillar method. Both of these methods present significant problems in regard to water use, waste disposal and revegetation. Clearly, the major need for research in the oil shale area is _in situ_ retorting.

Retorting is the process by which the shale is heated to the temperature (about 900°F) to convert the solid organic material in the shale (kerogen) to oil. <u>In situ</u> retorting involves underground heating by combustion in the crushed or broken formation, introduction of hot gas, or introduction of superheated steam.

<u>In situ</u> processing requires the crushing -- or rubblization -- of the rock. Various methods have been tried. One technique is to mine a cavity and then induce a cave-in with explosives or hydraulic pressure. Heat can be applied either through underground combustion of a proper mixture of fuel and air pumped from the surface, by circulating surface-heated gases through the deposit, or by circulating steam. Use of both conventional and nuclear explosives with minimal pre-mining are also promising possibilities.

In order to meet the far term goals for shale oil production, at acceptable environmental quality standards, accelerated research into <u>in situ</u> retorting will be needed.

NUCLEAR POWER

Nuclear power is expected to be a major contributor to U.S. energy needs in the near future. Its emergence vividly demonstrates the long lead times necessary to deal with an essentially new technology. After decades of research and development, only relatively recently have reactors contributed significantly to power generation. Currently, there are 53 reactors generating 36,000 megawatts supply -- about 2% of our energy.
 -76 being built @ 77,000 MW-* (112 being planned @ 130,000 MW)*
Further R&D is needed to assure that nuclear power can come into full production and become a major long term energy source. Effort is nec-

essary in the area of safety, environmental R&D, and the fuel cycle, including waste disposal, and new, more advanced reactor designs, such as the high temperature gas-cooled reactor and the breeder reactor.

Current reliance is placed primarily on the light water reactor for energy prodution in the next 15 years. Breeder reactors will not come into widespread commercial use until after 1990. There are two primary safety-oriented research needs to improve current nuclear technology -- disposal of high-level fission products and waste heat disposal.

The fission wastes contain plutonium with a half life of over 20,000 years. A novel solution being considered is chemical separation of transuranic elements and then burning them out in fission reactors. This could reduce storage time significantly. These can be converted into solid, glass-like pellets and stored temporarily in hug bunker-like structures. Long-range disposal of this waste is needed, however.

The problem of waste heat disposal is essentially the same for light water reators as for fossil fuels. But the heat is 50% higher because of lower thermal conversion. Development of environmentally benign methods of disposing of this heat is necessary.

The light water reactor can use only U-235 uranium, which amounts to only about .7% of natural uranium ore. Providing enough enriched fuel supply for Project Independence timetables would require tripling current uranium mining production and enrichment capability. The breeder reactor could solve the uranium supply constraint in the decades which follow. It converts non-fissionable U-238 and thorium to fissionable material, in essence producing more fuel than it consumes. Current enrichment facilities employ a gaseous diffusion process. This involves a lengthy series of steps to separate the U-235 fuel from the more abundant

U-238. The search for a cheaper, more efficient system has resulted in experimentation with the gaseous centrifuge system.

Centrifuge plants would be smaller, require 1/10 the electricity and would reject less heat, promoting total efficiency. However, reliability, maintenance costs and anticipated life of the proposed plants are uncertain. Full development of these plants would have a positive effect on the Project Independence goals by supplying nuclear fuel more efficiently. Laser separation is also being investigated in the laboratory. Only a few weeks ago, this technique was advanced considerably at the Livermore Lab, using a dual laser system.

There are several types of breeder reactors under study for the Nation's long-term future -- the liquid metal fast-breeder reactor, light water breeder, gas-cooled fast breeder, and the molten salt breeder. The liquid metal fast breeder is the main effort in the breeder field. These reactors will not be available within the 1975-1985 time frame, but the blueprint planners must consider the long-range potential of nuclear energy in order to plan effectively.

GEOTHERMAL

Geothermal energy is the natural heat of the earth, constantly radiating from the earth's core. Under present conditions geothermal power is a limited resource. It will very likely account for no more than 1% of total energy requirements by 1985. Technical problems which require the attention of planners include resource exploration and management, conversion of geothermal steam to electric power, and environmental protection. Accelerated R&D is needed, particularly in areas such as exploration techniques, heat transfer, corrosion, etc., to determine the extent of commercial geothermal resources, how best to develop them, and what role it will serve in meeting future energy needs.

SOLAR ENERGY

Energy from the sun is a virtually inexhaustible, pollution-free resource. The technical feasibility of using it on earth has been proven in several small-scale applications such as space and hot water heating. Solar energy could also be used to produce electricity. The basic problem with this resource is that it is a diffuse, low-density energy source, so that size and capital cost of equipment to use it is presently high in comparison with fossil and nuclear fuels.

The technology for solar heating and cooling is rapidly approaching commercial stages, but still requires a good deal of development and marketing. Several other aspects of solar power are now drawing the attention of scientists and engineers. Wind power and organic waste conversion are currently at states where commercial availability by the 1980's is possible. Exploitation of wind power requires large machines in order to produce as much as 1 to 10 megawatts of energy. They must operate over a wide range of wind velocities to produce electric power for a grid. This requires designs which can operate at wind speeds as low as ten miles per hour. Variable pitch blades are needed to maintain constant generator speed at different power levels. Once these technical problems are overcome, the problem is one of producing the required machinery and operating many of them in large, integrated grids. And like Solar Power, energy storage may well determine is long term viability as an energy source.

The combustion of plant and animal wastes is a form of solar energy use. It is already being used commercially. Other fuels may be produced by photosynthesis conversion.

FUSION

Although fusion will not be a commercial source of energy during the near term, its development will have a great impact on the energy future of the country. The development of this highly desirable energy source has been pursued for two decades, but its scientific feasibility remains to be proven.

Fusion reactions occur when two light nuclei such as Deuterium, Tritium, or Hydrogen collide and rearrange themselves so as to form other nuclei of smaller total mass with a consequent release of energy. The reaction releases sufficient energy to satisfy the balance between mass and energy required by Einstein's equation $E=mc^2$. The necessary conditions occur only in the sun and stars. No sustained fusion reaction has been achieved in laboratories as yet.

Three different approaches are being pursued; magnetic pinch in which large magnets are used to confine the very hot plasma, the Tokomak and laser fusion which uses high powered lasers to compress and ignite fuel pellets.

ENERGY CONSERVATION

One of the primary thrusts of the nation must be the conservation of energy to reduce demand for scarce energy supplies. Research into this area will provide useful information about the most efficient methods for reducing energy demand, particularly in those areas in which significant amounts of energy are wasted. R&D can also develop new, more-efficient technologies for energy use.

Many areas of energy conservation do not lend themselves to massive, centralized R&D efforts. They involve many relatively small improvements -- in processes, in construction, in individual habits. One exception is the automobile. The automobile is a large energy consumer, using 13% of all energy consumed in the US and 30% of all petroleum. Moreover, fuel substitutability in auto transport is very low. In addition, fuel economy has been dropping in recent years, reaching a low of 13 miles per gallon. It may be necessary for the government to take the initiative in promoting new automotive energy systems, such as using fuel cells, stirling and rankine cycles, etc.

Many important steps can be taken in the construction of building to conserve energy resources. These include improved insulation to minimize loss of heat; improved heat transfer systems to minimize the temperature difference required to transfer heat from one fluid to another; new equipment to control heat more precisely, combined with new overall systems configurations to make use of each level of potential energy.

A comprehensive energy R&D program is a vital component for the planning of the Nation's energy future. Decisions made to accelerate self-sufficiency will have a great impact on the long term outlook for energy supplies. Technologies developed in our space program are finding effective use in virtually all of the energy development programs.

We have just taken a thoughtful look at the options available and the difficult choices to be made during the months and years ahead.

REFERENCES

1. ERDA figures, July 1975
2. ERDA figures, July 1975

AAS 75-281

THE SATELLITE SOLAR POWER STATION — A FOCUS FOR FUTURE SPACE SHUTTLE MISSIONS

PETER E. GLASER*

The option for using satellite solar power stations for large-scale power generation on Earth, collecting and converting solar energy into microwave energy, transmitting it to the Earth's surface, and transforming it into electricity, is reviewed. The current state of technology and the necessary developments for accomplishing these functions are discussed, and the results of recent microwave transmission and rectification demonstration tests are mentioned. The requirements for Earth-to-orbit transportation are presented. Considerations are given to cost projections, resource use and economic comparisons. Environmental issues, including impact of waste heat release, space vehicle exhaust, noise pollution and location of antenna sites are listed. Biological effects and radio frequency interference are explored. The time frame for accomplishing the operational system is outlined.

I. INTRODUCTION

The successful accomplishment of the Apollo-Soyuz Space Mission represents a step towards future space activities oriented towards achievement of goals perceived as beneficial to society, as worthy of a large measure of public acceptance and support, in consonance with the "spaceship Earth" philosophy and based on international cooperation.

As we consider future space activities, we must recognize that our ability to predict the future is inadequate and yet we need to develop a future oriented perspective and select among alternatives those activities which address societal needs and problems of long-term relevance. High on the list of such societal needs and problems is the availability of energy resources to meet future global requirements.

For the first time in history, we are conscious of the possibility that there are limits to the continuous growth of human activities, if we rely solely on the Earth resources. Efforts are being made to examine the availability of Earth resources, their extent and economical development and the long-term effects which their increasing use may have on the Earth's ecology. Society has experienced the disruptive effects which are caused by the sudden shortages of energy resources. Although these shortages at present can be traced to political causes, it is becoming clear that

---

*Vice President, Engineering Sciences, Arthur D. Little, Inc., Cambridge, Massachusetts.

energy shortages and fuel price escalation are not short-term or one-time phenomena. The overwhelming consensus is that alternative sources of renewable, non-polluting energy must be developed initially as a supplement to the present non-renewable energy sources and eventually as a replacement for them.

One of the few such alternative energy sources available on a scale substantial enough to have a significant worldwide impact is the Sun. Today, the development of solar energy applications is perceived to be a promising approach to meeting future energy demands. Solar energy is certainly abundant enough to provide a significant portion of future energy demands and clean enough to satisfy the most ardent environmentalists. The challenges lie in finding methods for efficiently and economically converting this inexhaustible source of energy into useful forms. Although solar energy falls on Earth in tremendous quantities, it is not easily convertible and certainly is not "free." In a sense, it can be considered to be a widely distributed resource, somewhat akin to low-grade mineral resources, which will require technological sophistication to "mine" economically.

There are two obvious obstacles to harnessing solar energy: one, it is not constantly available on Earth during nighttime or when local weather conditions obscure the Sun; the other, it is diffuse. Although the total amount of energy available is far beyond all conceivable future needs, the collection and conversion of solar energy into useful forms must be carried out over a large area, entailing a large capital investment for the conversion apparatus. The challenge lies in making the best use of this capital investment. Solar energy most likely will be developed not because it is cheaper than alternative energy sources, but because these alternative energy sources sooner or later will be exhausted, will become increasingly more expensive, will continue to be subject to political and economic control by the nations possessing them, and will produce undesirable and, as yet, incompletely understood environmental consequences, especially on the huge scale which will be required to meet projected demands even with controlled growth.

The potential payoff of research and development in solar energy applications has been recognized and is for the first time being given significant support. The solar energy conversion methods which are being investigated include the power of the wind, the use of bioconversion to produce gaseous, liquid or solid fuels, the conversion of solar energy into heat for use in buildings or for conversion to power, the use of temperature differences of sun-warmed oceans for power production, and finally, the direct conversion of sunlight into electricity.

The applications of solar energy on Earth are restricted by the day and night cycle and by weather and will require suitable forms of energy storage. Terrestrial solar-generated power on a substantial scale will be economical in only a few favorable geographical locations where the capital-intensive solar energy conversion systems can be used to best advantage. These Earth-bound obstacles can be overcome when the solar energy conversion system is placed in orbit around the Earth where solar energy is nearly constant for 24 hours a day.

It is fortunate that space technology can provide the stepping stones for achieving solar energy conversion in synchronous orbit. Today, it is commonplace to use solar energy conversion to provide electrical power for the large number of satellites which are or have been orbiting the Earth or exploring the solar system. These unique technological achievements may open up a new path for human evolution because we can gain a new dimension to tap the inexhaustible resources represented by the Sun.

The concept of a satellite solar power station (SSPS) first proposed in 1968 (Ref. 1) and presented at Hearings on Energy Research and Development and Space Technology (Ref. 2) represents an extension of existing technology and is based on the successful start of the development of an effective space transportation system. As with the unique contributions of present satellites to worldwide communications, satellite solar power has the potential to provide an economically viable and environmentally and socially acceptable option for power generation on a scale substantial enough to meet a significant portion of future world energy demands.

An assessment of the feasibility of the SSPS concept performed by Arthur D. Little, Inc., Grumman Aerospace Corporation, Raytheon Company, and Spectrolab Inc. (Ref. 3), and continuing investigations being supported by NASA (Ref. 4) are showing that the SSPS is worthy of serious consideration as an alternative energy production method. Its development can be realized by building on scientific realities, on an existing industrial capacity for mass production, and on demonstrated technological achievements.

II. RATIONALE FOR AN SSPS

Since 1968, the configuration of the SSPS has gone through an evolution (Figure 1) which resulted in the adoption of the baseline design shown in Figure 2 as a basis for technical and economic analyses. While the satellite is maintained in synchronous orbit around the Earth, two symmetrically arranged solar collectors convert solar energy to electricity. The electricity is fed to microwave generators incorporated in a transmitting antenna located between the two solar collectors. The antenna directs the microwave beam to a receiving antenna located in direct line of sight on Earth, where the microwave energy is efficiently and safely reconverted to electricity. Additional complete SSPS systems can be established to deliver power almost anywhere on Earth by placing the receiving antennas either on land or on water in the desired geographical areas and selecting appropriate orbital locations for the solar collectors. The power generated at the receiving antenna interfaces with appropriate transmission networks.

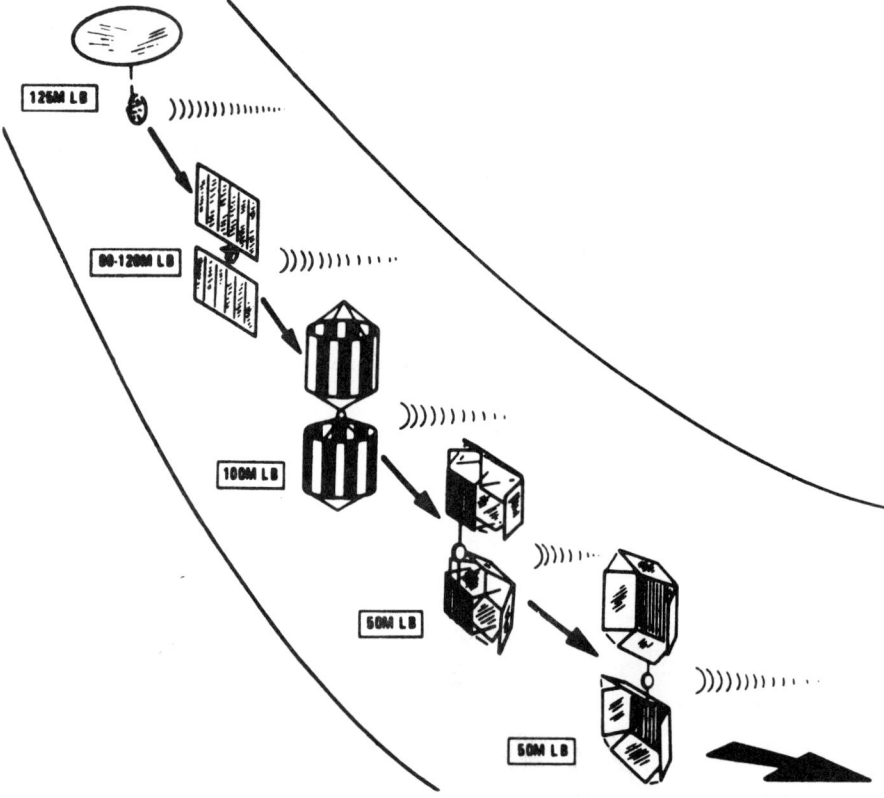

FIGURE 1 CONFIGURATION EVOLUTION

FIGURE 2   DESIGN CONCEPT FOR A SATELLITE SOLAR POWER STATION

The design of the SSPS is based on the following rationale:

1. The amount of solar energy available in synchronous orbit ranges from 6 to 15 times that available on Earth. The solar energy in orbit is available nearly continuously except for short periods around the equinoxes, at which time the satellite will be shadowed by the Earth (Figure 3). Averaged over a year, this results in a 1% reduction of the energy that would be available if the SSPS were continuously exposed to sunlight. The time period when an SSPS will be shadowed can be predicted accurately, allowing for load management of the terrestrial transmission network. The SSPS will be shadowed near midnight at the receiving antenna site, a time when energy storage is unlikely to be called upon.

2. Synchronous orbit is a favorable operational environment for the SSPS because zero-gravity conditions and the absence of wind, rain, and snow permits the deployment of large-area structures with minimum weight, leading to a marked reduction in material used per unit of delivered power. In addition, the space vacuum permits the operation of microwave generators and other components without the evacuated enclosures required on Earth.

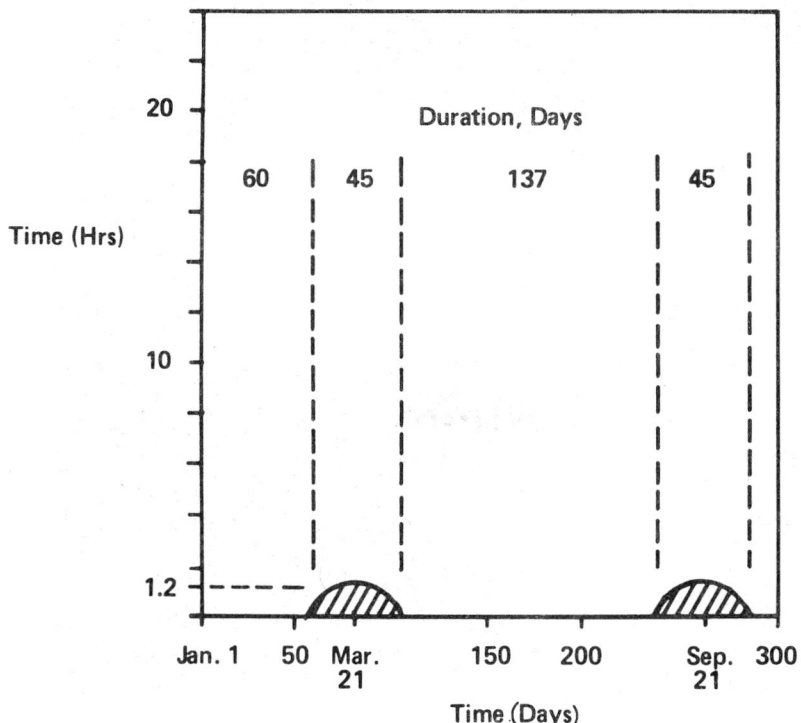

FIGURE 3 ECLIPSE DURATIONS FOR 0 DEG GEOSYNCHRONOUS ORBIT

3. The microwave beam can be directed to most desired receiving antenna sites which may be located in the vicinity of power users. These sites can be established on low-value land or offshore. A receiving site on land also provides opportunities for multiple land use.

4. The SSPS has the potential for large-scale power generation, ranging from about 2,000 to 20,000 megawatts on Earth. The lower output — 2,000 megawatts — is the smallest that can be effectively generated because of the necessary geometric relationships between the transmitting and receiving antennas. The upper limit — 20,000 megawatts — is fixed by the capacity to radiate waste heat from microwave generators to space at an acceptable power-to-weight ratio. Because of the broad range of power outputs available on Earth, the SSPS can be designed to be placed in service to first meet incremental energy demands and to subsequently replace conventional power-generation plants.

5. The environmental effects of the SSPS and the associated space transportation system are projected to be within acceptable limits, particularly because all waste heat associated with solar energy conversion and microwave generation can be rejected to space, no other waste products are generated, the microwave beam densities can be designed to meet international standards, and the system can be made inherently safe. Although a similar satellite could be used to convert nuclear

energy instead of solar energy, such a system would contravene all of the basic advantages inherent in the large-scale use of solar energy.

6. Solar energy can be converted by several types of processes: e.g., thermionic, thermoelectric, thermodynamic and photovoltaic. Photovoltaic conversion of solar energy directly to electricity is of interest because:

- a. Solar cells represent demonstrated technology as a result of widespread use in the space program.

- b. The United States program to develop low-cost mass-produced silicon solar cells addresses the same technology.

- c. The process is a passive one which could reduce maintenance requirements and increase reliability during the desired operational lifetime of the SSPS — 30 years. Present communication satellites (e.g., INTELSAT IV) already have a project lifetime of 10 years. A lifetime potential in excess of 30 years should be achievable.

- d. Sun-following requirements for solar cell arrays are modest, thus minimizing propellant requirements for station-keeping.

- e. Several solar cell materials are being investigated which promise to have significantly higher efficiencies and lower weight compared to silicon, e.g., gallium arsenide.

Thermodynamic conversion is of interest because:

- a. machinery operated on the Brayton cycle has had a long and successful history of terrestrial applications;

- b. gas-bearing technology has reached the point where operation for an extended period, particularly in space, appears to be feasible; and

- c. the very large concentration factors required to achieve elevated temperature levels for high thermodynamic efficiency may be approached by novel techniques to achieve the geometric perfection of solar concentrators and high-pointing accuracies.

These two approaches as well as others are now under investigation (Ref. 5). Although considerable technical and economic analyses will be required to establish which of the potential approaches for solar energy conversion will be optimum, the fact that there are several promising approaches to accomplish this objective indicates that the appropriate technology could be developed.

7. The microwave method for transmitting the power from orbit to Earth uses state-of-the-art or achievable technology to obtain high efficiency in generation, transmission, and rectification, promises to satisfy environmental requirements, and can be operated safely. Microwave generation, transmission and rectification technology is based on demonstrated results stemming from commercial use. Existing mass-production capability of upward of one million microwave devices in 1974 serving a half a billion dollar per year market in the United States alone is indicative of the state of the art. Successful demonstration of microwave transmission and the achievement of an 85% rectification efficiency of microwave power directly into DC indicate that the goal of high system efficiency is within reach. In the 10-centimeter wavelength region, atmospheric absorption of microwaves under most adverse weather conditions will be less than 4%. The maximum power density within the microwave beam can be chosen to reduce interactions with the ionosphere and to obviate any potential for damage at the receiving antenna site. Within a short distance from the receiving antenna the microwave power density reaches such low values that even the most stringent international microwave exposure standards can be met. The microwave power density in the vicinity of the transmitting antenna is low enough so that other satellites in lower orbits or aircraft flying through the beam would not be damaged.

III. SSPS DESIGN CONSIDERATIONS

A. STRUCTURES

Based on the above rationale for selection of specific technology, the characteristics of an SSPS can be established. The SSPS baseline configuration (Figure 4) selected to achieve a 5,000-megawatt output was subjected to structural and dynamic analyses to determine the elastic characteristics (natural frequencies, generalized masses, and mode shapes) of the structure so the elastic coupling between the SSPS attitude control system and the spacecraft structural modes could be investigated. Deflection and internal member loads under the various flight loading conditions were determined to verify structural integrity. This permitted an evaluation of the flight control performance of the baseline design. Parametric sizing studies determined the influence of structural flexibility upon the performance of attitude control systems. Although the very large area lightweight structures required for the SSPS represent new challenges, present analytical tools can be used to predict performance.

Materials being considered for space flight structures can meet structural requirements for the SSPS. Dielectric materials can be selected for the continuous support structure to avoid interference with the microwave beam as the transmitting antenna, which always faces the Earth, rotates once a day with respect to the solar collector, which always faces the Sun.

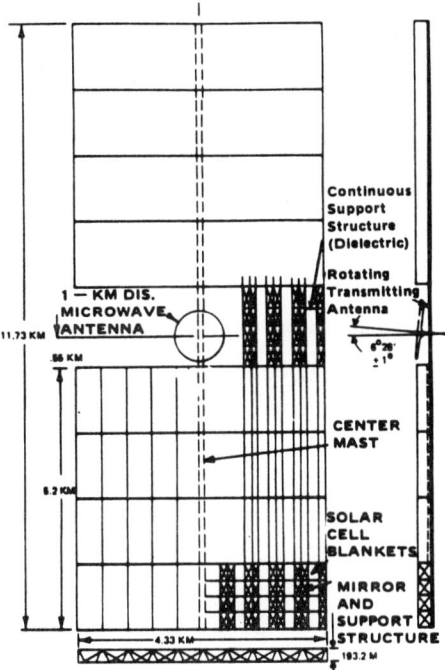

FIGURE 4 SSPS BASELINE CONFIGURATION

B. PHOTOVOLTAIC CONVERSION

The structural components of the solar collector array (Figure 5) have a dual role: provide support, and serve as electrical conductors. The solar cell blankets are fastened to the structure and a thin-film mirror reflects sunlight onto the solar cell to achieve a concentration factor of 2. This design approach is projected to result in a solar collector weight of about 3 pounds per kilowatt if 2-mil-thick silicon cells are used in the full-scale SSPS. By contrast, a current experimental solar cell array weighs about 10 lb/kw (Ref. 6) (arrays weighing 30 lb/kw are in orbit). The efficiency goal for the 2-mil-thick silicon solar cell is 18%. Recently a 16% efficiency has been achieved for solar cells of 8-mil thickness for use in communication satellites. The solar cells are to be encapsulated in plastic covers which are being developed to withstand the space environment. The challenge lies in being able to produce thin solar cells with high efficiency at low cost.

Since these goals are difficult to achieve simultaneously, the degree to which they can be achieved will have to be established in an integrated development program, which includes the following areas:

- Improvement of solar cell performance,
- Reduction of solar cell cost,
- Reduction of solar cell array blanket weight,
- Reduction of solar cell array blanket cost,
- Improvement of solar cell concentration technique,
- Lengthening of solar cell array life in space environment,
- Minimization of energy inputs,
- Development of high-voltage, high-power switching,
- Design of high-voltage circuit,
- Development of high-level DC power distribution,
- Development of power control and circuit protection, and
- Development of solar cell array assembly techniques in orbit.

An important step toward these goals is the United States Energy Research and Development Administration program to produce low-cost silicon solar cells at 50¢ per peak watt in 10 years. Figure 6 shows solar cell cost projections based on present achievements. One SSPS would be of sufficient magnitude to obtain the projected solar collector costs if a 70% slope experience curve were to apply.

Damage to solar cells from space radiation will cause a logarithmic decay of their effectiveness. However, improvements in solar cell radiation resistance are expected to result in a 30-year operating lifetime for the SSPS, with minimum degradation, after which normal cell effectiveness could be restored by adding a small area of new solar cells. This would indicate that there is no absolute time limit on effective SSPS operation.

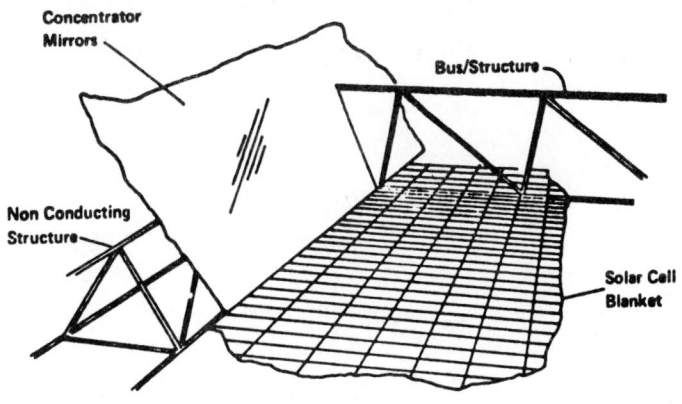

FIGURE 5 DETAIL OF SOLAR COLLECTOR ARRAY

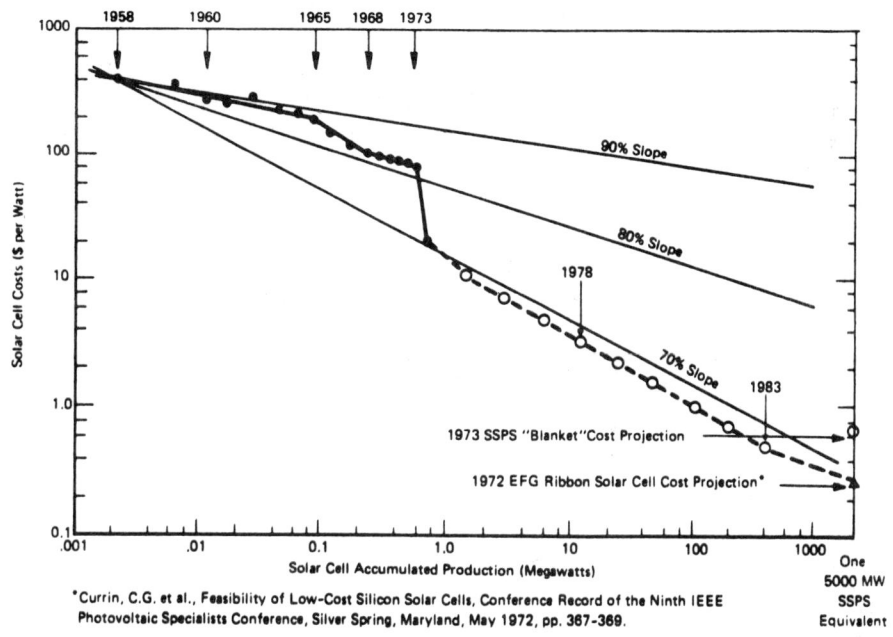

*Currin, C.G. et al., Feasibility of Low-Cost Silicon Solar Cells, Conference Record of the Ninth IEEE Photovoltaic Specialists Conference, Silver Spring, Maryland, May 1972, pp. 367-369.

FIGURE 6 SILICON SOLAR CELL COST PROJECTIONS

Another space environment influence on solar cell operation is micrometeoroid impacts. In synchronous orbit on the basis of the probability of damage-causing impacts by micrometeoroids during a 30-year period, the SSPS is expected to suffer a 1% loss of solar cells.

C. MICROWAVE POWER GENERATION, TRANSMISSION AND RECTIFICATION (REF. 7)

The power generated by the SSPS in synchronous orbit must be transmitted to a receiving antenna on the surface of the Earth and then rectified into DC at a voltage acceptable to transmission networks. The microwave power densities received on Earth must be at levels which will not produce undesirable environmental or biological effects. Finally, the microwave power must be in a form that can be converted, transmitted and rectified with very high efficiency by known devices. These conditions can be met by a wavelength of about 10 centimeters in the microwave part of the spectrum. At this wavelength, induced radio-frequency interference can be limited so that an internationally agreed-upon frequency could be assigned to the SSPS.

The efficiencies of microwaves power transmission will be high when the transmitting antenna in the SSPS and the receiving antenna on Earth are large. The dimensions of the transmitting antenna and the receiving antenna are governed by the distance between them and the choice of wavelengths.

The mast on which the transmitting antenna is mounted (Figure 7) is the primary current-carrying element from the solar collectors to the microwave generators. A rotary joint, which allows the antenna to rotate once a day with respect to the solar collectors, is the only major active element in the SSPS. A gimbal mount can be used to direct the microwave antenna towards a desired receiving antenna site on Earth.

The size of the transmitting antenna is influenced by the microwave generators, which because of their approximately 90% efficiency can use passive radiators to reject waste heat to space, and the structural considerations to accommodate the arrangement of the individual microwave generators. The size and weight of the transmitting antenna can be reduced by increasing the size of the receiving antenna and transmitting a higher-frequency microwave. The size of the receiving antenna is influenced by the choice of microwave power flux densities required for efficient microwave rectification, and the illumination patterns across the antenna face.

The microwave generator design is based on use of a crossed-field device, such as an Amplitron, which has the potential of achieving a high efficiency, high reliability, and very long life. The design goals for the Amplitron are a 90% efficiency for a 5-kilowatt output and a weight of a fraction of a pound per kilowatt.

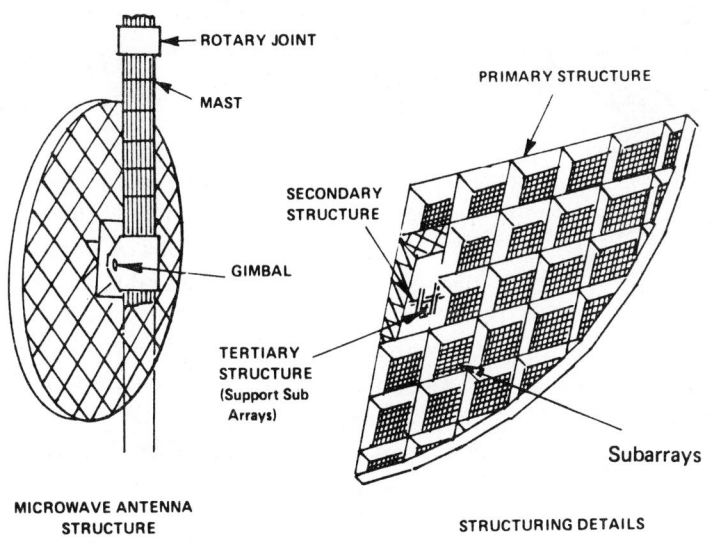

FIGURE 7 MW ANTENNA STRUCTURE

A series of microwave generators will be combined in a subarray (e.g., about 15 meters square) which forms part of the transmitting antenna. Each subarray must be provided with an automatic phasing system so that the individual radiating elements of the antenna will be in phase. These subarrays will be assembled into a slotted-wave-guide phased-array transmitting antenna, about 1 kilometer in diameter, to obtain a microwave beam of a desired distribution. The distribution within the beam can be designed to range from uniform to Gaussian. (Figure 8).

The receiving antenna is designed to intercept, collect, and rectify the microwave beam into DC which can then be fed into a high-voltage DC transmission network or converted into 60-Hz AC. Half-wave dipoles distributed throughout the receiving antenna capture the microwave power and deliver it to solid-state microwave rectifiers. Schottky barrier diodes have already been demonstrated to have an 80% microwave rectification efficiency at 5 watts of power output. With improved circuits and diodes, a rectification efficiency of about 82% has been recently demonstrated at the Goldstone antenna site in California by JPL (Ref. 4).

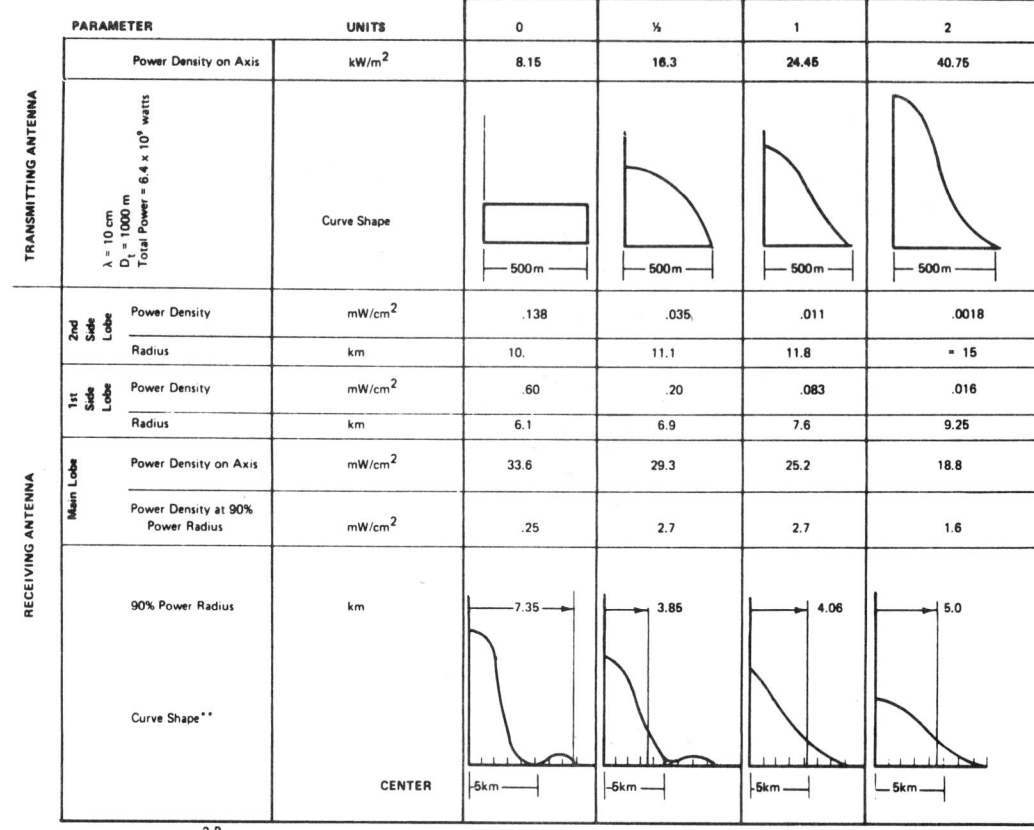

FIGURE 8 POWER DENSITY DISTRIBUTIONS

The diodes combined with circuit elements which act as half-wave dipoles are uniformly distributed throughout the receiving antenna, so that the microwave beam intercepted in a local region of the receiving antenna is immediately converted into DC. A receiving antenna based on this principle is fixed and does not have to be pointed precisely at the transmitting antenna; thus, its mechanical tolerances do not need to be severe. Furthermore, the density distribution of the incoming microwave radiation need not be matched to the radiation pattern of the receiving antenna; therefore, a distorted incoming wavefront caused by non-uniform atmospheric conditions across the antenna will not reduce efficiency.

The amount of microwave power that is received in local regions of the receiving antenna can be matched to the power-handling capability of the solid-state-diode microwave rectifiers. Any heat resulting from inefficient rectification in the diode circuits can be convected by the antenna's ambient air, producing atmospheric heating similar to that over urban areas. Only about 15% of the incoming microwave beam would be lost as waste heat. The low thermal pollution achievable by this process of rectification cannot be equaled in any known thermodynamic conversion process.

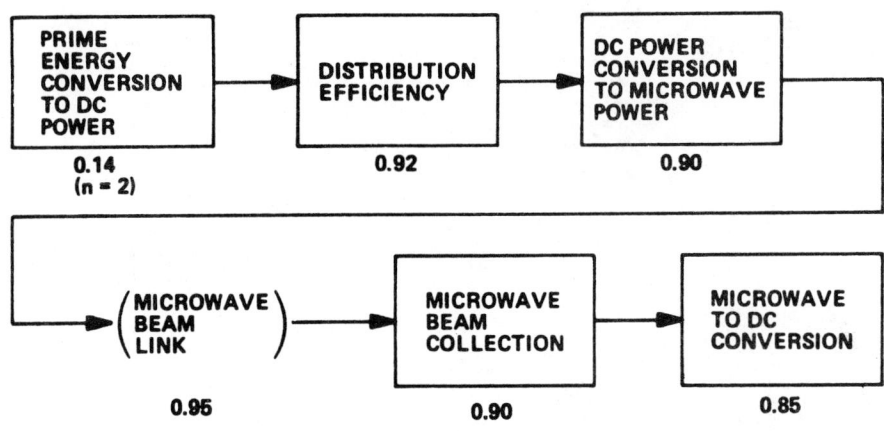

FIGURE 9   EFFICIENCY CHAIN

D. SSPS WEIGHTS

The solar collector is the heaviest orbiting component of the SSPS (Figure 10). The goal is to achieve a weight of about 4.5 pounds per kilowatt for the orbiting portion of the SSPS.

The rectifying elements in the receiving antenna can be exposed to local weather conditions. The antenna can be designed so that sunlight will still reach the land beneath it, with only a fraction lost due to shadowing, thus enabling the land to be put to productive use.

The solar energy conversion is the least efficient part of the system (Figure 9). Improvements in solar cell efficiencies through the use of other materials or alternative solar energy conversion processes with higher efficiency should have a significant effect on system performance.

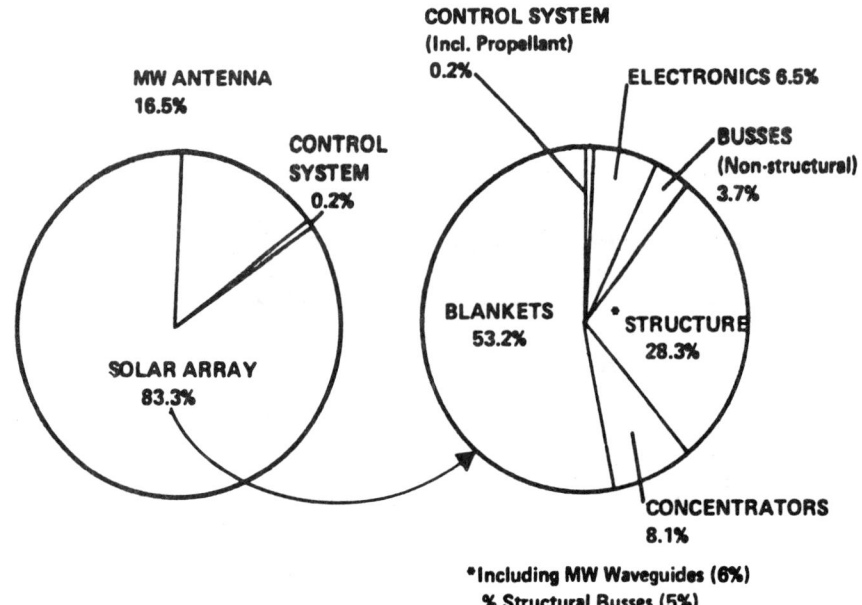

FIGURE 10 SSPS WEIGHT BREAKDOWN

E. ORBIT PERTURBATIONS

The SSPS structure in space — because of its size and weight — will be subject to orbital perturbations (Figure 11), of which the gravity gradient is the most important. Thus, a reaction control system will be required to keep the SSPS in the appropriate orbit and the solar collectors pointing toward the Sun within one degree. These systems will require about 30,000-100,000 pounds of propellant per year, depending upon specific orbit characteristics selected; Argon is the preferred propellant. Longitudinal drift of the SSPS can be minimized by deploying systems in the vicinity of the Earth's minor axes, one suitable location being at about 123 degrees West longitude, the other at 57 degrees East longitude. In these two locations, the SSPS system would be able to supply most of the prospective large-scale power users on Earth.

F. SPACE TRANSPORTATION AND ASSEMBLY MODES

The SSPS will require a high-volume two-stage transportation system: i.e., a low-cost stage capable of carrying high-volume payloads to low Earth orbit; and a high-performance stage capable of delivering partially assembled elements to synchronous orbit or possibly to an intermediate orbital altitude for final assembly and deployment.

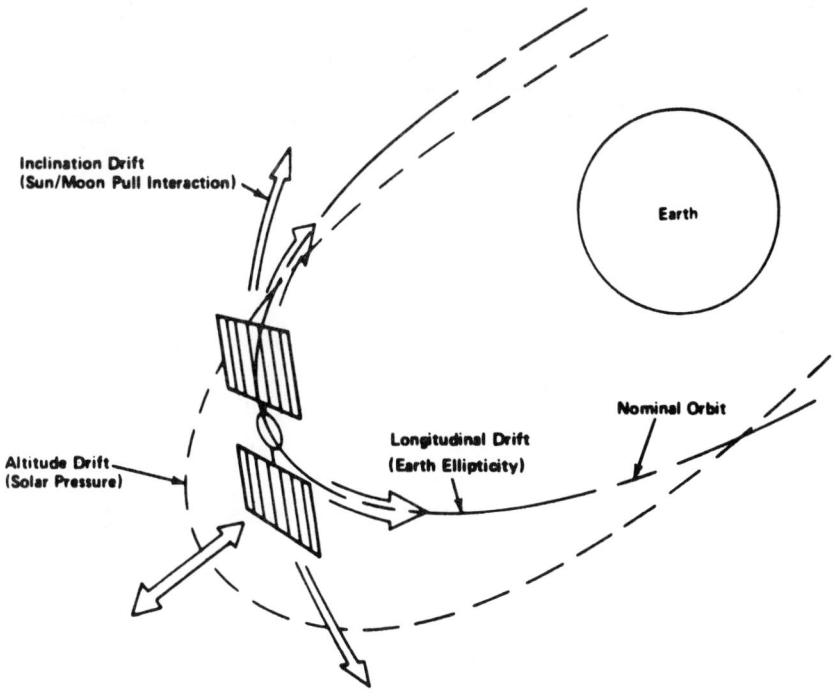

FIGURE 11 NOMINAL ORBIT PERTURBATIONS

Flight mode selection will be influenced by payload element size, payload assembly techniques, desirable orbit locations for assembly, time constraints, and requirements for man's participation in the assembly, which most likely will be primarily in a control function. The transportation system could be one of the currently planned propulsion stages, or an advanced concept could be developed specifically for SSPS systems. Space transportation combinations will have to be identified which can minimize the cost required for SSPS delivery, assembly, and maintenance.

The space Shuttle now under development provides the necessary first step towards the space transportation system required for the operational SSPS. This Shuttle can be used for SSPS technology verification and flight demonstration, and for transporting elements of the prototype SSPS into orbit. Operational experience with a prototype SSPS will be essential to permit an orderly evolution to the very-high-volume transportation system needed for an operational SSPS. Design considerations being given to a space transportation system capable of carrying a 150,000-pound payload to low Earth orbit and an upper-stage tug indicate that such a system could be realized at a cost of about $100 per pound placed in synchronous orbit. Given the capabilities of such a space transportation system, nominal component sizes and orbital assembly factors for the SSPS payloads can be identified. The large volume of required payloads could entail as many as 150 flights for each

SSPS assembled in synchronous orbit, based on a modified space Shuttle, or as few as 60 flights for an advanced space transportation system based on heavy lift launch vehicles. The development of a dedicated space transportation system will be essential to the success of an SSPS. Figure 12 indicates a number of candidate launch systems (Ref. 8).

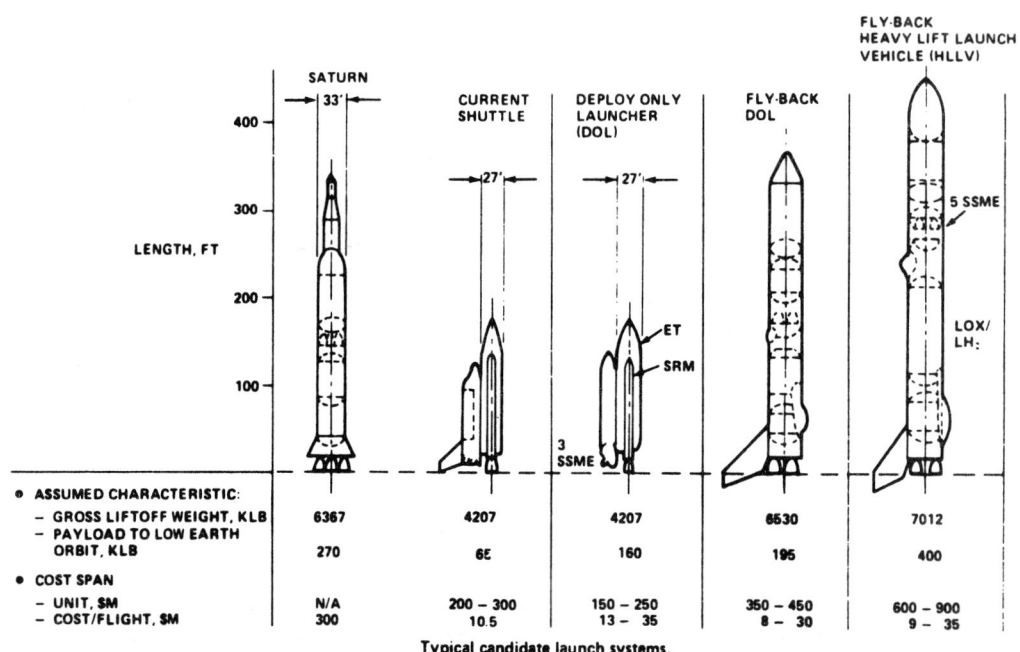

Typical candidate launch systems.

Source: Grumman Aerospace Corporation, Reference 8.

FIGURE 12 TYPICAL CANDIDATE LAUNCH SYSTEMS

IV. SSPS EVALUATION CRITERIA

The results of the technical feasibility investigations permit projection of the directions that future technical developments would have to take. But other criteria also need to be considered to determine whether an SSPS is worth developing: i.e., economic attractiveness, environmental impacts, social desirability, political implications, and public acceptance. As a start, consideration has been given to the first two, but a detailed assessment of the SSPS technology should be accomplished.

A. ECONOMIC ATTRACTIVENESS

The SSPS engineering cost projections (Figure 13) indicate that prototype system costs of $1,600-2,500 per kilowatt are within the range of competing energy production prototypes. These engineering cost projections are based on estimates of the individual SSPS subsystems and represent desirable cost objectives. The engineering costs need to be transformed into economic costs, which include the cost incurred in the implementation of the project, the required cash flow, and uncertainties in cost estimates in the R&D and operational phases. The social costs associated with the use of the SSPS, which include environmental impacts, safety, energy requirements during construction and operation, and demographic effects of any large-scale power project, need to be compared with those for competing energy production methods.

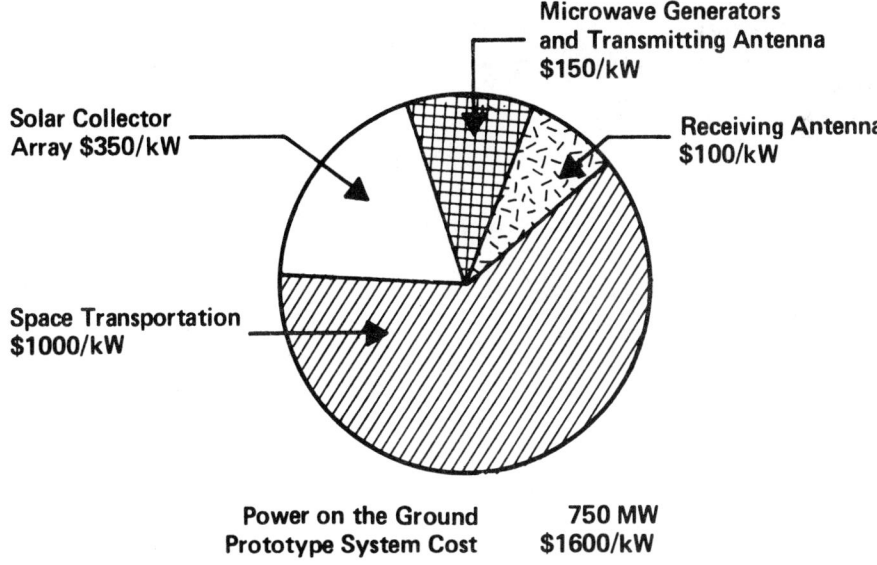

FIGURE 13 PROTOTYPE SSPS COST PROJECTION

B. ENVIRONMENTAL IMPACTS

The operational phases of the SSPS will produce various possibilities for environmental impacts:

- *Waste Heat.* Waste heat released at the antenna site would amount to the equivalent of about 15% of the rectified microwave power, which is substantially less than the waste heat released from energy production methods based on thermodynamic cycles.

- *Land Despoilment.* Land despoilment would be limited because the receiving antenna is a lightweight structure 80% transparent to sunlight and unobstructive to rain, microwave radiation can be excluded from beneath the antenna, maintenance will be minimal, and transportation of supplies to the site will be infrequent compared to conventional power plants.

- *Resource Consumption.* Materials requirements will be largely limited to those which are in plentiful supply, such as silicon and aluminum. The SSPS would require less than 2% of the yearly supply of critical materials, such as platinum, available to the United States.

- *Energy Consumption.* The energy required to produce the materials required during SSPS construction as well as the propellants to place it into orbit would be repaid — i.e., regenerated — in one to two years of SSPS operation.

- *Atmospheric Pollution.* According to present estimates of space vehicle emissions, the multiple launches required to orbit the SSPS are not expected to add any significant amount of pollution to the atmosphere, and will be within allowable goals.

- *Microwave Exposure.* Exposure to the microwave beam can be controlled by providing suitable enclosures for the maintenance crew to work on the receiving antenna. Within 10 kilometers from the beam center the microwave power density would meet the lowest international standards for continued exposure to microwaves (Figure 14). The microwave beam directional system and the phase control achieved by means of a pilot signal beamed from the center of the receiving antenna preclude the deviation of the microwave beam beyond allowable limits. In case of failure of the microwave beam pointing system, the coherence of the microwave beam would be lost, the energy dissipated, and the beam spread out so that the microwave beam density would approximate communication signal levels on Earth. The effects on birds exposed to microwave power flux densities within the beam at the receiving antenna and the effects on aircraft accidentally flying through the beam are projected to be negligible, but they should be determined experimentally.

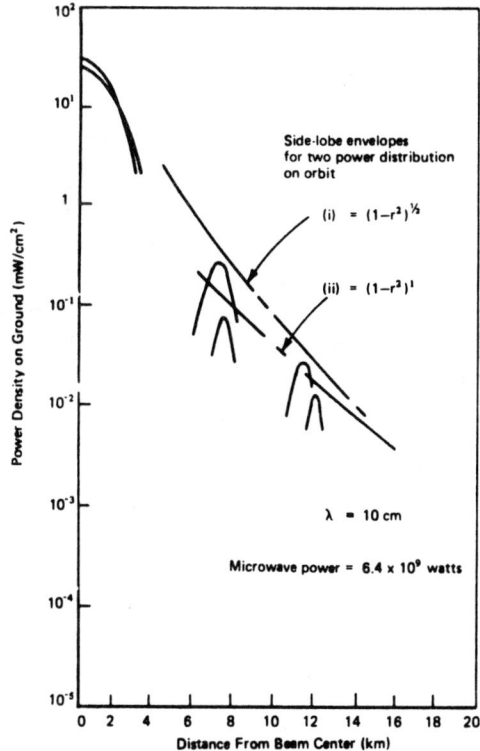

FIGURE 14 MICROWAVE POWER DENSITY DISTRIBUTION ON GROUND

- *Radio-Frequency Interference.* RF interference by the fundamental microwave frequency and its harmonics, turn-on and shutdown sequences, random background energy, and other superfluous signals resulting from specific design approaches for the microwave-generation devices could be controlled by filters, choice of frequency, and narrowband operation. Detailed and specific effects and impacts on radio astronomy services, shipborne radar, and communication systems will have to be determined before the international acceptability of specific frequency allocations can be established. Preliminary investigations indicate that, for example, amateur sharing, state police radar, and high-power defense radar may suffer substantial interference if the 3.3 GHz frequency is chosen as the fundamental frequency.

- *Noise Pollution.* The noise from launch operations will be of concern in the immediate vicinity of the launch facility. Suitable location of the launch facilities and noise control during launches could reduce noise pollution to acceptable levels.

V. SSPS PROGRAM PHASING

Three major steps are required to achieve a commercial version of an SSPS before the year 2000. The first, the technology development and verification phase, would culminate in a series of space Shuttle flights to low-Earth orbit to check the functions of various components and to obtain experience in component deployment and assembly. This experience will be applied during the second step, the development of a prototype SSPS that would deliver limited power from low-Earth orbit. A pilot-size SSPS would be built to deliver useful power from synchronous orbit to Earth. The development of a second-generation space transportation system, the third step, would be completed in time to make the full-scale SSPS operational before the year 2000. Integration of these three steps will assure the maximum usefulness of the SSPS components during various stages of SSPS development.

VI. CONCLUSIONS

Since the concept of a Satellite Solar Power Station was first proposed in 1968, considerable work has been carried out on various aspects of the SSPS. Assessments are being carried out by academic, industrial, and government groups in the United States and abroad. The results of these investigations have indicated so far that the option represented by the SSPS for the large-scale generation of power on Earth continues to be promising and that further work is required to confirm the validity of this option. The conclusions which can be drawn today are:

- The SSPS is technically and economically promising.

- The SSPS has potential to be environmentally acceptable.

- An orderly incremental SSPS development program can preserve this energy option.

- Critical SSPS technology developments can contribute to other worthwhile endeavors in space.

- Developments being carried out on advanced space transportation systems, solar energy conversion systems, and other related technology tend to be supportive of SSPS development.

- No one energy production method, whether based on nonrenewable or renewable fuels, can be relied on to meet future world energy demands on the scale required.

- Several options, particularly for solar energy conversion, appear to be promising, but only very few options possess the potential for the large-scale power generation represented by an SSPS.

The near-term objective is to investigate the SSPS option for power generation at a level of effort and on a time scale which is meaningful in terms of steps which must be taken in developing other options capable of meeting future energy demands. Any new concept such as the SSPS has to be poised on the uncomfortable edge between advocacy of the potential represented by its successful accomplishment and the calm dispassionate assessment of the benefits that it may represent to society.

The SSPS represents an undertaking which, because of its magnitude, worldwide implications on energy availability, and potential for the industrial use of space, could benefit from the combined efforts of many countries. Such an international undertaking could assure that the benefits of the SSPS will be widely available without political control by any one nation, and that each country would find it advantageous to participate in such an endeavor. For these reasons it should be in the common interest to obtain agreements on such aspects as suitable frequency assignment for the microwave beam, favorable launch sites for the space transportation system and sites for the receiving antennas. An international project can also assure that the functions of the SSPS are those which are in keeping with its peaceful character. Although the SSPS will be international in scope, national control will be exercised on the receiving antenna sites on Earth in keeping with the evolving body of international space law.

The SSPS represents a possible achievement which cannot be duplicated on Earth on the scale envisaged. The successful implementation of the objectives of the SSPS should sustain a highly energy-dependent civilization and remove energy-related concerns from society's agenda. But even beyond this, it represents an opportunity to enter not only a new era of energy resource development but, in a broader sense, it represents the first steps towards the industrialization of space and the extension of civilization beyond the confines of the Earth's surface.

VII. REFERENCES

1. Glaser, P.E., "Power from the Sun: Its Future," SCIENCE, Vol. 162 (November 1968), pp. 857-886.

2. Glaser, P.E., "Power from the Sun via Satellite," Hearings before the Subcommittee on Space Science and Applications and Subcommittee on Energy of the Committee on Science and Astronautics, U.S. House of Representatives, Ninety-Third Congress, First Session, May 7, 22 and 24, 1973. pp. 258-262 and pp. 281-327.

3. "Feasibility Study of a Satellite Solar Power Station," NAS 3-16804, Arthur D. Little, Inc., NASA CR-2357, February 1974. NTIS N74-17784.

4. "Microwave Power Transmission System," NASA, Lewis Research Center, Contract No. NAS 3-17835; "Reception Conversion Subsystem (RXCV) for Microwave Power Transmission System." JPL, Contract No. 953968, NAS 7-100; "Orbital Assembly and Maintenance Study," NASA, Johnson Space Flight Center, Contract NAS9-14319; "Space Based Solar Power Conversion and Delivery Systems," NASA, Marshall Space Flight Center, Contract NAS 8-31308.

5. Woodcock, G.R., and Gregory, D.L., "Derivation of a Total Satellite Energy System," AIAA paper 75-640, AIAA/AAS Solar Energy for Earth Conference, Los Angeles, April 24, 1975.

6. Ralph, E., Spectrolab, Sylmar, California, Private Communication, April 1975.

7. Brown, W.C., and Maynard, D.E., "The Adaptation of Free Space Power Transmission Technology to the SSPS Concept," AIAA/AAS Solar Energy for Earth Conference, Los Angeles, April 24, 1975.

8. Kline, R., and Nathan, C.A., "Overcoming Two Significant Hurdles to Space Power Generation: Transportation and Assembly," *ibid*.

AAS 75-283

NUCLEAR POWER IN THE SHUTTLE ERA

S. R. Ross

Public Service Company of Colorado

Nuclear power is expected to represent about 25% of electric capacity in the United States by 1985. Light water reactors will be the predominant type installed in the country by that time. The High Temperature Gas-cooled Reactor (HTGR) is expected to assume a significant role in nuclear power production by the latter part of the 1980's. The Fort St. Vrain Nuclear Generating Station of Public Service Company of Colorado described in this paper is the first commercial nuclear plant with an HTGR. The thermal light water reactors and HTGR's are expected to be supplemented by fast breeder reactors in the future.

INTRODUCTION

Nuclear power is assuming an ever increasing role in the generation of electric power in the United States. Whereas nuclear plants currently represent about 7% of the electric generating capacity in the country, it is generally projected that nuclear plants will represent about 25% of the country's electric generating capacity by 1985, in the middle of the space shuttle decade of the 1980's.

Nuclear energy also must assume a significant role in meeting the overall energy requirements of the country. The energy flow pattern for all types of energy for the United States as it actually took place in 1970 is illustrated in Figure 1.[1] You will notice that nuclear energy represented an insignificant part of the total.

In the same report,[1] which was prepared in 1973, the energy flow pattern for 1985 as illustrated in Figure 2 was projected. You will note in this case that nuclear energy was projected to represent the equivalent of 9.4 million barrels of oil per day, or about 16% of the energy requirements of the entire nation. While projections made subsequent to 1973 may have been modified by several factors such as the foreign oil embargo, energy conservation, economic recession, and slippage in power plant

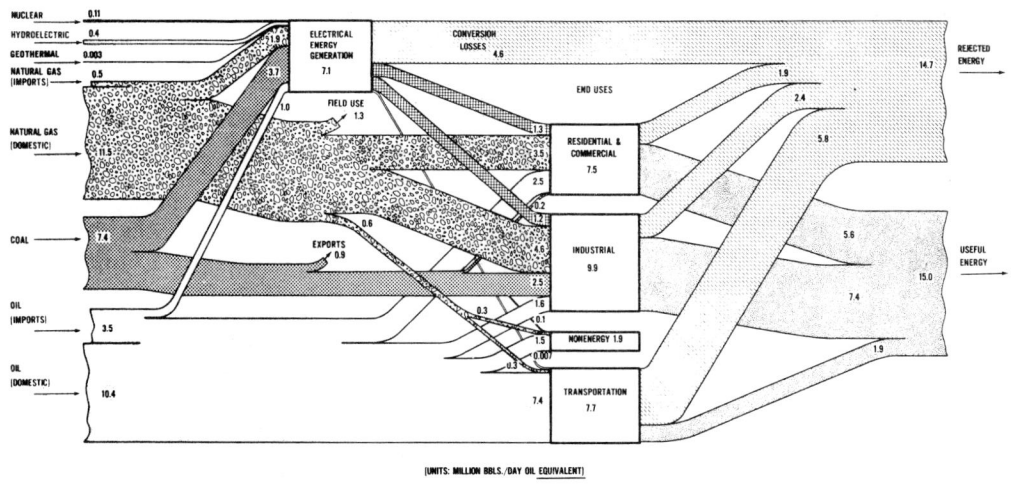

Fig. 1 Energy Flow in the U.S. in 1970

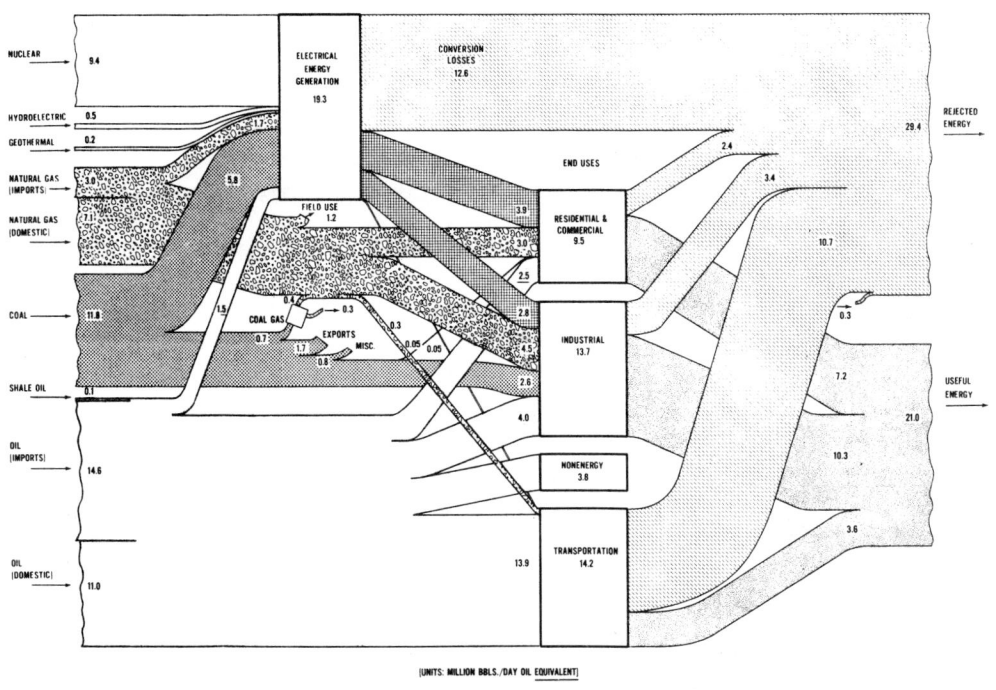

Fig. 2 Projected Energy Flow in the U.S. in 1985

construction schedules, I believe that it is clear that all available energy resources will be needed to meet the dramatic increase in energy requirements that is envisaged for the mid 1980's. Nuclear power clearly must play a significant role in meeting these requirements.

NUCLEAR POWER TODAY

As of December 31, 1974, there were 235 nuclear power plant projects in the United States. Of these, 55 plants were operable, 73 plants were under construction and 107 plants had been announced. Figure 3 shows the geographical location of these projects. It is interesting to note that

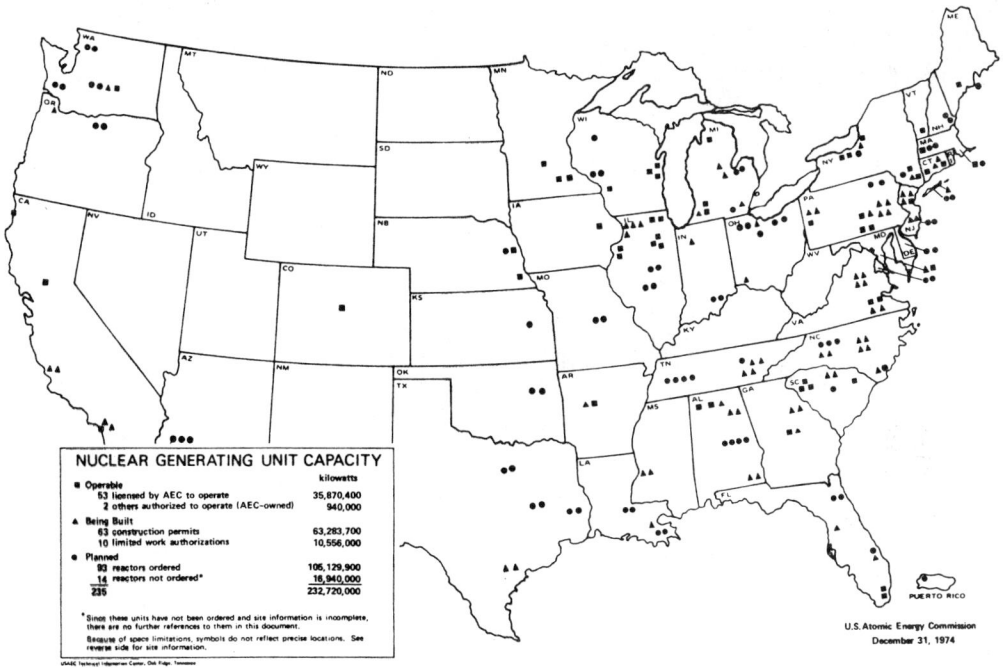

Fig. 3   Nuclear Power Reactors in the United States

the Public Service Company of Colorado Fort St. Vrain Nuclear Generating Station is the only plant presently located in this part of the country. One point which may seem somewhat surprising is the considerable number of plants which are located in the South and Southwest where there are large resources of petroleum and natural gas. You will also note that there are a large number of plants located in the Midwest where abundant resources of coal are located. The location of nuclear plants in these areas where fossil fuels are readily available is indicative of the competitive position which has been achieved by nuclear power.

There are several reactor types which are predominant in the United States today. Figure 4 depicts in cartoon form the types which are most common for plants in the country at the present time. The light water reactor

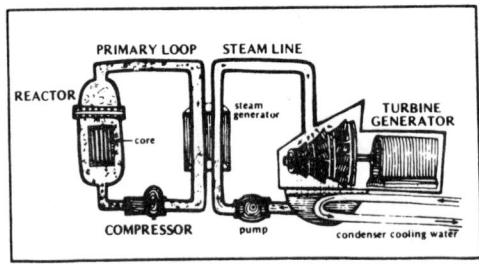

Fig. 4 Predominant Reactor Types in the United States

is by far the most common type in U.S. plants at the present time. There are two types of light water reactors--the boiling water reactor and the pressurized water reactor. In both types, water is circulated through the reactor core as the heat transfer medium or reactor coolant. The objective is to utilize the heated reactor coolant to generate steam which is used to drive a steam turbine and thus generate electricity. The boiling water reactor shown in the upper right hand corner is characterized by the direct use in a steam turbine of the steam which is generated by circulation of water through the reactor core. On the other hand, the pressurized water reactor which is shown in the upper left hand corner utilizes an intermediate heat exchanger or steam generator to generate steam for the steam turbine. The water in the closed circuit circulating through the reactor core is at a higher pressure than in the boiling water reactor and hence the name pressurized water reactor. The reactor type depicted at the bottom is the high temperature gas-cooled reactor. This is the same reactor type as that used at the Public Service Fort St. Vrain Station. You will note that the configuration is somewhat similar to the pressurized water reactor. The main difference is that helium gas is used as the reactor coolant instead of water. The chief advantage of this system, as compared to the water reactor systems, is its ability to produce steam at a higher temperature. Whereas the typical steam temperature achieved by boiling water and pressurized water reactors is about 550°F, the steam produced by the high temperature gas-cooled reactor is at a temperature of 1000°F. Since the high temperature gas-cooled reactor, or HTGR as it is commonly called, is the type used at our Fort St. Vrain Station, I would like to talk about it in more detail later.

Of the 225 reactors in nuclear power plants listed as either operable, under construction, or on order in the U.S. as of December 31, 1974, 73 were boiling water reactors, 144 were pressurized water reactors, 7 were HTGR's, and 1 was a liquid metal fast breeder reactor. Thus, it is evident that the HTGR is somewhat of a latecomer among reactor types. However, the potential for higher plant efficiencies from the advanced converter HTGR makes it a promising reactor concept for the future. Boiling water reactors are manufactured by General Electric Company. There are three manufacturers of pressurized water reactors--Westinghouse, Babcock & Wilcox, and Combustion Engineering. General Atomic Company is the manufacturer of the HTGR.

Nuclear plants are quite complex; however, it is worth noting that the turbine generator portion of nuclear plants is very similar to the turbine generator portion of fossil fuel plants. The significant difference is that a nuclear reactor is used as the heat source to generate steam in nuclear plants rather than a fossil fired steam generator.

FORT ST. VRAIN STATION

Now I would like to talk in more detail about our own Fort St. Vrain Station, the first commercial application of the high temperature gas-cooled reactor.

The key to any reactor system is the nuclear fuel. In the HTGR the fuel is fabricated in the form of tiny spherical particles. Photomicrograph cross sections of two of these particles are shown in Figure 5. In actual size these particles are only about ten thousandths of an inch in diameter.

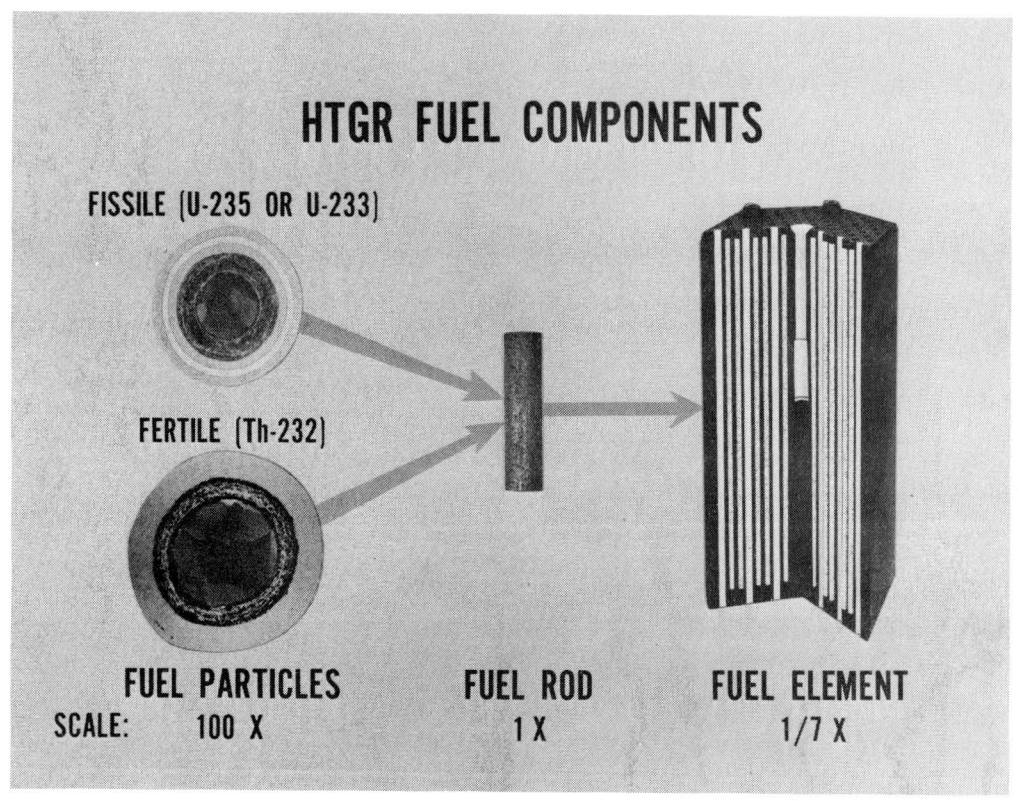

Fig. 5 HTGR Fuel Components

The central portion is the fuel kernel which contains uranium highly enriched in the isotope uranium-235 and thorium. In the present design it is made in the carbide form. The kernel is coated with several layers of pyrolitic carbon, and in the fissile fuel particles, an additional layer of silicon carbide is added. In the Fort St. Vrain design, a silicon carbide layer is added to the fertile particles as well. Each tiny particle forms its own pressure sphere and as long as the coatings on the particles remain intact, the fission products, or nuclear wastes, which are produced by the chain reaction remain inside the particles until the fuel is sent to a reprocessing plant for recovery of the remaining nuclear fuel. Thus, the coatings on the fuel particles form the first line of defense against the inadvertent release of radioactive materials from a plant and, therefore, are basic to the plant design from the standpoint of the safety of the public.

As I mentioned previously, the reactor fuel is composed of highly enriched uranium and thorium. This is a departure from the fuel cycle used in most other plants which are fueled with uranium slightly enriched in the isotope uranium-235. When thorium is incorporated in the fuel of a nuclear reactor, the thorium is converted into another nuclear fuel, uranium-233 during operation. Thus, this advanced converter concept produces a significant amount of new fuel while it is operating. The ultimate extension of this phenomena is the breeder reactor in which more fuel is actually produced during operation than is consumed. I will talk a bit more about the breeder reactor later.

In order to be able to assemble nuclear fuel into a reactor core, it is necessary that it be fabricated in a shape that can be handled conveniently. In the HTGR the tiny fuel particles are first fabricated into cylindrical fuel rods as shown in Figure 5. These fuel rods are then inserted into hexagonal blocks made of graphite. An HTGR fuel element is also shown in Figure 5. One of these blocks is about 14 inches across the flats and about 31 inches tall, and weighs over 300 pounds. The fuel rods are shown pictorially in one set of holes. The alternate set of open holes is the coolant holes through which the helium coolant flows to transfer heat from the fuel to the steam generators.

This graphite fuel element illustrates another important feature of the HTGR. Graphite is not only an excellent moderator material for nuclear reactors, but has the added advantage that it is an excellent structural material at high temperatures. This is the key to the ability of the reactor cycle to produce steam at a temperature of 1000°F resulting in plant efficiencies comparable to those of the most modern coal fired power plants.

The reactor core is then inserted in a prestressed concrete reactor pressure vessel which is characteristic of the high temperature gas-cooled reactor. A cross section of the prestressed concrete reactor vessel, or PCRV, and its internals is shown in Figure 6. The reactor core of graphite-body fuel elements may be seen in the upper half of the vessel interior. The

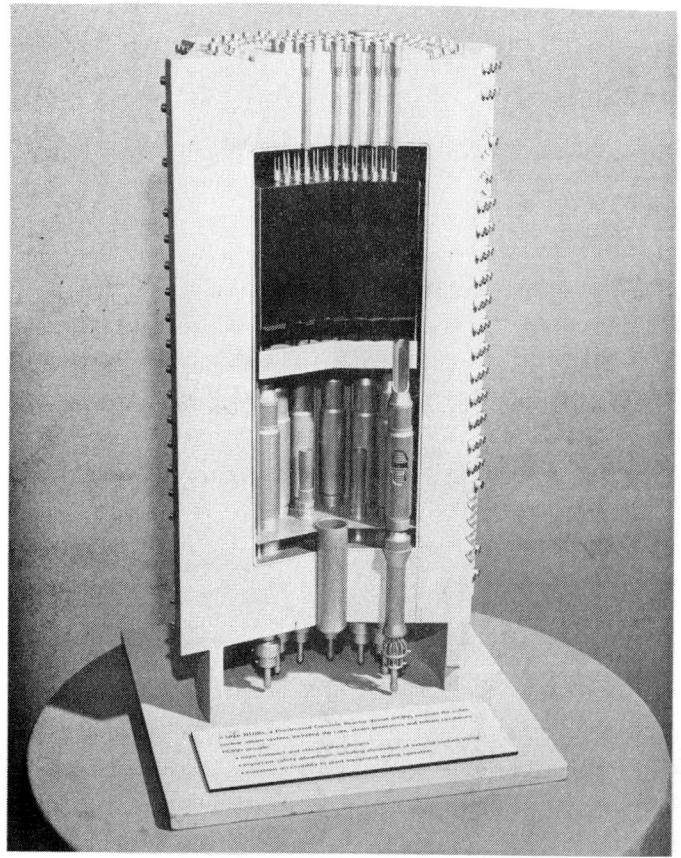

Fig. 6 Fort St. Vrain PCRV Cutaway

core is supported by a concrete support floor. The control rod drive mechanisms which control the reactor power level are visible in the top head of the vessel. In the bottom half of the vessel interior may be seen the cylindrical steam generator modules and the steam driven helium circulators which circulate the helium coolant through the reactor system inside the vessel.

As you can see, one of the important advantages of this design is that all of the major components of the reactor system are located inside of this one vessel, including the steam generators and the helium circulators.

The prestressed concrete vessel is also advantageous from the standpoint of field fabrication and provides several feet of concrete shielding which is required around any nuclear reactor.

The PCRV is located in the reactor building in the power station as shown in the cutaway model of Fort St. Vrain in Figure 7. The turbine generator at Fort St. Vrain has a capacity of 330,000 KW. An exterior view of the completed Fort St. Vrain Station is shown in Figure 8. The Station is now in the final phase of startup and is expected to be on line by this coming winter.

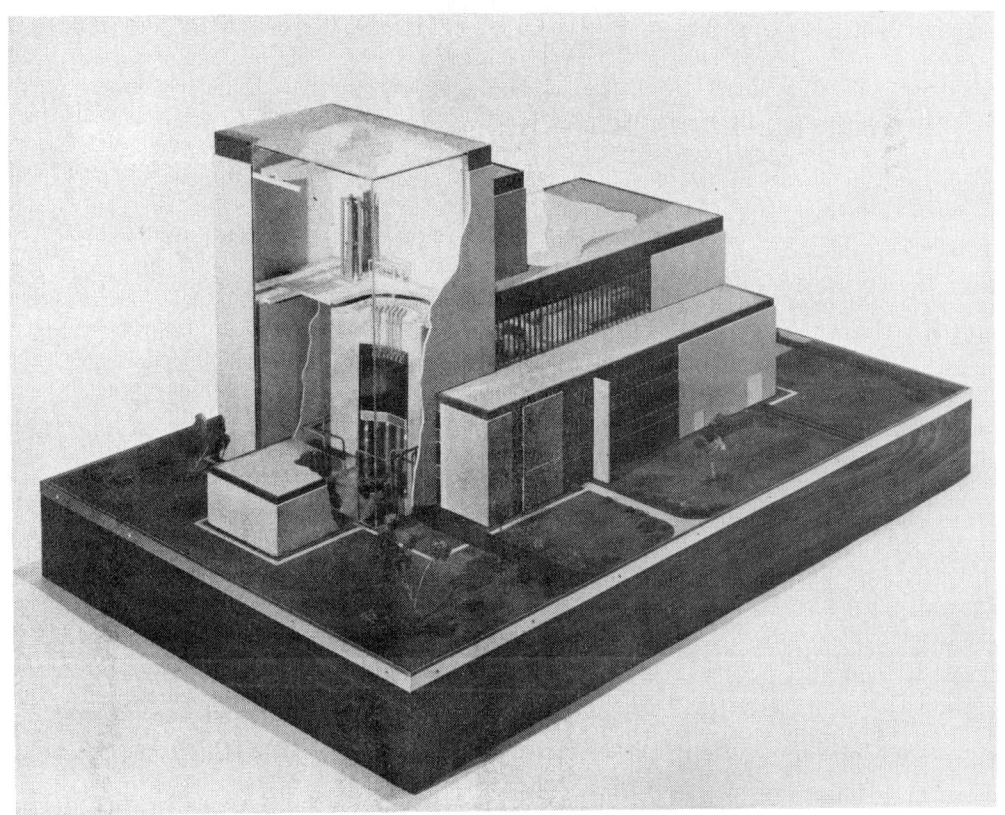

Fig. 7 Fort St. Vrain Plant Cutaway

NUCLEAR POWER IN THE FUTURE

Larger power stations with HTGR's are scheduled to come on line on other utility systems in the next few years. The standard sizes offered for

Fig. 8 External View of Fort St. Vrain

these larger plants are 770 MWe and 1160 MWe. A cutaway of the Fulton Station ordered by Philadelphia Electric Company, which will have two 1160 MWe units, is shown in Figure 9.

General Atomic also is proceeding with the design of a gas turbine HTGR power plant. This is a concept in which the helium coolant would be directly cycled through gas turbines coupled to the shaft of the electric generators. This advanced concept would result in a compact plant design. A cutaway of the essential components of this system is shown in Figure 10.

The nuclear plants I have discussed thus far have been the so-called thermal reactors which do convert some fertile material to fissionable material. However, there are even more advanced projects now underway. The most notable of these at the present time are the breeder reactor programs. A breeder reactor produces more fissionable material than it consumes.

The national breeder reactor program is a country-wide effort to construct a sodium-cooled fast breeder reactor in Tennessee. The Clinch River Breeder Reactor is supported by all segments of the electric industry, both public and investor-owned. Public Service is one of the investor-

Fig. 9 Philadelphia Electric 1160 MWe HTGR's

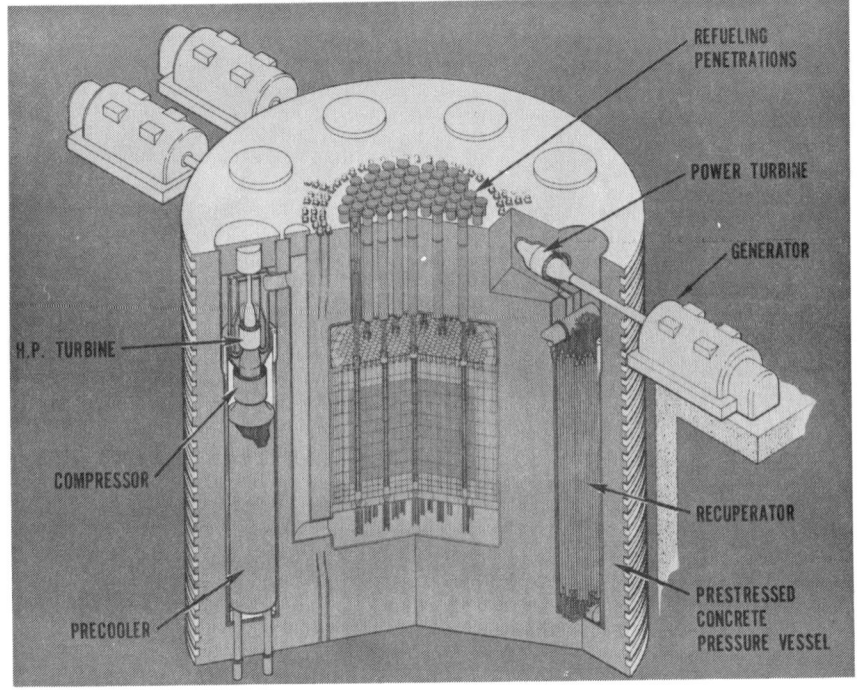

Fig. 10 Gas Turbine HTGR

owned participants in this nationwide effort. Even though the costs are very high and have been affected dramatically by inflation, the objective is very important in the long term because, as I mentioned previously, the breeder reactor has the potential to actually produce more fuel than it consumes. Thus, the breeder reactor is very important from the standpoint of long-term economic nuclear fuel reserves.

Even though the sodium-cooled breeder reactor is the favorite concept for the national breeder program, there are other breeder reactor concepts which deserve consideration. One of these which Public Service is particularly interested in is the gas-cooled fast breeder reactor, as illustrated pictorially in Figure 11. Public Service is a member of a group of utilities who are also supporting the design of this concept. General Atomic

Fig. 11 Gas-cooled Fast Reactor Station

Company, the designer of the HTGR, is also the designer of the gas-cooled fast reactor, or GCFR. Much of the experience gained from the development of the HTGR can be applied to the GCFR in contrast to the sodium breeder which is entirely different than other reactors in existence today in the

United States. It must be recognized that breeder reactors involve new
design problems as compared to the present thermal reactors and therefore,
their commercial development will be a long-term effort. It is quite
possible that breeder reactors will not generally be commercially available
until the late 1980's or 1990's. Therefore, the present generation of
reactors will be required to fill the gap until that time and more advanced
concepts such as the HTGR can play an important role in more efficient
utilization of our nuclear fuel resources until the fast breeder reactors
are commercially developed.

NUCLEAR PLANT OPERATING RECORD

What is the operating record of nuclear plants so far? It must be remembered that until now, only a few nuclear plants have been in operation
and most of these for only a short period of time. Thus, a well-developed
statistical base has not yet been established for nuclear plants as compared to the large number of fossil plants in operation. However, the
statistics that are available indicate that nuclear plants are not faring
badly, as illustrated in Figure 12. The left hand side of this figure
illustrates the availability of nuclear plants as compared to fossil fuel
plants.[2] You will note that in general, the availability of nuclear plants

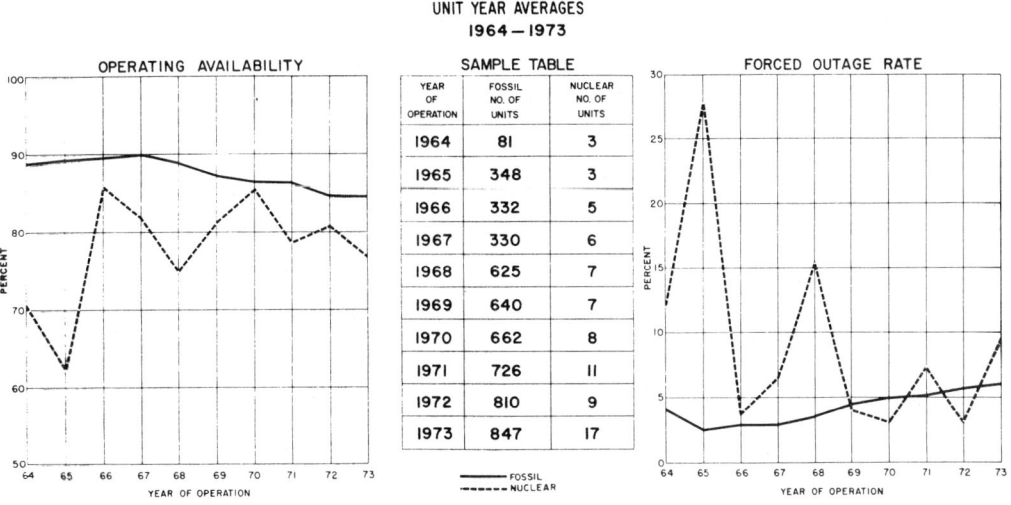

Fig. 12 Comparison of Fossil and Nuclear Operating Statistics

has been only slightly lower than that of fossil fueled plants. The right hand graph illustrates the forced outage rate of nuclear plants as compared to fossil plants. Again, the forced outage rate of nuclear plants is comparable to that for fossil plants. Another important statistic of interest, which is not illustrated in the figure, is plant capacity factor, which is an expression indicating the percent of generation produced from a plant during a year as compared to its maximum potential if it were operating at 100 percent capacity during the entire year. During the years of 1964-1973, the average capacity factor for 20 nuclear plants was 64.2% as compared to an average capacity factor of 68.9% for 894 comparable fossil fired units.

NUCLEAR PLANT SAFETY

How safe are nuclear plants? During the past year, the results of the most complete assessment of the safety of nuclear power reactors ever undertaken were released.[3] This study was performed under the independent direction of Professor Norman B. Rasmussen of the Massachusetts Institute of Technology. The graph shown in Figure 13 illustrates some of the major findings of this study. This figure illustrates the relative frequency of fatalities due to a number of man-caused events including the estimated frequency for nuclear plants. It may be seen that the actual frequency of fatalities based on actual experience for all types of man-caused calamities is considerably higher than that calculated for nuclear plants. Since there has not yet been a fatality resulting from commercial nuclear power plant operation, the projections for nuclear plants had to be based on very rigorous mathematical methods. For example, the figure illustrates that there is about a one in five chance each year that 100 people will be killed by air crashes and there is even one chance in 100 that 100 persons on the ground will be killed by being struck by a crashing aircraft. By comparison, it is predicted that there is only about one chance in 10,000 that 100 fatalities would result from the operation of 100 nuclear power plants in any given year. The probability that 100 fatalities would occur is even much smaller and is predicted to be only about one chance in 1,000,000. The study also investigated the predicted frequency of natural events. Again, it was found that the risk from natural events is considerably greater than that from nuclear power plants and is in a range not

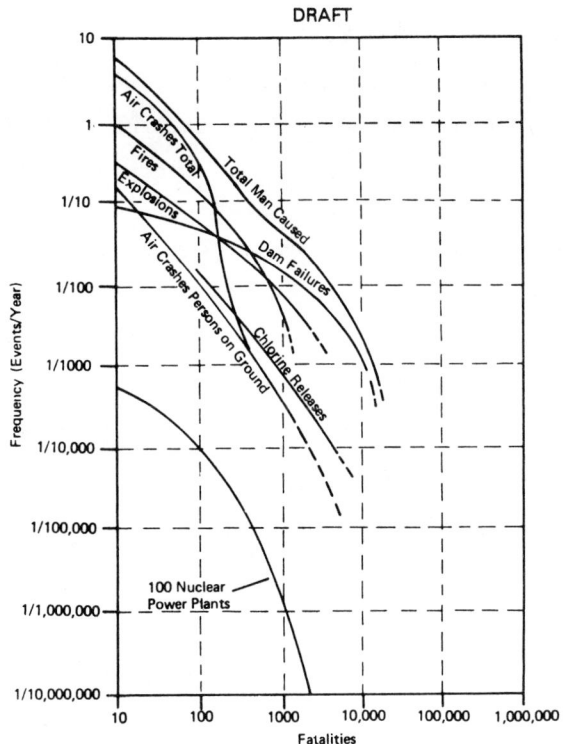

Fig. 13 Frequency of Fatalities Due to Man-caused Events

much different than that for man-caused events as just discussed. As a matter of fact, the only natural event which was identified as having a predicted frequency for fatalities roughly comparable to that from 100 nuclear plants was the very small probability that persons would be hit by a falling meteor. On an individual basis, it was estimated that the chance that you, as an individual, might be killed as a result of the operation of 100 nuclear plants, would be only one chance in 300,000,000 per year. Thus, it can safely be concluded that the risk from nuclear plants is very small indeed.

CONCLUSION

What then is the future for nuclear power? President Ford indicated in his energy message that he was setting a national goal to have 200 nuclear plants on line by 1985. This is a big order considering the many delays which are currently being experienced in the construction of nuclear plants and the ever accelerating costs and problems of raising capital funds to build plants. The typical schedule for building a nuclear plant is presently about ten years. Nuclear plant costs have risen from $240 per KWe for typical plants commissioned in 1969 to about $750-800 KWe in 1975 for plants scheduled for completion in the early 1980's. There are many obstacles to placing nuclear plants on line in an expeditious manner, but it is clear that nuclear energy presently is one of only two realistic alternatives for producing large amounts of electric power in the near term; the other is coal. Therefore, it is essential for our nation's well-being to recognize the importance of nuclear power and continue its development in a timely manner.

REFERENCES

1. Joint Committee on Atomic Energy Staff, "Certain Background Information for Consideration When Evaluating the 'National Energy Dilemma,'" U.S. Government Printing Office, 1973.

2. "Report on Equipment Availability for the Ten-Year Period, 1964-1973," Edison Electric Institute, Report EEI No. 74-57, December 1974.

3. "Reactor Safety Study, An Assessment of Accident Risks in U.S. Commercial Nuclear Power Plants," draft report, U.S. Atomic Energy Commission, August 1974.

AAS 75-285

AVAILABILITY AND VARIABILITY OF SOLAR
ENERGY IN THE ROCKY MOUNTAIN REGION

Richard C. Burriss*

Empirical values for attenuation of solar
radiation in the Rocky Mountain region due
to latitude, altitude, water vapor and cloud
cover are presented. These values are then
used to calculate insolation for Colorado
Springs, a location for which no long
historical record of observed insolation
is available.

This report is an attempt to combine readily available climatic data with geographical data to account for variation in solar radiation. All radiation values reported include direct and diffuse radiation from sun and sky at all wavelengths. They do not include radiation reflected or scattered from the earth or objects and structures visible to a radiation receptor surface. Climatic inputs[1] are mean daily total cloud cover from sunrise to sunset and mean daily surface dew point temperature, both interpolated from mean monthly values. Geographic inputs are latitude and elevation above sea level. Solar inputs are the solar constant (1.94 Langleys per minute), declination and radius vector (sun to earth distance).

The Rocky Mountain area was chosen because of assumed homogeneous type and density of clouds over the area, the variation in elevation and a lack of widespread air pollution. An empirical set of equations was developed to include attenuation of radiation by the clear dry atmosphere, by moisture and by clouds for five observation sites: Albuquerque, New Mexico; Boise, Idaho; El Paso, Texas; Grand Junction, Colorado; and Phoenix, Arizona.

* Kaman Sciences Corporation, Colorado Springs, Colorado.

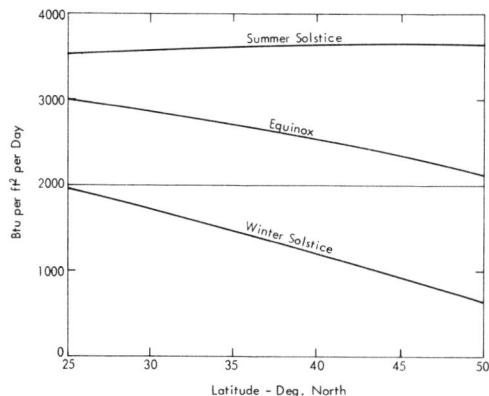

Fig. 1. Variation with Latitude of Daily Insolation on a Horizontal Surface at 6300 ft Above Sea Level, with a Surface Dew Point Temperature of 30°F and no Clouds.

Fig. 1 shows the variation of insolation with latitude at 6300 ft. above sea level with a surface dew point temperature of 30°F and no cloud cover. Notice that mid-summer insolation is nearly constant throughout the latitude range. Mid-winter insolation decreases by a factor of 3 from 25°N to 50°N due to latitude alone.

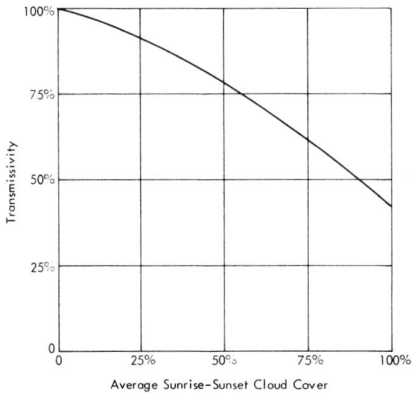

Fig. 2. Variation with Cloud Cover of Transmissivity of Clouds with no Atmospheric Attenuation.

Fig. 2 shows the variation of insolation with average sunrise to sunset cloud cover. There is no atmospheric attenuation in this calculation. With no clouds, there is 100% insolation; with 100% cloud cover, it reduces to 42%. Cirrus cloud overcasts transmit 85% of solar radiation, Altostratus 41% and Nimbostratus 15%[2]. Since this study does not include cloud type or density, we might assume that the clouds of the Rocky Mountain area average near midway between the extremes of transmissivity.

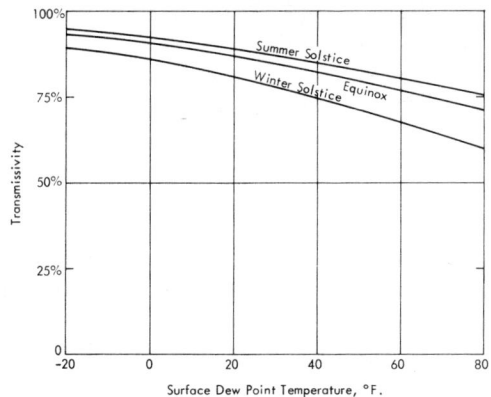

Fig. 3. Variation with Surface Dew Point Temperature of Transmissivity of the Atmosphere at 38.8°N, 6300 ft Above Sea Level and no Clouds.

Fig. 3 shows the variation of transmissivity of the atmosphere with surface dew point temperature at 6300 ft. elevation, 38.8°N, and no clouds. Note the increase in attenuation as the optical path length increases from summer to winter. Similarly, increasing attenuation due to optical path length is caused by increasing latitude or decreasing elevation.

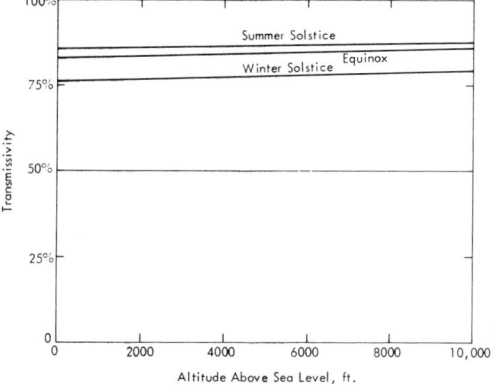

Fig. 4. Variation with Altitude of Transmissivity of the Atmosphere at 38.8°N, Dew Point Temperature 30°F and no Clouds.

Fig. 4 shows the variation of transmissivity with altitude at 38.8°N, no clouds and 30°F surface dew point temperature. The differences between sea level and 10,000 ft. are only about 2%.

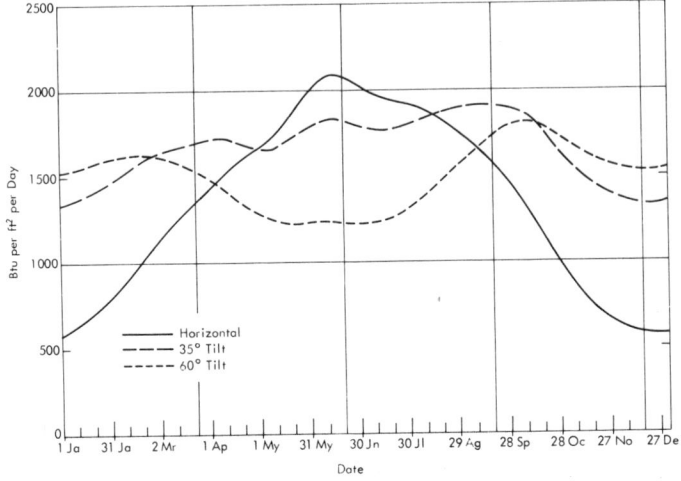

Fig. 5. Variation Throughout the Year of Insolation on a Receptor with 2 Glass Cover Plates, Facing South at Colorado Springs.

The distribution of insolation throughout the year for horizontal, 35° tilt and 60° tile receptors with two glass cover plates is shown in Fig. 5. The asymmetry about the summer solstice is caused by moist cloudy spring weather and dry, clear autumn weather.

CONCLUSIONS

Readily available meteorological data may be used to calculate solar insolation values within a homogeneous region at locations where radiation observations are not available. These calculated values may then be used to determine the orientation of solar collectors and the expected insolation to be received by the collectors.

REFERENCES

1. ESSA, Climatic Atlas of the United States, Government Printing Office, June 1968.
2. List, R.S., Smithsonian Meteorology Tables, Smithsonian Institution Press, D.C., 6th Rev. Ed., 4th Reprint, 1968.

AAS 75-286

MODELING THE WESTERN COAL INDUSTRY

William Ganter, Claude McMillan, Fred Glover[*]

INTRODUCTION

Today's energy problem is difficult to fully understand, partly because of its many complex supply-demand interrelationships. Coal reserves in the Western states have been pointed to as a short term solution to a portion of our energy problem. MSC, Inc., under contract to the U.S. Department of the Interior, has developed a model to accommodate some of the critical mine to market relationships of Western coal. The insights to be derived from this model, and from its solution by recent parallel innovations in solution methodology, promise to yield significant gains in our understanding of key interrelationships in the Western coal mining and distribution operations, and to identify ways to coordinate and implement these operations more advantageously.

MODEL STRUCTURE

The model allows the most cost-effective supply-demand relationship to be determined subject to the constraints imposed. The underlying framework constitutes a multi-stage representation of the total coal cycle, including the mining, transportation, conversion, distribution and final use stages. The input and output parameters of each stage are expressed in equivalent energy units, where the units employed are billions of BTU's. The final use stage is the demand in a market. Each individual element in a stage is capable of including the dimensions of cost, efficiency, and capacity in its characterization.

---

[*] Drs. Ganter, McMillan, and Glover are affiliated with MSC, Inc., Boulder, Colorado, and Drs. McMillan and Glover are also Professors at the University of Colorado.

MATHEMATICS

In mathematical terms, the model is formulated as a generalized capacitated network flow problem. The power of this type of formulation to handle complexities embodied in the model is the result of important recent innovations in mathematical and computer modeling techniques. A further significant advantage of this model is that MSC has a very highly effective computer software system for solving this type of problem. Our code is a couple of orders of magnitude (i.e., 100 times) better in terms of speed and problem size than available linear programming computer codes, which constitute the only competing alternative solution method.

The technical innovations underlying our procedure were made by Drs. Fred Glover and Darwin Klingman (see ACM Transactions on Mathematical Software, Vol. 1, No. 1, March 1975, pp. 47-55). Their modeling approach and solution methodology have recently been applied to a number of significant large scale applications in government and industry. The computer code is implemented as a FORTRAN language application program.

COST-EFFECTIVENESS

The model is cost minimizing across all stages. However, it also has the unique feature of imposing a penalty cost at each stage on the loss of energy during the various succeeding stages from mine to end use. This cost is always fairly and uniformly applied, thus making it an important measure to compare such trade-offs as locating a steam plant at mine mouth with its greater transmission line loses verses hauling the coal to a plant located in the utility's service area.

The model functions by pulling the end BTU demand at each market through the distribution, conversion, transportation and mining stages, all simultaneously in a cost minimizing competition. Each mine (or other energy resource) is assigned a capacity in BTU units that it can produce during the year. Its production is not allowed to exceed this capacity.

The capacity of a mine is strategically adjusted upward above the BTU content of its maximum coal production to allow for energy resources that will be consumed in the mining and transportation stages only. Once the coal is converted to another energy form, this is no longer a factor. The cost of these fuel oils and other energy sources is included in the unit cost at each stage. The BTU units required to move the coal energy along its cycle are a multiplier in the artificial penalty cost imposed at certain later stages. For example, if a mining cost was 21¢ per million BTU's and the rail efficiency was .95 to a final market, and 2,000 million BTU units were shipped, then the artificial penalty cost would be

$$.21 \ (2000/.95 - 2000) = \$22.10$$

If another mine producing at the same cost, but closer to this market and having a .98 transportation efficiency, could have supplied this demand, the artificial penalty cost would be

$$.21 \ (2000/.98 - 2000) = \$8.57$$

EXAMPLE

We can consider a simple hypothetical example drawn from the 1974 Wyoming coal industry which illustrates the kind of computations possible with the model. Let us assume that a midwestern electric utility has an annual demand for an equivalent of 5 billion BTU's at the customer meters. Wyoming mine A is located near a large generating station with a power grid connection to this utility. Wyoming mine B has a good direct rail route to a steam plant located in the utility company's service area. The data in the following table applies. Capacity is assumed to present no problem. Costs are in dollars per billion BTU's.

	Mine A	Mine B
Mining Cost	200	160
Mine Efficiency	.971	.978
Rail Cost	--	230
Rail Efficiency	--	.964
Conversion Cost (on generated output)	30	40
Conversion Efficiency	.33	.29
Transmission cost (at receiving meter)	260	--
Transmission Efficiency	.60	--
Distribution Cost (at receiving meter)	50	50
Distribution Efficiency	.89	.89
Final Demand	5	5

Cost using Mine A:

	Cost	BBTU Loss
Mining	$ 5,674	.85
Conversion	281	19.01
Transmission	1,461	3.74
Distribution	250	.62
	$ 7,666	24.22

There is no externally assessed penalty cost in this figure. However, the equivalent BTU's mined are paid for. The overall efficiency of alternative A is considerably worse than that of alternative B.

Cost using Mine B:

	Cost	BBTU Loss
Mining	$ 3,216	.45
Rail Hauling	4,455	.73
Conversion	225	13.75
Distribution	250	.62
	$ 8,146	15.55

This figure contains an external penalty cost for the extra coal mined (in the model) to account for the rail efficiency; it is .73x160 = $117. The electric costs are all priced (scaled) at the receiving meter.

CONSTRAINTS

A great variety of constraints can be imposed which can allow different issues to be studied. A forced amount of flow between a mine and a market can be required, as in honoring a supply contract. Regional production limitations due to environmental factors can be set. Other possibilities include rail shipping route limitations, regional demand quotas, and fuel substitutability requirements.

The cost variable can be made to incorporate a variety of additional components, such as a state mineral severence tax, or a labor contract increment. Costs associated with environmental impact, land reclamation, and the purchase of water rights are other possibilities.

POLICY IMPLICATIONS

Many kinds of policy decisions can be studied by using this type of model. Of particular importance are questions of long term energy facilities planning and conservation measures. The model will identify the best markets for the introduction of new energy technologies, and also offer guidelines for cost verses efficiency trade-offs in new energy technologies.

AAS 75-287

MINIMUM COST SOLAR THERMAL ELECTRIC POWER SYSTEMS: A DYNAMIC PROGRAMMING BASED APPROACH

William S. Duff*

A dynamic programming methodology is presented for finding minimum cost solar thermal electric power systems (STEPS). This methodology was developed as part of an 18-month study for NSF/RANN performed by Colorado State University and Westinghouse Electric Corporation. [1,2,3]. The objective of the study was to identify designs of STEPS concepts that were most promising in terms of projected cost per kilowatt hour of electricity generated.

There are two basic STEPS concepts. The first is the distributed system shown in Figure 1. In this system individual solar collectors ranging in size up to about seven meters in diameter are used to collect the sun's energy and convert it to thermal energy in a heat transport fluid. The thermal energy from collectors is transported in ever increasing pipe sizes to the center of the field where it is converted to electricity by a turbine-generator.

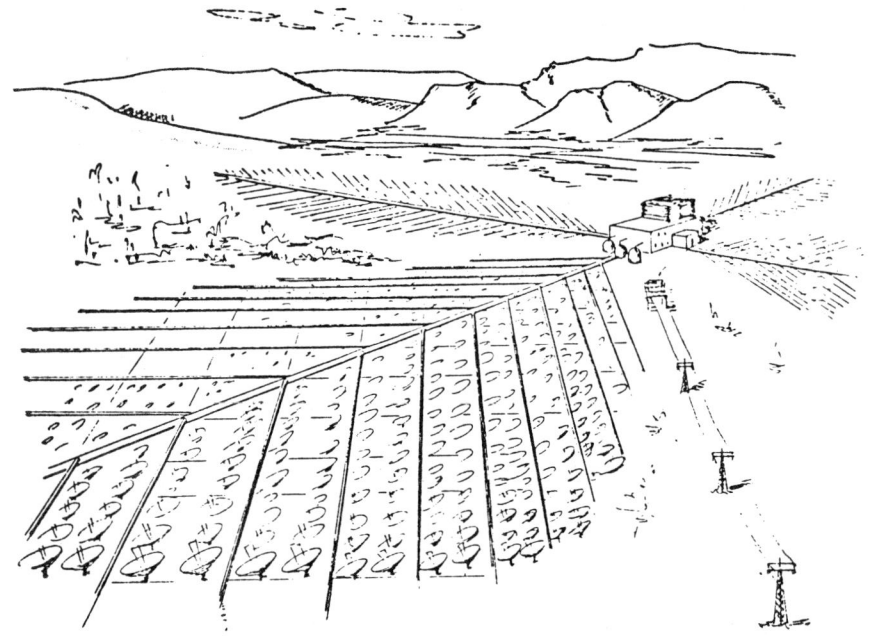

Figure 1. Distributed System

---

*Assistant Professor, Mechanical Engineering, Industrial Engineering Program and Solar Energy Applications Laboratory, Colorado State University, Fort Collins, Colorado.

Two types of distributed systems were examined in this study. The first, shown in part a of Figure 2, uses steam as the heat transport medium. The direction of the net flow of energy is shown. The energy flows optically from the concentrator subsystem to an absorber-heat exchanger subsystem where it is converted to steam. Concentrating solar collectors, the combination of a concentrator and absorber-heat exchanger, are then arrayed in a field. Then the thermal energy is transported to some point in the field where a turbine-generator-cooling tower converts the steam to electric energy. In some types of collectors, flat plate collectors, there is no concentration of solar radiation and the absorber-heat exchanger subsystem is the collector. Subscript designators corresponding to each of the subsystems that will be used in subsequent parts of the paper are C for the concentrator, A for the absorber-heat exchanger, F for the field and T for the turbine-generator-cooling tower.

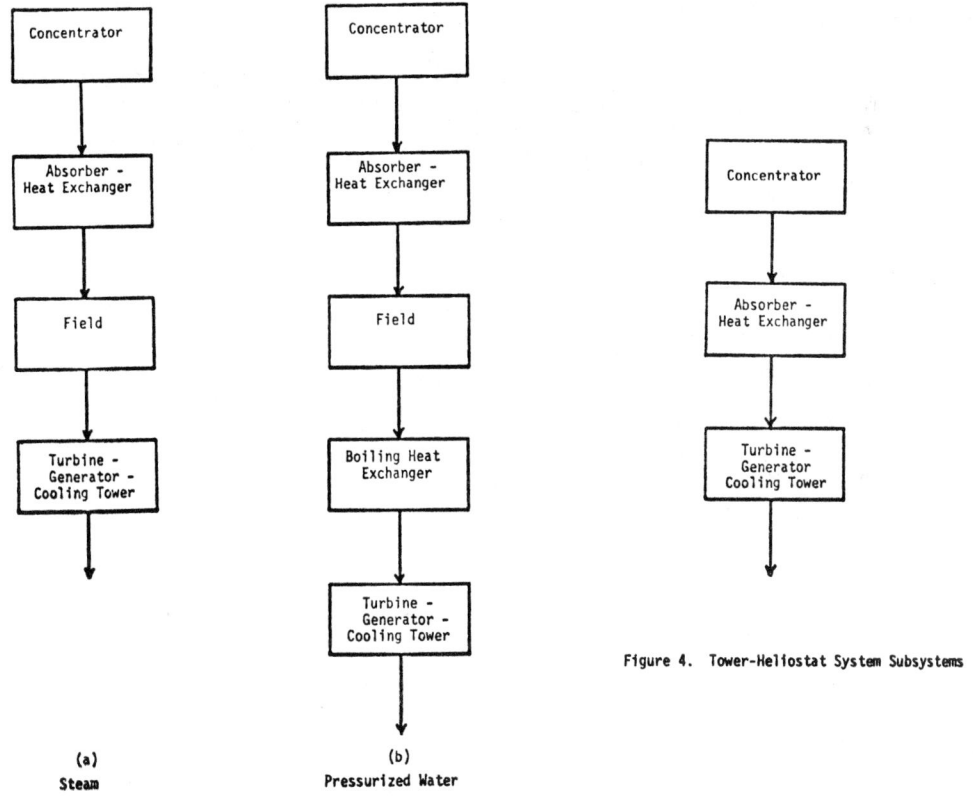

Figure 4. Tower-Heliostat System Subsystems

Figure 2. Distributed System Subsystems

Part b of the figure is similar to part a except that pressurized water is used as the heat transport fluid instead of steam. At the center of the field the pressurized water is put through a heat exchanger where steam, the working fluid of the turbine, is generated.

The second basic STEPS concept is the tower-heliostat system shown in Figure 3. In contrast to the distributed system, the principal

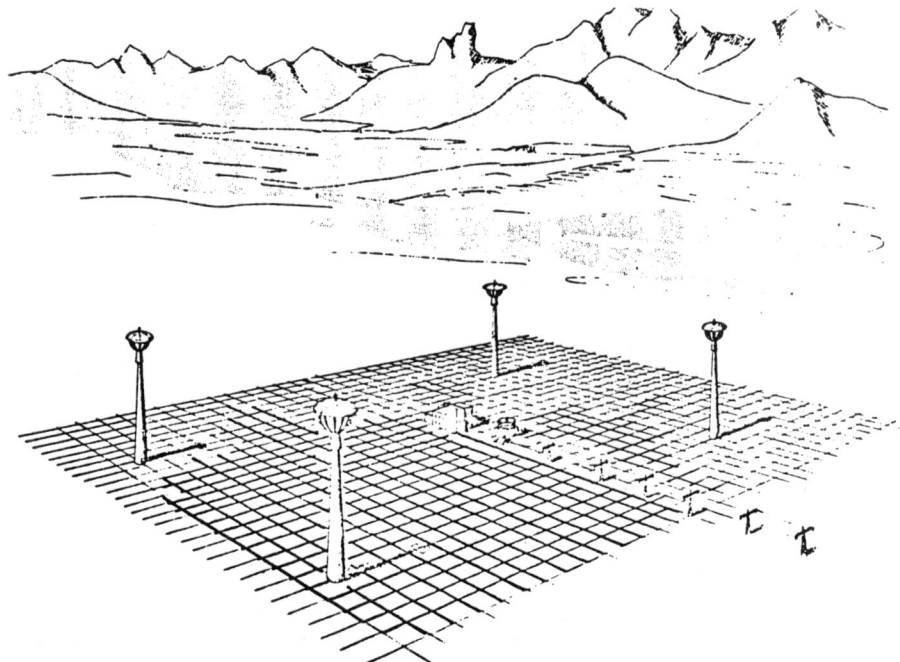

Figure 3. Tower - Heliostat System

mechanism for energy transport is optical rather than thermal. This avoids the thermal losses inherent in distributed systems. Solar energy is concentrated by the heliostats to the top of a tall tower onto an absorber-heat exchanger. Here it is converted to thermal energy in the form of steam or pressurized water. Thermal losses are virtually negligible since the mass flow rate of the heat transport fluid is so great and the distance to the turbine is relatively small. As in the distributed system, the thermal energy is then used to generate electricity. The subsystem breakdown for the tower-heliostat concept is shown in Figure 4. Heat transport fluids other than steam or water can be used. However, in this part of the study only steam and pressurized water were considered.

In the distributed system many different kinds of concentrating collectors or flat plate collectors could be used, and the distributed system and the tower-heliostat system appear to be quite different. Thus it seems that it would be difficult to find a common basis for systems optimization. The remainder of this paper describes such a methodology. This methodology utilizes parametric cost and performance models, derived using internally consistent approaches, as inputs to the optimization.

Problem Structure

The notation used in this paper follows Nemhauser [4]. If r_A is the stage return, X_A is the stage input state variable and D_A is the stage design variable for stage A, where X_A and D_A are vectors and r_A is a scaler, then the equation for the total system cost in terms of subsystem costs can be written as

$$N_F(D_F, X_F)[r_C(D_C, X_C) + r_A(D_A, X_Z0]$$
$$+ r_F(D_F, X_F) + r_T(D_T, X_T) \tag{1}$$

where N_F is the number of collectors in the field.

CONCLUSIONS

A general methodology has been presented which permits the choice of most promising STEPS concepts for further study in specific hardware research. The cost in cents per kilowatt hour (KWH) for the minimum cost STEPS plant was found to be about 3.5 after plant operation and maintenance costs were added to the costs. This is competitive with peaking power plants that STEPS plants would displace. Depending on the plant and the price of the fuel, peaking power plants generate electricity at from 2 to 6 cents per KWH. STEPS plants compete with peaking plants rather than base load plants because the solar insolation levels match up with peak electric power demands in most locations.

A disadvantage to the STEPS concept is the fact that solar energy is unreliable. Thus, a STEPS plant would not be a very attractive investment for utilities, given limited investment capital, since utilities must provide capacity. Because of the unreliability of STEPS plants no credit, or only very limited credit in an aggregated statistical sense, can be achieved in terms of capacity purchased. If additional investment capital could be provided, utility companies could achieve this required capacity by buying fossil fuel peaking plants to back up STEPS plants. This would add about 1.5 cents per KWH of capital investment costs to the 3.5 cents per KWH cost of the STEPS plants. This is still competitive with some fossil fuel peaking plants.

Because of the invarient embedding properties of the analysis, various other direct applications of this methodology are possible. For instance, a single collector could be used for individual residential electric power generation and heating and cooling. Another possibility is that a field of collectors could centrally supply thermal energy directly for heating and cooling of groups of buildings.

REFERENCES

1. "Solar Thermal Electric Power Systems," Final Report, Colorado State University, November 1974, Vol. 1, Executive Summary.

2. _____, Vol. 2, Systems Studies and Economic Evaluation.

3. _____, Vol. 3, Appendicies.

4. Nemhauser, G. L., <u>Introduction to Dynamic Programming</u>, Wiley and Sons, 1966.

AAS 75-288

100 MWe SOLAR POWER PLANT DESIGN CONFIGURATION AND PERFORMANCE*

F. A. Blake**
Martin Marietta Aerospace
Denver, Colorado

Abstract

Point design of a 100 MWe solar energy conversion power plant was a major task element of the Solar Power System and Component Research Program conducted jointly by Martin Marietta and Georgia Institute of Technology with support from the National Science Foundation RANN Grant AER74-07570. The design configuration of the 100 MWe solar power plant, a discussion of the dominant influences on the design, the projected performance of the plant, and a discussion of the substantiating evidence supporting the projection are presented in this paper.

I. Introduction

Solar Power System and Component Research was authorized under National Science Foundation Grant AER74-07570, with a period of performance from January 15, 1974 to January 31, 1975. Martin Marietta Corporation (grantee) and Georgia Institute of Technology (major subcontractor) led the team of cooperating organizations in the performance of the program. System analysis was supported by data and consultation by the Salt River Project of Phoenix, Arizona and the Bonneville Power Administration of Portland, Oregon. Collaboration on the cavity receiver steam generator design and test planning was provided by the CNRS+ Solar Energy Laboratory of Odeillo, France and the Babcock and Wilcox Research and Development Division of Alliance, Ohio.

Historical benchmarks in the technology of solar thermal energy conversion bound the technology required for a utility scale solar power plant in the vital areas of;
High performance concentrators (Fig. 1 and 2)
High quality steam generation (Fig. 3)
High temperature electrical generation (Fig. 4)
Large area solar energy collection (Fig. 5)

Figure 1. Concentrators of solar powered irrigation installation at Meadi, Egypt, 1913, which demonstrated peak output of 55.5 brake horsepower from 13,269 ft² collector-concentrator system installed on 0.88 acre site. [1] (Smithsonian Photo)

*Work supported by National Science Foundation RANN Grant AER74-07570, Mr. George M. Kaplan, Program Manager.
**Program Manager, Solar Power Research, Martin Marietta Aerospace, P. O. Box 179, Denver, CO 80201.

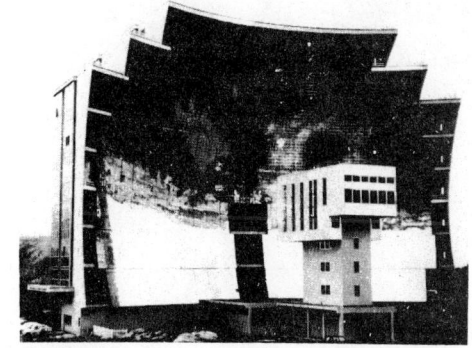

Figure 2. Concentrator of C.N.R.S. Solar Energy Laboratory, Odeillo, France. Output = 1000 KWth to Focal Plane from 1923 m² mirror area [2].

Figure 3. Solar Plant No. 3 of St. Illario-Nervi, Genoa, Italy. Capability = 150 Kg/Hr of Superheated steam at 150 atm, 500°C from 270 mirrors 1 m in diameter aimed into central receiver cavity. [3]

+ Centre National de la Recherche Scientifique

Figure 4. Solar Thermionic Research Installation of NASA-JPL Solar Electric Thermionic (SET) Generator Program. Output = 137.5 watts at 1754°C from 9.5 ft. dia. Electroformed Nickel Mirror.(4)

Figure 6. Artist concept of power station and North Heliostat Field of 1 MWe Solar Energy Conversion Demonstration Plant Proposed as Candidate for Research Applied to National Needs by Martin Marietta Corporation in 1973.

II. Baseline Features of Solar Plant Subsystems

Selection of the basic system characteristics to be optimized in the "point design" central receiver/optical transmission solar energy conversion power system resulted from the in-house study and proposal analyses that led to the granted program. These included establishing the following subsystem baselines.

Solar Collector/Concentrator

The concentrating heliostat was established as the baseline solar collector to accomplish both the large-area energy collection and the necessary concentration required to convert the solar energy to steam energy at temperatures consistent with modern power plants. The optical performance of the concentrating heliostat falls between the classical parabolic cylinder and the parabola of revolution. Considerable design flexibility exists with the heliostat, in that concentrations as low as 200 could be used without jeopardizing the 1000°F steam generation capability. The experimental prototype concentrating heliostat built by Martin Marietta to demonstrate the concept and develop optical accuracy data cost effectively is shown in Figure 7. The design point concentration for the collector-concentrator is 1840 overall, which results in a net range based on intercepted area between 1433 (June 21) and 1736 (December 21).

Figure 5. Large Area Solar Collector Benchmark-Field of Heliostats of C.N.R.S. Solar Energy Laboratory. Collector area = 63 Mirrors x 6.5 m x 7.5 m = 3071 m².(5)

Referencing the established technology generated in these benchmark programs against the requirements of modern power generation equipment led to conceptual design of the candidate 1 MWe central receiver power plant shown in the artist's concept of Figure 6 in early 1973. This concept and a research program which would lead to a demonstration plant of it were submitted for review by representatives of the scientific community, utility industry, and government research organizations. From this review the framework of the Solar Power System and Component Research program (NSF Grant AER74-07570) was established. Sizing of the point design system was increased to 100 MWe to accommodate the increased efficiency afforded in both larger equipment performance and reduced operational centers and to enable evaluation of the effects on configuration which would result from the scale up to utility size.

Figure 7. Concentrating Heliostat Optical Accuracy Development Prototypes (Martin Marietta Corporate Research Program)

Central Cavity Configuration Heat Receiver

Historical precedent and the potential for attaining highly efficient thermal energy conversion combined to establish the cavity configuration for the heat receiver. All the solar thermal programs undertaken during the power research programs of the space program used the cavity for its performance benefits. Additionally, the only solar powered steam generators used to obtain steam temperature-pressure conditions suitable for modern turbines--those of Prof. Francia at Genoa, Italy--used cavity configurations[3] as shown in Figure 8. Among the most desirable features of the cavity are its near-black-body absorption, the flexibility of designing the interior to attain nearly uniform flux at the heat conversion interface, and its ease of insulation. Figure 9 shows a 1/32-scale model of the boiler cavity for the point design plant which was used to empirically verify the convection loss projection.

Figure 8. Cavity Configuration Central Receiver Steam Generator of Prof. G. Francia's Solar Pilot Plant No. 3 at Genoa, Italy, in operation. (Photo Courtesy of Georgia Tech High Temperature Materials Laboratory).

Figure 9. 1/32 Scale Model of Solar Power System Steam Generator Cavity Used to Verify Convection Loss Projection. Northward Facing Tower Mounting Illustrates Appearance of Installation.

Air-Cooled Condensing

To maximize the environmental desirability of the solar-conversion power plant, and in acknowledgement of the extreme water shortage normally accompanying abundant solar insolation sites, the baseline design will use an air cooled condenser and carry the economic and performance penalties that result. (This does not preclude the use of a water cooled condenser should industrially allocated water such as that of the Central Arizona Project, be available at the solar site.)

Single Expansion Turbine

Based on preliminary evaluation of the turbines suitable for cycling operation, and the problems attendant with out-of-plant steam transmission, the thermodynamic cycles considered were limited to those featuring a single expansion from superheated steam. The efficiency improvement from reheating would largely be negated by transmission losses.

Sizing of the Point Design Solar Power Plant

100 MWe output capacity of the design to be defined during the granted program was established based on information fed back from electrical utility representatives on the NSF Review Board and present at presentations covering the solar plant concept. For smaller plants the operational cost per kWh increases sharply due to higher manpower costs.

III. Design Configuration of the 100 MWe Solar Plant

The overall layout for the 100 MWe solar-energy-conversion power plant is shown in Figure 10. Eight solar collector fields are located around a central steam-turbine-generation power plant and control center. Appearance of the plant would be similar to the air cooled condensing plant of the Black Hills Power and Light Company at Wyodak, Wyoming shown in the inset. Each solar collector field is 1400x1400-ft square (45.0 acres) on a site having a natural slope to the south of 10 to 13 degrees to accommodate terracing with minimum excavation. The contour profile and position of the tower mounted steam generator are shown in Figure 11.

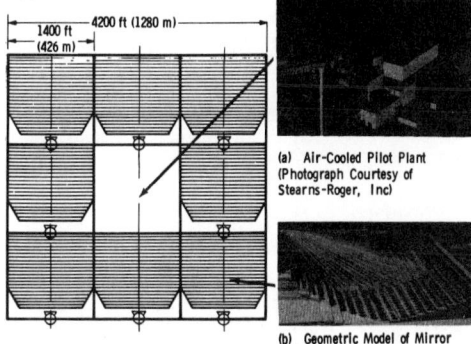

Figure 10. 100 MWe Solar Plant Layout. Insets of Air-cooled power plant and Geometric Model of Solar Collector Field Illustrate General Equipment Layout for Respective 45-Acre Fields.

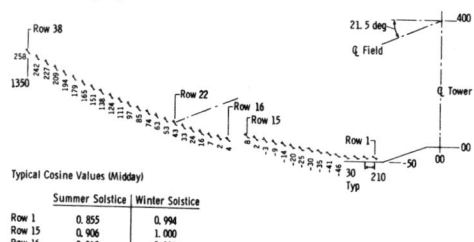

Figure 11. Heliostat Field Elevation Profile Established to Eliminate Row-to-Row Shadowing and Blocking for Solar Elevation Angles Above 20°, the Angle at which Design Threshold Direct Insolation (1.234 Langleys, 80 w/ft^2, 861 w/m^2) is Obtained in Southwestern U. S.

The design parameters of the 100 MWe solar plant are shown in Table 1. The most striking of the features is the 14,720 individual heliostat mirrors required to collect and concentrate the incident solar energy. Each of the mirror fields contains 1840 heliostats, the current baseline design, but could accommodate up to 2002 within the current boundaries, should growth be necessary. For perspective, the CNRS solar furnace has operated a coordinated field of 63 similar-sized heliostat mirrors since 1970.

Table 1. Parameters for 100 MWe Solar Power Plant

Layout Area – 405 Acres

Solar Energy Collection	- 8 Modules of 45 Acres (1400 x 1400 ft) Each
Power Plant Site	- 1 Module of 45 Acres Centrally Located

Solar Collector Subsystem

Heliostat Assemblies	- 14,720 Units of 400 ft^2 Each for Total Area of 5,870,000 ft^2
Heliostat Deployment	- 1840 Heliostats per Collector Module for Each of 8 Modules

Power Generating Plant

Turbine-Generator	- One Unit Rated at 100,000 kWe (141,000 SHP) Operating on 1250 psig, 950°F Steam
Waste Heat Rejection	- Air-Cooled Condenser

Solar Insolation Threshold

	- 1.234 Langleys = 80 W/ft^2; = 861 W/m^2

Tower-Mounted Heat Exchanger System - 8 Units Mounted Along South Border of Modules

Tower Height	- 400 ft (450 ft Above Lowest Elevation Mirror)
Boiler/Superheater	- Cavity Configuration with 20 ft^2 Apertures
Superheater Area	- 820 ft^2 per unit
Boiler Area	- 1,890 ft^2 per unit
Preheater Area	- 3,020 ft^2 per unit
Steam Rate	- 120,000 lb/hr
Steam Conditions	- 1,300 psig, 955°F at Outlet

Based on recommendations of turbine manufacturers, three thermodynamic cycles were evaluated during the preliminary system configuration design task. The overriding requirement during the evaluation was to demonstrate the capability to perform in the cycling operation attendant with solar applications. Identifying points of the cycles and the resultant sizing parameters imposed on the solar collector field are shown in Table 2. The sensitivity of the system to the thermodynamic cycle is evident in the reflecting surface requirement. Size of the mirror field ranges between 8.22 million ft^2 for the lowest (600 psig, 750°F) efficiency cycle and 5.87 million ft^2 for the highest (1250 psig, 950°F) efficiency cycle.

Table 2. Sizing Effects of Thermodynamic Cycle

Pressure, psig	Temperature, °F	Cycle Efficiency	Resultant Parameters		
			We/ft^2	ft^2/KW	ft^2/100 MWe
600	750	0.25	12.2	82.2	8.22 x 10^6
850	900	0.32	15.5	64.4	6.44 x 10^6
1250	950	0.35	17.0	58.7	5.87 x 10^6

For perspective, the 17.0 w/ft^2 attainable from the mirror powered solar thermal system is nearly double the 8.67 w/ft^2 attainable from current high quality photovoltaic panels (based on the terrestrial testing of the Skylab panels).

The 100 MWe system selected uses a single 100 MWe regenerative nonreheat turbine-generator. The steam inlet is at 1250 psig, 950°F, and the conversion efficiency is 35 percent. Steam temperature and pressure are lower than used in most present conventional steam plant design to be compatible with the thermal cycling required in a solar plant application. Key points of the system cycle are identified in the schematic of Figure 12.

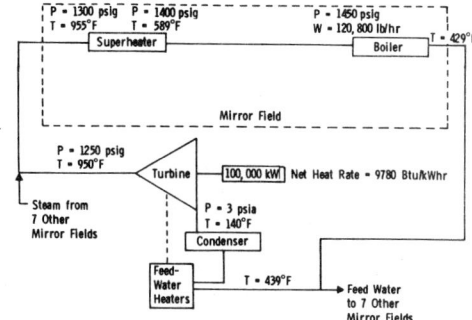

Figure 12. 100 MWe Solar Power Plant Schematic

Converting the concentrated solar energy into thermal energy in the working fluid occurs in eight tower-mounted cavity configuration boiler/superheaters. Each of the eight cavities for the 100 MWe system has a 20x20-ft. opening, 30-ft. depth, and is 40-ft. wide and 40-ft. high at the heat exchange interface in the rear of the cavity. A side-cutaway view of the steam generator is shown in Figure 13. An insulated door protects the cavity from the environment during shutdown periods and drastically reduces the temperature drop during shutdowns. For example, during a one-hour shutdown due to cloud cover the cavity average temperature will only drop about 10°F. During an overnight shutdown of 16 hours, the cavity temperature will drop about

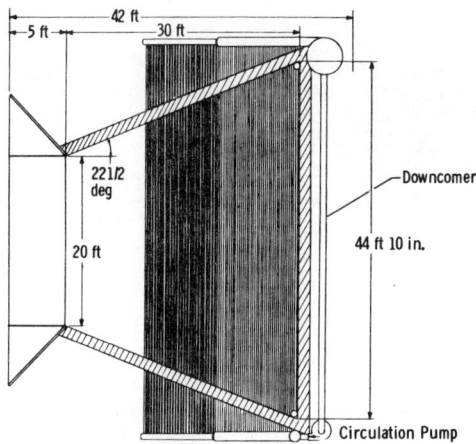

Figure 13. Cut-Away View of 100 MWe Steam Generator

160°F. The overall efficiency of the cavity is 95% without using glass in the aperture.

IV. Projected Performance - 100 MWe Solar System

Thermal collection efficiency for the solar-energy-conversion power system has been defined as "the output from the heat exchange subsystem divided by the direct insolation intercepted by the full mirror area". Because this efficiency directly contributes to the sizing of the mirror field for any design, it is of primary importance that each factor contributing to it be optimized. The factors currently established for the optical loss group (i.e., the cosing of the tracking angle, specular reflectivity of the mirrors, and optical accuracy of the concentration) and the thermal group (i.e., the convection, IR and reflection through the aperture, and insulation) are shown in Table 3. The data are shown for the winter solstice, equinoxes, and summer solstice. Two factors are affected by time of year, the area loss cosines and the optical accuracy which is partially dependent on the cosines. Attainable thermal collection efficiency values range from 0.614 at the summer solstice to 0.763 at the winter solstice. The waterfalling effect of the energy collection losses is illustrated in Figure 14.

Area Loss Factor (Cosines)

The area loss factor varies with both time of day and time of year. Values shown in Table 3 and Figure 14 are the daily averages for the respective three days and illustrate the time-of-year effect. A finer breakdown of the time-of-day variation is illustrated for the equinox day in Table 4. Row to row variations within the field range from 5% at the ends of the day to 10% near noon.

Specular Reflectivity

The reflectivity parameter of concern to the designer of a concentrating power system is the specular reflectivity, rather than the total reflectivity normally obtained with spectrophotometers. Additionally, the reflectivity needs to be referenced to a terrestrial solar spectrum in the geographical zone where the installation is to be located.

Table 3. Time of Year Effect on Performance Factor

	Day 355 Winter Solstice		Day 80/264 Equinox		Day 172 Summer Solstice	
	η	MW	η	MW	η	MW
Threshold Potential	-	471.8$_{th}$	-	471.8$_{th}$	-	471.8$_{th}$
Area Loss (Cosines)	0.944	444.5$_{th}$	0.895	421.6$_{th}$	0.779	367.2$_{th}$
Reflectivity	0.85	377.8$_{th}$	0.85	358.3$_{th}$	0.85	312.1$_{th}$
Optical Accuracy	1.0	377.8$_{th}$	1.0	358.3$_{th}$	0.975	304.3$_{th}$
Aperture Radiation	0.985	372.1$_{th}$	0.985	352.9$_{th}$	0.985	299.7$_{th}$
Convection Loss	0.968	360.2$_{th}$	0.968	341.6$_{th}$	0.968	290.2$_{th}$
Insulation Loss	0.998	359.5$_{th}$	0.998	340.9$_{th}$	0.998	289.6$_{th}$
Turbine-Alt Mech Eff	0.35	125.8$_E$	0.35	119.3$_E$	0.35	101.3$_E$

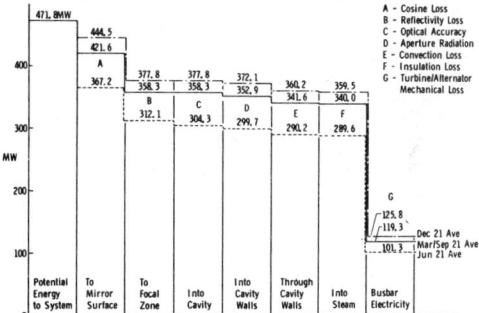

Figure 14. Projected Heat Balance-100 MWe System

Table 4. Cosine Pattern for Equinox Days for Solar Plant Location at 33.6°N. 111.3°W (Horse Mesa, Arizona)

Heliostat Row	Average of Cosines of All Heliostats in Designated Row									Average of Cosines for the Day
	8 am	9 am	10 am	11 am	Noon	1 pm	2 pm	3 pm	4 pm	
38	0.797	0.838	0.869	0.888	0.895	0.891	0.875	0.847	0.809	0.857
31	0.806	0.851	0.885	0.906	0.914	0.909	0.891	0.861	0.819	0.872
23	0.816	0.866	0.903	0.926	0.935	0.930	0.910	0.877	0.830	0.888
16	0.824	0.877	0.917	0.941	0.951	0.945	0.924	0.889	0.839	0.901
15	0.833	0.886	0.924	0.947	0.954	0.947	0.924	0.886	0.833	0.904
8	0.840	0.895	0.935	0.959	0.967	0.959	0.935	0.895	0.840	0.914
1	0.854	0.911	0.952	0.977	0.985	0.977	0.952	0.911	0.854	0.930
Average of Field	0.824	0.875	0.912	0.935	0.943	0.941	0.916	0.881	0.832	0.895

In an effort to obtain comparative data on several candidate mirror surfaces with specularity collimation similar to that of the solar monitoring instrumentation, the reflectivity test rig shown in Figure 15 was assembled. Reflected energy from the test mirror is monitored by the reverse-mounted normal-incidence-pyrheliometer. To avoid shadowing, the mirror is tested with a 10-degree tilt (Cos = 0.98481). Reference insolation input to the mirror is monitored by a pair of independently tracked pyrheliometers.

Results of specular reflectivity tests completed to date are shown in Table 5. The design value used in the system performance projection is 85%. From the samples tested it would appear that this is nearly attainable with current production glass laminated mirrors. The small improvement needed is available from either less tinted glass or

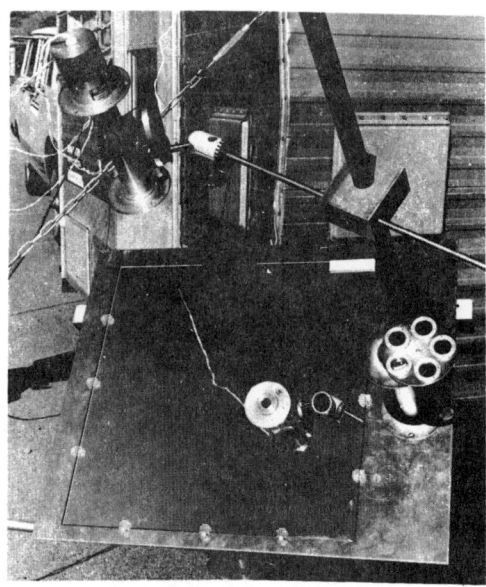

Figure 15. Specular Reflectivity Test Rig Using Reverse Mounted Normal Incidence Pyrheliometer to Monitor Direct Reflection Beam from Mirror

Table 5. Terrestrial Spectrum Reflectivity Test Data for Sample Mirror Surfaces.

Description of Mirror Sample	Test Date	Specular Reflectivity, %		
		High	Low	Average
First Surface Aluminized Teflon (Reference Surface)	10-1-74	86.5	86.1	86.27
	10-8-74	86.6	83.5	85.11
Second Surface Silvered Commercial Float Glass Laminated Mirror, 3 mm Front to Mirrored Surface	9-20-74	83.1	82.7	82.9
	9-23-74	83.9	83.3	83.56
	10-8-74	84.1	82.7	83.47
Second Surface Silver on Acrylic, 3.2 mm Thick	9-23-74	79.3	78.3	78.7
Second Surface Silvered Commercial Float Glass, 6 mm Thick	9-20-74	72.2	70.5	71.27
	10-8-74	73.3	72.7	73.05

decreasing the thickness of the mirrored segment of the laminated mirror from 3 to 2 mm.

Optical Performance

The optical pattern from a warped glass mirror operating as a heliostat (helio = sun, stat = stationary), such as the 4 x 4-ft mirror assembled for curvature control by a "wheel" structure shown in Figure 16, exceeds the size of the solar image that would be obtained by "on-axis-tracking" due to the combined effects of spherical aberation and manufacturing tolerances. By correlation of the theoretical image with test images such as shown in Figure 17, it was possible to evaluate candidate curvature control structures for accuracy. The image shown had a concentration of 5.32:1 at the test range distance of 108 feet, and a manufacturing error equivalent to 3.7 minutes tangential error for the surface.

Figure 16. Wheel Curvature Control Structure Installed on 4x4-ft Second Surface Glass Mirror, 108-ft Focal Length.

Figure 17. Optical Image from Wheel Structure Controlled 4x4-ft. Mirror, Concentration = 5.32:1.

Optical accuracy of an assembled heliostat was evaluated against the computer predicted patterns using the equipment shown in Figure 7 and the target of Figure 17. By focusing for a tight pattern at noon, it was possible to partially compensate for spherical aberation enough to offset the manufacturing inaccuracy. The result was endorsement of the analytical computer model as being suitable for realistic system design. The optical image patterns have been the controlling influence in sizing the cavity apertures of all boiler, superheater, and steam generation cavities analyzed. Overall the mirror curvature compensates for the divergent angle of solar rays, enabling use of an aperture size equal to the mirror size for fields as large as 1400x1400 ft. Rationale for the major design modification which shifted the location of the tower from the center of the field to the center of the south border was based on the performance gain defined by these patterns.

Thermal Performance of Steam Generator Cavity

A very substantial example of Space Program Technology being applied to terrestrial engineering needs is the "Thermal Analyzer Computer Program" which was generated and refined during the course of Skylab and Viking programs. Detail thermal analysis of the cavity and working fluid was obtained through its use. Input to the program is the radiant energy flux pattern keyed to the analysis nodes established for the interior of the cavity, illustrated in Figure 18. Resultant surface temperatures and energy distribution form the primary output. Energy losses established include the reflected solar radiation, the IR radiation from the surfaces which is lost through the aperture, and the insulation loss. The energy pattern reaching the working fluid at each node is established as well as the resultant temperature of the working fluid along its flow path. Convection, not normally being a part of spacecraft thermal analysis was treated separately based on empirical data developed from the model cavity shown in Figure 9.

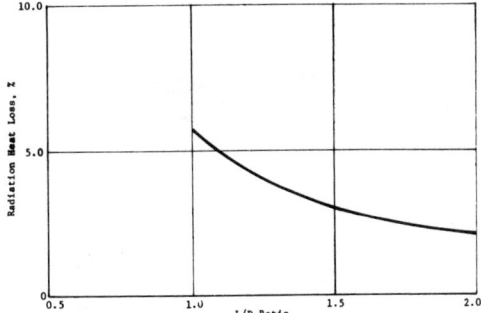

Figure 19. L/D Ratio Effect On Radiation Losses

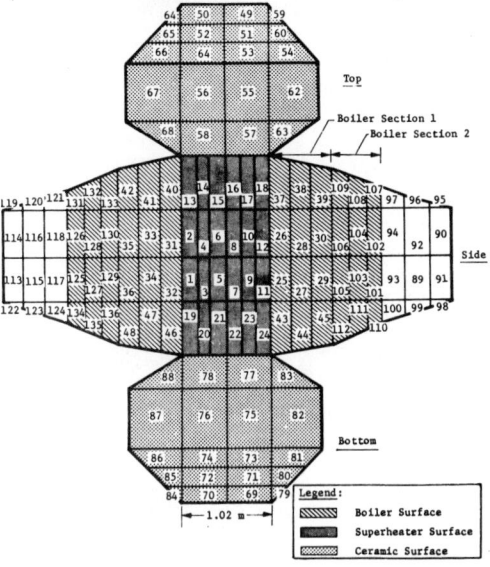

Figure 18. Developed Surface of Cavity Interior Showing Node Identification for Thermal Analysis and Zones of Superheating, Boiling, Preheating, and Insulation.

The shape of the cavity was selected to minimize the amount of direct flux impinging on uncooled surfaces. Results of a preliminary analysis to evaluate losses as a function of L/D are shown in Figure 19. As a result, we selected a cavity depth of 30-ft for the required 20-ft aperture (L/D=1.5). The sides of the cavity are at a 45° angle from the cavity axis and the top and bottom surfaces are are 22.5° from the axis to accommodate the incoming concentrated sunlight pattern from the heliostat field.

Thermal losses and resultant cavity efficiency values established by the Thermal Analyzer Program for the finalized full scale design are included in Table 6. A segment of the analyzer output, the cavity surface temperatures and working fluid temperatures for the superheater zone, for a specific control mode case which produced a flux gradient in the zone is shown in Figure 20. Steam flows through the superheater in two parallel sections in both downward and upward passes with mixing occurring at both the bottom header and the outlet. For the example shown, peak surface temperature was 1040°F, peak steam temperature was 960°F, and the mixed outlet steam temperature was 920°F.

Table 6. Full Scale Cavity Performance

Losses	Percent
Reflected Solar	0.8
IR	1.2
Convection	
No Wind	2.2
10-mph Wind	3.2
15-mph Wind	3.7
Insulation	0.2
Cavity Efficiency	
(Heat in Steam/Energy into Cavity)	
No Wind	95.6
10-mph Wind	94.6
15-mph Wind	94.1

V. Summary

Technology required for development of a solar energy conversion power plant yielding in excess of 17 watts per square foot of collector area in terrestrial installations in the Southwestern U. S. is available. A straightforward hardware development program directed at economic optimization and augmentation type applications into systems with existing storage is warranted to demonstrate that solar power has near term potential as an alternative power source.

Key features of the candidate system are:
    Central receiver energy conversion,
    Tower mounting of central receiver,
    Multiple solar collector fields,
    Single-central power generation station,
    Concentrating heliostats of laminated glass mirrors,
    Southward sloping site.

6. NSF/RANN/SE/AER74-07570/FR/75/1, Final Report "Solar Power System and Component Research Program (January 15 - November 15, 1974)," January 1975.

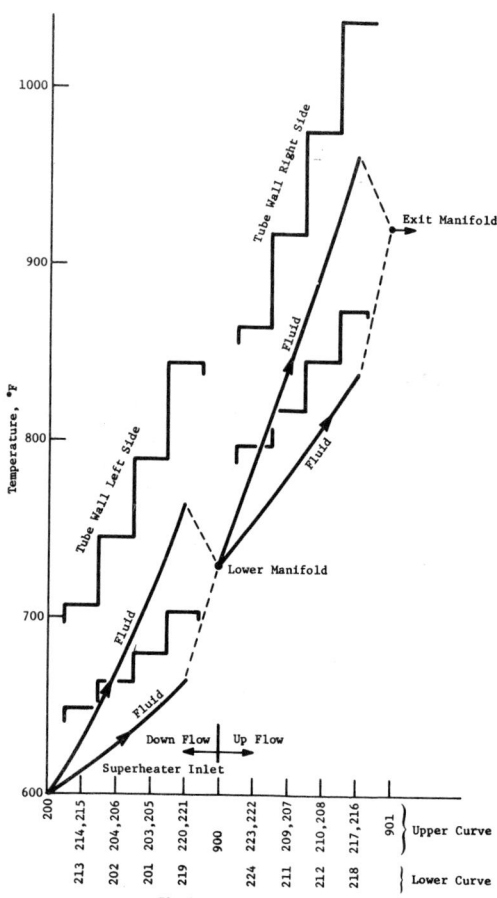

Figure 20. Superheater Zone Temperature Profiles for Full Scale Steam Generator.

VI. References

1. C. G. Abbot, "Utilization of Solar Energy". Annals of Astrophysical Observator (Smithsonian) Vol. IV, Chapter IX, 1922.

2. F. Trombe, C. Royere, J. F. Robert, "Premiers Resultats Obtnus Avec le Four Solaire de 1000 KW". Paper E-144 presented at International Congress "The Sun in the Service of Mankind", Paris, 1973.

3. G. Francia, "Pilot Plants of Solar Steam Generating Stations". Solar Energy, Vol. 12, No. 1, September 1968, pp 51-64.

4. P. Rouklove, F. A. Blake, "Performance Testing of a Solar Thermionic Generator System", Solar Energy Society Conference Paper, March 1965.

5. F. Trombe, L. Phat Vinh, "Thousand KW Solar Furnace Built by the National Center of Scientific Research in Odillo, France", Solar Energy, Vol. 15, No. 2, June 1973.

1300

MINIMIZATION OF OVERTIME SHIFT HOURS WITH AN L-P MODEL

Gerald L. Kaes[*]

It was desired to find the most economical work schedule to complete a construction project on a "crash" basis. Seven work schedules were possible, each having unique costs. Manpower loading could also be varied, with the necessity that the work force "buildup" be "staggered" during the early weeks. Further complications arose from labor union restrictions concerning overtime hours.

The problem may be stated in mathematical notation as follows:

Minimize: $\sum_{i=1}^{7} C_i M_i H_i W_i$ (Cost, $)

Subject to: $\sum_{i=1}^{7} W_i \leq 16$ (Total Weeks)

$\sum_{i=1}^{7} M_i H_i W_i = 26{,}500$ (Total Manhours)

$W_5 \leq 2$ (Union Restraint)

$W_1 + W_5 \leq 13$ (Union Restraint)

Maximum men/shift = 35
Maximum men/shift (week 1) = 15
Maximum men/shift (week 2) = 25

Where: C_i = Hourly cost, schedule i

M_i = Total men worked, schedule i

H_i = Productive hours/man/week, schedule i

W_i = Weeks worked, schedule i

---

[*]Technical adviser for computer aided design with Stearns-Roger Incorporated, Box 5888, Denver, Colorado, 80217, (303) 770-6400.

Table 1 presents additional data for the schedules.

Table 1
WORK SCHEDULE DATA

Schedule	Hourly Cost ($)	Productive Hours/Week/Man	Overtime Hours/Week/Man
1	15.46	40	0
2	19.69	48.2	14
3	21.32	50.8	20
4	20.71	52.2	20
5	15.84	75.5*	0
6	18.08	90.2*	6.5
7	20.16	100.9*	12.5

*Note – two shifts worked

The possibility of varying manpower loading within each schedule introduced a degree of complexity over a simple linear program solution. The need to restrict manpower during weeks one and two also increased the complexity.

To overcome these difficulties, each work schedule was restructured into several subvariables of fixed manpower loading.

Thus: $\quad C_1 H_1 W_1 M_1 = C_1 H_1 W_{11} M_{11} + \ldots C_1 H_1 W_{17} M_{17}$

Where: $\quad M_{11} \ldots M_{17} = 5, 10, 15, 20, 25, 30, 35 \quad (\text{men/shift})$

And: $\quad W_1 = W_{11} + \ldots W_{17}$

And so on for all the schedules.

The problem was now easily solvable with conventional L-P methods. Furthermore, it could be reasoned from inspection of Table 1 that schedule 1 would be present in the solution at the maximum activity possible. Therefore, "forcing variables" were added to the matrix to insure that variables W_{13} and W_{15} would be in the solution with activities of at least unity, simulating the work buildup during weeks one and two. The matrix details are given in Figure 1.

Several runs were made determining the effects of various labor alternatives on the job cost and completion schedule. Partial results are given in Tables 2 and 3.

Figure 1
MATRIX DETAILS

Row	W_{11}	W_{12}	W_{13}	W_{14}	W_{15}	...	W_{51}	W_{52}	W_{53}	...	Row Type	R.H.S.
Hours	200	400	600	800	1,000		378	756	1,134		$\geq$	26,500
Weeks	1	1	1	1	1		1	1	1		$\leq$	16
Sch. 1 + Sch. 5	1	1	1	1	1		1	1	1		$\leq$	13
Sch. 5	0	0	0	0	0		1	1	1		$\leq$	2
Week 1	0	0	1	0	0		0	0	0		$\geq$	1
Week 2	0	0	0	0	1		0	0	0		$\geq$	1
Cost ($)	3,092	6,184	9,276	12,378	15,460		6,000	12,000	18,000		Minimum	
No. of Men	5	10	15	20	25		5	10	15			

Note: Missing columns were constructed in similar fashion using Table 1 data.

Table 2

L-P RESULTS—COST DATA

Case	Cost ($)	Total Weeks
1 (Base)	433,623	15.2
2 (No Schedule 5)	439,912	16.0

Table 3

L-P RESULTS—WORK SCHEDULES*

Case	W_{13}	W_{15}	W_{17}	W_{57}	W_{67}	Total
1	1	1	9	2	2.2	15.2
2	1	1	11	0	3	16.0

*Those schedules not present in the solution are omitted.

Structuring the schedules in this fashion does introduce mild discontinuities in the manpower variables; however, as can be seen by the results it had little or no effect on the solution.

In conclusion, a simple technique of structuring was used to reduce a rather complex problem into a small, easily solvable problem. The simplified problem could be solved quickly and economically using a conventional linear program algorithm. This technique has obvious applications to a wide class of similar problems.

ACKNOWLEDGEMENT

The author wishes to thank Mr. Frank J. Mazanec, Jr. of Stearns-Roger Incorporated for his help in preparing this paper.

AAS 75-290

DEPOSITED TODAY - CONSUMABLE TOMORROW

SEWAGE AND SOLID WASTE HYDROGENATION

Clyde W. LaGrone*

Certain segments of our industrial and social nation are doomed to extinction unless industry and government begin to recognize that continual small methodical steps of technology can equal, and even surpass, a heralded scientific breakthrough, the practical or economic application of which may take several decades to implement. Currently, a breakthrough of simple technology with immediate application must usally struggle to obtain recognition, acclaim, and financial support commensurate with its potential impact on our environment.

This is strikingly exemplified in the case of the various techniques of converting municipal sewage and solid wastes into usable products such as methane, crude oils, combustible chars, fertilizers, soil conditioners, etc. The techniques for conversion such as pyrolysis, "wet oxidation", hydrogenation, and others are not new to the technical world. Yet none have received the publicity and, in turn, the financial support necessary to take them from the conceptual stage to a wide application. Several companies are actively engaged in the design and installation of these systems at various locations around the world, but their techniques, although being implemented, are faced with the reality of marginal economic return.

---

*Technical consultant to Energid, Inc., P.O. Box 11416, Denver, Colorado, 80211, (303) 825-2882.

The Bureau of Mines has conducted many in-depth studies of the techniques employed in sewage conversion over the last two decades, and found that all of these processes either require, or are improved by, the ability to operate at relatively high pressures. High pressures normally result in high costs.

Ironically, a technological breakthrough in applications engineering of these various techniques was made in the mid 1960's. A patent was applied for in 1967 and was finally issued in 1971, providing the missing link in mass application on an economically stable basis of conversion techniques developed earlier. The invention was based upon the realization that deep boreholes, similar to oil wells and thousands of feet deep, suitably lined, can be utilized as a particularly efficient apparatus for physical, chemical, and/or thermal processes, thereby overcoming one of the major problems associated with the application of various conversion techniques, i.e., the economical generation and maintenance of fairly high pressure over a wide range of temperatures.

The hydrostatic pressures created when the boreholes are filled with liquids or fluidized solids are of sufficient magnitude to promote a great many useful chemical reactions. Other advantages also result when deep boreholes are regarded as a primary component of the process equipment, eliminating the necessity for the expensive high pressure pumps associated with a conventional conversion process by the utilization of "off the shelf" hardware and equipment in conjunction with the unique injection and collection design to initiate and control chemical reactions at elevated temperatures. The "borehole" technique requires only the low pressure raw material feed pump, coupled with the activating heat source, to initiate the reaction.

The "borehole" technique relies on the combined hydrostatic head and chemical reaction to generate the high pressures required to extract the desired by-products. The raw material enters at low pressure. Pressure increases due to hydrostatic head as the material continues down the hole and approaches the reaction zone. The reaction zone begins at the point where the hydrostatic head reaches the minimum pressure required to sustain the desired reaction.

The pressure continues to increase due both to increasing hydrostatic head and chemical reaction. Pressure decreases gradually as the material is allowed to change direction and return to the surface. Exit pressure can be controlled within workable limits, using the standpipe height above ground as a guage. Thus, both entry and exit pressures can be relatively low, while the reaction pressure is high.

Complex and expensive above ground high pressure controls are eliminated since the high pressure exists only in the underground reaction, which also provides an obvious safety advantage. Heat losses, too, are minimized due to the insulation properties of the surrounding earth. Moreover, the balanced pressures existing throughout the "borehole" pressure vessel permit the pipes used in the system to be thin-walled, thereby enhancing their heat transfer capacity and reducing the cost.

The process associated with sewage and solid waste conversion varies considerably, depending upon the method, in time required to accomplish the conversion. Investigations showing acceptable degrees of conversion have ranged from one minute to several days. Those methods employing long periods for conversion are inefficient because of the energy consumed during the process. The Energid process has a short but efficient reaction time, allowing the generation of consumable liquid and gas by-products from sewage and solid waste materials that have been collected within the previous 24-hour period. Since the raw material must be in the reaction zone less than ten (10) minutes, high conversion with low energy consumption is possible.

Also, sewage need only pass through primary treatment prior to processing. Solid waste is accumulated by normal trash and garbage collection techniques, then shredded prior to dilution and injection into the "borehole" pressure vessel.

CONCLUSIONS

1) Conversion of organic waste materials created in the United States to synthtic fuels offers an annually replenished potential of one billion plus barrels of oil, or ten trillion cubic feet of gas per year.

2) The "borehole" pressure vessel technique offers the applications engineering breakthrough which will allow the economic implementation of various physical, chemical and thermal processes.

3) It is possible that, in addition to converting sewage and solid wastes to consumable by-products, this invention may be used in an efficient, economically viable process for the conversion of relatively hard and abrasive materials, such as oil shale, into similar synthetic fuels.